Design
and Estimating
for Heating,
Ventilating,
and
Air Conditioning

Design
and Estimating
for Heating,
Ventilating,
and
Air Conditioning

Ennio A. Rizzi

VNR VAN NOSTRAND REINHOLD COMPANY

NEW YORK   CINCINNATI   ATLANTA   DALLAS   SAN FRANCISCO
LONDON   TORONTO   MELBOURNE

Van Nostrand Reinhold Company Regional Offices:
New York   Cincinnati   Atlanta   Dallas   San Francisco

Van Nostrand Reinhold Company International Offices:
London   Toronto   Melbourne

Library of Congress Catalog Card Number: 79-10917
ISBN: 0-442-26952-8

Manufactured in the United States of America

Published by Van Nostrand Reinhold Company
135 West 50th Street, New York, N.Y. 10020

Published simultaneously in Canada by Van Nostrand Reinhold Ltd.

15 14 13 12 11 10 9 8 7 6 5 4 3 2 1

Library of Congress Cataloging in Publication Data

Rizzi, Ennio A
   Design and estimating for heating, ventilating, and
air conditioning.

   Includes index.
   1. Heating.   2. Ventilation.   3. Air conditioning.
I. Title.
TH7222.R55   697   79-10917
ISBN 0-442-26952-8

*To my wife, Marie, who patiently encouraged my writing and so
efficiently typed the entire manuscript, and to my children
Ennio Jr. and Remo who suffered hours of "fatherless" attention.*

PREFACE

The primary purpose of this book is to enable the reader to gain fundamental knowledge in HVAC design applications and their related cost estimates. There are so many handbooks, manuals, and guides currently available that emphasize overly theoretical approaches to the subject. This leaves unanswered many practical questions on the applications of HVAC design and estimating. For this reason and others, I decided to write this book to encourage the reader to acquire a background in heating, ventilating, and air conditioning using simple methods without resorting to extensive mathematical expressions beyond the use of algebra. This background will give an overall approach to the multitude of simple and sophisticated systems in use today.

This book is intended to provide a description of HVAC principles that can be understood by both beginners and professionals, be they draftsmen, designers, sheetmetal workers, or estimators. The HVAC applications that are described include residential, commercial, institutional, and industrial systems.

New technology is being devised each day. Prices of materials and labor rates fluctuate. However, HVAC design, takeoff, and estimating principles stay fairly constant. Knowing these principles, it becomes a matter of applying current market costs for materials and labor in order to design and estimate heating, ventilation, and air conditioning systems.

It is only after a project is built that its actual costs are known. Before it is constructed, that is, while it is in the design and specification stage, its cost can only be approximated. A systemmatic estimate is the best technique for understanding construction costs and is a prerequisite of every design job. This book will help you understand the scope of many HVAC applications and how best to approximate their costs.

INTRODUCTION

This book, Guideline Handbook for Design and Estimating in Heating, Ventilating, and Air Conditioning, is developed in four sections.

The sections contain sources of technical information and data available only from many publications scattered throughout the industry. These sections consolidate the most useful updated guidelines in simple and progressive steps and with constructive information, thus creating a pathway into the field of HVAC.

Section I covers the basic concepts of HVAC applications exploiting such areas as: heating principles; heating and cooling load estimates for residential and commercial applications; psychrometrics; refrigeration theory; energy conservation; air conditioning and heating with air, steam, hot water; and electricity; and application of solar energy.

Section II amplifies the fundamental subjects covered in Section I. However, the emphasis is placed upon the actual apparatus available to the HVAC designer and contractor. Equipment is described individually and in systems. Examples of commercially available units are illustrated, and finally, the estimated time to install a wide range of HVAC equipment is given.

Section III details guidelines for energy conservation that can be achieved by selecting proper HVAC systems. Other important factors that affect the selection of HVAC systems are also discussed; these include owning and operating cost, initial cost, return of invest-

ment, insurance, and others. Estimating procedures for HVAC system takeoff and budget estimates are also enumerated along with the preparation of design drawings.

Section IV is a collection of more than fifty useful reference tables that are most commonly used by HVAC engineers, estimators, and contractors. These tables are the basis of the calculations made in other sections of the book.

The abbreviations and symbols used in this book (those preferred for HVAC specifications and drawings) follow this introduction.

In addition to the general table of contents, each section begins with a more detailed list of contents.

This book could not have been written without the contributions of product data and literature from numerous organizations. Thus, grateful appreciation is expressed to the following for permission to use portions of their publications.

- American Society of Heating and Air Conditioning Engineers (ASHRAE) for data used in the chapters on automatic control and owning and operating costs, and in many reference tables.
- Joy Manufacturing Compnay for descriptions used in the chapter on air pollution control equipment.
- York Division of Borg-Warner Corporation for the many illustrations used throughout the book.

- Sheet Metal and Air Conditioning Contrators' National Association, Inc. (SMACNA) for various plates from their Low Velocity Duct Standard, 4th ed., 1969.
- Anemostat Products Division, Dynamics Corporation of America for portions of the Anemostat Engineering Manual E-70.
- Thrush Products Inc. for material from Hydronic System Design Manual E-15.
- American Air Filter Co., Inc. for description used in the chapter on air filters.

The author gratefully acknowledges the help of Mr. J. B. Godel, P.E. for his assistance in technical editing and guidance, Mr. M. Kass, P.E. for his continuous support and the organizations listed below for giving their permission to use data and illustrations from their literature.

- Aladdin Heating Corporation
- Baltimore Aircoil Company, Inc. Subsidiary of Merck and Co., Inc.
- Cleaver-Brooks Division of Aqua-Chem, Inc.
- Dynaforce Corporation
- EDPAC-AC. Manufacturing Co.
- Dowco Corporation
- Dunham Bush Incorporated
- The Fairbanks Company
- Heat Recovery Corporation
- Honeywell, Inc.
- The Marley Company
- Owens-Corning Fiberglas Corp.
- Singer Company Climate Control Division
- Van-Packer Company
- Vibration Mountings and Controls, Inc.

ABBREVIATIONS AND SYMBOLS

A	Area		CG	Ceiling grille
a	Thermal conductance of air space		CHT	Ceiling heater
AAV	Automatic air vent		CKV	Check valve
AC	Number of air changes per hour		CR	Ceiling register
ACP	Air compressor		CT	Cooling tower
ACU	Air conditioning unit		D	No. of degree days
AD	Air dryer		DB	Dry-bulb temperature in °F
ADP	Apparatus dewpoint		DD	Annual degree days
AF	Air filter		DH	Duct heater
AGV	Angle gate valve		DHWT	Domestic hot water tank
AP	Access panel		DU	Dehumidifier unit
ASHRAE	American Society of Heating Re-frigerating and Air-conditioning Engineers, Inc.		DV	Diaphragm valve
			E	Ejector or eductor
			EDR	Equivalent direct radiation
ASME	American Society of Mechanical Engineers		EF	Exhaust Fan
			EJ	Expansion joint
AV	Angle valve		EWF	Eye wash fountain
B	Air leakage per hour per foot of crack		F	Fuel consumption
			f_o	Air film coefficient (outside)
BD	Backdraft damper		f_i	Air film coefficient (inside)
BFP	Back flow preventer		FD	Fire damper
BFV	Butterfly valve		FF	Final filter
BG	Bottom grille		FH	Flexible hose
BJ	Balljoint		fpm	Feet per minute
BR	Bottom register		FTV	Foot valve with strainer
Btu	British thermal units		GLV	Globe valve
Btu/hr	British thermal units per hour		GV	Gate valve
BV	Ball valve		GVQC	Gate valve, quick closing
C	Thermal conductance		GVT	Gravity ventilator
cal	Calorie unit of heat		h	Enthalpy
CC	Cooling coil		H_L	Latent heat
Cf	Temperature correction for energy consumption formula		H_S	Sensible heat
			H_T	Total heat
cfm	Cubic feet per minute		HC	Heating coil

HCV	Hand control valve	R_t	Sum of thermal resistances
HL	Heat loss	RU	Refrigeration unit
HPT	Hydro-Pneumatic tank	SCKV	Stop check valve
Hs	Heat required to raise temperature of air leaking into space	SD	Splitter damper
		SE	Sewage ejector
HTW	High temperature water heating system	SF	Supply fan
		SG	Sight glass
HU	Humidifier unit	SHF	Sensible heat factor
HVAC	Heating ventilating and air conditioning	SJ	Swivel joint
		SLV	Sleeve valve
HX	Heat exchanger	SMACNA	Sheet Metal and Air Conditioning Contractors' National Association, Inc.
k	Thermal conductivity		
kWh	Kilowatt hour		
L	Length	SP	Sump pump
LCV	Level control valve	T_1	Temperature, entering air
LTW	Low temperature water heating system	T_2	Temperature, leaving air
		TCV	Temperature control valve
MBH	Thousand Btu per hour	TD	Temperature differential
MFD	Motorized fire damper	ΔT	Difference between indoor and outdoor design temperature in °F
mph	Mile per hour		
MTW	Medium temperature water heating system		
		TF	Transfer fan
N_b	Number of building load units	TG	Top grille
NV	Needle valve	TK	Tank
OA	Outside Air	T_m	Temperature of mixture of air in °F
OAL	Louver, outdoor air	TR	Top register
P	Pump	TXV	Thermal expansion valve
PA	Pipe anchor	U	Coefficient of heat transmission
PAC	Package air conditioner	u	Unit fuel consumption quantity per degree day
PCV	Pressure control valve		
PG	Pipe guide	UH	Unit heater
psi	Pounds per sq. in.	UV	Unit ventilator
PSV	Pressure or temperature safety valve	V	Volume
PV	Three-way plug valve	VAV	Variable air volume
Q	Heat transferred	VB	Vacuum breaker
QD	Quick disconnect	VD	Volume damper
QOV	Quick opening valve	VP	Velocity pressure
Q_w	Heat transferred as moisture	W	Grains of water per pound of dry air
R	Thermal resistance		
REF	Roof exhaust fan	WB	Wet-bulb temperature in °F
RH	Relative humidity	WG	Water gauge
$RSHF$	Room sensible heat factor	WH	Wall heater

CONTENTS

SECTION III   SYSTEM COSTS AND ENERGY CONSERVATION   331

SECTION IV   HVAC AND ENGINEERING TABLES   367

Design
and Estimating
for Heating,
Ventilating,
and
Air Conditioning

SECTION I

Principles of Heating, Ventilation, and Air Conditioning

1
INTRODUCTION

Air conditioning is the field of engineering dealing with the design, construction, and operation of equipment used to establish and maintain desired, atmospheric conditions within an enclosure, whether it be a large commercial building or a small box containing delicate equipment. This can be accomplished in one or more of the following ways.

1. Maintain the air at the desired temperature.
2. Control the moisture content of the air.
3. Hold contamination to an acceptable level.
4. Circulate the air properly in order to have the correct proportion of oxygen (without which the human body cannot long maintain a high level of efficiency).

Human comfort is not the only purpose of air conditioning. Mechanical cooling, ventilation, or heating must also be provided to prevent deterioration of machinery components in general storage spaces, to prevent excessive pressure buildup in containers, to stop contamination from leaking out of a confined space, and, to maintain specified temperatures for electrical and electronic equipment.

In heating, ventilating, and air conditioning work (HVAC), there is a distinction between cooling and dehumidifying, ventilating, and heating. Although all of the above functions are related, it is important to understand these distinctions.

1. *Cooling and Dehumidifying.* These terms are generally used to indicate the process where the sensible and latent heat are removed from the air or controlled to create a comfort condition in a given space. The terms *sensible* and *latent* heat are defined in the next section.

2. *Ventilation.* A mechanical or natural circulation of air in a given space, with or without treatment other than heating to desired temperature.

3. *Heating.* The primary function of heating in HVAC work is to raise the temperature of a given space and maintain it at some prescribed level during winter weather.

In this Section we will examine the underlying principles of heating, ventilation, and air conditioning, including load calculations for residential and commercial applications and an explanation of comfort levels as applied to HVAC. Actual air conditioning and heating systems are described so that the reader can become familiar with their design and use prior to using the application data found in Section II.

2
HEATING PRINCIPLES

Before attempting a discussion of specific heating systems, the reader must first understand a number of heat concepts. Heat is a form of energy, and every substance on earth contains some form of heat, even ice. It is believed that heat is actually the rapid vibration of the molecules, atoms, or electrons in a substance, be it solid, liquid, or gas. This motion slows as the temperature is lowered and finally stops at about 460° below zero on the Fahrenheit scale or the equivalent of 273° below zero Centigrade. These theoretical temperatures are known as absolute zero.

A basic law of physics states that heat will always flow or transfer in the direction of lower temperature. Thus, heat can be thought of as a ball which always run downhill to the lowest level. Heat will flow down the temperature hill warming the colder objects along the way. The rate at which the heat will flow depends upon the steepness of the temperature hill and the property of the material that it contacts. This type of heat transfer is by *conduction*.

Heat can also be transferred by *convection*. For example, a radiator in a room warms the surrounding air which decreases in density and rises. Cold air takes its place which in turn is heated, setting up a circulation of air that warms the room. The same type of transfer can occur in liquids as in air, i.e., heat is transferred in a hot water boiler by setting up convection currents that circulate the cold water through the heat source.

Radiation is the third method of heat trans-

fer. Unlike conduction and convection, it does not depend upon a medium (like a metal, gas, or liquid) to act as a carrier. The sun heats the earth by radiation. This heat is transferred by electromagnetic waves and is dependent upon the temperature of the source and the properties of the surface upon which the radiant heat falls. For example, a highly reflective surface, such as a light colored roof, will absorb less radiant heat than a dark one.

The heat (or energy) transferred is measured using two common units. The first is called the British thermal unit (Btu), which is defined as the amount of heat required to raise the temperature of 1 lb of water 1°F. The second unit is the calorie (cal.), which is the amount of heat required to raise the temperature of 1 gram of water 1°C. (One Btu equals approximately 252 cal.) At present, Btu's are more commonly used in HVAC work, but with the rising acceptance of the metric system calories may soon become the common unit of measurement of heat transfer.

2. HUMIDITY IN THE ATMOSPHERE

The earth's atmosphere contains about 77% nitrogen, 22% oxygen, plus traces of other gases. Water vapor also exists in atmospheric air in varying amounts, from almost neglible quantities in the case of dry air, to the point where it is saturated with moisture.

The water vapor content of air is called humidity. It has a great influence on human comfort. The common expression "It isn't the heat it's the humidity" recognizes the discomfort-producing effects of moisture-laden air in hot weather. At the other end of the scale, extremely low humidity also has undesirable effects on the human body. The measurement and control of the moisture content of the air is an important phase of air conditioning engineering. The reader must understand the concepts of saturated air, absolute and specific humidity, and relative humidity. *Psychometrics* is a branch of physics that deals with the properties of moist air. This subject is covered later in this Section.

Saturated Air.   The air holds varying amounts of water vapor, and as temperature rises, the amount of moisture that the air can hold increases. But for every temperature there is a definite limit to the amount of moisture that the air is capable of holding. When air absorbs the maximum amount of moisture which it can hold at a specific temperature, it is said to be saturated.

The saturation point is usually called the dew point. If the temperature of saturated air falls below its dew point, some of the water vapor in the air must condense. The dew that appears on foliage in the early morning when there is a drop in temperature is an example of condensation. The "sweating" of cold water pipes is the result of water vapor from the air condensing on the cold surface of the pipes.

Absolute and Specific Humidity.   The amount of water vapor in the air is expressed in terms of the weight of the moisture. The weight is usually given in grains (7000 grains equals 1 lb). Absolute humidity is the weight in grains of water vapor per cubic foot of air. Specific humidity is the weight in grains of water vapor per pound of air. (The weight of water vapor refers only to moisture in the vapor state, and not in any way to the moisture that may be present in the liquid state, such as rain, or dew.)

Relative Humidity.   Relative humidity is the ratio of the weight of water vapor in a sample of air to the weight of water vapor that same sample of air would contain if saturated, at the given temperature. This ratio is usually stated as a percentage. For example, when air is fully saturated, its relative humidity is 100%. When air contains no moisture at all, it's relative humidity is zero %. If the air temperature is 70°F and the relative humidity is 50%, it means that the amount of water vapor present in the air is 50% of what the air would contain at 70°F if it were saturated.

As far as comfort and discomfort resulting from humidity are concerned, it is the relative humidity and not the absolute or specific humidity that is the important factor. This can be easily understood from the discussion that follows.

Moisture always travels from regions of greater wetness to regions of less wetness, just as heat moves from higher to lower temperature regions. If the air above a liquid is saturated, the two are in equilibrium and no moisture can migrate from the liquid to the air; simply stated, the liquid cannot evaporate. If the air is only partially saturated, some evaporation can take place.

If the specific humidity of the air is 120 grains/lb, it is defined as the actual weight of the water vapor in the air. When the temperature of this air is 76°F, the relative humidity is nearly 90%, or nearly saturated. At such a relative humidity, a person may perspire freely but the perspiration does not evaporate rapidly; thus, a general feeling of discomfort results.

However, if the temperature of this air were 86°F, the relative humidity would then be only 64%. Although the absolute amount of moisture in the air is the same in both examples, the relative humidity in the second case is lower, because at 86°F the air is capable of holding more water vapor than at 76°F. The body is now able to evaporate its excess moisture and the general feeling is much more agreeable, even though the temperature of the air is 10° higher. Heat is absorbed in the process of evaporation. Hence the cooling effect.

In both examples, the specific humidity is the same, but the ability of the air to evaporate liquids is quite different at the two temperatures. This ability to evaporate moisture is directly measured by the relative humidity. It is for this reason that the control of relative humidity is of extreme importance in air conditioning.

2.2. HEAT OF THE AIR

The heat of the air is considered from three standpoints:

- sensible heat (H_S)
- latent heat (H_L)
- total heat (H_T)

Sensible heat is the heat that changes the temperature of a substance (air) when added to or removed from it. Sensible heat changes are measured by the household, or dry-bulb thermometer. *Latent heat* is a measure of the change of a substance's internal energy as it goes through a change of state. When water boils it undergoes a change of state from liquid to gas. The internal energy of a substance is measured by a thermodynamic term called *enthalpy. (h)*

Air always contains some water vapor. Any water vapor in the air contains the latent heat of vaporization. Remember that the amount of latent heat present has no effect upon the temperature of the air, as read on a dry-bulb thermometer. Any mixture of dry air and water vapor contains both sensible and latent heat. The sum of the sensible heat and the latent heat in any sample of air is called the *total heat* of the air. Put into a mathematical form

Total Heat (H_T) = Sensible Heat (H_S)
$$+ \text{Latent Heat } (H_L) \quad \text{(Eq. 1)}$$

We have defined the measurement of heat transferred in terms of the British thermal unit. For heating and cooling calculations, it is convenient to put this in the form of a rate of heat transfer, such as the number of Btu's per hour (Btu/hr).

The following equations are used to determine the total, sensible, and latent heats for samples of flowing air.

For cooling

$$H_T = \text{cfm} \times 4.5 \times (h_1 - h_2) \quad \text{(Eq. 2)}$$
$$H_S = \text{cfm} \times 1.08 \times (T_1 - T_2) \quad \text{(Eq. 3)}$$
$$H_L = \text{cfm} \times 0.68 \times (W_1 - W_2) \quad \text{(Eq. 4)}$$

For heating

$$H_T = \text{cfm} \times 4.5 \times (h_2 - h_1) \quad \text{(Eq. 5)}$$
$$H_S = \text{cfm} \times 1.08 \times (T_2 - T_1) \quad \text{(Eq. 6)}$$
$$H_L = \text{cfm} \times 0.68 \times (W_2 - W_1) \quad \text{(Eq. 7)}$$

where

$H_T, H_S,$ and H_L are in Btu/hr
cfm = airflow, in cu ft/min.
h_1 = enthalpy or total heat of entering air, in Btu/lb
h_2 = enthalpy or total heat of leaving air, in Btu/lb
T_1 = temperature of entering air, in °F
T_2 = temperature of leaving air, in °F
W_1 = grains of water/lb of dry air @ entering condition
W_2 = grains of water/lb of dry air @ leaving condition
1.08 = constant, derived as follows:

$$\frac{\text{Specific heat of standard air Btu/lb/(°F)}}{\text{Specific volume of standard air (cu ft/lb)}}$$

or

$$\frac{0.24 \times 60 \text{ min./hr}}{13.33 \text{ cu ft/lb}} = 1.08$$

4.5 = constant, derived as follows:

$$\frac{60 \text{ min./hr}}{13.33 \text{ cu ft/lb}} = 4.5$$

or

$$60 \times 0.075 = 4.5 \text{ lb/cu ft}$$

where

0.075 = density of standard air, in lb/cu ft
0.68 = constant derived as follows:
13.33 = specific volume of standard air in cu ft/lb

$$\frac{60}{13.33} \times \frac{1060}{7000} = 0.68$$

where

1060 is the average latent heat of water vapor (Btu/lb)
7000 is the number of grains in 1 lb of water vapor

2.3. TEMPERATURE RELATIONSHIPS

As stated above, the dew point occurs when condensation appears, and for a given amount of moisture in the air this phenomenon begins at a specific temperature. When testing the effectiveness of air conditioning equipment and when checking the humidity of a space, two other temperatures are generally considered. These are the dry-bulb (DB) and wet-bulb (WB) temperatures. The dry-bulb temperature is the temperature of the sensible heat of the air as measured by an ordinary thermometer. Such a thermometer in air conditioning engineering is referred to as a dry-bulb thermometer because its bulb is dry, in contrast with the wet-bulb type. Wet-bulb temperatures are indicated by a thermometer having its bulb covered by a film of water; this is usually accomplished with a loosely woven cloth sleeve or wick placed around the bulb and then wet with water. The air moving past the thermometer must be moving rapidly (in turbulent flow). Often, wet- and dry-bulb thermometers are combined into a single instrument called a psychrometer. Data from these thermometers, used in conjunction with formulas or charts will give the user an indication of the relative humidity.

The relationships between the three temperatures should be clearly understood. These relationships are:

1. When the air contains some moisture but is not saturated, the dew-point temperature is lower than the dry-bulb temperature, and the wet-bulb temperature lies between them.
2. As the amount of moisture in the air increases, the differences between the dry-bulb temperature and wet-bulb temperature become less.
3. When the air is saturated, all three temperatures are the same.

2.4. HUMAN COMFORT

Ordinarily the human body remains at a fairly constant temperature of 98.6°F. It is very important that this body temperature be maintained, and since there is a continuous heat gain from interior processes, there must also be a continuous out-go to maintain heat balance. Excess heat must be absorbed by the surrounding air or be lost by radiation. As the temperature and humidity of environment vary, the body automatically regulates the amount of heat which it gives off. However, this ability to adjust to varying environmental conditions is limited. Furthermore, although the body may adjust to certain atmospheric conditions, it may do so with a distinct feeling of discomfort. The discussion which follows will help you understand how atmospheric conditions affect the body's ability to maintain a heat balance.

The body gains heat by:

1. radiation
2. convection
3. conduction
4. as a by-product of physiological processes that take place within the body

The heat gain by radiation comes from our surroundings, but since heat always travels from regions of higher temperature to regions of lower temperature, the body receives heat from those surroundings that have a temperature higher than body surface temperature. The greatest source of heat radiation is the sun. Indoor heat radiation is gained from heating devices, operating machinery, hot steam piping, etc. The heat gain by convection comes from currents of heated air only. Such currents of air may come from a hot plate or a motor. The heat gain by conduction comes from objects with which the body is in contact from time to time.

Most body heat comes from within the body itself. Heat is being continuously produced inside the body by the oxidation of foodstuffs and by other chemical processes, by friction and tension within the muscle tissues, and by other causes as yet not completely identified.

There are two types of body heat losses, namely, loss of sensible heat and loss of latent heat. Sensible heat is given off by radiation, convection, and conduction. Latent heat is given off in the breath and by evaporation of perspiration.

The body is usually at a higher temperature than that of its surroundings, and therefore radiates heat to the surrounding area. This action is called heat radiation loss. The temperature of the air does not influence this radiation, except as it may alter the temperature of such surroundings. The heat loss by convection occurs when the heat is carried away from the body by convection currents, both by the air coming out of the lungs and by exterior air currents. The heat loss by conduction is caused by bodily contact with colder objects. Since the body is usually at a higher temperature than that of its surroundings, it gives up heat by conduction through physical contact with its surroundings.

When the air temperature and relative humidity are not too high and when the body is not too active, the body gets rid of its excess heat by radiation, convection, conduction, and by a slight amount of perspiration. When engaged in work or exercise, the body develops much more internal heat and perspiration increases. If the relative humidity is low, perspiration rapidly evaporates. As the perspiration evaporates, the body loses additional heat (latent heat of vaporization). However, if the relative humidity of the air is high, the moisture cannot evaporate, or it does so at a slow rate; hence, the excess heat cannot be removed by evaporation, and discomfort follows.

The amount of heat given off by the body varies according to the body's activity. When seated at rest, the average adult male gives off about 380 Btu/hr. Doing work he gives off an average of 500 to 600 Btu/hr.

In perfectly still air, the layer of air adjacent to the body absorbs the sensible heat given off by the body and increases in temperature. This layer also takes up the water vapor given off by the body, and increases in relative humidity. The body is thus surrounded by an envelope of air which is at a higher temperature and relative humidity than the ambient air; therefore, the amount of heat which the body can lose to this envelope is less than that amount it can lose to the ambient air. If the air is set in motion, the envelope is continually broken up and replaced by the ambient air, and the heat loss from the body is increased. When the increased heat loss improves the heat balance, the sensation of a "breeze" is felt; when the increase is excessive, the sensation of a "draft" is felt.

From the foregoing discussion, it is evident that the three factors (temperature, humidity, and air motion) are closely interrelated in their effects upon comfort and health. In fact, a given combination of temperature, humidity, and air motion will produce the same feeling of warmth or coolness as a higher or lower temperature, providing the humidity and air motion are varied accordingly. The term given to the net effect of these factors is known as the "effective temperature." This temperature cannot be measured by any instrument, but may be found on a special psychrometric chart when the dry-bulb and wet-bulb temperatures and air velocity are known.

Though all of the combinations of temperature, relative humidity, and air motion of a particular effective temperature may produce the same feeling of warmth or coolness, they are not all equally comfortable. It has been found that a relative humidity below 15% produces a parched condition of the mucous membranes of the mouth, nose, and lungs, and increases susceptibility to disease germs. A relative humidity above 70% causes an accumulation of moisture in clothing. For best health conditions, a relative humidity ranging from 40 to 50% for cold weather, and from 50 to 60% for warm weather is desirable. An overall range from 30 to 70% is acceptable.

3

HEATING LOAD ESTIMATE

The heating load estimate is a computational procedure which accounts for the probable heat loss occurring in a room or space to be heated, during the winter weather. It is usually necessary to obtain drawings or sketches of the space involved and extract the following information which will make the heating computations possible and accurate.

- General construction of the space or building (roof, walls, floor, ceiling, etc.).
- Orientation and location of the building.
- Window types and quantities.
- Ceiling height and construction profile.
- Machinery used or intended to be used within the space.
- Lighting layouts.
- Utility services available.
- Occupancy, that is, the number of people and their degree of activity within the given space.
- Code requirements that deal with the eventual heating system for the building or space involved.

When the above data are recorded we are ready to proceed with the actual heating load calculation.

3.1. HEATING LOAD CALCULATION

This calculation is usually made using the format shown in Figure 1. However, different manufacturers, engineering offices, or contrac-

tors have their own standard which may vary in format but end with the same result.

Let us go through a step-by-step computation of a heating load using the form shown on Figure 1, and describe each entry and term in order to better understand the procedure. When you are familiar with the terminology refer to Chapter 6 for a worked out example.

STEP 1: Design Conditions.
Indoor Dry-Bulb Temperature (DB°F). The ideal solution for a heating system is to design an apparatus with a capacity to generate enough heat to offset the load needed in the most severe weather conditions possible at a particular locality. In applications where the comfort of the occupants is of primary importance such factors as: the temperature of the air, the air motion, the relative humidity, and physiological factors (age, sex, health, etc.) must be considered before selecting an indoor dry-bulb temperature. A guideline of recommended indoor conditions for human comfort is presented in Table 22, Section IV.

Note that an occasional failure of a heating plant to maintain a preselected indoor design temperature during brief periods of severe weather is not a critical matter. However, the incidence of such "occasional" failures to meet indoor design temperatures is a matter of customer tolerance to the infrequent discomfort and must be compared to the additional cost of a system that can maintain the indoor temperature for all outdoor conditions. This becomes a

Location ....   Date ....

HEATING LOAD CALCULATIONS

1. Design Conditions

	Dry-Bulb Temp. °F	Humidity (% RH)	grains/lb	
Indoor	....	....	....	(X) = Insulation factor from tables.
Outdoor	....	....	....	(Y) = Heat loss from tables.
Temperature Diff. .... (ΔT)	....	....	(Z) = Total length of duct, in linear feet (LF)

2. Transmission Loss

	Area (sq ft) gross	net	U factor	ΔT (°F)	Transmission Loss (Btu/hr)
Walls	....	.... ×	.... ×	.... =	....
	....	.... ×	.... ×	.... =	....
	....	.... ×	.... ×	.... =	....
	....	.... ×	.... ×	.... =	....
Glass	....	.... ×	.... ×	.... =	....
	....	.... ×	.... ×	.... =	....
	....	.... ×	.... ×	.... =	....
	....	.... ×	.... ×	.... =	....
Ceiling	....	.... ×	.... ×	.... =	....
Roof	....	.... ×	.... ×	.... =	....
	....	.... ×	.... ×	.... =	....
Floor	....	.... ×	.... ×	.... =	....
Slab	....	.... ×	.... ×	.... =	....
Other	....	.... ×	.... ×	.... =	....

Total Transmission = ...

3. Infiltration–Ventilation

Ventilation .... cfm   .... cfm × ΔT × 1.08 = ...

Infiltration .... cfm   .... cfm × ΔT × 1.08 = ...

4. Humidification (if added to Heating Equipment Load)

.... cfm × .... grains/lb diff. × 0.68 = ...

5. Duct Heat Loss

Supply duct .... (X) × .... (Y) × .... (Z) ÷ 100 = ...

Return duct .... (X) × .... (Y) × .... (Z) ÷ 100 = ...

TOTAL HEATING LOAD   ....

FIGURE 1. Heating load calculations form.

matter of customer choice and judgement of the designer which is developed through experience.

Outdoor Dry-Bulb Temperature (DB°F).   The outside design temperatures, shown in Table 41, Section IV, are representative of general practice, which may vary due to local conditions such as altitude and exposure. Before selecting an out-

door design temperature from Table 38, the designer should consider if, the structure has a low-heat capacity,* is not insulated, has more

*The heat capacity is the amount of heat needed to raise the temperature of the structure 1°F. Low-heat capacity implies that the temperature of the structure can be raised more quickly than a high-heat capacity structure, assuming the same heat source.

than normal glass area, or is occupied during the coldest part of the day.

The "median of extremes" temperature should probably be selected as the outdoor design temperature. A moderate heat capacity, some internal load, and daytime occupancy would indicate that the 99% value was a reasonable choice. Massive, institutional buildings with little glass can usually be designed using the 97.5% value.

Relative Humidity (in grains of water per pound of air). Humidification is usually not considered in many heating applications. However, for comfort conditions, winter heating without humidification may produce extremely dry conditions within the heated space. Relative humidity ranges between 30 and 60% within the comfort zone, but in most winter climates, a humidity of over 30 to 40% may cause excessive window frosting as well as condensation in certain types of wall and roof construction. Indoor design humidities for comfort heating applications are given in Table 22, Section IV.

STEP 2: Transmission Losses.
The heat loss by transmission can be divided in two general groups:

1. Transmission losses or heat transmitted through the confining wall, floor, ceiling glass, or other surface.
2. Infiltration losses or heat required to warm outdoor air which leaks in through cracks around doors and windows or through open doors and windows, or heat required to warm outdoor air used for ventilation.

Note that before proceeding with the calculation of transmission losses it is necessary to know the rate of heat transfer through a building structure, such as a wall, floor, or ceiling. This rate of heat transmission is called the *U* factor*. This is expressed in Btu/(hr)(sq ft) of surface (°F) for the difference between the air temperature on each side of the structure.

The heat transmission coefficient can be calculated for any combination of layers of differ-

*Most of the heat transmission coefficients shown throughout this book have been taken with permission of ASHRAE. Others have been calculated.

ent materials used in building structures if the thermal conductivity (*k*) or thermal conductance (*C*) of each material is known (see Table 6, Section IV).

The following equations are used for calculating overall coefficients of heat transmission *U*, in Btu/(hr)(sq ft)(°F)

$$U = \cfrac{1}{\cfrac{1}{f_i} + \cfrac{1}{a} + \cfrac{1}{C} + \cfrac{x}{k_1} + \cfrac{x}{k_2} + \cdots + \cfrac{x}{k_n} + \cfrac{1}{f_o}}$$

(Eq. 8)

$$U = 1/R_t \qquad \text{(Eq. 9)}$$

where

k = thermal conductivity in Btu/(hr)(sq ft) (°F/in.) for a homogeneous material.

x = thickness of homogeneous material in in.

C = thermal conductance in Btu/(hr)(sq ft) (°F) for the thickness or type of material stated.

a = thermal conductance of air space (heat transmitted by radiation, conduction convection in Btu/(hr)(sq ft)(°F) for the thickness of air space under consideration). Normally $a = 1.1$ for air spaces exceeding $\frac{3}{4}$ in. in thickness where the surfaces have no reflective insulation such as aluminum foil.

f = air film coefficient or surface conductance (heat transmitted by radiation, conduction, and convection in Btu/(hr)(sq ft)(°F) difference between the surface and surrounding air, or vice versa).

f_i = surface conductance on inside of wall, floor, or ceiling.

f_o = surface conductance on outside wall, floor, or roof.

R_t = the sum of thermal resistances (R_1 + $R_2 + R_3 \cdots$), in (hr)(°F)(sq ft)/Btu.

The basic formula for the loss of heat by transmission through walls, roofs, ceilings, floors, glass, or other exposed surfaces is given by the equation

$$HL = A \times U \times (T_i - T_o) \qquad \text{(Eq. 10)}$$

where

HL = heat loss (in Btu/hr)

A = area of exposed surface (in sq ft)

U = coefficient of transmission, air-to-air in Btu/hr/sq ft/°F temperature difference

T_i = indoor temperature near the surface involved in °F

T_o = outdoor temperature, or temperature of an adjacent unheated space or of the ground in °F

The following is an example of a calculation of a transmission loss. Find the transmission loss of a 150 sq ft, 8-in. brick wall, assuming an indoor temperature (T_i) of 72°F and an outdoor temperature (T_o) of 0°F.

Using Equation 10 and substituting:

$$HL = 150 \text{ sq ft} \times 0.29 \frac{\text{Btu}}{(\text{hr})(\text{sq ft})(\text{F})}$$

$$\times (72°F - 0°F)$$

$$= 150 \times 0.29 \text{ Btu/hr} \times 72$$

$$= 3132 \text{ Btu/hr}$$

The heat loss is expressed in Btu/hr and is the product of the area of the surface, a transmission factor giving the rate of heat transfer for the material in question, and the effective temperature difference. The overall transmission factor, U for most types of wall, and glass construction can be found by the summation of all resistances by using the equation $U = 1/R_t$.

STEP 3: Infiltration and Ventilation.

Infiltration. Infiltration is the cold air which leaks in through windows, doors, and walls because of wind pressure against the building and by differences in air density between the warm and cold air. The amount of air entering a building is generally estimated in two ways: the air change method and the crack method.

Air Change Method. The air change method may be used with reasonable accuracy for calculating heat loss of residential construction. The amount of air leakage is estimated by assuming a certain number of air changes per hour. The following equation is used:

$$\text{cfm} = \frac{AC \times V}{60} \quad \text{(Eq. 11)}$$

or solving for the number of air changes:

$$AC = \frac{\text{cfm} \times 60}{V} \quad \text{(Eq. 12)}$$

where

AC = number of air changes/hr
cfm = cu ft/min.
60 = min./hr
V = volume of enclosed space in cu ft

For rooms of unusual size or proportion, the air change method is not accurate. Such an example would be a store with one long narrow room and no windows or doors on the side walls. Heat loss through infiltration would occur only at the front and back (through windows or doors). The air change method, based upon the air volume of this space, would indicate a heat loss much in excess of that which might actually occur.

With the air change method, an arbitrary allowance for exposure is often made. Opinions vary on the percentage of allowance according to geographical location and therefore, no rule covers the subject. Table 18, Section IV, indicates the air changes commonly used, but, they should be taken as a guide only.

Crack Method. As the name implies, the crack method is based on air leakage through cracks around windows and doors, according to the wind velocities expected. When the crack method is used for estimating leakage, it is expressed in terms of crack length using the following equation:

$$H_s = B \times L \times (T_i - T_o) \quad \text{(Eq. 13)}$$

where

B = air leakage/hr/ft of crack for the wind velocity and type of windows or door cracks involved, multiplied by 0.018.*

L = length of window or door crack in ft

$(T_i - T_o)$ = temperature difference or ΔT in °F

*Constant derived using the specific heat of air at constant pressure, 0.24 Btu/lb (°F) and density of standard air, 0.075 lb/cu ft hence: 0.20 × 0.075 = 0.018.

H_s = heat required to raise the temperature of the air leakage into the space from T_o to T_i in Btu/hr

For designers who prefer to use the crack method, the basis of calculation is as follows:

1. The length of crack used for computing the infiltration heat loss should be not less than half of the total length of crack in the outside walls of the room.

2. For a building having no partitions, air entering through the cracks on the windward side must leave through the cracks on the leeward side. Therefore, use one-half the total length of the crack when computing each outside building wall.

3. In a room with one exposed wall, use the full crack length. With two, three, or four exposed walls, take the wall having the crack that will result in the greater air leakage, but in no case use less than half of the total crack length.

4. In small residences the total infiltration loss is generally considered to be equal to the sum of the infiltration losses of the various rooms. However, this may not be accurate, since at any given time infiltration will take place only on the windward side and not on the leeward side. For determining the total heat requirements of larger buildings, it is more accurate to base the total infiltration loss on the wall having the longest crack length, but in no case on less than half of the total crack in the building.

An example in the use of the crack method is as follows: What is the infiltration heat loss per hour through the crack of a 3 × 5 ft double hung, nonweather stripped, wood window, based on a wind speed of 15 mph. Assume indoor and outdoor temperatures to be 70°F and 0°F, respectively.

Solution: According to Table 19, Section IV, the air leakage through a window of this type (based upon $\frac{1}{16}$-in. crack and $\frac{3}{64}$-in. clearance) is 27 cu ft/ft of crack/hr. Therefore, $B = 27 \times 0.018 = 0.486$ cu ft/hr. The length of crack (L) is $(2 \times 5) + (3 \times 3)$, or 19 ft. $T_i = 70°F$ and $T_o = 0°F$. Substituting in Equation 13:

$$H_s = 0.486 \times 19 \times (70°F - 0°F) = 646 \text{ Btu/hr}$$

*Ventilation.** When outdoor air is required for ventilation purposes, the quantity usually depends on the number of occupants, their activity, and the volume of the space. The recommended minimum requirements for the amount of outdoor air may be found in Table 25, Section IV. These amounts are frequently based on local and state laws.

When calculating the amount of outside air required, best results are obtained by taking no credit for infiltration. This is because on windless days, the infiltration is practically zero. It is common practice not to supply outside air for ventilation for industrial applications except where toxic gases, smoke, or fumes are present. Where outside air is required it is usually more satisfactory to bring it in through the heating apparatus in order to reduce cold draft to a minimum.

Therefore, analyze carefully the ventilation requirements, keeping in mind the possible infiltration. Normally, the total outside air allowance is equal to the quantity of infiltration or ventilation air (whichever is greater), but further consideration is necessary to determine the quantity of heat to be delivered to an enclosed space. The following example is illustrative:

Calculate the heating load from outside air in Btu/hr for a room which measures 50 ft long by 60 ft wide by 12 ft high. The inside design conditions are 72°F dry-bulb with 50% RH. The outside design conditions are 10°F and 80% RH. The ventilation at 10 cfm/person must be provided for 100 people. An exhaust fan rated at 1000 cfm operates continuously. The room is fairly air tight and crackage is not judged a serious factor. Infiltration, determined from the air change method, is 1.5 cfm/hr.

Solution:

Ventilation = 150 people @ 15 cfm/person

= 2250 cfm

$$\text{Infiltration (air change)} = \frac{1.5 \times 50 \times 60 \times 12}{60}$$

= 900 cfm

*Although this subject is part of the section on heating load applications, it is equally applicable for cooling load calculations.

Exhaust air = 1000 cfm

Total infiltration = 1900 cfm

Ventilation is greater than the total infiltration and 2250 cfm is therefore used as the total outside air in the heat load calculation.

Using Equations 6 and 7

$$H_S = 2250 \times 1.08 \times (72 - 10)$$

$$= 150,660 \text{ Btu/hr}$$

$$H_L = 2250 \times 0.68 \times (58 - 7.2)$$

$$= 77,724 \text{ Btu/hr}$$

From Equation 1

$$HT = H_S + H_L$$

$$= 228.384 \text{ Btu/hr}$$

STEP 4: Humidification

The amount of moisture that must be supplied to indoor air in winter is dependent on the absolute humidity difference between indoor and outdoor air and the amount of infiltration. A humidifier is rated according to the pounds of water evaporated per hour.

The equation used for this calculation is:

$$\text{lb/hr} = \frac{(\text{cfm OA.} + \text{cfm infiltr.}) \times \text{grains/lb diff.}}{7000 \times 13.33/60}$$

$$= \frac{\text{cfm} \times \text{grains/lb}}{1580} \qquad \text{(Eq. 14)}$$

Since the evaporation of moisture in the air may require that heat be added to maintain the air temperature, this should be included in the heat estimate after calculating the infiltration loss. It is obtained from the following equation:

$$\text{Btu/hr} = \text{lb/hr of moisted air} \times 100$$

$$\text{(Eq. 15)}$$

Note that this equation is not needed where the heat of evaporation is supplied by a source other than the air being conditioned, such as a steam humidifier.

STEP 5: Duct Heat Loss (or Gain).

Air leakage from an air heating system may effect the total heating requirement if the ductwork used to supply the heating areas is not located in the area to be heated. For this case the following should be known:

1. Air temperature difference between the inside and the outside of the duct.
2. Volume of air being circulated in cfm.
3. Length of duct installed in the unheated area.
4. Thickness of duct insulation, if any.

The following equation is used to determine the heat loss for heating or cooling applications:

$$Q = \frac{X \times Y \times Z}{100} = \text{Btu/hr}$$

$$\text{(Eq. 16)}$$

where

X = insulation factor from Table 20, Section IV

Y = heat loss found in Table 20, Section IV

Z = total length of duct (in ft)

4

AIR CONDITIONING

The approach in designing an air conditioning system is somewhat similar to that of the heating system previously described. Additional investigations should be made for the following steps.

1. Evaluate the system to be used.
2. Select such a system.
3. Select the equipment.
4. Design the air distribution system.
5. Design the piping distribution system.
6. Calculate the cooling load requirements.

Of all the above steps, the cooling load requirement is probably the most important.

4.1. COOLING LOAD CALCULATION

Cooling load calculations for individual areas are required to determine the peak air conditioning loads and the required supply of air or water for each area or zone of the building that is being designed.

For load estimates on residential construction, individual rooms must be separately calculated. Multistory buildings will have repetitive loads, and load estimates are required only for each typical room or area for each exposure of the building. For now, we will be concerned only with load estimates for institutional and commercial facilities. Residential cooling load calculations are given in Chapter 5.

The air conditioning loads for the top floor of a multistory building are usually obtained by adding the roof load to the corresponding lower floor loads. Normally the peak loads of individual exposures do not occur simultaneously and an overall or *block cooling load* calculation is needed to determine the maximum simultaneous cooling load that will be imposed on the system.

Cooling load calculations should be made using data found in the *ASHRAE Handbook of Fundamentals* and by using the format shown in Figure 2. An explanation of the terms used on the cooling load calculations form (Figure 2) follows. A worked out example of a commercial air conditioning design is given in Chapter 6.

STEP 1: Design Conditions.
The block load determines the size or capacity of the air handling apparatus and the refrigeration equipment. The block load consists of a bird's eye view of the building envelope at the time of peak refrigeration demand. The block load is *not* the sum of the peak demands at each area but is determined by calculating the total refrigeration load at different times of the day. These calculations will be influenced by the solar heat gain, the outside air temperature and the internal heat gain if it is a variable.

The selection of the indoor room design conditions may be dictated by a particular application or the specific requirements established by the owner, architect, or engineer. Since most applications are essentially related to the comfort of building occupants, design conditions of 72° to 78°F DB, and a relative humidity

Location ....   Proj. No. ....

Latitude ....   Prepared by ....

Hour   day   month   Date ....

....

COOLING LOAD CALCULATIONS

1. Design Conditions (Summer)

	Dry Bulb °F	Wet Bulb °F	Specific Humidity grains/lb
Outdoor	....	....	....
Indoor	....	....	....
Temperature Diff.	....	....	....

2. Solar Heat Gain

Exposure	Area sq ft	Solar factor	Shading factor	Sensible Heat Gain Btu/hr	Latent Heat Gain Btu/hr
Glass	.... ×	.... ×	.... =	....	
....	.... ×	.... ×	.... =	....	
....	.... ×	.... ×	.... =	....	
....	.... ×	.... ×	.... =	....	

3. Transmission Heat Gain

Exposure	Area sq ft	U factor	Temp. Diff. °F	
Glass	.... ×	.... ×	.... =	....
....	.... ×	.... ×	.... =	....
....	.... ×	.... ×	.... =	....
....	.... ×	.... ×	.... =	....
Walls	.... ×	.... ×	.... =	....
....	.... ×	.... ×	.... =	....
....	.... ×	.... ×	.... =	....
....	.... ×	.... ×	.... =	....
Ceiling	.... ×	.... ×	.... =	....
Roof	.... ×	.... ×	.... =	....
Floor	.... ×	.... ×	.... =	....
Partition	.... ×	.... ×	.... =	....
....	.... ×	.... ×	.... =	....

4. Internal Heat Gain

	No.	Factor	
a) People	.... ×	.... (Sensible Heat)	= ....
b) People	.... ×	.... (Latent Heat)	= ....

c) Lighting

	watts/sq ft	area sq ft	
	.... ×	.... × 3.4 (i)	= ....
	.... ×	.... × 4.1 (ii)	= ....

d) Motors

	hp.	Eff. Factor	
	.... ÷	.... × 2545 (iii)	= ....

FIGURE 2. Cooling load calculations form.

e) Appliances   No. H_s or H_L

.... × .... = ... = ...
.... × .... = ... = ...
.... × .... = ... = ...

5. Ventilation

Ventilation (Outside Air)   DB Temp. Diff.

cfm .... ×   1.08   × .... = ...
                                        grains/lb Diff.
cfm .... ×   0.68   × ....   = ...

6. Infiltration

Infiltration   DB Temp. Diff.

cfm .... ×   1.08   × .... = ...
                                        grain/lb Diff.
cfm .... ×   0.68   × ....   = ...

7. Duct Heat Gain

(b)   (c)   (d)

cfm .... × .... × .... ÷   100   = ...
cfm .... × .... × .... ÷   100   = ...

Total Sensible Heat = ...
Total Latent Heat   = ...

Total Cooling Load (Sensible plus Latent Heat) ....
Tons of Refrigeration ....

(i)   3.4 Btu/watt–Heat of Incandescent Lamp
(ii) 4.1 Btu/watt–Heat of Fluorescent Lamp
(iii) 2545 Btu/hr where
      1 hp = 746 watts × 3.4 Btu/hr/watt
(a)   Total Sensible Heat where Supply Air in cfm is
      determined
(b)   Duct Insulation Factor
(c)   Duct Heat Gain Factor
(d)   Total Ductwork Length

Check Figures:
Outside Air   = ... %
Btu/sq ft (Sensible) = ...
Btu/sq ft (Total)   = ...
No. Air Changes/hr   = ...
cfm/sq ft   = ...

FIGURE 2. (*Continued*)

range of 40 to 50% are currently accepted as satisfactory indoor design criteria.

STEP 2: Sun Load.

One of the largest single components of the room cooling load is the solar heat gain. Because of the hourly, daily, and seasonal variations in intensity of this load, a careful investigation of the proper design time and design conditions should be made. The maximum cooling load requirements for each exposure usually occur on or near the time of peak solar heat gain to the space. Cooling loads should, therefore, be based on this peak solar gain. Peak solar heat gains for East exposures occur during early or midmorning hours in July in northern latitudes and January in southern latitudes. West exposure will peak from mid to late afternoon in July in northern latitudes and January in southern latitudes. North exposures in the northern hemisphere and South exposures in the southern hemisphere will peak shortly after the maximum outdoor design temperature at 3 P.M. in July in the North and January in the South.

Peak loads for South exposures in the northern hemisphere and North exposures in the southern hemisphere, particularly those with large glass areas, should be based on the maximum loads which occur at or shortly after noon, sometime between September and December in the North and between March and June in the South.

Standard solar heat tables are based on data for absolutely clear days which usually occur at outdoor temperatures that are 15 to 20°F. below the normal design temperatures. Because of the smoke and dust laden atmosphere encountered in all but a few areas of the country, it is recommended that tabulated peak solar values for all but South exposures be reduced by 15 or 20% for buildings located in the larger metropolitan areas.

Maximum solar intensities for South exposures in the northern hemisphere and North exposures in the southern hemisphere, occurring sometime between September and December in the North and between March and June in the South, will occur at times when outdoor dew points are below the tabulated data.

In these cases, solar heat values should be increased in accordance with corrections given for the standard tabulated data. When the heating system is controlled by the outdoor temperature, these corrections should be made for dew points occurring at minimum outdoor design temperatures.

Additional corrections should be made to the solar heat values to adjust to applicable shading factors. Corrections should be made for external or internal shading factors that may also apply.

These corrections should be made for external overhangs, louvers, awnings, screens and for internal blinds, shades or draperies. Additional corrections may also be necessary for various types of heat absorbing or heat reflecting glass. Another frequently overlooked factor is the shading effect produced by nearby buildings as well as offsets and reveals of the building itself.

Solar heat gains through glass do not become instantaneous heat gains to the space. This radiant solar energy is first absorbed by the surfaces of the building and its contents. As the radiant heat is absorbed, the surface temperatures are increased. Part of the heat is removed by convection, while the remainder is absorbed by the structure. The heat absorbed by the structure is gradually removed by convection over an extended period and is, therefore, not an instantaneous load.

STEP 3: Transmission Heat Gain.
Transmission heat gain is, in an opposite sense, similar the description of transmission losses that was given as Step 2 in the heating load estimate, Chapter 3. Now the heat is being transmitted through the building structure and must be cooled by the air conditioning system. Equations 8 and 9 covering U factors are valid for this application, as is Equation 10 for the heat loss by transmission.

The transmission load is independent of the load from lights, people, and sun. It varies directly with the difference between outdoor and indoor temperatures. Generally speaking, when outdoor temperatures are less than the indoor temperatures, it should be regarded as a heating load.

STEP 4: Internal Heat Gain.
The air conditioning load in building interiors includes:
  1. heat generated by people
  2. lights
  3. miscellaneous equipment, such as motors, hoods, stoves, etc.

This is a consistent year-round cooling load and varies in magnitude depending upon occupancy and use. Interior zones with roof loads generally require cooling whenever lighted and occupied. However, due to heat loss through the roof, heating is required to maintain prescribed room temperatures during prolonged unoccupied periods with low outdoor temperatures.

The heat gain from people consists of both sensible and latent heat. Consideration should be given to the maximum number of people that may be occupying the space when both the maximum and minimum room sensible heat loads occur. Reduced occupancy may result in overall load reduction. Occupancy at minimum sensible loads may require a downward adjustment of the supply air dew point to provide proper dehumidification.

Total building occupancy is seldom equal to the individual room totals. Proper diversity factors may be used to adjust the total block cooling load for the anticipated total building occupancy.

In the interior zones of a building, the lighting load is by far the most important source of a room's sensible heat. In perimeter areas, if not the most important, it is second only to the solar heat gain. Careful attention should be given to accurately estimate the effect of this important heat source. Additional heat load from the ballast of fluorescent fixtures should not be overlooked.

Ceiling return air plenums in combination with return air grilles or return air lighting troffers may be used to effectively remove some of the lighting load. It should be recognized, however, that the increased plenum temperature will result in increased convection and radiation through the ceiling and the floor above. About 25% of the total lighting load is the practical heat removal limit of most ceiling plenums. While ceiling plenums may effectively reduce the room's sensible heat load, because of increased return air temperatures, the total load of the air handling and refrigeration equipment remains unchanged.

Normally, on overall building loads, a diversity factor ranging from 0.85 to 0.95 may be applied to the total lighting load to account for fixtures that may not be in use or for burned out bulbs.

In addition to the people and lighting loads already discussed, various miscellaneous loads may have to be included. These include all types of electrically operated equipment including typewriters, bookkeeping machines, data processing equipment, communications equipment, etc. Laboratories may present special types of loads, such as steam heated or gas burning equipment.

Careful evaluation should be made to determine both the degree and usage of all types of miscellaneous internal loads. Frequently, it is possible to apply rather large usage or diversity factors to these loads. Where specialized equipment loads may be encountered, such as in computer rooms, it may be advisable to treat these as separate supplemental systems.

STEP 5: Ventilation.

Description and data for determining ventilation requirements have been previously presented in Chapter 3.

For cooling and/or heating, ventilation is required for supplying fresh air to occupants and for removing odors generated within the area.

STEP 6: Infiltration.

The principles of infiltration have been discussed in Chapter 3.

In cooling applications, the infiltration calculation is normally omitted because it is desirable to introduce sufficient outdoor, air through the air conditioning equipment to maintain a constant outward escape of air.

To do so a minimal pressure, capable of offsetting the wind pressure, is maintained within the area to be conditioned. Doing so, makes it unnecessary to account for the infiltration load. Nevertheless, when this condition does not prevail, the infiltration load is counted as a heat gain.

STEP 7: Duct Heat Gain.

The same procedure explained in Chapter 3, Heat Load Calculation is applicable for cooling load where the heat loss is counted as a heat gain within the space and included in the total refrigeration load computation.

In addition to the accounting of heat gains from various sources, the designer must remember that thermal storage is present in the structure and its contents. This is an inertia factor that can affect comfort in an air conditioning (or heating) system because it will take time to change the air temperature in a building when the building and its inanimate contents must also change temperature.

Thermal storage is present whether the design engineer takes it into account or not. Accounting for storage will produce a properly sized system, capable of balancing the actual loads and will result in a better, more satisfactory performing installation. Failure to

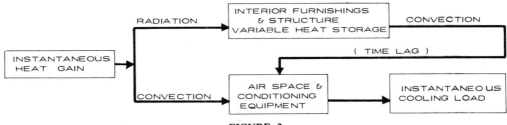

FIGURE 3.

account for storage results in uneconomical, oversized, poor performance systems with high operating costs.

An example of thermal storage is given in Chapter 6.3, which describes how part of the heat gain from solar heat is absorbed by objects and results in thermal lag.

4.2. INSTANTANEOUS HEAT GAIN VS INSTANTANEOUS COOLING LOADS

The instantaneous heat gain from solar heat can be divided into two parts, radiation and convection. The radiative part of the instantaneous heat gain is first absorbed by solid objects within the conditioned space, and is not encountered by the cooling equipment until some time later when it appears in the air stream. The heat storage of the objects are in direct proportion to their mass. The convective portion of the heat gain can be considered to be an instantaneous cooling load, i.e., the heat enters the air stream with only a slight time lag and is conditioned by the cooling equipment.

Figure 3 illustrates the breakdown of the instantaneous heat gain.

Heat lag should be carefully considered in the cooling load calculations. In certain types of buildings the effect of solar radiation is still apparent several hours after the sun has shifted from the exposure. In other types of buildings having a much lighter construction, the heat gain due to solar radiation decreases markedly by the passing of the sun. Some walls, warmed by the sun, may radiate heat long after the passing of the sun, thus requiring lower indoor temperatures to offset the radiant energy.*

*Extracted from Ashrae's *Handbook of Fundamentals*, p. 495 (1972).

5

RESIDENTIAL COOLING LOADS

Residential air conditioning is based on the same principles as the air conditioning of commercial buildings. However, peculiar differences between heat gain and cooling loads which occur in residential buildings are due to the nature of the service and the type of structure. This is due mainly to the following:

1. Residences, unlike many other structures, can be assumed to be occupied, and conditioned, for 24 hr/day, every day of the cooling season. The indoor design temperature is therefore a constant value, usually 75°F.
2. Equipment should be of the smallest possible capacity commensurate with performance in order to minimize initial equipment and distribution system costs as well as operating costs.
3. Most residential systems employ units of relatively small capacity (20,000 to 60,000 Btu/hr) which have no means for controlling output except by cycling the condensing unit. Since the cooling load is largely affected by conditions outside the house and that only a few days of each season are "design" days, it becomes apparent that a partial load situation exists during most hours of the season.
4. Most residences are cooled as a single zone with no means to redistribute cooling capacity from one area to another as loads change during the course of the day.
5. The loads on residential cooling systems are primarily those imposed by heat flow through structural components and by air leakage or ventilation. Internal loads, particularly those imposed by occupants and lights, are small compared to commercial or industrial installations.

6

EXAMPLES OF LOAD CALCULATIONS FOR COMMERCIAL AIR CONDITIONING

This Chapter contains two, step-by-step worked out calculations for a one-story office building in a suburb of New York City. The first is for the heating load, the second for the cooling load. These are typical design problems that begin with information taken from the building specifications and a drawing (or drawings) that are part of a bid document.

Imagine that you are sizing a HVAC system for this one-story office building prior to estimating the cost of the system. As before, the heating load calculations form is used, and this time it contains data taken from the drawing and specifications. Each step in the heating load calculations are explained on an accompanying sheet. The same process is repeated for the cooling loads. A standard cooling load calculations form is filled out with back-up explanations which follow.

Using the building specifications and design conditions given in Table 1, the floor plan in Figure 4, and Figure 1 in Chapter 3, the heating load calculations form was completed.

6.1. HEATING LOAD CALCULATION

STEP 1: Winter Design Conditions.
From Table 41, Section IV, the outdoor design condition is found to be 15°F DB for New York City, New York. (New York City is taken

as the nearest geographical point for Franklin Square data.)

STEP 2: Transmission Loss.
The U values for glass, found in Table 11, Section IV, are somewhat greater for winter than summer. Single glass has a U value of 1.13 and the U for glass block is 0.56.

The difference in dry-bulb temperatures (ΔT) for the heat loss calculations is equal to 57°F. (72°F indoor – 15°F outdoor = 57°F.) Also there is a heat loss through a slab. To calculate this loss, find the linear feet of the exposed edge from the floor plane (Figure 3), which equals 84* linear ft. Then refer to Table 16, Section IV, where the heat loss per foot of exposed slab edge is given as 38 Btu/hr. This loss was determined by using known conditions of 15°F outdoor temperature and 12-in. wide insulation with an R factor of 30.

STEP 3: Ventilation and Infiltration.
The estimated infiltration for the winter can vary with the load from changing wind velocities. By following the personal judgement of 0.75 air changes/hr, a calculation can be made (see Table 18, Section IV). Using Equation 11

---

*Two edges of the slab are exposed (58 LF + 26 LF = 84 LF). The other two edges abut existing buildings.

TABLE 1. Design Conditions and Construction Data for a One-Story Office Building

Building:	One-story office building
Location:	Franklin Square, New York
Indoor Condition:	72°F DB Winter
	75°F DB−64°F WB Summer
Construction:	
Windows	North: 3 each, 4 ft × 6 ft glass block (8 × 8 × 4 in.), no side shading
	West: 1 each, 4 ft × 6 ft fixed type single plate, shading with venetian blinds
Door	North: 3 ft × 7 ft glass swinging door, $\frac{1}{4}$-in. single plate type
Walls	Northwest: 4-in. face with 8-in. common brick and $\frac{5}{8}$-in. plaster
	Southeast: partitions 2 × 4-in. studs with metal lath and plaster both sides (adjacent building)
Roof	Height 12 ft 6 in., construction: 2-in. gypsum deck slab under built-up roof, deck slab on $\frac{1}{4}$-in. asbestos cement board, 2-in. preformed insulation
Ceiling	Acoustical hung ceiling @ 10 ft 0 in. throughout the space
Floor	6-in. concrete slab with perimeter insulation 12-in. wide (R = 3.30)
Overhang	Northwest corner 3 ft 0 in. wide

to determine the cfm of infiltration,

$$\frac{0.75 \text{ (air changes/hr)} \times (1508 \text{ sq ft} \times 12.9 \text{ ft})}{60 \text{ (min.)}}$$

$$= 235 \text{ cfm}$$

Infiltration through doors, due to their being opened, should be added to the above infiltration rate. From Table 17, Section IV, it is found that 15 cfm/door opening (in a 1 hr period) is the approximate infiltration rate through a 36-in. swinging door.

Assuming 6 window openings × 15 cfm/window = 90 cfm the total estimated infiltration will be:

$$235 \text{ cfm} + 90 \text{ cfm} = 325 \text{ cfm}$$

The ventilation requirement, is 325 cfm and 235 cfm is the infiltration estimate. In the heating load form use the larger rate to calcu-

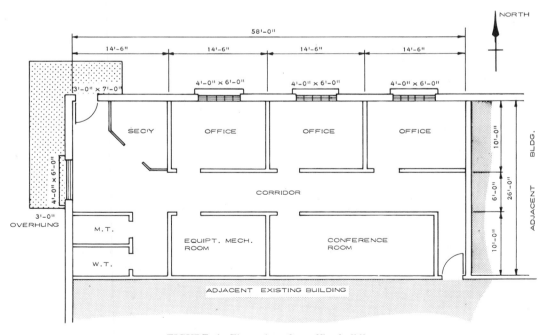

FIGURE 4. Floor plan of an office building.

Location ..... Franklin Square, N.Y. .....   Date ...6/6/78...

HEATING LOAD CALCULATIONS

1. Design Conditions

Dry-Bulb Temp.	°F	Humidity (% RH)	grains/lb
Indoor ....72.		....	....
Outdoor 15		....	....
Temperature Diff. ....57. (ΔT)		....	....

(X) = Insulation factor from tables.
(Y) = Heat loss from tables.
(Z) = Total length of duct, in linear feet (LF)

2. Transmission Loss

	Area (sq ft) gross	net		U factor		ΔT (°F)		Transmission Loss (Btu/hr)
Walls	N. 725	632	×	0.33	×	57	=	11,887
	W. 200	176	×	0.33	×	57	=	3,310
			×		×		=	
			×		×		=	
Glass	N. 24 ×3	72	×	0.56	×	57	=	2,298
	W. 4 ×6	24	×	1.13	×	57	=	1,545
			×		×		=	
Door	7×3	21	×	1.13	×	57	=	1,352
Ceiling			×		×		=	
Roof	58 ×26	1,508	×	0.12	×	57	=	10,314
			×		×		=	
Floor			×		×		=	
Slab	58 + 26	84	×	38	×	—	=	3,192
Other			×		×		=	

Total Transmission = ...33,898......

3. Infiltration–Ventilation

Ventilation ....235......... cfm   ....325....... cfm × ΔT × 1.08 = ...2,007......
Infiltration ....325......... cfm   .... cfm × ΔT × 1.08 = ...

4. Humidification (if added to Heating Equipment Load)

........325........ cfm × ......42...... grains/lb diff. × 0.68 = ...9,282......

5. Duct Heat Loss

Supply duct .... (X) × .... (Y) × .... (Z) ÷ 100 = ...—......
Return duct .... (X) × .... (Y) × .... (Z) ÷ 100 = ...—......

TOTAL HEATING LOAD   ......53,905......

FIGURE 1. Heating load calculation form.

late the heating load imposed by the outdoor air.

At this point, we can sum up the transmission losses and ventilation/infiltration losses. The total heat loss is found to be 53,905 Btu/hr. This does not include the heat required to in-

crease the specific humidity of the infiltrating outdoor air to indoor design conditions.

STEP 4: Humidification.

The ventilation and infiltration of outdoor air is found to be 325 cfm. The difference in

Location ....   Proj. No. ....

Latitude ....   Prepared by ....

Hour   day   month   Date   ....

....

COOLING LOAD CALCULATIONS

1. Design Conditions (Summer)

	Dry Bulb °F	Wet Bulb °F	Specific Humidity grains/lb
Outdoor	91	76	112
Indoor	75	64	72
Temperature Diff.	16		40

2. Solar Heat Gain—

Exposure	Area sq ft		Solar Heat Gain Factor		Shading factor		Sensible Heat Gain Btu/hr	Latent Heat Gain Btu/hr
Glass N.	72	×	30	×	0.40	=	864	
W.	24	×	170	×	0.55	=	2244	
N.	21	×	30	×	0.95	=	599	
		×		×		=		

3. Transmission Heat Gain

Exposure	Area sq ft		U factor		Temp. Diff. °F			
Glass N.	72	×	0.54	×	16	=	622	
W.	24	×	0.81	×	16	=	311	
N.	21	×	1.06	×	16	=	356	
		×		×		=		
Walls		×		×		=		
N.	629	×	0.33	×	6	=	1,245	
W.	176	×	0.33	×	8	=	465	
		×		×		=		
Ceiling		×		×		=		
Roof	1,508	×	0.12	×	68	=	12,305	
Floor		×		×		=		
Partition	1,260	×	0.40	×	5	=	2,520	
		×		×		=		

4. Internal Heat Gain

	No.		Factor		
a) People	10	×	250	(Sensible Heat) =	2,500
b) People	10	×	200	(Latent Heat) =	2,000

c) Lighting	watts/sq ft		area sq ft			
		×		× 3.4 (i)	=	
	5	×	1,500	× 4.1 (ii)	=	30,750

d) Motors	hp.		Eff. Factor			
assumed 50%—	3/4	÷	60%	× 2545 (iii)	=	1,590

e) Appliances   No. H_S or H_L

.... × ....   = ...

.... × ....   = ...

.... × ....   = ...

5. Ventilation

Ventilation (Outside Air)   DB Temp. Diff.

cfm ......_250_...... ×   1.08   × ...._16_....... = ..._4,320_....

grains/lb Diff.

cfm ......_250_...... ×   0.68   × ...._40_......... =   = ..._6,800_.....

6. Infiltration

Infiltration   DB Temp. Diff.

cfm .... ×   1.08   × .... =   = ...

grain/lb Diff.

cfm .... ×   0.68   × .... =   = ...

7. Duct Heat Gain

(b)   (c)   (d)

cfm .... × .... × .... ÷   100   = ...

cfm .... × .... × .... ÷   100   = ...

Total Sensible Heat = ..._60,691_....

Total Latent Heat   = ..._8,900_.....

Total Cooling Load (Sensible plus Latent Heat) ...._69,491_...........

Tons of Refrigeration ...._5.79 say 6 TONS_...........

(i)   3.4 Btu/watt–Heat of Incandescent Lamp   Check Figures:

(ii)   4.1 Btu/watt–Heat of Fluorescent Lamp   Outside Air   = ... %

(iii) 2545 Btu/hr where   Btu/sq ft (Sensible) = ...

1 hp = 746 watts × 3.4 Btu/hr/watt   Btu/sq ft (Total)   = ...

(a)   Total Sensible Heat where Supply Air in cfm is   No. Air Changes/hr = ...

determined   cfm/sq ft   = ...

(b)   Duct Insulation Factor

(c)   Duct Heat Gain Factor

(d)   Total Ductwork Length

FIGURE 2. Cooling load calculation form.

specific humidities between outdoor and indoor air at design conditions is found to be 42 grains/lb. Therefore, the humidification load will be:

325 cfm × 46 grains/lb × 0.68

= 282 Btu/hr (latent heat)

A separate humidifier to supply the latent heat of 9282 Btu/hr will be a better application than to incorporate this load into the primary heating equipment.

STEP 5: Duct Heat Loss.
It's assumed that a duct system is located within the heated space and, therefore, no duct heat losses are being considered.

6.2. COOLING LOAD CALCULATION

Using the building data given in Table 1, the floor plan in Figure 4, and Figure 2 in

Chapter 4 the cooling load calculations form was completed.

STEP 1: Summer Design Conditions.

From Table 41, Section IV, the outdoor design conditions are found to be 91°F DB, 76°F WB for New York City (Central Park). The latitude is 40°. The specific humidity in grains of water vapor per pound of air is found by plotting the indoor and outdoor design conditions on a psychrometric chart. There are 112 grains/lb of air at 91°F DB and 76°F WB, and 72 grains/lb of air at 75°F DB, 64°F WB. The difference in grains/lb (the amount of water the cooling equipment must remove from the outdoor air to maintain indoor design conditions) is: 112 grains/lb − 72 grains/lb = 40 grains/lb.

STEP 2: Solar Heat Gain.

The building shows two different types of glass construction, smooth colorless $8 \times 8 \times 4$ in glass block and $\frac{1}{4}$ in. plate of single weight. The entrance door is a swinging door, 36-in. wide and 84 in. in height, with $\frac{1}{4}$ in. plate single

glass. The single glass window is shaded with a light colored venetian blind. The glass block windows and doors are without shading. From Figure 3, the exposures and window areas are:

North exposure = (4 ft × 6 ft) × 3 = 72 sq ft

and 7 ft × 3 ft = 21 sq ft

West exposure = (4 ft × 6 ft) × 1 = 24 sq ft

It is important to assign a peak cooling load for the solar load on the glass and roof, as well as for internal loads. To arrive at this selection, emphasis must be placed on solar loads, roof loads, and internal loads. Since all of these loads increase and decrease independently of each other during the day and month, a sample of all three must be made for the period of occupancy. The time at which the greatest heat gain occurs should be used to compute the cooling load. In this example, internal loads such as people, lights, and equipment are assumed to be constant, and therefore, it will be necessary to make calculations of solar and roof transmission loads only. An example of this time selection is shown below.

PEAK COOLING LOAD–TIME SELECTION

TIME–12 p.m. (noon)

Material	Exposure	Area sq ft	× Shade Factor	× Solar Factor =	Btu/hr
Glass–block	North	72	.40	30	864
Glass–single	West	24	.55	34	448
Door–glass	North	21	.95	30	598
Roof		1500	.13(U)	30(TD)[a]	5850
				Total heat gain	7760

TIME–2 p.m.

Material	Exposure	Area sq ft	× Shade Factor	× Solar Factor =	Btu/hr
Glass–block	North	72	.40	33	950
Glass–single	West	24	.55	102	1346
Door–glass	North	21	.95	33	658
Roof		1500	.13(U)	53(TD)	10,335
				Total heat gain	13,289

TIME–4 p.m.

Material	Exposure	Area sq ft	× Shade Factor	× Solar Factor =	Btu/hr
Glass–block	North	72	.40	30	864
Glass–single	West	24	.55	170	2244
Door–glass	North	21	.95	30	598
Roof		1500	.13(U)	70(TD)	13,650
				Total heat gain	17,356

[a]Temperature differentials (TD) is the difference between indoor and outdoor design temperatures (in °F).

In the foregoing example the peak load calculations at the various hours are usually done on separate sheets. Although the hours of occupancy in a office building are from 8 A.M. to 5 P.M. the example uses only the hours 12 noon, 2 P.M. and 4 P.M. These times were selected using Tables 31 and 32, Section IV, for the larger solar factors and Table 8A for the greatest equivalent *TD*'s (temperature differentials). The hours of 8 and 10 were intentionally omitted because after inspection of the solar tables and the equivalent temperature differential table, these two times were insignificant by comparison. Therefore, the solar factors selected for 4 P.M. are 30 for North 170 for West.

Once the solar factor for glass is found (Tables 31 and 32), the *U* factor for the roof can be found by using Table 6, Section IV, ($U = 1/R_t$). Shade factors for glass are found in Tables 29 through 34 and Tables 35 and 36, and equivalent temperature differential (*TD*) for the roof in Table 7A.

At this point consideration must be given to the outdoor minus room *TD*. The equivalent *TD*'s are based on an outdoor minus room temperature difference of 20°F. If the actual *TD* varies from the base of 20°F, corrections must be made. When the difference is greater than 20°F, add the excess to the equivalent *TD* found in Table 7A. If the *TD* is less than 20°F, subtract this amount from the equivalent *TD* found in Table 7A.

The result of the calculations in the example, shows the time 4 P.M. is highest and therefore selected for the load calculation form.

Use North solar factor for shaded portion of window. The area of the shaded portion of window must be subtracted from the total window area. Use the solar factor for the unshaded portion of the window that agrees with the direction it faces.

The basic equation for solar heat gain is:

Solar Heat Gain (in Btu/hr)

= Area of Glass (in sq ft) × Solar Factor

× Shading Factor   (Eq. 17)

STEP 3: Transmission Heat Gain

In Figure 3 the windows were comprised of:

72 sq ft glass block (without indoor shading)

24 sq ft fixed single plate (with indoor shading)

21 sq ft single plate, door (without indoor shading)

The coefficient of transmission, *U*, for a glass block without indoor shading can be found in Table 10, to be 0.54. The *U* value for single plate with indoor shading can be found in Table 10 to be 0.81. The *U* value for single glass without indoor shading is found in the same table to be 1.06. The temperature difference for glass is the difference between outdoor and indoor design temperatures, 91°F - 75°F = 16°F.

The following applies to wall, ceiling, roof, floor and partition calculations.

The space between the ceiling and roof will be used as the return air plenum. Therefore, this space must be calculated as part of the conditioned space. This will necessitate the use of a 12 ft 6 in. rather than a 10 ft 0 in. wall height to obtain wall area for heat gain. For this case the *U* factor for a roof without ceiling will be used. The volume of the entire space, from floor to roof, should be used to calculate infiltration. The *U* factors for each building component can be found by using Table 6, Section IV, and are listed on the cooling load calculation sheet.

STEP 4: Internal Heat Gain.

People. There are 10 people in the office area. Table 29, Section IV, gives the rates of heat gain from occupants of conditioned spaces. For an office, it can be assumed that the occupants will be moderately active. The sensible heat factor is given as 250 Btuh and the latent heat factor is given as 200 Btuh. These factors are multiplied by the number of people in the area.

Lights. The office is designed for a lighting density of 5 W/sq ft for a total of 7540 W. The lighting used is of the fluorescent type, and therefore, the ballast of the lights should be accounted for. The 4.1 factor on the calculation form should be used.

Motors. It must be assumed that electric motors from office equipment totaling $\frac{3}{4}$ hp will be located in the space to be conditioned. Referring to Table 30, Section IV, it is found that fractional hp motors range between 50% and 60% efficiency. The equipment is assumed to be. used 50% of the time. These figures are entered on the load calculation form.

Appliances. Where there is significant heat gain from appliances, see Table 28 which lists their sensible and latent output in Btuh. The values found in Table 24 should be entered on the load calculation form.

STEP 5: Ventilation

Table 25 (Section IV) lists the recommended and minimum outdoor air requirements for occupants for a general office where no smoking will occur. It is advisable to use the recommended figure, for this will help assure a first-rate installation. In some cases the figure to use is dictated by local codes. For our example the recommended figure of 25 cfm/person will be used.

Recommended ventilation

= ventilation/person in cfm

× no. of persons

= 25 cfm × 10

= 250 cfm

STEP 6: Infiltration

Table 18, Section IV, suggests various air changes/hr for use in estimating infiltration.

The example building has two sides exposed with fixed windows, and one door. The infiltration through the fixed windows can be considered negligible. Table 21 suggests 0.90 air change/hr for ordinary windows, and 0.45 air change/hr with weatherstripping. Good judgement would indicate that the lower figure of 0.45 air changes/hr should be used as the most

representative of the circumstances. Use Equation 11 to arrive at cfm of infiltration where

$$\frac{\text{sq ft} \times \text{ceiling height} \times AC}{60 \text{ min.}} = \text{cfm}$$

and the infiltration through walls and windows;

$$\frac{1508 \times 10 \times 0.45}{60} \cong 114 \text{ cfm}$$

Prior calculations have been concerned with normal infiltration through walls, windows, and closed doors. At this point, it is necessary to determine infiltration through doors due to persons entering and leaving the conditioned area.

Figure 3 shows one 36-in. swinging door which has been estimated as opening six times per hour. Referring to Table 17, Section IV, it is found that 7.5 cfm/door opening (in a 1 hr period) is the approximate infiltration rate.

Infiltration through swinging door:

6 openings/hr × 7.5 = 45 cfm

Total infiltration:

114 + 45 = 159, say 160 cfm

The next step is to enter both the ventilation requirement and infiltration estimate on the cooling load form. The larger of the two rates is used to calculate the cooling load imposed by the outdoor air. In this example the ventilation rate exceeds the infiltration rate; therefore, the 250 cfm rate is used.

STEP 7: Duct Heat Gain.

This addition is intended to account for the heat entering the supply air stream through the air duct walls from spaces outside of the air conditioned space. For this example we assume that the air conditioning equipment with the ductwork distribution system will be located within the conditioned space. Therefore, the duct heat gain is negligible.

Adding all of the individual loads from the steps noted above, the total cooling load is found to be 69,491 Btu/hr, or 6 tons* of refrigeration.

Note that the total cooling load should be

*One ton is equal to 12,000 Btu/hr. See Chapter 8.2 for an explanation.

used as a guide for initial equipment selection only. The actual equipment selection should be based on the individual sensible and latent requirements vs. the specific sensible/latent capacity of the air conditioning system under consideration.

7

PSYCHOMETRICS

7.1. THE PSYCHROMETRIC CHART

The psychrometric chart is the air conditioning designer engineer's basic tool to determine the thermodynamic properties of air. This chart's importance is such that it can be the connecting link between the load calculation and the type of system and system operation that is required.

The psychrometric chart is a graphical presentation of the thermodynamic properties of air. This chart is for the normal temperature range common in air conditioning practice and based on standard atmospheric pressure 29.29 Hg. By use of this chart, it is possible to determine many properties of moist air when two of these properties are known.

Such a chart is shown in Figure 5. There are many intersecting lines for each set of conditions, and the properties for a specific moist air mixture are shown and described. Each property and its units are keyed to Figure 5 by encircled numbers as follows:

① *Dry-Bulb Temperature* is the temperature of the air indicated by an accurate thermometer expressed in degrees Fahrenheit. On the psychrometric chart these are vertical lines, originating from the dry-bulb side at the bottom of the chart.

② *Wet-Bulb Temperature* is the temperature of the air recorded by an accurate thermometer, whose bulb has been covered by a wetted wick and placed in the path of a current of rapidly moving air, expressed in degrees Fahrenheit. On the psychrometric

chart these lines are sloping, originating at the saturation line. (Saturation line also equal 100% relative humidity curve.)

③ *Dew Point Temperature* is the saturation temperature corresponding to the vapor pressure and relative humidity of moist air, expressed in degrees Fahrenheit. On the psychrometric chart, these temperatures are read on the saturation curve. At saturation, the dew point temperature is equal to dry-bulb temperature and is also equal to wet-bulb temperature.

④ *Specific Humidity* (*Humidity Ratio*) is the weight of water vapor per pound of dry air in a given air mixture, expressed in grains of water per pound of dry air. On the psychrometric chart these are horizontal lines.

⑤ *Relative Humidity* is the ratio of the partial pressure of the water vapor in an air mixture to its saturation pressure at the same dry-bulb temperature, expressed in percent. On the psychrometric chart these are curved lines.

⑥ *Specific Volume* is the cubic feet of the air mixture per pound of dry air. On the psychrometric chart these are shown as sloping lines.

⑦ *Enthalpy* is the summation of sensible and latent heats in 1 lb of dry air and the grains of moisture contained in it between assigned datum points ($0°F$ for air and $32°F$ for water), expressed in Btu/lb of dry air.

The enthalpy values given on the psychrometric chart are for saturated air. In any air

1. DRY BULB TEMPERATURE

2. WET BULB TEMPERATURE

3. DEW POINT TEMPERATURE

4. SPECIFIC HUMIDITY

5. RELATIVE HUMIDITY

6. SPECIFIC VOLUME

7. ENTHALPY SCALE

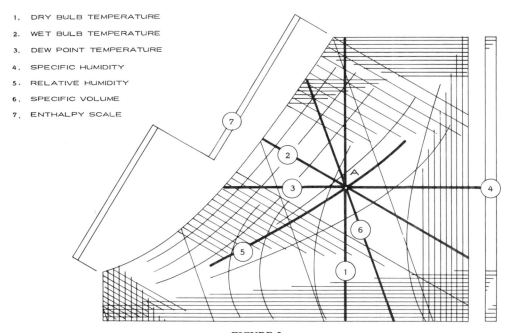

FIGURE 5

conditioning process where moisture is either being removed from or added to the air, it is sufficiently accurate to base the calculation on "enthalpy at saturation" and to assume that this enthalpy is constant for a given wet-bulb temperature regardless of the dry-bulb temperature. Where greater accuracy is required, refer to the Chapter on thermodynamics in the ASHRAE Guide. On the psychrometric chart the enthalpy values are read on a scale corresponding to the wet-bulb temperature.

Other terms associated with air properties are sensible heat factor, standard barometric pressure, and standard air. These are defined below.

Sensible Heat Factor is a ratio of the sensible heat to the total heat in the process. (See Equations 2 through 7 in Section I, Chapter 2.) Heat removed from a room by an air conditioning unit consists of room sensible heat (H_S) and room latent heat (H_L). To properly maintain a given design condition a proportion of the sensible and latent heat must be removed. This proportion or psychrometric process is called sensible heat factor and expressed as:

$$SHF = H_S/(H_s + H_L) \qquad \text{Eq. (17a)}$$

Standard Barometric Pressure is the standard sea level barometric pressure which is 29.29 in. Hg absolute.

Standard Air is air having a density of 0.075 lb/cu ft and a specific volume of 13.33 cu ft/lb. This is substantially equivalent to dry air at 70°F and 29.92 in. Hg, absolute barometric pressure.

Figures 6 and 7 correlate a typical air conditioning system and the corresponding affects on the air as shown on a psychrometric chart. In Figure 6, outdoor air (1) is mixed with return air (2) from the room and enters the apparatus (3). Air flows through the conditioner and is supplied to the space (4). The air supplied to the space picks up heat (Q), and moisture (Q_w), and the cycle is repeated.

On the psychrometric chart note that warm, outside air moved from (1) to (3) it mixes with the relatively cooler return air, slightly lowering both the dry-bulb and wet-bulb temperature as well as the humidity ratio. It is the air conditioner that brings the air to its design conditions at (4). As the air picks up the heat load from the room it begins its return to the point where it mixes with outside air.

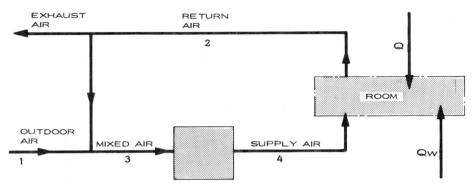

FIGURE 6

7.2. EXAMPLES OF PSYCHROMETRIC CHART USAGE

The following examples of processes of most interest to HVAC designers are: *mixed air temperature, dehumidification, sensible heat factor, and ventilation.*

7.2.1. Mixed Air Temperature

To select cooling equipment, the entering dry- and wet-bulb temperatures or relative humidity must be determined. If outside air is provided for ventilation, it will have different properties than room air. There is the need to determine the properties of the mixture.

Example:
    If 7200 cfm of room air at 75°F DB and 50% RH is mixed with 800 cfm of outside air at 95°F DB and 75°F WB, find the DB and WB temperature of the mixture (T_m).

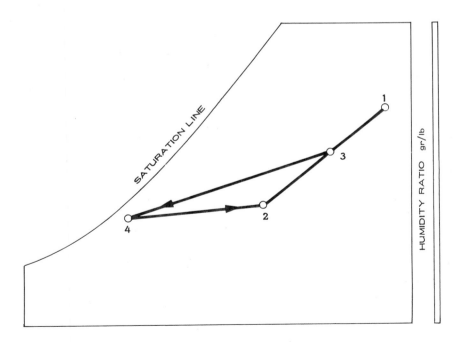

FIGURE 7

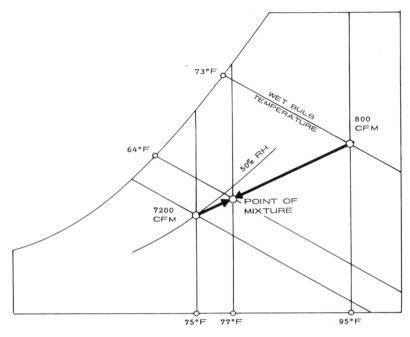

73°F

WET BULB
TEMPERATURE

800
CFM

64°F

50% RH

POINT OF
MIXTURE

7200
CFM

75°F   77°F   95°F

DRY BULB TEMPERATURE   °F

FIGURE 8

There are two ways of doing this. First by calculation:

$$T_m = \frac{(7200 \times 75) + (800 \times 95)}{7200 + 800}$$

$$= 77°F \; DB$$

The second method is with the aid of the psychrometric chart as shown below in Figure 8; the conditions of the room air and outside air are plotted on the chart and a line is then drawn between them. The mixture will be on this line which represents the locus of all possible mixtures having these end points.

With the addition of outside air, 8000 cfm flows through the system, the temperature change of the mixture will be proportional to the volumes of outside to inside air multiplied to the difference in their temperatures. On this basis, the temperature of the mixture is:

$$T_m = 75 + \left[\frac{800}{8000} (95 - 75) \right]$$

$$T_m = 77°F \; WB$$

By locating the 77°F DB point on the line connecting the properties of the two constituents, the wet-bulb temperature may be read as 64°F WB, see Figure 8.

7.2.2. Dehumidification

To cool and dehumidify a space in summer, the supply air must remove the sensible heat gain from the sun, lights, people, transmission, equipment, etc. To do this, the supply air must be cooler than the room, so that as it warms up to room temperature, it will absorb the heat gain in the space. The supply air must therefore, have a lower specific humidity or moisture content than the air in the room if there is a latent heat gain in the space. This drier air can absorb the moisture given off by the sources above and carry it back to the coil where the moisture is condensed. The process appears on the psychrometric chart shown in Figure 9 in the following way:

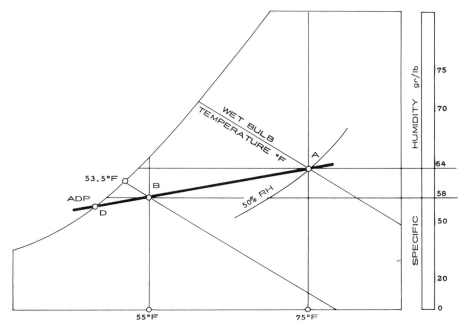

FIGURE 9

Assume that all air passing over a coil is recirculated room air, with properties commensurate with Point A. To get air from the coil at 55°F DB and 53.5°F WB, Point B, the average surface temperature of the coil must be at Point D. This temperature is often called the apparatus dew point or ADP. This would be the saturation temperature of the air having the same ratio of sensible to total heat. Not all the air passing through a coil actually contacts the coil surface. Point B represents the mixture of air, part of which has contacted the coil and is at condition D, and the smaller quantity of air at condition A which bypassed the coil surface.

If 5000 cfm of air at condition B in Figure 7 maintains a room at 75°f DB and 50% RH, how much sensible, latent, and total heat is the air absorbing in the room?

SOLUTION (from Equations 2, 3, and 4):

Sensible Heat Removed
= 1.08 × 5000 × (75 – 55)
= 108,000 Btu/hr

Latent Heat Removed
= 0.68 × 5000 × (64 – 58)
= 20,400 Btu/hr

Total Heat Removed
= 108,000 + 20,400
= 128,400 Btu/hr

Sensible Heat Factor.   Referring again to the data used in the example, on Figure 9 what is the ratio of sensible heat to total heat?

SOLUTION:

$$SHF = \frac{\text{Sensible heat } (H_s)}{\text{Total heat } (H_T)}$$

and

$$SHF = \frac{108,000 \text{ Btu/hr}}{128,400 \text{ Btu/hr}} = 0.84$$

This ratio is called the Sensible Heat Factor (SHF) and the condition is shown in Figure 10.

Looking at dehumidification another way is by means of a psychrometric chart as shown in Figure 10.

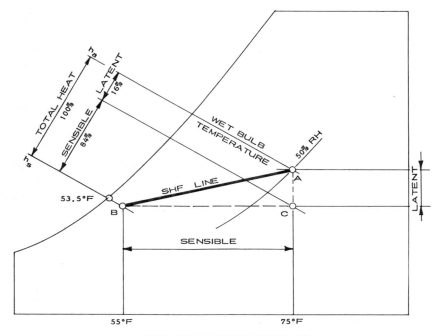

FIGURE 10

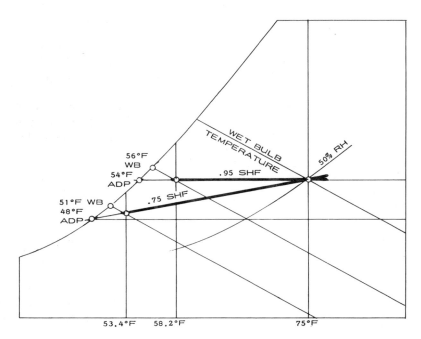

FIGURE 11

Each pound of air supplied to the room removes a finite amount of total heat which is equal to the difference in enthalpy between the room and supply air, or $(h_a - h_s)$. (The letter h is the accepted symbol for enthalpy.) The slope of line AB between the WB lines for the room and the supply air varies the amount of sensible and latent heat. In this case, the slope of line AB is such that 84% of the total is sensible heat and the difference between 100 and 84% equals 16% latent heat capacity.

The slope of line AB varies with the application. In office, apartment and motel rooms, the line is almost flat. For these cases the sensible heat capacity nearly equals the total heat capacity. The resulting sensible heat factor is above 0.90 to 0.95 for the latent load is small. As the slope of the SHF line becomes steeper, the sensible heat capacity decreases in relation to the total, and the latent heat increases. In restaurants, auditoriums, or classrooms, the ratio of sensible to total load would normally give an SHF in the 0.70 to 0.80 range.

Using the psychrometric chart shown in Figure 11 we see that the application with the 0.75 SHF requires that the coil operate at a lower ADP than one with a 0.95 SHF. As the SHF varies, different air quantities are required per ton of refrigeration (see Section I, Chapter 8) of total room load.

Example 5:

1. Find the air quantities per ton of refrigeration required for an office assuming a 0.95 SHF and 0.2 BF factor:

$$T_m = 0.2 \, (75°F - 54°F) + 54°F$$

$$= 58.2°F$$

$$\text{cfm/ton} = \frac{12000 \text{ cfm/ton} \times 0.95 \, SHF}{1.08 \, (75°F \text{ DB} - 58°F \text{ DB})}$$

$$= 620 \text{ cfm}$$

2. Find the air quantities per ton of refrigeration required for a classroom assuming a 0.75 SHF and 0.2 BF factor:

$$T_m = 0.2 \, (75°F - 48°F) + 48°F$$

$$= 53.4°F$$

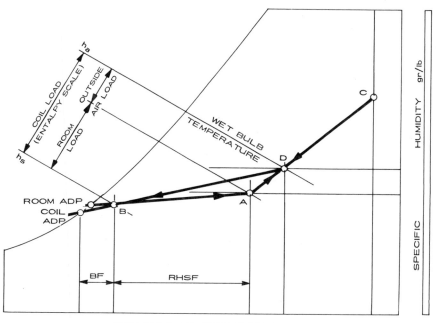

DRY BULB TEMPERATURE °F

FIGURE 12

$$\text{cfm/ton} = \frac{12000 \text{ cfm/ton} \times 0.75 \text{ } SHF}{1.08 \text{ } (75°F - 53.4°F)}$$

$$= 386 \text{ cfm}$$

Note that the higher the *SHF*, the more air that is delivered per ton of refrigeration.

Ventilation. Considering to add ventilation and using Figure 12 then the process will be as follows:

Air entering the coil, point D, is now different from room conditions, point A. The condition at point D may be determined if the cfm of outdoor air and the air conditioner cfm are known. The load characteristics of the applica-tion establishes the slope of the room sensible heat factor, (RSHF) line AB.

Ideally, the air leaving the dehumidifying coil at condition B should fall on the RSHF line. Air at condition B is delivered to the room and as it absorbs, the sensible and latent load, it warms up along line AB, maintaining the room at condition A.

An assumed "bypass" factor of the coil and the entering and leaving conditions at D and B will establish a point called ADP, this is the point at which the coil operates.

With outside air for ventilation, the load on the coil is made up of the room load (sensible and latent) plus the heat required to cool the outside air from its condition at C to the room condition at A.

8

REFRIGERATION CYCLE

Refrigeration is the general term used to describe the process of removing heat from spaces, objects, or materials, and to maintain them at temperatures below the temperature of their surroundings. In order to produce a refrigeration effect, it is merely necessary to expose the material to be cooled to a colder object of environment and allow heat to flow in its natural direction, i.e., from warmer material to colder material. The term is usually applied to an artificial means of lowering the temperature.

Mechanical refrigeration may be defined as a mechanical system or apparatus so designed and constructed that, through its function, heat is transferred from one substance to another. The refrigeration process is a mechanical cooling based on the use of a low boiling point, heat transfer media called *refrigerants*. Commonly used refrigerants such as R-12 and R-22 are put through a continuous process of compression, expansion, evaporation, and liquefaction called a refrigeration cycle.

Figure 13 gives a general idea of this type of refrigeration cycle which is the basic building block of an HVAC system.

8.1 REFRIGERATION CYCLE

Again referring to Figure 13, the liquid refrigerant enters the thermostatic expansion valve from the high-pressure side of the system and passes through an orifice, which reduces the pressure of the refrigerant. Due to its reduced pressure, the liquid refrigerant begins to boil and flashes into vapor. From the expansion valve, the refrigerant passes into the cooling coil (or evaporator). The boiling point of the refrigerant at the low pressure in the evaporator is about 20°F less than the temperature of the space surrounding the coils in the evaporator. As the liquid boils and vaporizes, it picks up its latent heat of vaporization from the surroundings, thereby removing heat from the space. The refrigerant continues to absorb latent heat of vaporization until all the liquid has been vaporized. By the time the refrigerant is ready to leave the cooling coil, it has not only absorbed this latent heat of vaporization, but has also picked up some additional heat, i.e., the vapor has become superheated. As a rule, the amount of superheat is 4° to 12°F.

The refrigerant leaves the evaporator as a low-pressure super-heated vapor having absorbed heat and thus cooled the space to the desired temperature. The remainder of the cycle is concerned with disposing of this heat and converting the refrigerant back into a liquid state so that it can again vaporize in the evaporator and repeat the cycle.

The low-pressure superheated vapor is drawn out of the evaporator by the compressor, which also keeps the refrigerant circulating through the system. In the compressor, the refrigerant is changed from a low-pressure, low-temperature vapor to a high-pressure vapor, and its tempera-

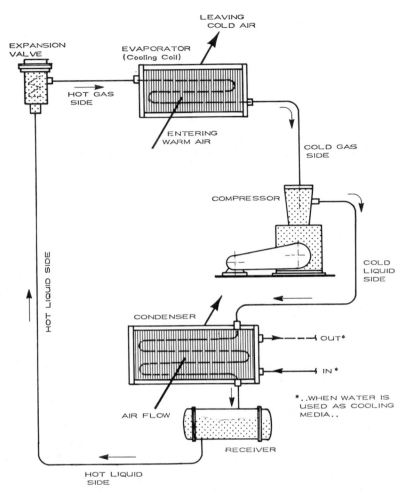

FIGURE 13. Schematic representation of refrigeration cycle.

ture rises accordingly. The high-pressure refrigerant is discharged from the compressor into the condenser. Here the refrigerant condenses, giving up its superheat (sensible heat) and its latent heat of vaporization to the ambient air in an air-cooled condenser, or to the cooling water in a water-cooled condenser. The refrigerant, still at high pressure, is now a liquid again. From the condenser, the refrigerant flows into a receiver, which serves as a storage place for the liquid refrigerant in the system. From the receiver, the refrigerant goes to the thermostatic expansion valve, and the cycle begins again.

8.2. MEASURE OF REFRIGERATION

In air conditioning work there is a convenient unit of measure called ton of refrigeration. One ton of refrigeration gives the same amount of "cooling" as the melting of one ton of ice over a 24-hr period. When 1 lb of ice at 32°F melts, it absorbs 144 Btu. One ton of 32°F ice absorbs 2000 times as many, or 288,000, Btu when it melts. Therefore, one ton of refrigeration will absorb 288,000 Btu over a 24-hr period. This amounts to 12,000 Btu being removed per hour, or 200 Btu/min. It is most convenient to remember the rate as being 12,000 Btu/hr for one ton of refrigeration.

9

WEATHER DATA ANALYSIS FOR HVAC ENERGY CONSUMPTION

9.1. DATA ANALYSIS METHODS

Engineers, designers, contractors, and others resort to a great variety of methods for predicting system operating conditions and for calculating energy consumption in commercial and industrial buildings. Today, the ones most commonly used are:

- degree day method
- equivalent full load hours method
- temperature frequency occurrence (Bin) method

9.1.1. Degree Day Method

The degree day method is a tool used in the HVAC industry to estimate the consumption of energy by heating plants during winter weather. The method was developed for residential heating, where the fuel oil, gas, and electricity consumption began the heating cycle as the outside temperature reached 65°F. This concept of "65°F base" was adopted as the average condition when the mean daily outdoor temperature is below 65°F. There are as many degree days as there are degrees difference in temperature between the mean temperature for the day and the 65°F base. The yearly degree days are therefore, the sum of all days in the year on which the mean temperature falls below 65°F, multiplied by the number of degrees difference. New York City for example, has an average of 4871 degree days per year.

To estimate energy consumption for heating only, the following equation was adopted by the ASHRAE.

$$F = U \times N_b \times D \times C_f \qquad \text{(Eq. 18)}$$

where

F = fuel consumption for the estimated period.

U = unit fuel consumption, or quantity of fuel used per degree day. This value is a unit of fuel consumption per degree day obtained from operating information listed in Table 4, Section IV.

N_b = number of "building load units." This value depends on the particular building for which the estimate is being prepared, and must be found by surveying plans, or by observation, or by measurement of the building.

D = number of degree days for the estimate period–Table 5, Section IV.

C_f = temperature correction as per Table 3, Section IV.

9.1.2. Equivalent Full Load Hours Method

This method (the energy consumption for each of several buildings of a given type) is expressed as the number of hours of operation at full cooling load which would produce the same energy consumption. These figures are averaged to arrive at the equivalent full load hours for a given type of building in a given climate. Annual cooling energy requirements are then estimated for a specific building by multiplying the peak cooling load of the building by the equivalent full load hours for that type of building. Because of the great diversity in the physical and operational characteristics of even similar types of buildings, this method is used only for a "ball park" estimate and not applied to design criteria for energy consumption.

9.1.3. Temperature Frequency Occurrence (Bin) Method

This method of estimating has become more frequently used in recent years. In the basic version, the calculated heat gain or loss of a building is expressed as a function of the prevailing ambient dry-bulb temperatures. Then a count is made of the number of times each month or year that the average temperature for each hour of the day was within a certain range or bin. A bin is merely a group of temperatures with upper and lower limits. For example:

$$\underset{45.1-46-47-48-49-50}{\overset{45 - 50}{\text{Bin}}}$$

Since temperatures can change frequently by one, two, or more degrees within an hour, it is difficult to establish the exact length of time a temperature was at a precise degree point. Therefore, temperatures are grouped into small bins and length of occupancy can be measured more precisely.

These temperature bins are usually in increments of 5° to 10°F, and are further categorized into occupied and unoccupied periods. The Air Force prints a manual used as a source for this type of temperature frequency data. There is also much data available from the Environmental Science Service Administration of the U.S. Department of Commerce, and from the *ASHRAE Handbook.*

The basic bin method provides for load calculation at various temperatures. It assumes that the internal load and solar radiation load are constant during the operating period covered by the temperature bins. This drawback can be overcome somewhat by establishing temperature bins of 5°F and 1-hr increments instead of 8-hr increments. This allows the internal load to be varied each hour of the day and matched to the solar load for that hour, but would not make allowance for various types of occupancy and operational days during the month.

Therefore, individual variations incorporating occupancy, sunshine percentage, etc., can be worked out to provide a fairly accurate energy consumption estimate.

9.2. GENERAL CHARACTERISTICS OF HOURLY METHODS

The greatest flexibility and accuracy in predicting energy requirements is achieved by those procedures which account for the various influencing factors on an hour-by-hour basis. One of the factors which greatly affects the building thermal load is the changing external environment (ambient weather and solar radiation). A calculation which can handle this factor on an hour-by-hour basis can best predict the thermal requirement pattern for a typical year's operation. While an exact calculation of the instantaneous heat gain or loss for each hour is an unreasonable expection because of inexact data, treating a large number of factors in great detail will best predict actual thermal load patterns. Some of the variable factors which can be handled in detail are:

1. The solar load as a function of building construction, location, and orientation, time of day and year, internal and external shading devices and patterns, and cloud cover.
2. The hourly variations in building occupancy and the operation of lights and other internal heat sources for different types of days during the year.
3. Infiltration.
4. A control temperature, and night setback or system shutoff during unoccupied periods.
5. The effects of lag and heat storage in de-

termining instantaneous heat gain and carryover loads.

All of these features permit a calculation procedure to predict yearly room loads and energy consumption with some degree of precision.

9.3. EXAMPLES OF DEGREE DAY CALCULATIONS

9.3.1. Estimating Gas Consumption

Estimate the gas needed in cubic feet for a heating system located in New York City where the heating season has 4871 degree days (see Table 5, Section IV). The gas heating value is 800 Btu/cu ft. The heating system uses a gas furnace and is designed for 55,000 Btu/hr with an outdoor temperature of 10°F and an indoor temperature of 70°F.

Solution:
Given: Heat Loss = 55,000 Btu/hr or 55 MBH
 Degree days = 4871 (this is D in Equation 18)
From Table 1, Section IV, an efficiency of 70% is assumed. From Table 4, Section IV, using 70% efficiency for gas usage, a factor of 0.0049 is given for 0°F outside and 70°F inside conditions (this is U in Equation 18). From Table 3, Section IV, a correction factor of 1.167 is used to suit outdoor and indoor conditions (this is Cf in Equation 18).
Therefore:

$$\mu \times C_f = \text{correct } \mu \text{ value}$$

$$0.0049 \times 1.167 = 0.00572$$

Accounting for correct unit values where:

$$1 \text{ therm} = 100,000 \text{ Btu}$$

$$\text{Heating value of gas} = 800 \text{ Btu/cu ft}$$

Applying Equation 18 we find:

$$F^* = 0.00572 \times 55 \times 4871 = 1532 \text{ therms}$$

$$= 1532 \times 100,000/800$$

$$= 191,500 \text{ cu ft}$$

*Equation 18 lists the term N_b, which is the number of building load units. When available, this term is used to calculate the hourly requirements for installed radiation. In this example N_b units are neglected.

9.3.2. Estimating Oil Consumption

Using the same heating system in the previous example and assuming the heating system uses an oil-fired boiler with an oil heating value of 144,000 Btu/gal., find the oil consumption for the heating season.

SOLUTION:
Given: Heat Loss = 55,000 Btu/hr or 55 MBH
 From Table 1, Section IV, an efficiency of 70% is assumed. From Table 4, Section IV, using a 70% efficiency for oil, a factor of 0.00347 is given for a 0°F outside condition and a 70°F inside condition, and a heating value of 141,000 Btu/gal. From Table 3, at the end of the book, a correction factor of 1.167 is used to suit outdoor and indoor conditions. Therefore:

$$0.00347 \times 1.167 = 0.00405$$

Accounting for the correction for the unit heating values of oil used:

$$0.00405 \times 141,000/144,000$$

$$= 0.00397 \text{ gal. of oil}/1000 \text{ Btu/hr}$$

Applying Equation 18 we find:

$$F = 0.00397 \times 4871 \times 55 = 1064 \text{ gal.}$$

9.3.3. Estimating Coal Consumption

Use the same heating system as in previous examples and assume the heating system uses a coal furnace with the coal heating value of 13,000 Btu/lb. Find the coal consumption used during the heating season.

SOLUTION:
Given: Heat Loss = 55,000 Btu/hr or 55,000 MBH
 From Table 1, Section IV, an efficiency of 80% is assumed. From Table 4, using an 80% efficiency for coal usage, a factor of 0.0357 is given for a 0°F outside condition and a 70°F inside condition, and a heating value of 12,000 Btu/lb. From Table 3, a correction factor of 1.167 is used to suit indoor and outdoor conditions.
Therefore:

$$0.0357 \times 1.167 \times 12,000/13,000 = 0.03846$$

Applying Equation 18 we find:

$$F = 0.03846 \times 55 \times 4871 = 10,304 \text{ lb}$$

9.3.4. Estimating Electrical Energy Consumption

With increased numbers of electric heating installations, operating cost estimates and experience records are accumulating. Experience indicates that Equation 18 cannot be used to estimate seasonal power consumption for electrical heating. Even using a utilization efficiency of 100%, all data available indicate that the use of Equation 18 will yield kilowatt-hour consumptions above those observed for residences that are built and insulated in accordance with NEMA requirements. This does not mean that the efficiency must be greater than 100%, but rather that calculated heat losses for these residences are higher than actual.

SOLUTION:
The following formula,* taken from the NEMA Manual for Electric House Heating, is the best

*To be applicable, the method of calculating the heat loss must be in accordance with the procedure specified in the National Electrical Manufacturers Association (NEMA) manual.

available for estimating the annual energy use of residential electric heating systems:

$$\text{kWh} = \frac{(H_L) \times (DD) \times C}{(TD)} \quad \text{(Eq. 19)}$$

where

kWh = annual kilowatt-hour consumption
C = constant (see note below)
H_L = heat loss of residence in kilowatts (Btu/hr divided by 3415)
DD = annual degree days
TD = difference between indoor and outdoor design temperature (in °F)

Note: The constant C depends on a number of variables such as weather conditions in the locality, orientation, design and construction of the residence, living habits of the occupants, design of heating system, and internal sources of heat. To be applicable, the method of calculating the heat loss must be in accordance with the procedure specified in the *NEMA Manual*. The *NEMA Manual* reports that experience with millions of electric heating installations has shown a conservative figure of 15 for C when three-quarters of an air change per hour is figured for infiltration and exhaust, and 17 when half an air change is figured.

10

AIR CONDITIONING SYSTEMS

10.1. GENERAL

Continuous changes in HVAC design applications have taken place since the origin of the basic air conditioning system about 55 years ago. This basic system consisted of a centrally located forced warm air heating and ventilating furnace which distributed tempered air through ducts. As more functional and economical demands for individual buildings developed, so did the science of HVAC. Demands for close control of the environment required new approaches for HVAC systems.

We see today a multitude of different systems. They vary according to application and are modular in many respects so that adding or subtracting different components can satisfy different design conditions. The main categories of the most commonly used systems and subsystems are:

All-Air Systems
    single duct—constant volume
    dual duct—high or low velocity
    single zone—with reheat
    multizone
    variable air volume (VAV)
Air—Water Systems
    induction
    fan coil with central air
    radiant panels with supplementary air
All-Water Systems
    2-pipe
    3-pipe
    4-pipe

Direct Expansion
(DX) system

10.2. ALL-AIR SYSTEMS

An all-air system is one in which the air is treated in a central refrigeration plant. The cold air is supplied to a space via ducts and distributed by means of terminal outlets or mixing terminal outlets. There are no additional cooling needs at the treated space. However, heating can be accomplished by a separate air, water, steam, or electric system.

The all-air system can be adopted for all types of comfort or process air conditioning systems. It can be used for individual control of conditions in systems having multiple zones, such as in office buildings, schools and universities, laboratories, hospitals, stores, hotels, and ships. All-air systems are also used for special applications where a need exists for close control of temperature and humidity, including clean rooms, computer rooms, hospital operating rooms, and textile industrial buildings.

10.2.1. Single Zone—Constant Volume

The simplest form of the all-air system is a single conditioner serving a single temperature control zone. The unit may be installed within or remote from the space it serves and may operate either with or without a distributing ductwork. Well-designed systems can maintain the temperature and humidity closely and efficiently

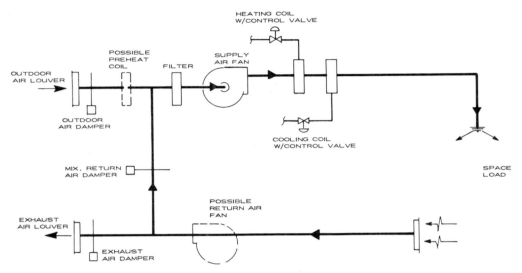

FIGURE 14. Single-duct system (flow diagram).

and can be shut down when desired without affecting the operation of adjacent areas.

This system responds to only one set of space conditions so its use is limited to situations where variations occur approximately uniformly throughout the zone served or where the load is stable. When installed in multiple, single zone systems can handle a variety of conditions efficiently.

This system would have applications for small department stores, individual shops in a shopping center, individual classrooms for a small school, computer rooms, etc. A rooftop unit, for example, complete with a refrigeration system, serving an individual space would be considered a single zone system. The refrigeration system, however, may be remote and serve several single-zone units in a larger installation. A schematic of the single-zone constant volume system is shown in Figure 14.

10.2.2. Dual Duct

A dual-duct system is one where all the air is centrally treated (cooled, heated, filtered, etc.), and distributed to the space by means of the two main ducts. One duct carries the cold air and the other the warm air, thus, providing cold or hot air at all times. In each conditioned space a mixing box or valve is responsive to a room thermostat and mixes the two airstreams

in proportions to satisfy the heat load of the space. The return air is usually handled in the conventional manner.

Schematics of dual-duct systems are shown in Figures 15 and 16. The difference between each system is that the first is used for high-velocity air, the second for low-velocity air. This means that the first uses high-velocity ductwork and therefore small duct sizes. Also, the air terminal box can serve as a mixing box and sound attenuator. The low-velocity system, or perhaps better stated as a normal velocity system, uses larger ducts and has no terminal boxes.

Some advantages of a dual-duct system, including those common to all-air systems are:

1. When 100% air systems are used, there is complete absence of water, steam and drain piping, electrical equipment, wiring, and filters in conditioned spaces.
2. The system has heating and cooling capabilities making it unnecessary to place additional mechanical equipment at the perimeter of the building, and thereby save valuable floor area.
3. Zoning of central equipment is seldom required. Under some design and operating conditions, however, it may be advantageous to provide separate air handling equipment for exterior and interior zones.

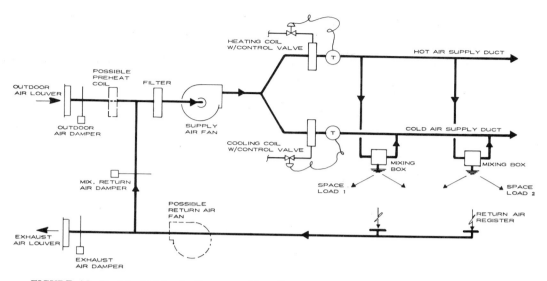

FIGURE 15. Dual-duct high velocity system (flow diagram). *Note:* T = thermostat, M = mixing valve.

4. Systems with terminal volume regulation are self-balancing.

Some of the disadvantages of a dual-duct system are:

1. The arrangement of two parallel ducts with crossovers to terminal points will require special attention on the part of designer and extra care by the installer.
2. Operating economy of constant volume, dual-duct systems is not as good as for variable volume systems.

3. Due to space limitations in the majority of installations, velocities and pressures are higher in the duct system than those required in other installations.
4. Humidity control without hot deck heat, during high ambient wet-bulb and high internal latent load conditions can be poor. Many packaged multizone systems are designed or operated in this fashion,

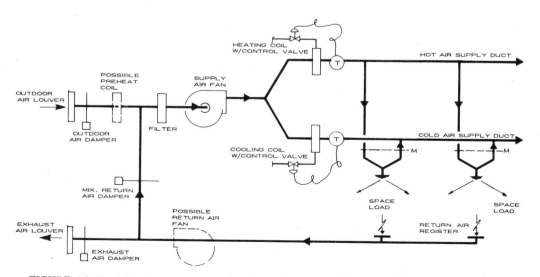

FIGURE 16. Dual-duct low velocity system (flow diagram). *Note:* T = thermostat, M = mixing valve.

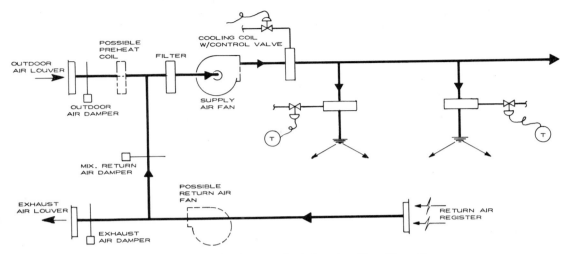

FIGURE 17. Single-zone reheat air system (flow diagram).

while few dual duct systems are treated that way.

10.2.3. Single Zone with Reheat

This system consolidates all major equipment in the machine room except for the reheat element which is located near the room or area to be conditioned. Primary treated air is supplied at a constant volume through a single duct to room units, each of which is equipped with a small steam or hot water reheat coil. Induction type reheat units provide convection heating when the primary air is turned off during nights and weekends.

The reheat system offers the designer great flexibility in providing zones in the initial design stages. These may be readily revised during construction of the building as changes are made. Field changes to accommodate zoning revisions require only the simple addition of a heating coil or terminal unit. A schematic of a single-zone with reheat system is shown in Figure 17.

10.2.4. Multizone Systems

The multizone system is used when a relatively small number of zones are fed from a single, central air handling unit. The temperature requirements of the different zones are met by mixing cold and warm air through zone dampers at the central air handler in response to zone thermostats. The mixed conditioned air is distributed throughout the building by a system of single-zone ducts. Either packaged units,* complete with all components, or field-fabricated apparatus casings may be used. The return air is usually handled in a conventional manner. The multizone system is similar in all other respects to the dual-duct system.

From an economic and practical standpoint, multizone systems can handle more than one room with a single duct. Although dual-duct systems often serve the same purpose, their incremental cost for room-by-room zoning (with parallel ducts already near all rooms) is appreciably lower and therefore easier to justify. Multizone packaged equipment is usually limited to about 12 zones, while built-up systems can have as many as can be physically incorporated in the layout. A schematic of a multizone system is shown in Figure 18.

There are a number of design considerations in a multizone system. These are:

1. Rooms with similar heat loads for areas in the same zone should be grouped so that the response of all areas from internal loading changes or external forces such as

*Packaged units are fabricated as one integral unit at the manufacturer's plant.

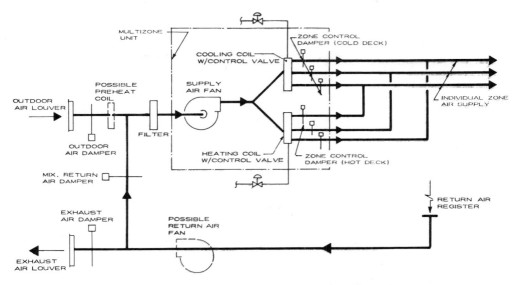

FIGURE 18. Multizone airstream system (flow diagram).

solar gain and thermal transmission (gain or loss) are approximately equal.

2. Any zone thermostat handling more than one room is limited by its ability to only sense one room's temperature.

3. Adequate provision for control of hot deck[†] temperature (including rescheduling) and cold deck temperature is recommended to avoid wide fluctuation in these temperatures.

4. Avoid combination of several large zones with only a few small zones. This can cause erratic behavior in the airflow to the small zones when the large zones are in a modulating condition.

10.2.5. Variable Air Volume (VAV)

The control of the dry-bulb temperature within a space requires that a balance be established between the space load and the medium supplied to offset the load. When air is this medium, the designer may choose between varying its temperature (constant volume), or varying its volume (variable volume) as the space load changes. Supply air temperature and air volume may be controlled simultaneously as a third

†Hot or cold decks are air chambers forming part of an air conditioning unit.

option as is the case for variable air volume systems.

Variable air volume systems may be used for interior or perimeter zones, with common or separate fan systems, common or separate air temperature control and with or without auxiliary heating devices. The variable volume concept may apply to the main system airstream or to the zones of control, or both. There are many combinations of VAV systems, limited mainly by the designer's imagination and his compliance with basic criteria. A schematic of one such VAV system is shown in Figure 19.

10.3 AIR–WATER SYSTEMS

As with the all-air system, the air apparatus and refrigeration plants in air–water systems are separate from the conditioned space. However, the cooling and heating of the conditioned space is affected in only a small part by air brought from the central apparatus. The major portion of the room thermal load is balanced by warm or cool water circulated either through a coil in an induction unit or through a radiant panel. In virtually all air water systems, both heating and cooling functions are carried out by changing the temperature of either the air or the water, or both.

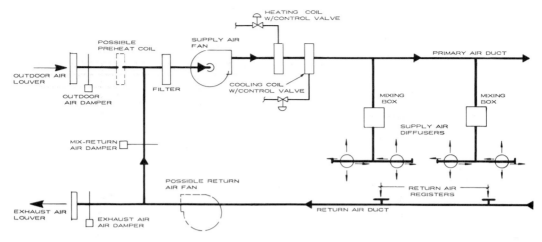

FIGURE 19. Variable volume system (flow diagram).

There are several reasons for the use of this type of system. Because of the greater specific heat* and the much higher density of water as compared to air, the air temperature can be controlled much more efficiently closer to the space to be conditioned. This means that the cross-sectional area required for the distribution pipes is significantly less than that required for a ductwork to accomplish the same cooling task since a substantial part of the heat gain can be removed directly from the conditioned space by a recirculating water system. Consequently, the quantity of air supplied can be low compared to an all-air system, and less building space need be allocated for the cooling distribution system.

Space requirements for ducts are further reduced by use of a high-velocity air distribution system. If the system is designed so that the air supply is equal to the air required for ventilation or to balance exhaust requirements, or both, the return air system can be eliminated.

These systems are used primarily for multiple perimeter spaces where a wide range of sensible load exists and where close control of humidity

is not required. Systems of this type have been commonly used for office buildings, hospitals, hotels, schools, better apartment houses, research laboratories, and other buildings. Their space-saving characteristic has made air–water systems especially useful in high-rise structures.

10.3.1. Induction System

This system uses a high velocity, high pressure, constant volume air supply to a high induction type of terminal. Induced air from the room is either heated or cooled within the terminal as required. This temperature control is by flow of water or air bypass. This system may use two pipes (one water circuit) or four pipes (two water circuits) for heating and cooling (see Figure 20).

Terminals come with or without a fan coil with supplementary air. The fan coil type of terminal provides direct heating or cooling of the room air. A supplementary constant volume air supply provides the necessary ventilation.

Induction units are usually installed at a perimeter wall under a window, but units designed for overhead installation are available. During the heating season, the floor-mounted induction unit can function as a convector during off hours with hot water to the coil and without a primary air supply. A wide variety of induction unit configurations is available, including units with low overall height

---

Specific heat is defined as the ratio of the number of Btu's required to raise the temperature of 1 lb of a substance $1°F$ to the Btu's required to raise the temperature of 1 lb of water a similar amount. The specific heat of water is 1.0; of air, 0.24. This means that the stored heat in 1 lb of water is about four times more effective in heating the air than a similar weight of air, barring transfer losses.

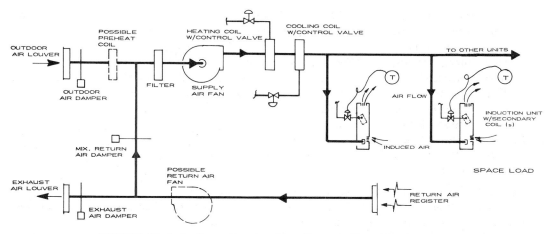

FIGURE 20. Induction unit system (flow diagram). *Note:* T = thermostat.

or with extremely large secondary coil face areas, to suit the particular needs of space or load.

Air–water systems employing induction units have the following characteristics:

1. The individual induction units contain no motors or other machinery. Routine service is generally limited to the temperature controls, cleaning the lint screen, and infrequent cleaning of the induction nozzles.
2. By use of high-velocity air supply ducts with parallel secondary water piping, the system takes a minimum of building space. The induction unit's compactness and variety of configurations are also space-saving features.
3. Located under the window, the induction unit can operate as a convector during off hours. It is not usually necessary to operate the air system to obtain off-hour heating.

10.3.2. Fan Coil System

The fan coil unit is a versatile room terminal used in both air–water and water-only systems. Fan coil units have been more commonly associated with water only systems than with air–water systems. Many of the standard features of air–water units are accordingly incorporated into it for water-only applications.

The fan coil type of terminal provides direct heating or cooling of the room air. A supplementary constant volume air supply provides the necessary ventilation.

Air–water systems employing fan coil units have the following characteristics:

1. Building management can reduce operating costs during partial occupancy by shutting unit fans off in unoccupied areas.
2. The variety of physical types of fan coil units offers a reasonable number of alternate layouts for the distribution piping and ductwork, and the equipment serving the conditioned space.
3. Where a three-or four-pipe concept is employed, a common air supply system can serve both interior and perimeter spaces.
4. Under the window fan coil units effectively distribute air for heating and cooling. Draperies which interfere with the free circulation of air will affect terminal performance. Light-colored draperies drawn across the return air passage will act as a filter and will discolor rapidly.
5. By proper selection, units can be obtained which will be adequately quiet for most applications. Wear and tear may cause some gradual increase in noise level.
6. Units can provide only limited heating capacity as convectors during unoccupied hours. If so employed, they must be

wired with central switching to discontinue fan operation conveniently during the unoccupied period.

7. Unit filters, motors, fans, and controls must be serviced in the conditioned space. Fan and motors in many models, however, can be quickly removed and replaced for other than routine maintenance.
8. The internal design of fan coil terminals makes them difficult to clean or decontaminate.

Figure 21 shows a vertical fan coil unit working in conjunction with a central air system. The central air system takes care of the ventilation requirements for the space by filtering and heating the air at a constant flow and temperature, while the cooling and reheating is done by the fan coil unit at the perimeter.

10.3.3. Radiant Panels with Supplementary Air

The radiant panel terminal in a ceiling or wall provides either radiant heating or cooling. A constant volume airstream is supplied to the terminal for dehumidification and ventilation.

Radiant panel systems are similar to other air–water systems in terms of the arrangement of the system components. Room thermal conditions are maintained primarily by direct transfer of radiant energy, rather than by convective heating and cooling.

The principal advantages of radiant panel heating and cooling systems are:

1. Comfort levels are better than in any other conditioning system because radiant heat is treated directly, and air motion in the conditioned space is at normal ventilation levels.
2. Cooling and heating may be obtained simultaneously, without central zoning or seasonal changeover, when three- and four-pipe systems are used.
3. Panel systems do not require any mechanical equipment such as fans and louvers at the outside walls, thus, simplifying the wall, floor, and structural systems.
4. All pumps, fans, filters, etc., are centrally located, simplifying maintenance and operation.
5. There is no mechanical equipment within the occupied space requiring maintenance or repair.
6. Draperies and curtains can be installed

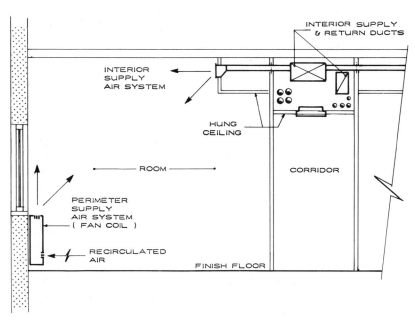

FIGURE 21. Air conditioned space by separate systems.

at the outside wall without interfering with the heating and cooling system.

7. Supply air quantities usually do not exceed those required for ventilation and dehumidification.

10.4. ALL-WATER SYSTEMS

All-water systems are those with fan coil room terminals to which are connected one or two water circuits. The cooling medium (such as chilled water or brine) may be supplied from a remote source and circulated through the coils of fan coil terminal, which is located in the conditioned space. These circuits may be either two-pipe, three-pipe, or four-pipe distribution. Ventilation is obtained through an opening in the wall, from bleed-off from an interior zone, or by infiltration.

The air side of this system is a constant supply air volume often referred to as "primary air" to distinguish it from room air, which is recirculated over the room coil. The primary air provides fresh clean air for ventilation. In the cooling season, the air is dehumidified in the central conditioning unit for comfortable, humidity conditions throughout the spaces.

The water side of the system consists of a pump and piping to convey water to heat transfer surfaces within each conditioned space. These heat transfer surfaces are in the form of a coil which may be an integral part of the air terminal (as with induction units), or be a completely separate component within the conditioned space (radiant panel) or have a dual capability (as can be the case with fan coil units). The water is cooled by direct refrigeration or, more commonly, by the introduction of chilled water from the primary cooling system, or by heat transfer through a water-to-water exchanger. To distinguish it from the primary chilled water circuit, the water side is usually referred to as the "secondary water" loop or system.

10.4.1. Two-Pipe Systems

A two-pipe system derives its name from the water distribution system that supplies secondary water to the induction units. In addition to

this water, induction units receive conditioned air from a central apparatus. The two pipes are for supply and return water. The design of the system and the control of the primary air and secondary water temperatures must be such that all rooms in the same system (or zone, if the system separated into independently controlled air and water zones) can be satisfied during both heating and cooling seasons.

Two secondary water arrangements applicable to two-pipe systems are illustrated in Figure 22 and Figure 23. In both systems, separate primary and secondary water circuits and circulation pumps are provided. In the arrangement shown in Figure 9, the return water is routed through both the hot water heater and the chiller. The water temperature is then controlled by a mixing valve. In the scheme shown in Figure 10, all the secondary water flows through the heat exchanger. The primary circuit remains separate. Its temperature is governed to control the temperature of the secondary water.

The two-pipe system was the first air water system developed and continues to be the most

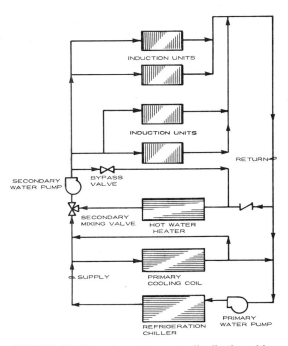

FIGURE 22. Two-pipe system water distribution with mixing control circuits.

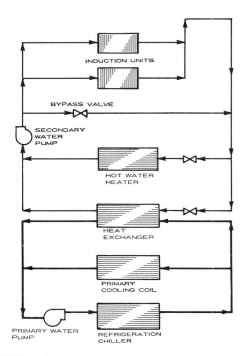

FIGURE 23. Two-pipe water distribution with separated circuits.

widely used. Characteristics of the two-pipe system are:

1. It is usually the least expensive system to install.
2. The seasonal changeover procedures are cumbersome, resulting in a greater need for competent operating personnel.
3. It is less capable of handling widely varying loads or controlling room temperatures than either three- or four-pipe systems.
4. Operating costs will probably be higher for two-pipe than for a four-pipe system.

10.4.2. Three-Pipe Systems

Three-pipe systems have three water pipes connected to each terminal unit; a cold water supply, a warm water supply, and a common return. Three-pipe systems have been classified into two categories, terminal mix and return mix systems. In both types of systems, hot and cold secondary water is provided at the units and, depending on the type of system, is either mixed before entering the unit or mixed in the

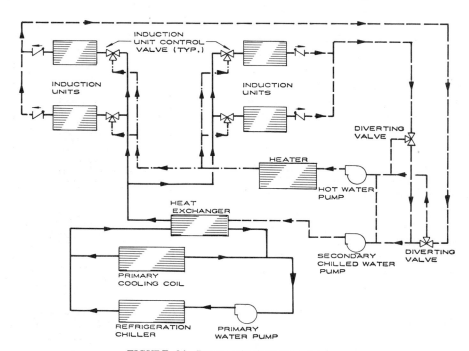

FIGURE 24. Return mix-three-pipe systems.

unit in accordance with temperature requirements. Both systems use a common return pipe. The return is switched at the diverting valve, and mixing takes place at the three-way valve. Figure 24 is a schematic diagram of a return mix three-pipe system.

The terminal mix three-pipe system utilizes a mixing valve at the inlet of the unit for temperature control. The hot and cold secondary water is mixed before it enters the unit. This form of three-pipe system is designed for operation as a part of a heat recovery system and should be used only when heat recovery is employed.

The three-pipe system satisfies variations in load by providing independent sources of heating and cooling to the room unit in the form of constant temperature primary and secondary chilled and hot water. The terminal unit contains a single secondary water coil. A three-way valve at the inlet of the coil admits the water from either the hot or cold water supply, as required. The water leaving the coil is carried in a common pipe to either the secondary cooling or heating equipment. The usual room control for three-pipe systems is a special three-way modulating valve which modulates either the hot or the cold water in sequence, but does not mix the streams. The primary air is cold and at the same temperature year-round.

Characteristics of the return mix three-pipe air–water system are:

1. The three-pipe system possesses most of the characteristics of a four-pipe system with respect to its ability to respond to load variations.
2. The system opeates without the summer to winter cycle changeover and primary air reheat requirements of the two-pipe changeover system.
3. The hot and cold secondary circuits are interconnected in a common return line. Because of the mixing that takes place, the system operating efficiency is less than that of a two-pipe system.
4. The hydraulic systems are complicated by the interconnected secondary circuit.

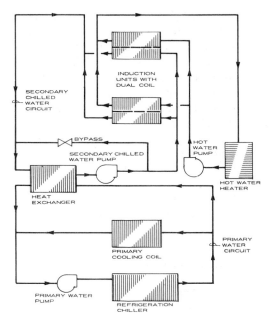

FIGURE 25. Four-pipe system water distribution with separated circuits.

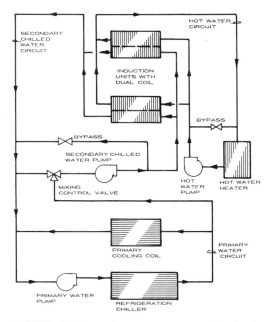

FIGURE 26. Four-pipe system water distribution with mixing control valve.

During off-peak operation, there is danger of overheating because of backflow from the hot to the cold secondary circuit.

5. Initial cost is higher for the return mix three-pipe system than for a two-pipe induction system. When the costs of secondary water system zoning, controls, and additional factors are evaluated and compared with the cost of the additional return pipe, the three-pipe system may not be less expensive than a four-pipe system.

6. The three-pipe system is more difficult to operate than a four-pipe system.

10.4.3. Four-Pipe Systems

Four-pipe systems derive their name from the four pipes that connect to each terminal unit: cold water supply, cold water return, warm water supply, and warm water return. The four-pipe system satisfies variations in cooling and heating load by providing such independent sources of heating and cooling to the room unit as constant temperature primary air, secondary chilled water, and secondary hot water. The terminal unit is usually provided with two independent secondary water coils; one served by hot water, the other by cold water. The primary air is cold and remains at the same temperature year-round.

Water system arrangements for four-pipe systems are shown in Figures 25 and 26. In both figures, separate water pumps and piping circuits are used. However, in Figure 25 the water is cooled by a water–water heat exchanger. In Figure 26 a three-way valve is used to maintain a constant water temperature.

10.5 DX SYSTEM

This system is widely used in a "packaged form," usually sized in small or intermediate tonnage ranges. Heat is removed by a simple method using a finned tube coil with direct expansion of the refrigerant into the coil, hence, the term DX.

In general terms, a DX system can be described as a refrigeration system where the cooling effect is obtained directly from the refrigerant (e.g., the refrigerant is evaporated in a cooling coil located in the airstream). A window air conditioner is an example of a DX system. Its capacity can range up to 25 or 50 tons. Centrifugal chillers and absorption chillers, are used for higher cooling demands. A more detailed explanation of these systems and their uses are developed in Section II, under Unitary Systems, Chapter 1.2.

11
HEATING SYSTEMS

Throughout this book, we define heating, ventilating and air conditioning as the process where temperature, humidity, and filtering of the air is controlled, thus, creating a preset environmental condition. Since heating is part of this process, it is considered to be part of the HVAC system. However, heating is usually treated separately so that it can later be (or not be) incorporated into the whole function of the HVAC system. Forced warm air, hot water, steam, and electric are major classifications for heating systems in today's industry.

11.1 FORCED WARM AIR SYSTEM

Forced warm air, considered a basic central system, consists of a furnace, containing a blower (centrifugal fan), a source of heat such as a gas or oil burner, and filters. In this package, the cold air is taken from the various spaces and some from outdoors, and passed through the heat source. The heated air is then channeled via ductwork to the space. The simplicity of the system in creating basic comfort during winter weather makes it widely used for residential, commercial and industrial applications.

The advantages of this system are:

- Air circulation is positive and when properly controlled creates a uniform ambient temperature.
- By properly designing the system it can be adapted for cooling application.

- Outdoor air can be channeled into the system in quanities to suit differing ventilation needs.

A forced warm air system is referred to by different labels according to use. It is known as a *package blower* (furnace) for residential use, a heating and ventilation unit for commercial applications, and a make-up air unit when used in industrial applications.

11.1.1. Airflow in Ducts

Unlike many HVAC components, duct systems are not available in catalogs and must be designed for specific applications. To design an air distribution system there must be an understanding of airflow in ducts, methods of design, and recommended air velocities. A poor design can result in a noisy, inefficient, or expensive system.

One of the first steps in designing an air distribution system is to compare the initial cost plus operating costs for a number of systems. This includes whether large ducts or a number of smaller ones that give the same air distribution are the most cost effective. Sometimes duct size is determined by architectural or structural limits on the space available.

The industry has adopted a set of standard nomenclature for air distribution systems which affect duct construction. These are:*

- Low pressure—where the maximum velocity of the air is up to 2000 fpm and the static pressure in the duct is not over 2 in. wg.
- Medium pressure—where the velocity of the air is greater then 2000 fpm or the static pressure in the duct is up to 6 in. wg.
- High pressure—where velocity of the air is greater than 2000 fpm or the static pressure is between 6 and 10 in. wg.

Velocity is related to the volumetric flow of air by the following equation:

$$cfm = A/v$$

where

v = velocity in feet per minute
A = cross-sectional area of duct in square feet
cfm = cubic feet per minute of air

The velocity to be used depends to the amount on the noise level** to be maintained or tolerated. Guides for initial design are shown in Figure 27.

Air flowing in a duct is based on a "potential energy" needed to deliver the air to different terminal points. This is characterized by three pressures:

Pressure is generally defined as force per unit area, and it is most commonly measured in pounds per square inch (psi). Other pressure units derived by this principle that are used in HVAC work are:

- *atomspheric pressure*—defined as the pressure of the earth's atmosphere at sea level, which is equal to:

    1 atm = 29.92 inches of mercury (in. Hg)
          = 14.7 pounds per square inch (psi)
          = 33.9 feet of water (ft of water)

- *gauge pressure*—defined by the principle of pressure obtained by measuring, in pressure gauges, the liquid in tubes. These tubes are U-tube or monometer type. They are measured in the unit of inches of mercury or inches of water.
- *absolute pressure*—defined as gauge pressure plus atmospheric pressure in pounds per square inch, absolute (psia).

**The noise generated in a duct system increases with duct velocity and system pressure.

- static pressure (s.p.)
- velocity pressure (v.p.)
- total pressure (t.p.)

Static pressure in a duct is the pressure (developed by a fan) necessary to overcome the frictional resistance to air flow by such components as heating coils, registers, branch takeoffs, etc. *Velocity pressure* is the pressure required to take the air at rest and accelerate it to the required velocity in the duct. *Total pressure* is the algebraic sum of static pressure and velocity pressure, and is the pressure which a fan must generate to deliver the air at desired points.

11.1.2. Duct Design Methods

The three most widely used methods of duct design are:

- equal friction†
- static regain
- total pressure†† methods

Equal Friction Method.   As implied by its definition, the design procedure for this method is based on maintaining the same air friction loss per 100 ft of length for the entire ductwork system. Knowing these losses makes possible the selection of a recommended initial duct velocity. The equal friction method has been the most commonly used because of its simplicity. Duct velocities are reduced in the direction of airflow, thereby, somewhat reducing noise problems. However, this method does not differentiate between various lengths of duct, nor does it provide for equal static pressure at succeeding terminals. Thus, balancing dampers must be used throughout the system and special provisions for sound attenuators may be required where high pressure losses occur. For these reasons, use of the equal friction method should be restricted to fairly simple, symmetrical, low velocity duct systems.

Static Regain Method.   The basic principle underlying the static regain method is that each

†Also called "constant pressure drop" method.
††Also called "velocity reduction" method.

Application	Typical Noise Level (decibels)	Main or Branch Ducts (average)	Velocities (fpm) Main Ducts		Branch Ducts	
			Supply	Return	Supply	Return
Residences	25–30	1000	1000	800	600	600
Apartments Hotel bedrooms Hospital bedrooms	24–45 25–35	1200	1500	1300	1200	1000
Private offices Director's rooms Libraries	30–45	1500	1800	1500	1200	1000
General offices Banks	30–45	1700	1800	1500	1400	1200
Stores Cafeterias Industrial	50–70 70–above	2000 2200	1800 2500	1600 1800	1600 2200	1200 1600

Return Air Openings		
Industrial	800	fpm
Commercial		
corridor	700	fpm
rooms	500	fpm
Residential	400	fpm
Relief Dampers		
Industrial	1000	fpm
Commercial	400	fpm
Outside Air Intakes (general application)	500–700 fpm	
Corridor and Stairwell (general application)	150	fpm

Louver Doors (or undercut doors)

s.p. of outlet, in.	0.005	0.025	0.05	0.10	0.25	0.50	1.00
Max, velocity, fpm		300	400	500	600	600	600

Note: in any application where a person is seated adjacent to a grille, reduce velocity to a maximum of 300 fpm.

FIGURE 27. Recommended air velocities in ducts, areas, and components.

section of duct is sized so that the increase in static pressure, due to the velocity reduction, offsets or balances the friction loss in the succeeding section of duct. Thus, the same static pressure is maintained at each branch and terminal throughout the system. With a system properly designed for no gain or loss, the total system friction loss is limited to the loss in the initial section of main and branch ducts, plus takeoff losses.

The duct size prior to distribution is determined from the initial air quantity required and from a suitable initial duct velocity. This duct size is maintained up to the first takeoff. The succeeding duct section is then sized by allowing a reduction in duct velocity to obtain a static regain equal to the friction loss of the duct to the second branch or terminal. Succeeding duct sections are similarly sized to the end of the run.

Total Pressure Method. The design procedure for this method is based on the selection of an arbitrary initial duct velocity, with progressive reductions in velocity made after each takeoff along the entire duct run. The system's friction loss is then calculated as the sum of the friction losses of the various duct sections, and the required terminal operating pressure, based on the corresponding air quantities and velocities.

Frequently, the static pressure required at the fan discharge is considered to be equal to the system's friction loss of the longest equivalent duct run. The friction loss for such a run is its actual length plus allowances for fittings.

This method requires knowledge and experience in duct design for reasonable accuracy without numerous repetitive trial and error selections.

11.1.3. Examples of Duct Design

An example of the equal friction method of duct design is given in Figure 28. This system is to serve an office area where the maximum space available for ductwork in the hung ceiling is 18 in., the noise level is not critical, and a pressure drop of 0.10 to 0.15 in. wg/100 ft is chosen for sizing the ductwork.

In examining Figure 28 we find that the main duct starting at the fan outlet of the air conditioning unit, carries 3000 cfm from point *A* to *B*. From point *B* to *C* the duct carries 2000 cfm, and so on. Each station, *A-B*, *B-C*, *C-D*, etc., is analyzed in Figure 29. The air volume and duct lengths are known. Sizing the ducts is accomplished as follows:

Enter the left scale of Figure 30 at 3000 cfm. Move horizontally to the right until the vertical, pressure drop line is intersected at 0.10 in. of wg/100 ft. This value for pressure drop is acceptable for a system of this type. We find that a 20 to 22 in. diameter duct at 1300 fpm velocity can be used. A check for velocity, in Figure 27, shows that 1300 fpm is acceptable and therefore can be used throughout the remaining ductwork. To convert round ducts to its rectangular equivalent, refer to Figure 31. For a 22-in. diameter (on diagonal line) we can select a duct of 38-in wide by 12-in deep, or whatever width to depth relationship is required for clearance, appearance, etc.

The friction loss has been set at 0.1 in. wg/100 ft of duct. To find the actual friction loss for a specific length, for example, section *A–B*

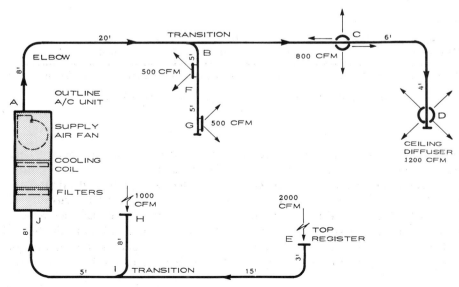

FIGURE 28. Office ductwork layout.

Duct Section	Air Volume (cfm)	Design Friction Factor (in. wg)	Velocities (fpm)	Length of Straight Duct	Elbow Fitting Eq. to ft of Duct.	Total Pres. Drop (in. wg)	Duct Sizes Diam.	Duct Sizes Rectangular
A–B	3000	0.10	1300	28	20 + 15	0.063	22	38 × 12
B–C	2000	0.10	1100	20		0.020	20	28 × 12
C–D	1200	0.10	1000	10	20	0.030	17	20 × 12
B–F	1000	0.10	980	5		0.005	14	14 × 12
F–G	500	0.10	800	5		0.005	11	8 × 12
E–I	2000	0.10	1100	15	20	0.035	20	28 × 12
H–I	1000	0.10	980	8		0.008	14	14 × 12
I–J	3000	0.10	1300	13	20 + 15	0.048	22	38 × 12

FIGURE 29. Duct element analysis.

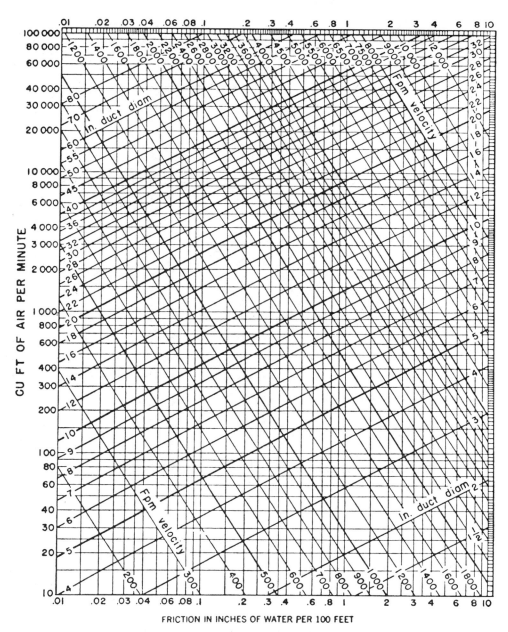

FIGURE 30. Friction of air in round ducts for volume 10 to 6000 cfm. (Based on standard air of 0.075 lb/ft density flowing through averaged galvanized duct.)

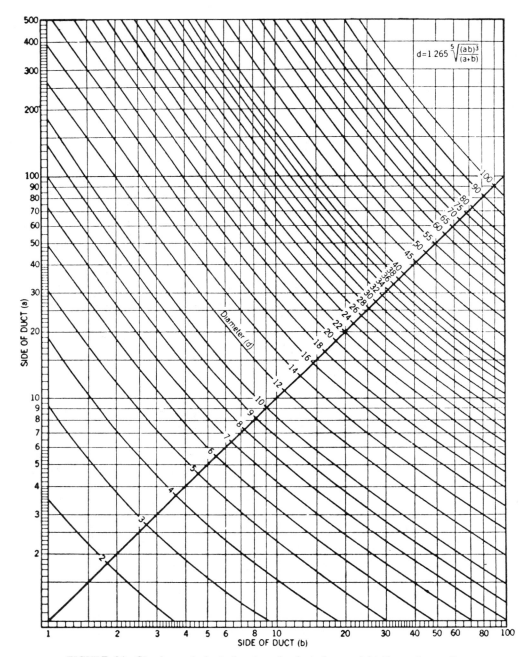

$$d = 1.265 \sqrt[5]{\frac{(ab)^3}{(a+b)}}$$

SIDE OF DUCT (a)

Diameter (d)

SIDE OF DUCT (b)

FIGURE 31. Circular equivalent of rectangular ducts for equal friction and capacity.

in Figure 29 which is 28 ft long:

$$\text{friction loss} = 0.10 \times \frac{28}{100}$$

$$= 0.028 \text{ in. of water}$$

Duct fittings add to the pressure drop which the fan must overcome. The friction in these

fittings is given in equivalent feet of straight duct. Returning to section A–B, we see that there is an elbow and takeoff transition. Figures 32 and 33 give equivalent straight duct lengths for rectangular elbows and takeoffs.

Using Figure 32 and assuming single thickness vanes in the elbow, a friction loss is found to be 20 LF (linear feet) of straight duct for a 12 × 36-in. duct, which is close in size to our 12 × 38-in. duct. In Figure 33, our takeoff to an almost square *B-F* duct section will give an equivalent of 15 LF of pressure drop.

These lengths are than added to the total length of straight duct. In addition, at the end of the run, a ceiling diffuser also adds to the total resistance of the system. The pressure drops of registers, diffusers, dampers, coils, filters, etc., are generally given in manufacturers' catalogs. For this example, assume a pressure drop of 0.05 in. wg at the diffuser at point *D*. The total pressure drop from *A* to *B* will be:

$$\begin{array}{rl} \text{length of duct:} & \text{28 LF} \\ \text{elbow:} & \text{20 LF} \\ \text{transition:} & \underline{\text{15 LF}} \\ & \text{63 LF} \end{array}$$

$$63 \text{ LF} \times \frac{0.10 \text{ in. wg}}{100 \text{ LF}} = 0.063 \text{ in. wg}$$

This is summarized in Figure 29.

In an equal friction (constant pressure drop) method, the *total pressure drop* in the system and the static pressure developed by the AC unit fan as in Figure 28 are obtained as follows:

total straight line supply ductwork
$$\begin{array}{rl} (A-D) & = 0.058 \\ \text{2 elbows } (A-B \text{ and } C-D) & = 0.04 \\ \text{1 transition} & = 0.015 \\ \text{1 branch takeoff } (B) & = 0.015 \\ \text{1 diffuser} & = \underline{0.05} \end{array}$$

total static pressure drop in the
supply air duct   = 0.178 wg

total straight line return
$$\begin{array}{rl} \text{ductwork } (E-J) & = 0.031 \\ \text{2 elbows } (E \text{ and } J) & = 0.04 \\ \text{1 transition} & = 0.015 \\ \text{filters (assumed dirty)} & = 0.25 \\ \text{return air register } (E) & = 0.15 \\ \text{coils} & = \underline{0.35} \end{array}$$
total static pressure drop
in the return system*   = -0.836 in. wg

*The static pressure produced by the return system is considered negative.

The minimum fan static pressure required is the difference between supply and return static pressure, or

$$0.178 - (-0.836) = 1.014 \text{ in. wg}$$

The fan will be selected on the basis of this static pressure requirement.

11.2. HOT WATER HEATING SYSTEM

Hot water heating or chilled water cooling systems are usually referred to as "hydronic systems." *Hydronics* is the science of heating or cooling with liquids. For heating applications a hot water system can be the most versatile. A basic hot water system consist of a boiler, pipes, and room heating units (radiators, convectors). The hot water is generated in the boiler, circulated through the pipes to the radiators or convectors where the heat is transferred to the room air.

Hot water systems may be classified by:

- pumping arrangement
- piping arrangement
- flow
- temperature
- pressure requirements

By flow designation hot water heating systems are of two types:

1. *Gravity Systems.* The circulation of hot water in a gravity system is obtained by the difference in density of the hot water in the supply pipes and the cooler water in the return pipes. This system was previously used in residences and other small buildings, but is no longer in common use, having been replaced by the forced flow system.

2. *Forced Systems.* In this system a small booster or circulating pump forces the hot water through the pipes to the room radiators or convectors.

Water heating systems may be classified according to their operating temperature such as:

1. *Low-Temperature Water Systems* (LTW). The limits for boiler pressure (160 psi) and maximum temperature (250°F) are set by the ASME Boiler Code. Boilers designed, tested and stamped for higher pressures may frequently be used in the system even though the 160 psi

Illustrations: (elbow types) · 1 Vane · 2 Vane

Equivalent Length of Straight Duct Feet

Duct Dimensions	R/B=.5	R/B=1.0	R/B=1.5	R/B=.5	R/B=1.0	R/B=1.5	No Vanes	Single Thickness Vanes	Double Thickness Vanes	R/B=.5 (1 Vane)	R/B=1.0 (1 Vane)	R/B=1.5 (1 Vane)	R/B=.5 (2 Vane)	R/B=1.0, 1.5 (2 Vane)
84 × 36	204	69	15	210	56	26	294	60	30	60	24	21	45	21
84 × 30	180	40	13	182	55	25	260	50	25	50	20	18	38	18
84 × 24	168	32	11	175	53	25	216	40	20	40	16	14	30	14
84 × 18	153	25	9	151	48	24	165	30	15	30	12	11	23	11
72 × 36	186	39	15	198	62	23	282	60	30	60	24	21	45	21
72 × 30	162	35	13	174	55	23	242	50	25	50	20	18	38	18
72 × 24	158	30	11	162	47	23	210	40	20	40	16	14	30	14
72 × 18	135	25	9	144	43	21	165	30	15	30	12	11	23	11
60 × 36	174	36	14	180	45	20	264	60	30	60	24	21	45	21
60 × 30	154	32	12	165	43	20	235	50	25	50	20	18	38	18
60 × 24	148	26	10	145	40	20	200	40	20	40	16	14	30	14
60 × 18	123	24	7	130	38	19	162	30	15	30	12	11	23	11
48 × 36	144	34	13	156	38	16	250	60	30	60	24	21	45	21
48 × 30	140	30	12	144	37	16	215	50	25	50	20	18	38	18
48 × 24	136	26	10	136	35	16	188	40	20	40	16	14	30	14
48 × 18	111	20	8	112	32	15	153	30	15	30	12	11	23	11
48 × 12	90	17	6	96	29	14	110	20	10	20	8	7	15	7
36 × 36	135	31	12	135	32	13	225	60	30	60	24	21	45	21
36 × 30	120	28	11	120	30	13	197	50	25	50	20	18	38	18
36 × 24	114	24	10	108	28	12	170	40	20	40	16	14	30	14

36 × 18	100	20	8	102	26	12	141	30	15	30	12	11	23	11
36 × 12	79	15	6	71	23	12	105	20	10	20	8	7	15	7
36 × 8	67	12	5	69	21	10	79	14	7	14	6	5	10	5
30 × 30	112	27	11	110	26	11	188	50	25	50	20	18	38	18
30 × 24	98	23	10	102	24	11	162	40	20	40	16	14	30	14
30 × 18	90	20	8	90	22	10	141	30	15	30	12	11	23	11
30 × 12	74	15	6	75	20	10	100	20	10	20	8	7	15	7
30 × 8	60	11	5	62	19	9	73	14	7	14	6	5	10	5
24 × 24	90	21	9	90	21	9	150	40	20	40	16	14	30	14
24 × 18	78	18	7	76	19	8	125	30	15	30	12	11	23	11
24 × 12	68	13	5	68	18	8	94	20	10	20	8	7	15	7
24 × 8	52	10	4	54	16	8	70	14	7	14	6	5	10	5
24 × 6	45	8	4	48	14	7	55	10	5	10	4	4	8	4
18 × 18	67	16	7	67	16	7	112	30	15	30	12	11	23	11
18 × 12	57	12	5	54	15	5	85	20	10	20	8	7	15	7
18 × 8	48	9	3	46	13	3	65	14	7	14	6	5	10	5
18 × 6	40	8	3	40	12	3	51	10	5	10	4	4	8	4
16 × 16	59	14	6	60	14	6	100	27	14	27	11	9	20	9
16 × 14	53	13	5	53	13	6	91	23	12	23	9	8	18	8
16 × 12	48	12	5	51	13	6	83	20	10	20	8	7	15	7
16 × 10	45	10	4	50	12	5	72	17	9	17	7	6	13	6
16 × 8	42	9	3	44	12	5	63	14	7	14	5	5	10	5
16 × 6	35	7	3	36	11	5	51	10	5	10	4	4	8	4
14 × 14	53	12	5	53	12	5	88	23	12	23	9	8	18	8
14 × 12	47	11	5	46	11	5	79	20	10	20	8	7	15	7
14 × 10	42	10	4	45	10	5	70	17	9	17	7	6	13	6
14 × 8	39	8	3	41	9	4	60	14	7	14	5	5	10	5
14 × 6	35	7	3	35	9	4	49	10	5	10	4	4	8	4
12 × 12	45	11	4	45	11	5	75	20	10	20	8	7	15	7
12 × 10	42	10	4	39	10	4	66	17	9	17	7	6	13	6
12 × 8	38	8	3	36	9	3	57	14	7	14	5	5	10	5
12 × 6	34	7	3	34	9	3	47	10	5	10	4	4	8	4
10 × 10	38	9	4	37	9	4	63	17	9	17	7	6	13	6
10 × 8	34	8	3	33	8	4	54	14	7	14	5	5	10	5
10 × 6	28	6	2	29	7	3	44	10	5	10	4	4	8	4
8 × 8	30	7	3	30	7	3	50	14	7	14	5	5	10	5
8 × 6	25	6	2	25	6	2	41	10	5	10	4	4	8	4

FIGURE 32. Static pressure losses of rectangular elbows. (*Courtesy* of York, Division of Borg-Warner Corp.)

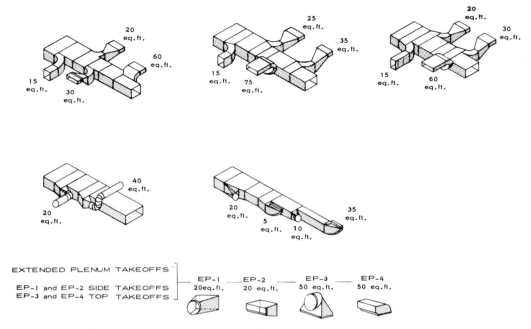

FIGURE 33. Static pressure losses for take-offs fittings in feet of equivalent length of straight duct.

limit will not be exceeded. Steam-to-water or water-to-water heat exchanges are also often used in LTW systems.

2. *Medium-Temperature Water System* (MTW). A hot water heating system operating at temperatures of 350°F or less, with pressures not exceeding 150 psi. The design supply temperature is approximately 250° to 325°F with a pressure rating for boilers and equipment of 150 psi.

3. *High-Temperature Water System* (HTW). A hot water heating system operating at temperatures and pressures of over 350°F and about 300 psi. The maximum design supply water temperature is 400° to 450°F. It is necessary that the pressure-temperature rating of each component be checked against the design characteristics of the particular system in order to assure a safe and efficient application.

According to piping-flow arrangements, hot water systems can be further classified as:

• series loop
• one-pipe
• two-pipe, reverse return
• two-pipe, direct return

In general, the selection of a particular hot water system for a specific building should be based on a careful analysis of such factors as:

• the nature of the heating load
• physical characteristic and layout of the building
• comfort and control requirements
• the economics of the complete system

Figure 34 is an isometric view of a residential forced hot water heating system showing the boiler, circulator, and convectors. This would be classified as a double circuit, one-pipe system.

The following are examples of hot water systems most commonly employed.

1. Single Circuit, One-Pipe. The hot water leaves the boiler and flows in one pipe around the building before returning to the boiler. The radiators or convectors are connected in series to the feed pipe.
2. Double Circuit, One-Pipe. The hot water leaves the boiler and splits after a distance into two circuits which come together before returning to the boiler.

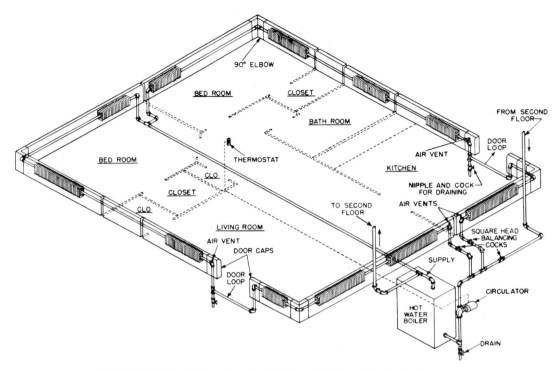

FIGURE 34. Typical baseboard installation with forced circulation hot water.

3. Two-Pipe, Direct Return. The separate return lines do not join until they reach the boiler inlet.

4. Two-Pipe, Reverse Return. The separate return lines join into a hot water return main that can be balanced.

We will go through a step-by-step procedure for designing the system and selecting its components.

11.2.1. Single Circuit, One-pipe System

To design a single circuit, one-pipe hot water system as shown in Figure 35, the following steps are usually followed:

1. Determine the heat loss of building.
2. Make a piping layout.
3. Determine the gallons of water required per minute.
4. Select the circulator pump.
5. Size the main.
6. Size the radiator branches.

7. Select the boiler.
8. Select the pressure tank.
9. Select the flow control valve.

STEP 1: Determine the Heat Loss of the Building.

Calculate the heat loss of the structure by using the format described in a previous chapter. For the example in Figure 35 a total heat loss of 65,000 Btu has been figured.

STEP 2: Make a Piping Layout.

First determine if a one- or two-loop circuit will be most adaptable. This will depend somewhat on the design of the structure and the number of radiators required. For an average size residence, design the system as a single circuit. The circuit should be as short and direct as possible with horizontal branches from main to risers not over 3 to 4 ft long. To simplify use of the following Tables, show radiator size in Btu's. This can be changed to square feet of radiation by dividing the Btu output of each radiator

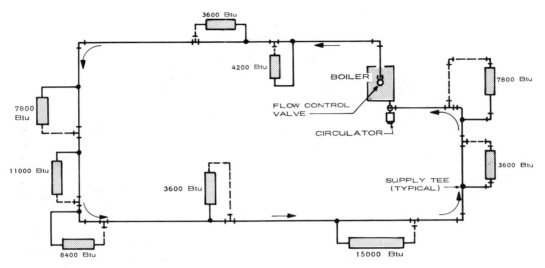

FIGURE 35. Single-circuit, one-pipe hot water system.

by a Btu emission corresponding with the type of radiation used and the average water temperature of the heating system.

Heat Emission Rates

Average water Temperature[a] (°F)	Approx. Btu Emission/hr/sq ft of Radiation, at Minimum Design Temperature
160	130
165	140
170	150
180	170
190	190
195	200

[a]Average water temperature is the sum of the maximum and minimum boiler temperatures, divided by two.

The Btu ratings shown above are for free standing, cast iron radiation. There are many different types of radiators, convectors, and baseboards. It seems no two types have the same output and it is not advisable to assume ratings based on assumed similarities. The manufacturer's ratings must be known to properly figure a job and to size the piping. If necessary, consult the maker for this data. This can reduce the design effort and avoid expensive additions that may be required after the job is installed if incorrect data is assumed during design.

STEP 3: Gallons of Water Required per Minute.

The flow rate for the system must be known so a properly sized circulator pump can be selected. Most tables are based on a 20°F temperature drop, as this has been found to be the most economical for this type system. It is assumed, therefore, that there will be a difference of 20° between the return and supply water temperature of the boiler. However, this difference will occur only when the system is operating at its maximum design temperature.

To find the gpm required for the given system, divide the total heat loss of 65,000 Btu's by 9,600.

$$\frac{65,000}{9,600} = 6.77 \text{ gpm}$$

The number 9,600 is established by multiplying the 20° temperature drop by 60 min. in one hr and by the 8 lb of water in one gal. at 215°F (20 × 60 × 8). To simplify, divide total heat loss by 10,000 which is accurate enough for this purpose.

STEP 4: Selecting a Circulator Pump.

The selection of a circulator pump is usually done by using pump manufacture performance data. For this example, use Thrush water circu-

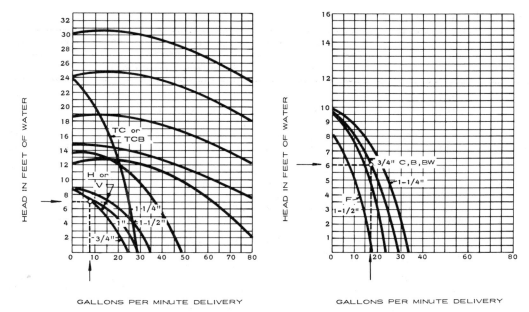

FIGURE 36. Performance Thrush water circulators. Performance curves for Thrush water circulators are shown in combined charts above. Capacities are based on 1750 rpm motors. Always select the circulator nearest capacity required for the job. Too much gallonage may result in a humming sound, but extra head over what is required will not cause any trouble.

lator data shown on Figure 36. Read to the right on the bottom line (chart for Model *H, V* or *B*) until 6.8 gpm is found. Draw a line upward until it intersects the first circulator capacity curve. From this point read at the left a head of 7 feet of water. Therefore, a $\frac{3}{4}$-in. H or $\frac{3}{4}$-in. V circulator will deliver 6.8 gpm against a head pressure of 7 ft.

From some installations it may be more practical to use a larger size circulator or a "hi-head" circulator because of a possible reduction in pipe sizes. For this example, a $\frac{3}{4}$-in. H or $\frac{3}{4}$-in. V Thrush circulator is selected.

STEP 5: Sizing the Main.
In a forced circulation system the piping design must be such that the pressure, or head, in feet of water, created by the circulator forcing the water from the boiler to any radiator, be equal to the friction loss, or head in the radiator circuit when the system is operating at its maximum design temperature.

The measured length of the main circuit in the example shown in Figure 35 is 120 ft. To

this must be added the resistance of the one-pipe tees.* Allow 10 equivalent feet of piping for each one-pipe tee. Ten tees times 10 will be 100 equivalent feet of piping to be added to the measured length of 120 ft, making a total equivalent length of 220 ft.

To size the main refer to Table 2A heading to the right from the 7-ft. circulator head, the closest equivalent length of piping to that required for the system (220 ft) is found to be 210 ft in the 400-milinch† column. Continuing

---

*One-pipe tee or venturi fitting is an engineered tee designed to operate on the principle of Bernoulli's theorem. It has a built-in restriction which causes an increase in water velocity and thereby a differential in pressure. This differential in pressure will always cause a flow of water from higher pressure to low pressure region, thus, causing the water to continuously enter at this point.

†Milinch—Water flowing through a pipe produces friction with a resultant loss of energy. This loss is called the pressure drop and is expressed in milinch/lineal ft of pipe or feet of water/lineal foot of pipe. A milinch is equal to 1/1000 in. One foot or 12 in. has 12,000 milinches.

TABLE 2A. Milinches per Foot of Pipe

Circulator Head in Feet	50	100	150	200	250	300	350	400	500	600	700
	Total Equivalent Length of Pipe in Feet—Longest Circuit										
1	240	120	80	60	48	40	34	30	25	20	17
2	480	240	160	120	96	80	69	60	48	40	34
2.5	600	300	200	150	120	100	86	75	60	50	43
3	720	360	240	180	144	120	103	90	72	60	51
3.5	840	420	280	210	168	140	120	105	84	70	60
4	960	480	320	240	192	160	137	120	96	80	68
4.5	1080	540	360	270	216	180	154	135	108	90	77
5	1200	600	400	300	240	200	172	150	120	100	85
5.5	1320	660	440	330	264	220	189	165	132	110	94
6	1440	720	480	360	288	240	206	180	144	120	102
6.5	1560	780	520	390	312	260	222	195	156	130	111
→7	1680	840	560	420	336	280	240	[210]	168	140	119
7.5	1800	900	600	450	360	300	257	225	180	150	128
8	1920	960	640	480	384	320	274	240	192	160	136
8.5	2040	1020	680	510	408	340	292	255	204	170	145
9	2160	1080	720	540	431	360	308	270	216	180	153
9.5	2280	1140	760	570	455	380	326	285	228	190	162
10	2400	1200	800	600	480	400	343	300	240	200	170
10.5	2520	1260	840	630	505	420	361	315	252	210	180
11	2640	1320	880	660	529	440	378	330	264	220	187
11.5	2760	1380	920	690	554	460	396	345	276	230	197
12	2880	1440	960	720	576	480	412	360	288	240	204
14	3360	1680	1120	840	672	560	479	420	336	280	240
16	3840	1920	1280	960	770	640	550	480	384	320	274
18	4320	2160	1440	1080	864	720	617	540	432	360	309
20	4800	2400	1600	1200	962	800	688	600	480	400	340
24	5760	2880	1920	1440	1152	960	825	720	576	480	410
28	6720	3360	2240	1680	1344	1120	960	840	672	560	480
32	7680	3940	2560	1920	1538	1280	1097	960	768	640	548
36	8640	4320	2880	2160	1728	1440	1234	1080	864	720	617
40	9600	4800	3200	2400	1920	1600	1370	1200	960	800	685
45	10800	5400	3600	2700	2160	1800	1543	1350	1080	900	771
50	12000	6000	4000	3000	2400	2000	1715	1500	1200	1000	857
55	13200	6600	4400	3300	2640	2200	1886	1650	1320	1100	943
60	14400	7200	4800	3600	2880	2400	2060	1800	1440	1200	1028

down this same column to Table *B*, find 66,000 Btu's or 66 MBH (closest and next largest) to the required 65,000 Btu. Reading to the left on the same line, a 1 in. pipe is indicated as being required for this circuit.

STEP 6: Sizing the Radiator Branches.
To size the radiator branches refer to Table 3. In Table 3, capacities are given for one or two, one-pipe tees for upfeed radiators and includes sizes for first, second, and third floor radiators as well as downfeed radiators. Downfeed radiators require two-pipe tees.

Use the same 400-milinch column as was used in Table 2B. Reading to the right on the 400-milinch line in the nonadjustable tee section, find that $\frac{1}{2}$-in. pipe will carry up to 11,000 Btu's and $\frac{3}{4}$-in. pipe will carry up to 26,000 Btu's.

Table 3 gives carrying capacities of Thrush nonadjustable one-pipe tees in various milinches of resistance. After determining milinch column

TABLE 2B. Milinches per Foot of Pipe

Nominal Pipe Size	50	100	150	200	250	300	350	400	500	600	700
	Heat Carrying Capacity of Pipes in Thousands of Btu's[a]										
$\frac{1}{2}$"	5.2	7.2	9	10.5	12	13	14	16	17.5	19.5	22
$\frac{3}{4}$"	11.1	15.2	19	22.5	25	27.5	30	34	37	40	44
→1"	21	30	38	44	49	55	60	66	72	80	89
$1\frac{1}{4}$"	45	65	82	95	105	120	130	138	155	175	182
$1\frac{1}{2}$"	68	97	125	148	165	180	190	215	240	270	283
2"	132	190	240	280	310	351	370	415	470	508	565
$2\frac{1}{2}$"	215	312	390	462	508	570	605	655	723	800	860
3"	390	565	700	805	900	1010	1140	1250	1350	1505	1710
$3\frac{1}{2}$"	570	830	1050	1200	1350	1500	1630	1800	2000	2250	2480
4"	800	1200	1400	1700	1900	2100	2300	2600	2950	3150	3475
5"	1400	2200	2700	3200	3500	4000	4300	4600	5400	5900	6400
6"	2500	3700	4600	5400	6000	6700	7000	8000	9000	10000	11000

[a]Based on 20° temperature drop

Tables A and B provide a very accurate means of selecting the proper circulator, the carrying capacity of mains and risers when only two factors are known. The first step is to determine the total Btu (radiation) to be carried by the main. Second, the total equivalent length of main must be found.

For greatest efficiency, design one-pipe systems between the 300 and 700 milinch columns.

Assuming an installation with a total of 115,000 Btu to be designed in the 300 milinch column; it is found by referring to Chart B that a $1\frac{1}{4}$-in. main will carry up to 120,000 Btu. Following up the 300 milinch column the equivalent length of the main can be found at, say 240 ft. Moving to the left from 240 to the first column shows a circulator with 6 foot head is required. Dividing 115,000 Btu by 10,000 gives 11.5 gal./min. A Thrush circulator to deliver 11.5 gpm at a 6-ft head would be selected.

TABLE 3. Nonadjustable One-Pipe Tees

Milinches per Foot of Pipe	Riser Capacities in Thousands of Btu per Hour[a]													
	One One-Pipe Tee						Two One-Pipe Tees							
	a		b		c		a		b		c		Down-Feed	
	Pipe Size		Pipe Size		Pipe Size		Pipe Size		Pipe Size		Pipe Size		Pipe Size	
	$\frac{1}{2}$"	$\frac{3}{4}$"	$\frac{1}{2}$"	$\frac{3}{4}$"	$\frac{1}{2}$"	$\frac{3}{4}$"	$\frac{1}{2}$"	$\frac{3}{4}$"	$\frac{1}{2}$"	$\frac{3}{4}$"	$\frac{1}{2}$"	$\frac{3}{4}$"	$\frac{1}{2}$"	$\frac{3}{4}$"
100	5	12	3.5	9	3	7	6	16	5	11	4	9	5	11
150	6	15	4	10	3.5	9	8	20	6	14	5	11	6	14
200	7	18	5	12	4	10	9	23	7	16	6	13	7	16
250	8	20	5.5	14	4.5	12	11	26	7	18	6	15	7	18
300	9	22	6	16	5	13	12	29	8	21	6.5	16	8	21
350	10	25	6.5	17	5.5	14	13	32	9	22	7	18	9	22
→400	11	26	7	18	6	15	14	34	9	24	8	19	9	24
500	12	29	8	21	6.5	17	16	38	11	26	9	22	11	26
600	14	32	9	22	7	18	18	43	12	30	10	24	12	30

[a]Based on 20°F. temperature drop.

in Tables 2A and 2B for selecting main size and circulator head the same milinch column is referred to in Table 3 for carrying capacity of Thrush one-pipe tees and risers.

Table 3 gives carrying capacities for the use of one up-feed one-pipe tee, and one common tee, also two up-feed one-pipe tees, one at supply riser and one at return riser of each heating unit.

All-down-feed radiators require two one-pipe tees and capacities are shown in the last column in Table 3.

Due to the friction loss thru piping, valves, fitting, and other component the above capacities are expressed in thousands of Btuh and based on the following riser length and fittings:

a) 10 ft of pipe, 4 elbows, 2 union ells (or 1 union ell and 1 radiator valve) and 1 C.I. radiator.
b) 26 ft of pipe, 10 elbows, 2 union elbows (or 1 union ell and 1 radiator valve) and 1 C.I. radiator.
c) 46 ft of pipe, 10 elbows, 2 union elbows (or 1 union ell and 1 radiator valve) and 1 C.I. radiator.
d) Downfeed, 20 ft of pipe, 4 elbows, 2 union elbows (or 1 union ell and 1 radiator valve) and 1 C.I. radiator.

For each additional 15 equivalent feet of riser piping deduct 20% from the supply tee capacities shown above.

In Item b and c, the riser capacities are for one or two radiators on the same riser. In other words, one radiator carrying the full load or two radiators having a combined capacity equal to the full load.

STEP 7: Selecting the Boiler.

Net boiler output should be used for selecting a proper size boiler. This may be net rating in Btu or installed square foot of water reduction at 150 Btu emission. To select a boiler on the basis of net installed radiation divide the total heat loss by 150 Btu emission. In the above example this would be 65,000/150 = 433 sq ft of installed radiation.

STEP 8: Selecting Pressure Tank Size.

Because water expands when heated, an expansion tank must be provided in the system. In an

TABLE 4. Size and Capacities of Pressure Tanks Selection.

Gallons Capacity	Size (in.)	Total Btu Requirement of System
8	12 × 16	50,000
12	12 × 24	70,000
15	12 × 30	85,000
18	12 × 36	100,000
24	12 × 48	200,000
30	12 × 60	400,000
36	12 × 72	500,000

"open system," the tank is located above the highest point in the system and has an overflow pipe extending through the roof. An open system means that the tank is literally open to the atmosphere. In a "closed system," the tank is placed anywhere in the system, usually near the boiler.

Half of the tank is filled with air, which compresses when the heated water expands. Note that higher water pressure can be obtained in a closed system than in an open one. Higher pressure raises the boiling point of the water. Higher temperatures can therefore be maintained in the radiators without steam and smaller radiators can be used.

For this example see Table 4 for tank selection, which is found to be 12 gal.

STEP 9: Select Flow Control Valve.

The flow control valve is to be the same size as the trunk or main branch in which it is installed.

11.2.2. Double Circuit, One-Pipe System

As with the single circuit layout, to design a double circuit, one-pipe hot water system as shown in Figure 37, the following steps are usually employed:

1. Determine the heat loss of building.
2. Make a piping layout.
3. Determine the gallons of water required per minute.
4. Select the circulator pump.
5. Size the main.
6. Size the radiator branches.
7. Select the boiler.

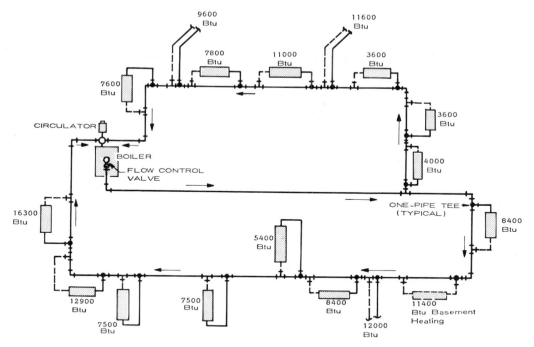

FIGURE 37. Double-circuit, one-pipe hot water system.

8. Select the pressure tank.
9. Select the flow control valve.

STEP 1: Determine the Heat Loss of the Building.

Calculate the heat loss of the structure by using the format described in a previous Chapter. For the example, in Figure 37, a total heat loss of 148.8 MBH or 148,000 Btu has been figured.

STEP 2: Make a Piping Layout.

First determine whether a one- or two-loop circuit will be most adaptable. This will depend somewhat on the design of the structure and the number of radiators required. For an average size residence, design the system as a single circuit. The circuit should be as short and direct as possible with horizontal branches from main to risers not over 3 to 4 ft long. To simplify use of the tables, show the radiator sizes in Btu's. This can be changed to square feet of radiation by dividing the Btu output of each radiator by a Btu emission corresponding with the type of radiation used and the average water temperature selected for the heating system. This is a good example of a one-pipe, two-circuit instal-lation. For heat emission rates see the Table on page (113).

STEP 3: Gallons of Water Required per Minute.

The flow rate for the system must be known so a proper size circulator pump can be selected. As with the previous example, a 20°F temperature drop will be used. It is assumed there will be a difference of 20° between the return and supply outlet water temperature of the boiler. However, this difference will occur only when the system is operating at its maximum design temperature.

To find the gpm required for the given system, divide the total loss of 148,800 Btu's by 9,600.

$$\frac{148,800}{9,600} = 15.5 \text{ gpm}$$

STEP 4: Selecting a Circulator Pump.

The selection of a circulator pump is made by use of the Thrush water circulator data given in Figure 36.

Referring to the Chart of Model *C*, in Figure 29, a 1-in. *C* Thrush circulator is found to deliver 15.5 gpm against a head pressure of 6 ft.

STEP 5: Sizing the Main.

In a forced circulating system the piping design must be such that the head in feet of water created by the circulator, be equal to the friction head in the radiator circuit when the system is operating at its maximum design temperature. The measured length of the longest circuit is 115 ft. This includes the trunk main and longest branch main circuit. The longest branch main circuit is used because it offers the most resistance. If the circulator can pump through the longest circuit, it will be more than ample for the shorter circuit. A measured length of 115 ft plus 100 (10 equivalent feet of pipe for each supply tee on the longest branch main) will give a total equivalent length of 215 ft for the longest circuit (trunk and longest branch main circuit).

From Table 2A, using a 6-ft circulator head, read across to find the closest equivalent length of piping to that required for the system (215) to be 240 ft (closest and next largest). This will be found to be in the 300-milinch column. Continuing down this same column to Table 2B find 180,000 Btu's (closest and next largest) to the required 148,000 Btu's. Reading to the left find that $1\frac{1}{2}$-in. pipe will be required to carry the total radiation. This will be the trunk main size.

Branch mains are sized in a like manner. In the 300-milinch column, Table 2B, we find 120,000 Btu's can be carried in $1\frac{1}{4}$-in. pipe and up to 55,000 Btu's in 1-in. pipe. The longest branch main circuit must carry 89,800 Btu's and the shorter branch circuit 95,000 Btu's. Therefore, both circuits for this system require $1\frac{1}{4}$-in. piping.

STEP 6: Sizing the Radiator Branches.

To size the radiator branches refer to Table 3. In Table 3, capacities are indicated for using one or two one-pipe tees for upfeed radiators and includes sizes for first, second, and third floor radiators as well as downfeed radiators. Downfeed radiators require two-pipe tees.

On the 300-milinch line, by using one one-pipe tee find that up to 9000 Btu's can be carried by $\frac{1}{2}$-in. pipe, and up to 22,000 Btu's on $\frac{3}{4}$-in. pipe for first floor radiators. Up to 6000 Btu's can be carried by $\frac{1}{2}$-in. pipe and up to

16,000 Btu's on $\frac{3}{4}$-in. pipe for second floor radiators. In the same 300-milinch line, downfeed radiators of up to 8000 Btu's can be carried on $\frac{1}{2}$-in pipe and up to 21,000 Btu's on $\frac{3}{4}$-in. pipe.

When two, one-pipe tees are installed, one in the supply and one in the return radiator branch, more water is diverted through the radiator and the carrying capacity of the radiator branches is thereby increased. To find the capacity of radiator branches in Btu's when two, one-pipe tees are used, read to the right on the same milinch line of Table 3 as was used previously in the column under the heading of one, one-pipe tees.

STEP 7: Selecting the Boiler.

Net boiler output should be used for selecting a proper size boiler. This may be as net rating in Btu or installed square feet of water radiation of 150 Btu emission. To select a boiler on net installed radiation basis divide the total heat loss by 150 Btu emission. In this example it would be $65,000/150 = 433$ sq ft of installed radiation.

STEP 8: Selecting Pressure Tank Size.

As with the previous example, use Table 4 to select a 24 gal. pressure tank.

STEP 9: Select Flow Control Valve.

The flow control valve is to be the same size as the truck main, or $1\frac{1}{2}$ in.

11.2.3. Two-Pipe Direct Return System

To design a two-pipe direct return hot water system as shown in Figure 38, the same steps are followed as in the two previous examples.

STEP 1: Determine the Heat Loss.

Calculate the heat loss of the structure by using the format described in a previous chapter. For the example in Figure 38, a total heat loss of 96,000 Btu has been calculated.

STEP 2: Make a Piping Layout.

The layout has been made with radiation divided as evenly as possible.

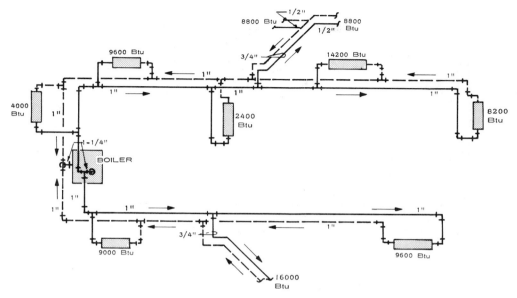

FIGURE 38. Two-pipe direct return hot water system.

STEP 3: Gallons of Water Required per Minute.

To determine the required gallons per minute to be circulated, the total Btu loss is divided by 9600 for 20°F temperature drop, or 96,000/9600 = 10 gpm.

STEP 4: Selecting a Circulator Pump.

Referring to the circulator performance charts in Figure 36 it can be found that a regular 1-in. H or V Thrush circulator will deliver 10 gpm at a 7-ft-head. 7-ft head × 12,000 (milinches in 1 ft) = 84,000 milinches.

STEP 5: Sizing the Main.

Main size is based on milinch resistance of piping in the longest circuit. This includes both supply and return mains. The measured length of the longest circuit is 136 ft. To this measured length add 50% (68 ft) for friction loss in boiler, fittings, and end radiator. The total equivalent length would then be 204 ft (136 + 68). Divide 84,000 milinches (as found in *step 4*) by 204 = 411 milinches/ft of piping in the longest circuit.

Referring to Table 2B, use the 400-milinch column (closest and next smallest to the required 411 milinches). The longest circuit supplies a total of 56,000 Btu's, and therefore, 1-in. supply and return will be required (1-in.

main in 400 milinch column will carry up to 66,000 Btu's). The shorter circuit having 40,000 Btu's will also require 1-in. supply and return. The combined total of 96,000 Btu's will require $1\frac{1}{4}$-in. flow riser. Also, $1\frac{1}{4}$-in. return drop to the 1-in. circulator is needed.

STEP 6: Sizing the Radiator Branches.

Radiator branches would also be sized from the same 400-milinch column on Table 2B.

STEP 7: Selecting the Boiler.

To select the boiler divide total Btu required for the job (96,000) by 150 = 640 sq ft net boiler load.

STEP 8: Selecting the Pressure Tank Size.

For this example see Table 4 for tank selection. An 18 gal. capacity is required.

STEP 9: Select Flow Control Valve.

The flow control valve is the same size as flow riser, $1\frac{1}{4}$ in.

11.2.4. Two-Pipe Reverse Return System

The term "reverse return" comes about in a system where a radiator or convector is fed first and returned last. To design a two-pipe reverse

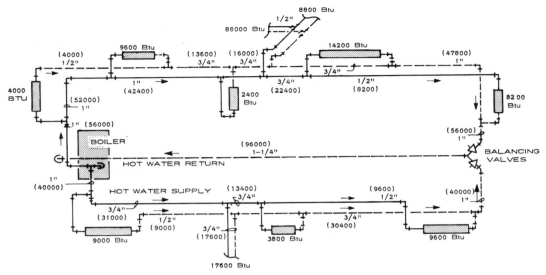

FIGURE 39. Two-pipe reverse return hot water system.

return system as shown in Figure 39, the same steps are followed as with previous examples.

STEP 1: Determine the Heat Loss.
Heat loss for the above job is calculated to be 96,000 Btu.

STEP 2: Make a Piping Layout.
Layout has been made with radiation divided as evenly as possible, as with the direct return system.

STEP 3: Gallons of Water Required per Minute.
To determine the required gallons per minute to be circulated, total Btu loss is divided by 9600 for 20°F temperature drop. Total Btu 96,000/9600 = 10 gpm.

STEP 4: Selecting Circulator Pump.
Referring to circulator performance charts, Figure 29, we find that a regular 1-in. H or V Thrush circulator will deliver 10 gpm at a 7-ft head. 7-ft head × 12,000 (milinches in 1 ft) = 84,000 milinches.

STEP 5: Sizing the Main.
Main size to be based on milinch resistance of the longest piping in the supply and return mains. The measured length of the longest cir-

cuit is 136 ft. To this measured length add 50% (68 ft) for friction loss in boiler, fittings, and end radiator. To total equivalent length would than be 204 ft (136 + 68). Divide 84,000 milinches (as found in *Step 4*) by 204 = 411 milinches/ft of piping in the longest circuit.

Referring to Table 2B, by use of the 400-milinch column (closest and next smallest to the required 411 milinches), find that $\frac{1}{2}$ in. pipe will supply up to 16,000 Btu's, $\frac{3}{4}$ in. pipe up to 34,000 Btu's, 1 in. pipe up to 66,000 Btu's, and $1\frac{1}{4}$-in pipe up the 138,000 Btu's. Using these data and starting with the last radiator the required Btu carrying capacity of the main is noted along the piping back to the boiler also see Figure 39. Sizes of the mains between radiator branches for the accrued Btu's are obtained from the sizes listed above or from the 400-milinch column of Table 2B.

STEP 6: Sizing the Radiator Branches.
Radiator branches would also be sized from the same 400-milinch column, Table 2B.

STEP 7: Selecting the Boiler.
To select the boiler divide total Btu required for the job, 96,000, by 150 = 640 sq ft net boiler load.

STEP 8: Selecting the Pressure Tank.
For this example see Table 4 for tank selection.

STEP 9: Selecting the Flow Control Valve.
The flow control valve is the same size as flow riser, $1\frac{1}{4}$ in.

11.3. STEAM HEATING SYSTEMS

A steam heating system is comprised of a boiler in which the steam is generated and a piping system which conveys the steam to such terminal outlets as radiators, fin tube elements, convectors, and others.

Due to its thermodynamic properties, steam can deliver a large quantity of heat in a small amount of circulating fluid. Steam can be maintained at a pressure which is ample enough to create a positive circulation. This thermodynamic property is that as water is heated until it boils, more and more heat energy is stored in the water–steam mixture even though its temperature may remain constant. This is called the heat of evaporation. Because of the fact that considerably more heat energy can be stored in a water-steam mixture as compared to a hot water heating system, steam can be a very efficient heating medium.

Steam heating systems are classified according to steam temperature, piping arrangement, and operating temperature. These are as follows:

One-pipe:	Steam and condensate are carried in one pipe.
Two-pipe:	Steam and condensate return are in separate pipes.
Vapor system:	A low pressure, two-pipe arrangement.
Upfeed:	Main is located below the radiators.
Downfeed:	Main is extended above the radiators.
Wet return:	A condensate return line that is below the water level in the boiler.
Dry return:	A return line above the boiler water level.
Gravity circulation:	A system wherein the condensate returns to the

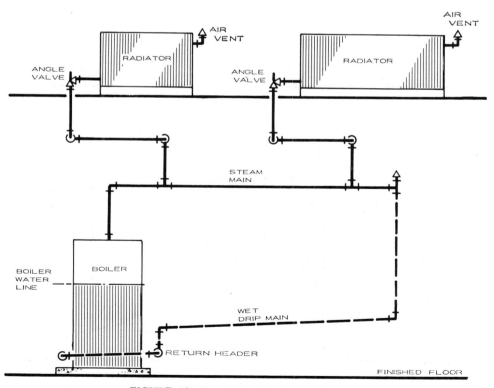

FIGURE 40. One-pipe gravity steam system.

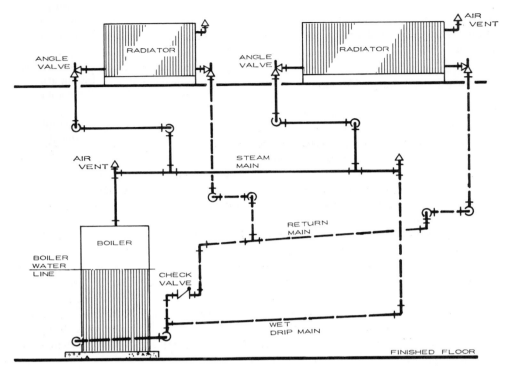

FIGURE 41. Two-pipe gravity steam system.

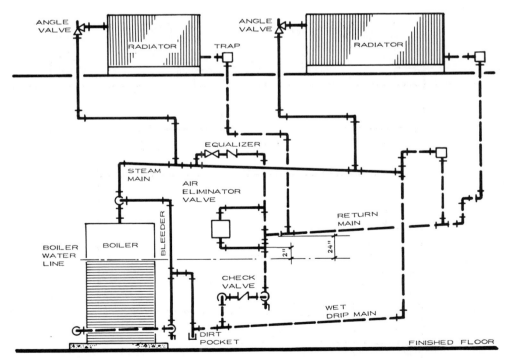

FIGURE 42. Vapor system.

High pressure:

boiler by gravity instead of by a pump.

This system operates between 50 psig and 150 psig, and is primarily used for industrial heating and for process work. The need for high pressure is to force the steam through long lengths of pipe or to drive machinery.

This system often requires expensive piping is dangerous by nature so that state and local codes require licensed operators to run it.

Medium pressure:

This system operates between 15 psig and 50 psig. It is used for institutional and industrial heating and for process work.

Low pressure:

Most commonly used system in HVAC, for heating and cooling applications. It operates between 0.5 psig and 15 psig.

Vacuum:

A vacuum system uses a pump to create a partial vacuum in a receiver tank so that the condensate return can be drawn back to the boiler.

Some of the systems above can be combined to create a new system such as a one-pipe, gravity, downfeed or a two-pipe low pressure, upfeed, etc. The three most commonly used systems are shown in Figures 40, 41, and 42, a one-pipe gravity system, a two-pipe gravity system, and a vapor system, respectively.

11.3.1. Estimating Steam Consumption

In estimating steam consumption the efficiency is generally assumed to be 100%. For low-pressure steam an average heating value of 1000 Btu/lb of steam is used.

Where the heat loss is calculated as the heat rate in Btu/hr per degree difference in temperature, the following Equation may be used:

$$F = \frac{H \times 24 \times D}{1000} \qquad \text{(Eq. 20)}$$

where

F = pounds of steam required for estimate period

H = calculated heat loss, Btu/hr (degree difference)

D = number of degree days for the period of estimation

1000 = Btu delivered per pound of steam condensed

11.3.2. Selecting Pipe Sizes for Steam Mains

The subject of pipe design for steam applications is complicated. This book is intended to only touch some of the highlights as a matter of information for those who are not involved in this work. For those who engineer steam pipe systems more extensive texts are recommended.

Factors which enter into the selection of steam line sizes are:

1. Maximum quantities of steam (in pounds per hour) to be delivered at terminal points.
2. Steam pressure in pounds per square inch available at source.
3. Allowable or desired pressure drop of the steam flowing through the line, usually expressed in pounds per square inch per hundred feet.

Before the first of these factors, maximum flow, can be obtained, it is necessary to secure or compute the maximum steam demands of all points to be served and the pressure required. Having obtained these figures, the maximum flow of steam in each section of the line can readily be figured. When the new line is to begin at the boiler plant, the second factor, initial pressure, is known from plant operating practice. When it begins elsewhere, either a pressure gauge test may be necessary at the point of connection or a calculation made of the theoretical pressure at that point.

Assume for the moment that an intermediate point on a steam line requires the same pressure steam as the most distant point on the line. The allowable pressure drop would be the difference between the initial pressure and the pressure required at the intermediate point. This total allowable pressure drop divided by the "equivalent" line length (which makes extra allowance for the greater friction in bends, valves and fittings) in hundreds of feet, gives the permissible unit pressure drop per hundred feet. For example, assume a low-pressure steam main with an initial pressure of 11 psig, service pressure at the end of a steam main of 10 psi, length of line 370 ft, plus an allowance for bends and fittings equal to 30 ft or a total equivalent length of 400 ft. The pressure drop will be

$$= \frac{\text{initial pressure} - \text{required pressure}}{\text{equivalent length}} \times 100$$

$$= \frac{(11 - 10)}{(370 + 30)} \times 100$$

$$= \tfrac{1}{4} \text{ psi/100 ft} \qquad \text{(Eq. 21)}$$

If several points are to be connected to the steam main, with different service pressure requirements, the pressure drop to each point should be calculated as a separate problem.

Where high-pressure steam is being transmitted through underground steam mains it is often desirable to set up a pressure drop per hundred feet that will permit the use of smaller steam mains to reduce initial costs. This pressure drop should be determined after a careful analysis of all factors entering into the installation. Such factors include:

- Initial costs
- steam pressure required at terminal points
- expansion and anchor points
- critical velocity of steam
- auxiliary apparatus for control, reducing pressures, drips, etc.

Once the above factors are known, one of several formulas can be used to determine the line size. The formula for the flow of steam in pipe, that is generally accepted in the district heating industry, has been established by Unwin. Unwins formula and the graphical solution of this formula is readily available in standard texts on this subject.

11.3.3. Recommended Velocities for Flow of Steam in Pipes

The table below gives reasonable velocities that are based on the average practice for steam lines. The lower velocities should be used for smaller pipes and the higher velocities for pipes larger than 12 in.

Condition of Steam	Pressure, (psi)	Use	Reasonable Velocity, (ft/min.)
Saturated	0 to 15	Heating	4000–6000
Saturated	50 and up	Miscellaneous	6000–10000
Superheated	200 and up	High pressure steam piping	7000–20000

11.3.4. Estimating Capacities of Condensate Return Piping

The various types of condensate return systems require distinctly different pipe sizes. No general recommendation is made as to the type of system as each problem needs individual analysis. However, it generally pays to return the condensate

- from large buildings located close to the boiler plant
- if a gravity flow can be obtained
- when the boiler feedwater makeup requires extensive treatment, providing practically all the condensate can be returned

Saturated steam mains should be dripped through a drip and trapped at intervals of 300 to 400 ft, and at all low points. Drip lines should be sized according to the required trap size and can be connected to condensate return mains at the nearest convenient point. High pressure drip lines cannot be connected to vacuum return lines.

Tables 5a, b and c give pipe capacities for three conventional methods of returning condensate:

- gravity flow
- vacuum return
- pumped return

TABLE 5. Pipe Capacities for Condensate Return

a) Gravity Flow, lbs of condensate per hour

Size of Pipe,[a] (in.)	Pitch of Pipe per 100 Ft				
	6	12	24	30	60
1	660	940	1,350	1,490	2,100
1¼	1,220	1,730	2,490	2,512	3,970
1½	2,000	2,840	4,080	4,500	6,350
2	4,350	6,150	8,830	9,730	13,760
2½	7,860	11,120	15,970	17,590	24,880
3	12,700	18,000	25,830	28,440	40,220
4	26,960	38,100	54,760	60,290	85,270
5	48,000	67,800	97,450	107,290	151,750
6	76,200	108,700	156,100	171,890	243,000

[a]Size of pipe should be increased if it carries any steam.
Size of pipe can be reduced if head pressure is available at source.

b) Vacuum Return, square feet equivalent radiation

Pipe Size (in.)	1	1¼	1½	2	2½	3	4	5	6
Sq Ft Equiv. Radiation	1,400	2,400	3,800	8,000	13,400	21,400	44,000	77,400	124,000

4″ grade per 100 ft in direction of flow.
Equivalent radiation divided by 4 equals pounds condensate per hour.

c) Pumped Return, pressure drop in psi/100 ft.

Velocity	2 Ft Per Sec.			3 Ft Per Sec.			4 Ft Per Sec.		
Pipe Size Inches	Gals. per Minute	Lbs. per Hour	Pressure Drop	Gals. per Minute	Lbs. per Hour	Pressure Drop	Gals. per Minute	Lbs. per Hour	Pressure Drop
1	6	2,880	1.50	8	3,840	2.50	10	4,800	4.00
1¼	10	4,800	1.00	14	6,720	1.80	18	8,640	2.60
1½	13	6,240	.78	19	9,120	1.60	25	12,000	2.40
2	20	9,600	.49	32	15,360	1.20	44	21,120	2.10
2½	30	14,400	.43	45	21,600	.91	60	28,800	1.55
3	45	21,600	.30	70	33,600	.72	95	45,600	1.24
4	80	38,400	.23	120	57,600	.50	160	76,800	.84
5	125	60,000	.16	190	91,200	.35	250	120,000	.58
6	180	86,400	.13	270	129,600	.29	360	172,800	.48
8	300	144,000	.09	475	228,000	.20	600	288,000	.30
10	500	240,000	.07	750	360,000	.15	1000	480,000	.26

Recommended Velocities: Sizes 1″ to 1½″ Incl.−2 ft per second.
                                        Sizes 2″ to 3″   Incl.−2 to 3 ft per second.
                                        Sizes 4″ to 10″ Incl.−2 to 4 ft per second.
Usually not much economy in using sizes under 1½″ diameter.

11.4. ELECTRIC HEATING SYSTEMS

When electricity passes through a resistance wire, the molecules are stirred up creating heat. This electricity is actually the flow of electrons through the conductor. The number of electrons that flow per unit time is called the current and is measured in amperes (A). The driving force, which is analogous to the pump head in a hydraulic system, is called voltage. The heat that is generated is due to the internal resistance of the wire which is measured in ohms. The wire heats up uniformly because of conductive heat transfer. Electric power is what is required to keep current flowing against this resistance. The power needed for 110 volts (V) to push 4 A through a conductor is determined by the formula

$$\text{Power (in watts)} = \frac{\text{Volts}^2}{\text{Resistance (in ohms)}}$$

$$= 110 \times 4$$

$$= 440 \text{ W}$$

Most commercial heating elements are designed with a margin of safety of rated wattage vs. the maximum watts that can be delivered. The type of application determines the life to a major extent. It is much more important that the proper voltage be applied to heating elements than to electric motors. A motor is built to operate satisfactorily at rated voltage plus or minus 10%. At 90% of the rated voltage, an electric motor will run slightly warmer, but will still carry its rated load. However, if a resistance heating element is operated at 90% of the rated voltage, it will produce only 81% of its rated heat output. For this reason, it is important that wiring to electric heaters be adequate, and that consideration be given to good voltage regulation.

The basic formula which applies to change of heat output with voltage change is:

$$\text{Power (in watts)} = \frac{\text{Volts}^2}{\text{Resistance (in ohms)}}$$

Since the resistance of the heating element remains constant, the heat output in watts varies as the square of the voltage is applied. If 230 V. were applied to a 115 V. element, the heat output would be four times the rated output, and the life of the heating element would be very short.

11.4.1. Heat Transfer with Electric Heaters

In Chapter 2, Heating Principles, it was stated that heat is transferred from higher temperature bodies to lower temperature ones. It flows like water in all directions unless controlled or directed. The rate depends on the temperature difference between the source and the receiving body. There was also a brief explanation of the three methods of heat transfer: conduction, convection, and radiation. The explanation is amplified below with the emphasis on heat transfer with electric heaters.

Conduction. Conduction takes place in solids, liquids, and gases and is the transfer of heat from one particle of matter to another. The heat source must be in direct contact with the object. A silver spoon used to stir a hot cup of tea soon becomes hot all over due to conduction.

Convection. Convection is the transfer of heat in a liquid or gas by the mixing of the warmer particles with the cooler. This mixing action may be the result of differences in temperature, as in natural convection. When an electric strip heater generates heat, the air around the heater is warmed and expands. Since the heated air weighs less than air above it, it rises. In so doing, it mixes with and warms the cooler air. A mixing action may also be produced by mechanical means, as in forced convection (or circulation). The warm air register in a home heats air in the room by convection. The heat from an electric forced air heater is convected heat.

Radiation. Radiation is the transfer of heat from one body to another as the result of the bodies emitting and absorbing a form of energy called radiant energy. Radiant energy may be regarded as being transferred through space in the form of waves, similar to light. Heat energy is transformed into radiant energy at the surface of a heater. It passes through the surrounding medium to other substances which convert the radiant energy back into heat energy. A substance transparent to radiant energy such as glass could not convert any of the radiant energy to heat energy as it passes through. A solid, nonreflecting substance can readily absorb radiant energy and convert it into heat.

The rate of heat transfer varies with the application, medium of transfer, and the temperature difference as noted above. As a transfer medium, air is slow. Liquids vary—water is fast and molasses is slow. Molten salts are usually very good. Metals also vary—aluminum is good and cast iron is less efficient.

For example, a heater operating in air may have a sheath temperature of 1000°F. The heat is transferred slowly, principally by convection, to the surrounding air. The same heater lightly clamped to a steel bar would have a sheath temperature of 920°F. Here the surface of the heater in contact with the bar is transferring heat to the bar faster than to the air. The same heater tightly clamped to the same bar would have a sheath temperature of 800°F. Better contact between the two surfaces increased the conduction of heat. Of course, there is still a lot of heat being transferred by convection through the air from the sides of the strip heater not in contact with the steel bar. The same heater immersed in water would operate at about 230°F. Here the heat would be conducted rapidly into the water which surrounds the heater intimately on all sides.

11.4.2. Heat Absorption

The life of an electric heating unit is influenced by its operating temperature for beyond a certain temperature, deterioration of a resistance wire increases rapidly. The sheath temperature of the heating element must be higher than the process temperature since heat moves from the higher to the lower temperature. If the material being heated does not readily absorb the heat from the heating element, the sheath temperature will increase, causing an increase in wire temperature. The faster heat is absorbed by a material, the lower will be the operating temperature of the sheath for the same process temperature.

If the strip heater is clamped to a bar only at its ends so that it can't expand lengthwise, the middle of the strip will bow away from the bar. The temperature in the midsection will increase over the temperature of the ends, because there is no conduction to the steel at the middle. Good contact is necessary for efficient heat transfer.

A pot on an electric range is a good example of conduction and absorption. If the pot does not cover the heating element, the element will be red in color in the uncovered section, and will be black where it is covered. This shows that the covered section is conducting heat to the pot and the pot is absorbing the heat. To absorb the most heat from the element, the pot should have a flat bottom so that good contact will be made with the element.

Different materials absorb heat at different rates (a physical property known as thermal conductivity). This is important to know because it is useless and even destructive to attempt to supply heat to a material faster than it can be absorbed. Silver will absorb heat rapidly, air very slowly. This list of materials is in the order of highest to lowest absorption rates.

Materials	Thermal Conductivity [Btu/(hr)(sq ft)(°F/ft)]
Silver	245
Copper	227
Aluminum	128
Steel	26.2
Ice	1.3
Glass	0.3–0.5
Water	0.348
Air	0.014

When heating a material that absorbs heat rapidly, an element that produces a lot of heat per square inch of sheath surface can be used. If the material absorbs heat slowly, a small amount of heat per square inch must be used. This rate of producing heat in the element is called "watt density." Heating elements are built with ratings ranging from 4 to 65 W density. In some special applications, much higher watt densities can be used.

The following table lists some maximum watt densities that are satisfactory for heaters immersed in these materials, provided there are no unusual sludge or sheath problems.

Materials Heated	Approx. Temp. (°F)	Allowable Watt Density (W/sq in.)
water (domestic)	212	50–65
water (industrial)	212	35
alkali cleaning solution	212	35
prestone	300	20–30
metal melting pot	500	25

Curves are available in manufacturers' catalogs to determine the allowable watt densities of strip heaters for both clamp-on and air heating applications. Allowable watt densities for other heating elements are usually obtained from the manufacturer.

11.4.3. Types of Electric Heating

Application of electric heat can be used for the following:

- Liquid heating
- Surface heating
- Snow melting
- Pipe line heating
- Air heating—oven and comfort

Liquid Heating. In industry, many types of liquids must be heated. The most common of these is water which is used for process needs and wash room purposes. Other liquids such as oil, plating solutions, caustic solutions, etc., are heated in a manner similar to water. Many processes require open vessels for liquid heating. Others use a closed vessel. Some states or municipalities require that heated water storage systems be inspected and certified by the state or an insurance company before being placed in operation. The tank must have a pressure relief valve with a trip lever and a pressure–temperature gauge.

The normal expansion and contraction of immersion elements in water heaters tends to remove any lime deposit, and no maintenance is required. Where extreme hard water conditions prevail, it is advisable to inspect after about three months use. If no lime deposit is found, it is safe to assume that no further inspection is necessary. A build up of lime acts as a heat insulator which may cause the heater to fail prematurely.

Where headroom is available, tanks should be installed vertically, as there is less mixing action when the cold liquid enters the tank at the bottom. Since heat rises, the liquid below a tank heater does not get hot. Therefore, heating elements should be placed low in a tank.

Surface Heating. Electrical heating elements can be used to heat surfaces by locating them in holes or grooves, by clamping them to a surface, or by mounting them spaced away from the surface. When heaters are clamped on to surfaces, intimate contact between the heater and surface is essential. A unit of proper watt density should be selected to ensure its long life. Heaters spaced away from the surface require a different watt density than clamped heaters. An over-temperature thermostat in the air space is desirable to protect the heaters from overheating.

Snow Melting. Snow melting can also be considered a surface heating problem, and the same basic heat fundamentals apply. However, much test work has been done on the amount of heat required to melt snow on concrete and asphalt surfaces. It is, therefore, not necessary to go through a complete calculation. With an air termperature of 10°F, 10 W/sq ft is enough heat to melt the ice immediately in contact with the surface. The ice, up to 1-in. thick, can then be removed by scraping. A large number of installations use about 33 W/sq ft. This density will melt a snowfall of 1-in./hr at an air temperature of 26°F.

Pipe Line Heating. In many manufacturing processes there exists the need to transport a viscous liquid through pipes. To accomplish this, the temperature of such liquids as tar, molasses, heavy oils, etc., is raised to make the liquid flow freely. It is then necessary to maintain the temperature along the entire length of the pipe by replacing the radiation losses. Another problem is that of preventing exposed water pipes from freezing. This too, can be done by supplying enough heat to replace radiation losses in order to maintain the water temperature above the freezing point. Tubular heaters which are strapped to the pipes, are ideal for such applications.

Air Heating. Of the three methods of transferring heat from electric heaters the primary method covered so far has been by conduction, although radiation and convection were present to some degree. In air heating, the principal method of heat transfer is convection, where the heat is transferred first from the heater to an air current and then from the air current to the object to be heated. The procedure previously outlined for the heating load calculation, Chapter 3, can be used. However, the heating load expressed in Btu/hr is converted to kilowatts by the divisor of 3413 Btu/hr kW.

11.4.4. Economics of Electric Heat

In almost any part of the United States, a dollar's worth of coal, gas, or oil contains more heat units than a dollar's worth of electricity. To justify using electric heat, some other factors must outweigh this difference in cost. Such factors which often outweigh the cost differential and may lead to the selection of electric heat are efficiency, maintenance, control, installation, and initial cost.

Efficiency. A kilowatt-hour contains 3412 Btu. It can always be converted to heat with an efficiency of up to 100%. In general, all of the heat contained in the kilowatt-hour is usable, although the conversion efficiency may not be 100%. In converting fuel to heat, however, the efficiency may range up to a maximum of 80% because of the losses in burning and the heat transfer losses.

Maintenance. Electric heat is usually controlled by automatic devices so that the labor of attendance can be entirely eliminated. Such devices are a thermostat that will maintain any desired temperature automatically, and a time clock to control the time of any temperature change in a heating application.

Control. Many control devices are available to fit the quantity of heat to the requirements of the job, giving a flexibility that can not be obtained in any other way. Some of these devices are:

1. An input controller, which permits power to flow at rated value for an adjustable percentage of each minute.
2. A double throw switch to connect heating elements in either series (low heat), parallel (high heat), or wye (low heat), delta (high heat) arrangements.
3. A variable automatic transformer to change the applied voltage and thereby control the heat input.

Installation. Heat can be applied locally in a production line, thus, eliminating transportation of the object to be heated to a separate furnace location. Electric equipment is light in weight and can be suspended from the ceiling, erected vertically, or fit into a minimum space. There is no need for chimneys or flues to handle products of combustion.

Initial Cost. The overall investment in heating equipment is generally lower if electric heat is used, because less floor space is usually required, auxiliaries are at a minimum, and controls are simple. Low investment with long equipment life means lower interest, taxes, and depreciation.

In summary, the lowest overall cost of the heating system is the ultimate objective. Therefore, all of the above factors should be evaluated in determining what source of heat should be used in addition to the cost of the electricity or other fuel.

12

SOLAR ENERGY

The topic of solar energy is currently receiving great interest in today's HVAC industry. By now everyone is painfully aware of the shortages of fossil fuel. HVAC manufacturers with the encouragement of government agencies are turning their technological skills in the direction of harnessing the sun's energy and converting it to usable heat. We are now seeing more prototype and "packages" of solar heating components mainly for domestic usage than ever before.

Solar energy is generally interpreted by the public to involve a solar collector which will provide energy to the home. Actually, solar energy has a broader meaning. It can involve: *photovoltaic energy*, the direct conversion of the sun's energy into electricity; *bioconversion*, the utilization of agricultural or municipal wastes to provide fuel; *ocean thermal*, providing power by harnessing the temperature difference between the surface waters and the ocean depths; *wind*, harnessing wind energy to generate power; and *solar thermal electric*, concentrating the sun's rays to obtain high temperatures and thus generate electric power.

For the purpose of this Chapter let's exploit the solar energy used in the HVAC field. A general approach is to begin with the sun-energy nomenclature which will aid in the understanding of solar energy usage.

12.1. SOLAR ENERGY NOMENCLATURE*

Absorber, or absorber plate: A surface, usually blackened metal, in a solar collector which absorbs solar radiation.

Absorptance: The soaking up of heat in a solar collector. Measured as percent of total radiation available.

Active solar system: Any system that needs mechanical means such as motors, pumps, valves, etc., to operate.

Ambient temperature: Another way of saying how cold or how hot it is outdoors.

Bioconversion: Utilization of agricultural or municipal wastes to provide fuel.

British thermal unit (Btu): A unit of energy defined as the amount of energy required to heat one pound of water one degree Fahrenheit.

Collector, or solar collector: A device for receiving solar radiation and converting it to heat in a fluid.

Collector efficiency: The fraction of incoming radiation captured by the collector. If your system captures half of the incoming radiation, you have a system that is 50 percent efficient. Efficiency is the capability of a collector to capture Btu's under various climatic conditions. Efficiency varies according to

*Extracted from *BUYING SOLAR*, Federal Energy Administration Publication FEA/G-76/154

outside temperatures, whether skies are clear or cloudy, whether it is windy or not, and, of course, the quality of the collector. There's no way a collector can be 100 percent efficient; that is, to capture all the Btu's that fall on the collector; 55 percent is good under desirable weather conditions.

Collector tilt: The angle measured from the horizontal at which a solar heat collector is tilted to face the sun for better performance.

Concentrator: Reflector or lens designed to focus a large amount of sunshine into a small area, thus increasing the temperature.

Conductivity: The ease with which heat will flow through a material determined by the material's physical characteristics. Copper is an excellent conductor of heat; insulating materials are poor conductors.

Convection: When two surfaces—one hot, the other cold—are separated by a thin layer of air, moving air currents (called convection currents) are established that carry heat from the hot to the cold surface.

Emittance: A measure of the heat re-radiated back from the solar collector. Measured as fraction of the energy which would be radiated by a totally black surface at the same temperature.

Fluid: Any substance such as air, water, or antifreeze used to capture heat in the collector.

Galvanic corrosion: If you have this in your system, you have problems. This is caused when different metals are not isolated properly and a liquid comes in contact with both metals. The result is galvanic corrosion and repair bills.

Heliostat: A mirror used to reflect the sun's rays into a solar collector or furnace.

Hybrid solar system: A system that uses both active and passive methods to operate (e.g., a solar system which uses pumps to heat and nocturnal cooling to cool).

Insolation: The rate of solar radiation received per unit area.

Kilowatt: One thousand watts of power; equal to about $1\frac{1}{3}$ horsepower.

Kilowatt-hour (kWh): The amount of energy equivalent to 1 kilowatt of power being used for 1 hour = 3,413 Btu.

Langley: A unit of measurement of insolation. (One langley equals one gram-calorie per square centimeter.) The langley was named for American astronomer Samuel P. Langley.

Ocean thermal: Providing power by harnessing the temperature differences between the surface waters and the ocean depths.

Passive solar system: A system that uses gravity, heat flows, evaporation or other acts of Mother Nature to operate without mechanical devices to collect and transfer energy (i.e., south facing windows).

Photovoltaic: Direct conversion of the sun's energy into electricity.

Pyranometer: An instrument for measuring solar radiation.

Radiation: Any object that is warmer than its surroundings radiates heat waves (similar to light waves, but invisible) and, thus, emits heat energy.

Reradiation: After an object has received radiation or is otherwise heated, it often reradiates heat back. Generally speaking, matte black surfaces are good absorbers and emitters of thermal radiation while white and metallic surfaces are not.

Selective surface: A special coating sometimes applied to the absorber plate in a solar collector. The selective surface absorbs most of the incoming solar energy and reradiates very little of it.

Solar cell: A device, usually made of silicon, that converts sunlight directly into electrical energy.

Solar constant: The average amount of solar radiation reaching the earth's atmosphere per minute. This is just under 2 langleys, or 2 gram-calories per square centimeter. This is equivalent to 442.4 Btu/hr/ft^2, 1395 watts/m^2 or .1395 watts/cm^2.

Solar rights: An unresolved legal issue involving who owns the rights to the sun's rays.

Sun tracking: Following the sun with a solar collector to make the collector more effective.

System efficiency: Btu's are lost from the time the sun's rays hit the collector to the moment they are used to heat the house or the water supply. The question is how many Btu's are used in comparison to the original number coming in. The answer is the efficiency of the whole system. This is a very important consideration.

Thermosyphon: The principle that makes water circulate automatically between a collector and a storage tank above it, gradually increasing its temperature.

12.2. SOLAR ENERGY SYSTEM

Solar energy is put into use by a basic system composed of a solar collector, a piping network, and a series of controls (relays, thermostat, valve etc.,) Figure 43 shows a typical application of solar energy.

12.2.1. The Solar Collector

The solar collector is the component whose main function is to capture the sun's energy. Many types of collectors are available such as:

1. High-performance collectors which focus and track the sun.
2. Vacuum-sealed collectors that have very low heat losses.
3. Flat collectors which are considered to be conventional.

Various other types of collectors are available in different shapes and efficiencies. To compare their quality is a very complex process.

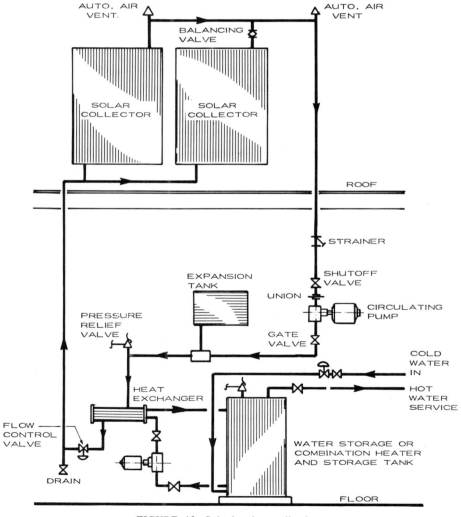

FIGURE 43. Solar heating application.

The designer must rely heavily on the reputation of the manufacturer when evaluating his data.

Basically, a collector is composed of a frame which contains an insulating blanket, an absorber or collector plate array, and a serpentine of channels (pipes). When the solar radiation passes through the transparant cover and hits the absorber plate, most of the radiant heat is absorbed by the plate and transferred through the pipe walls to the air, water, or other media. Some of the radiation is reflected off the plate and back to the cover. How much radiant solar heat is absorbed or reflected back depends on the physical characteristics of the collector plate.

An important design factor for the collector, depending on climate conditions, is its position with relation to the sun. The collector must face south, and its ideal inclination should be to bring it as perpendicular to the sun's rays as possible During the winter season, the optimum angle is the latitude of the area involved plus 10 to 15°; in the summer it is the latitude of the area minus 10 to 15°.

Next to the collector, another factor effecting the efficiency of a solar energy system is the flow of fluid through the collector. For exam-

TABLE 6. Economic Comparison of Solar Space Heating Plus Solar Hot Water Heating vs. Oil Heat Plus Electric Hot Water Heat.

Region	Solar System Cost ($)	Pay Out Time (yr)			
		[a]3¢/kWh [b]40¢/gal.	[a]3½¢/kWh [b]43¢/gal.	[a]4¢/kWh [b]45¢/gal.	[a]5¢/kWh [b]45¢/gal.
East Coast—Boston					
40% Solar	4875	12.1	11.4	10.7	—
50% Solar	6750	12.6	11.8	11.2	—
East Coast—New York					
40% Solar	4700	13.0	12.1	11.4	10.5
50% Solar	6800	14.4	13.5	12.8	11.9
East Coast—Washington					
40% Solar	3475	10.4	9.6	9.0	—
50% Solar	5300	12.0	11.2	10.5	—
Upper Midwest (Omaha—Chicago)					
40% Solar	3200	9.2	8.5	8.0	—
50% Solar	4825	10.6	9.8	9.3	—
Lower Midwest (St. Louis—Nashville)					
40% Solar	3275	10.8	10.0	9.3	—
50% Solar	4825	12.3	11.4	11.0	—
Southwest (Dallas)					
40% Solar	2200	9.2	8.3	7.7	—
50% Solar	3000	10.0	9.1	8.5	—
60% Solar	4875	12.4	11.5	10.8	—
Southern California (Los Angeles)					
50% Solar	1500	6.7	6.1	5.6	—
60% Solar	2175	7.8	7.2	6.7	—
70% Solar	3000	9.0	8.3	7.7	—

[a]Electricity.
[b]Oil.

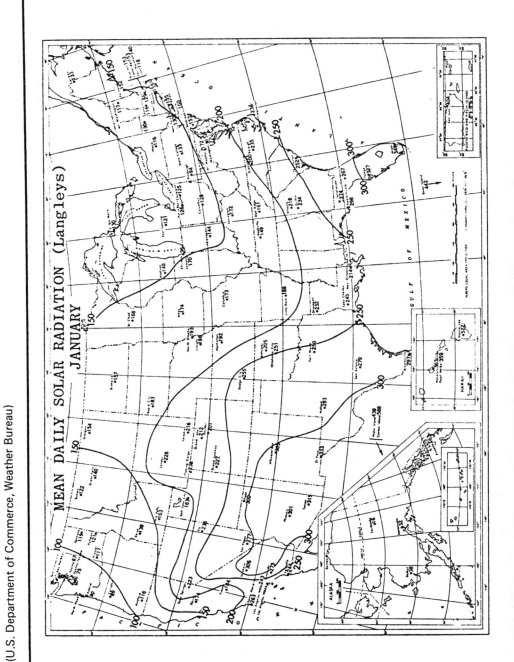

TABLE 7. Mean Daily Solar Radiation, Monthly and Annual (U.S. Department of Commerce, Weather Bureau)

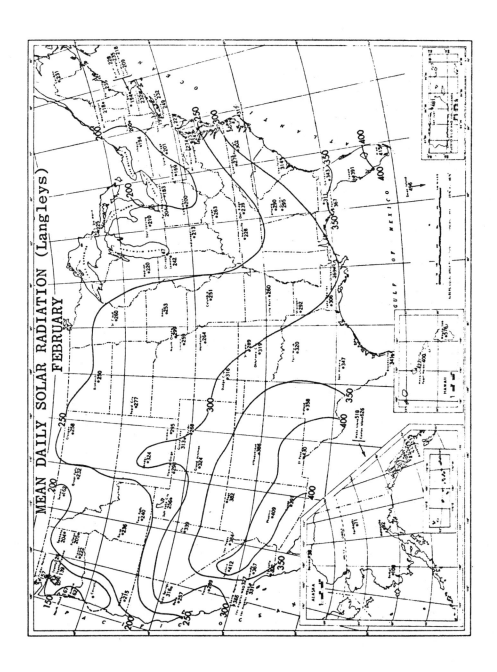

MEAN DAILY SOLAR RADIATION (Langleys)
FEBRUARY

TABLE 7. *(Continued)*

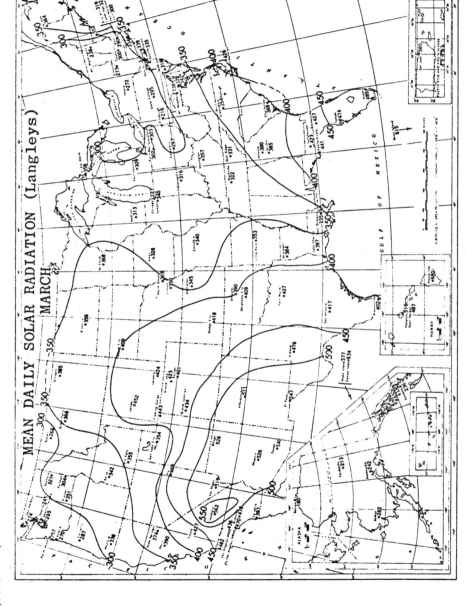

MEAN DAILY SOLAR RADIATION (Langleys)
MARCH

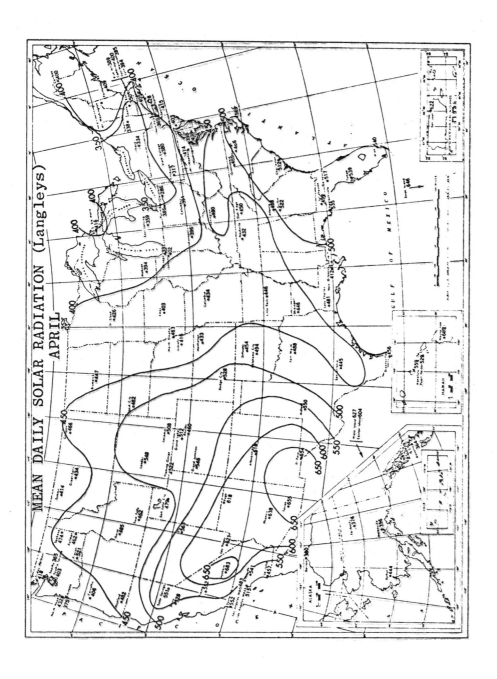

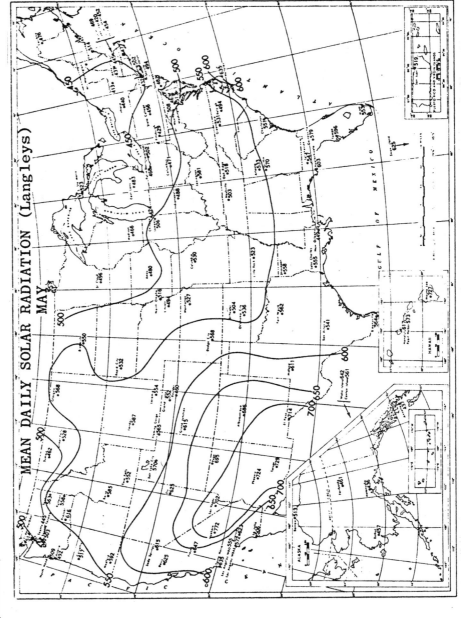

TABLE 7. *(Continued)*

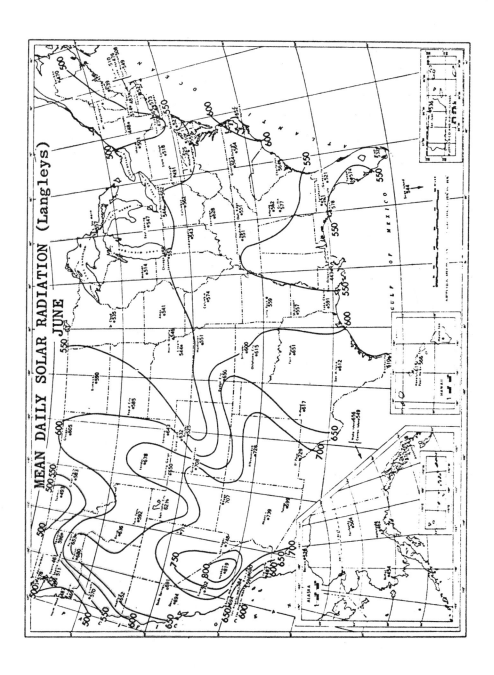

MEAN DAILY SOLAR RADIATION (Langleys)
JUNE

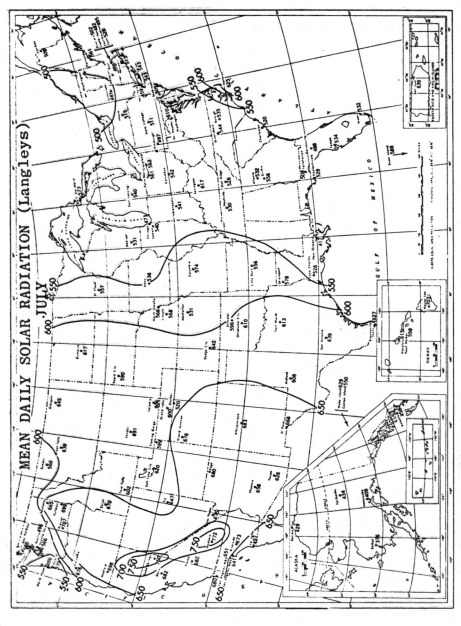

TABLE 7. *(Continued)*

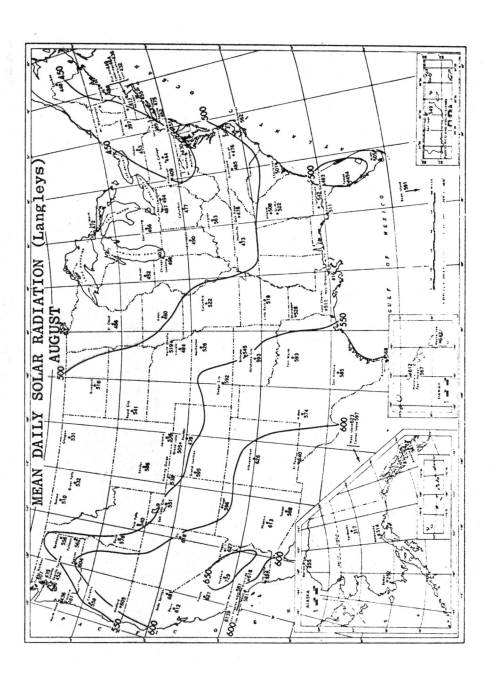

MEAN DAILY SOLAR RADIATION (Langleys)
AUGUST

TABLE 7. (Continued)

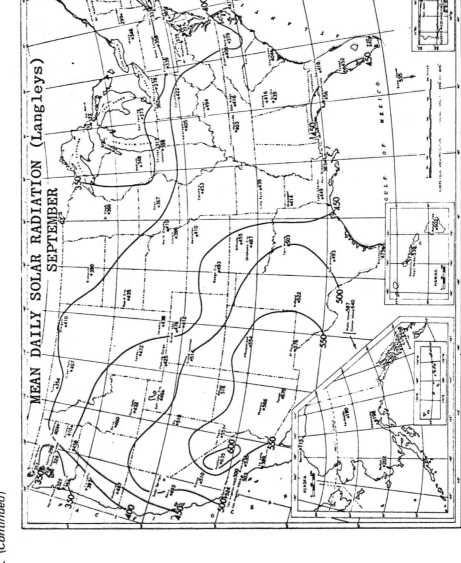

MEAN DAILY SOLAR RADIATION (Langleys)
SEPTEMBER

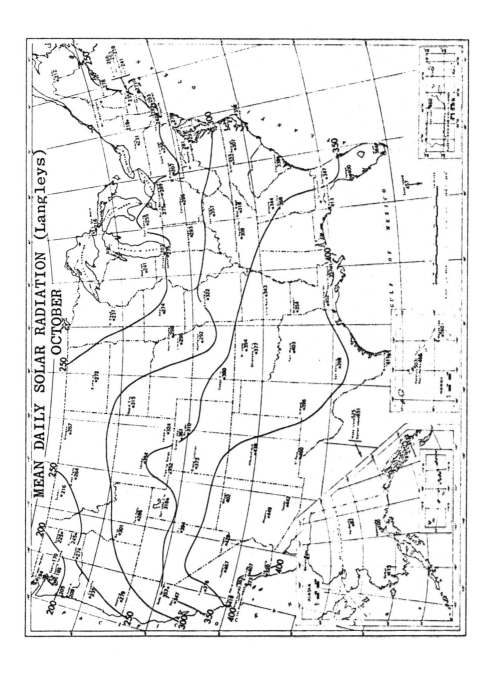

MEAN DAILY SOLAR RADIATION (Langleys)
OCTOBER

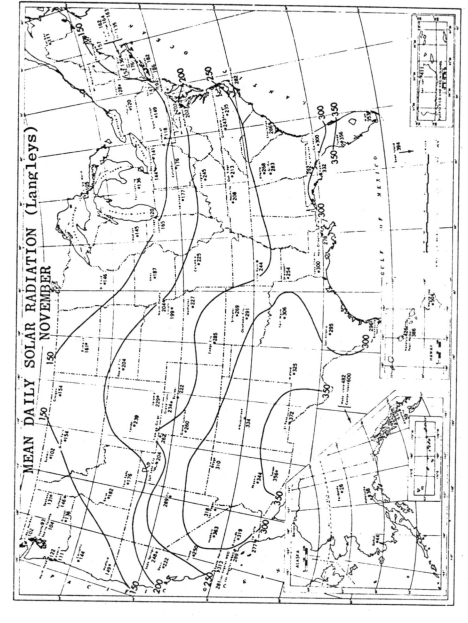

TABLE 7. (Continued)

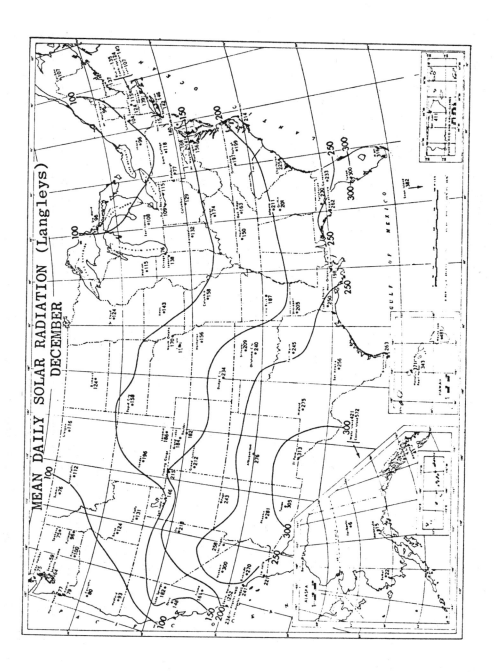

MEAN DAILY SOLAR RADIATION (Langleys)
DECEMBER

TABLE 7. (*Continued*)

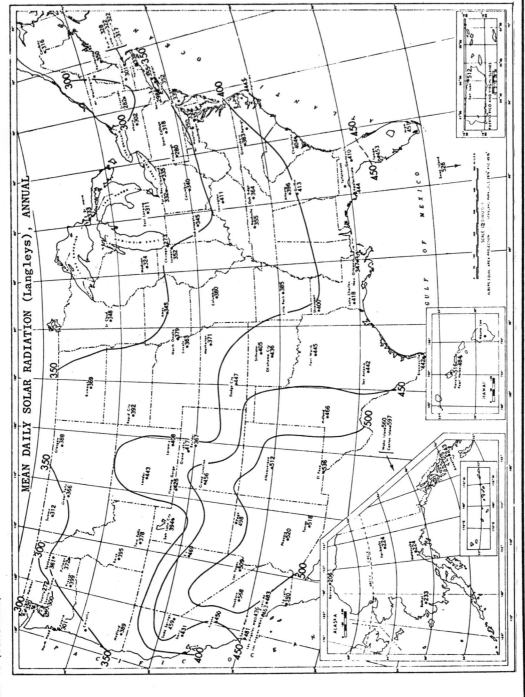

MEAN DAILY SOLAR RADIATION (Langleys), ANNUAL

These charts and table are based on all usable solar radiation data, direct and diffuse, measured on a horizontal surface and published in the Monthly Weather Review and Climatological Data National Summary through 1962. All data were measured in, or were reduced to, the International Scale of Pyrheliometry, 1956.

Langley is the unit used to denote one gram calorie per square centimeter (1 langley = 1 gm. cal. cm²).

ple, if there is no movement at all, the fluid in the collector will become very hot and will dissipate heat that becomes lost to the system, but as the fluid moves, less head dissipation will occur. Therefore, a close control of air or water flow will improve the efficiency of the system. The temperature at which the collected heat must be stored also effects the system's efficiency. It has been found that the hotter the system operates the lower it's efficiency.

12.2.2. Heat Storage

The heat captured by the collector is usually stored by three methods: water, rocks, and change-of-state storage. Each has its advantages and disadvantages as noted below.

Water Storage.   Water has the advantages of low cost and high heat capacity. Water storage tanks must be protected from freezing if located outdoors. Tanks must be protected from corrosion either through use of a corrosion inhibitor in the water or by constructing them of corrosion-resistant material. Storage tanks must be heavily insulated to prevent the stored heat from being lost.

Materials used for water storage have both advantages and disadvantages. When rust and corrosion inhibitors are added, care must be taken to ensure that the water in the system does not mix with the potable water because inhibitors are toxic. Plastic tanks, though corrosion proof are more expensive than steel but the types for storing large quantities of water at high temperature. Concrete is safe, durable, and economical, but if a leak developes it is hard to repair. Great care must be taken to make certain that the concrete tank is properly sealed when it is first built and installed. Some experts recommend bladders or diaphragms lining the tank to insure water-tightness.

Rocks and Pebbles Storage.   For air-type solar collectors, rocks or pebbles are the most cost-effective way to store the collected heat. Rocks are easily obtainable, economical, and are not subject to problems stemming from corrosion, freezing, or leaking. Further, there is no need for an expensive container. However, rocks and pebbles do have two disadvantages: they take up a lot of space and are much heavier than water. The space needed for storing heat in rocks is roughly three times that for water. This can be a major problem for some homes.

Another advantage of rocks is that they can be used to store heat at temperatures above 212°F, while water containers designed for such temperatures must be designed for more than atmospheric pressure, making the tank more expensive. The ability of the rocks to reach higher temperatures is important because such heat will not be wasted if the collector delivers it.

Change-Of-State Storage.   The third method of storing solar energy, change-of-state storage, uses materials or substances that change from solid to liquid when heated. This change permits the storage of more heat per pound than if the change did not occur. When a material such as Glauber's salts is cooled and transforms from a liquid to solid state, it gives off extra heat. This method of storage is still in the experimental stage.

The great advantage of change-of-state storage is that it can contain a great amount of heat in a limited space and at a limited weight. Let's say that the objective of a storage system is to hold 200,000 Btu's at 100° to 160°F. Water systems would need 53 cu ft at 3300 lb, rocks would need 175 cu ft at 17,500 lb, while Glauber's salts would require only 19 cu ft at 1740 lb. In addition, such salts are obtainable at relatively reasonable prices.

General Considerations.   In order to provide heat when the sun doesn't shine, conventional heat sources are used to augment the solar system. A conventional furnace (oil, gas, electricity) serves as a back-up energy source for most space heating and cooling, hot water, and some swimming pool solar systems. For reasons of cost, most solar systems are not designed to handle all of the anticipated demand. For example, if a solar system were designed to handle 100% of the heating load during the middle of winter, the collector and storage system would have to be large enough to provide energy for extended periods when the sun didn't shine. As

TABLE 8. Mean Daily Solar Radiation (Langleys) and Years of Record Used

STATES AND STATIONS	JAN	YRS	FEB	YRS	MAR	YRS	APR	YRS	MAY	YRS	JUNE	YRS	JULY	YRS	AUG	YRS	SEPT	YRS	OCT	YRS	NOV	YRS	DEC	YRS	ANNUAL
ALASKA, Annette	63	6	115	6	236	7	364	7	437	6	438	6	438	6	341	6	258	7	122	7	59	7	41	7	243
Barrow	#	8	38	8	180	8	380	8	513	8	528	8	429	9	255	8	115	10	41	10	#		#		206
Bethel	38	9	108	9	282	10	444	9	457	10	454	10	376	10	252	10	202	10	115	10	44	9	22	9	233
Fairbanks	16	25	71	27	213	25	376	28	461	28	504	29	434	28	317	29	180	29	82	30	26	26	6	26	224
Matanuska	32	6	92	6	242	6	356	7	436	7	462	6	409	6	317	6	198	6	100	6	38	6	15	7	224
ARIZ., Page	300	3	382	3	526	3	618	3	695	2	707	2	680	2	596	3	516	3	402	3	310	3	243	3	498
Phoenix	301	11	409	11	526	11	638	11	724	5	739	11	658	11	613	11	566	11	449	11	344	11	281	11	520
Tucson	315	5	391	5	540	4	655	5	729	5	699	5	626	6	588	6	570	6	442	6	356	6	305	6	518
ARK., Little Rock	188	9	257	9	353	10	446	10	523	9	559	9	536	8	518	8	439	9	343	8	244	10	187	10	385
CALIFORNIA, Davis	174	18	257	17	390	18	528	18	625	18	694	18	682	18	612	18	493	18	347	18	222	19	148	19	431
Fresno	184	31	289	31	427	31	552	31	647	31	702	32	682	32	621	31	510	31	376	32	250	31	161	32	450
Inyokern (China Lake)	306	11	412	11	562	11	657	11	772	11	819	11	772	11	729	11	635	8	467	8	363	11	300	12	568
LaJolla	244	19	302	18	397	19	457	20	506	19	487	21	497	22	464	22	389	22	320	21	277	20	221	20	380
Los Angeles WBAS	248	10	331	10	401	10	460	9	460	4	525	5	641	5	581	5	503	5	373	10	289	10	234	10	463
Los Angeles WBO	243	9	327	9	436	9	483	9	555	9	584	9	651	9	581	9	500	9	362	10	281	10	212	10	436
Riverside ‡	275	8	367	8	478	8	541	9	623	9	680	9	673	9	618	9	535	11	419	11	313	11	270	9	483
Santa Maria	263	11	346	11	482	11	552	10	635	11	694	11	680	11	613	11	521	11	419	11	313	11	252	11	481
Soda Springs	223	4	316	4	374	4	551	4	615	4	691	4	760	4	681	4	510	3	357	3	248	3	182	3	459
COLO., Boulder	201	5	268	4	374	4	460	4	460	4	525	5	532	5	439	5	412	5	310	10	222	10	182	4	367
Grand Junction	227	9	324	9	434	9	546	8	615	8	708	8	676	8	595	8	514	9	373	9	260	9	212	10	456
Grand Lake (Granby)	212	9	313	9	423	8	423	2	552	8	632	8	600	8	505	8	476	6	361	6	254	6	184	7	417
D. C., Washington (C.O.)	174	11	266	11	344	2	411	16	551	11	577	11	536	2	416	2	375	3	299	6	211	6	166	3	356
American University	158	39	231	39	322	39	398	39	467	39	510	39	496	39	440	38	364	38	278	38	192	39	141	39	333
Silver Hill	177	7	247	16	342	16	438	16	513	9	555	7	511	7	457	7	391	9	293	10	202	7	156	6	357
FLA., Apalachicola	298	10	367	10	441	10	535	10	603	10	578	9	529	9	511	9	456	9	413	10	332	10	262	10	444
Belle Isle	297	10	330	10	412	10	463	10	483	10	464	10	488	11	461	10	400	10	366	10	313	10	291	10	397
Gainesville	267	11	343	11	427	12	517	12	579	12	521	12	488	11	483	11	418	10	347	10	300	10	233	10	397
Miami Airport	274	2	415	19	489	19	499	21	553	6	532	10	532	10	505	10	440	10	384	10	353	10	316	10	451
Tallahassee	327	2	311	2	423	2	499	10	547	10	521	3	508	3	512	3	398	5	400		292		230	2	451
Tampa	327	8	391	8	474	8	539	8	596	8	574	8	534	8	494	8	452	9	*	9	356	9	300	9	453
GA., Atlanta	218	11	290	11	380	11	488	11	533	11	562	11	556	11	508	11	416	11	344	11	268	11	211	11	396
Griffin	234	17	295	9	385	10	522	11	570	11	577	11	556	11	522	11	435	11	368	11	283	11	201	11	413
HAWAII, Honolulu	363	4	422	4	516	4	559	5	617	5	615	5	615	5	612	5	573	5	507	5	426	5	371	5	516
Mauna Loa Obs.	522	2	576	2	680	2	689	3	727	3	*		703	3	642	3	602	3	560	3	504	4	481	4	---
Pearl Harbor	359	5	400	4	487	4	529	5	573	5	566	6	598	5	567	5	539	5	466	5	386	5	343	5	484
IDAHO, Boise	138	10	236	9	342	9	412	16	485	14	443	9	670	10	576	10	460	10	301	14	182	11	124	11	395
Twin Falls	163	20	240	20	355	20	462	21	552	21	557	20	602	20	540	19	432	19	286	20	176	20	131	19	378
ILL., Chicago	96	19	147	19	227	19	331	14	424	10	470	10	473	10	403	16	313	15	207	15	120	15	76	20	273
Lemont	170	6	242	7	340	7	400	10	476	10	539	9	540	10	498	10	398	10	275	5	165	8	138	5	352
IND., Indianapolis	144	10	213	10	316	10	396	10	488	9	543	11	541	11	490	11	405	11	293	10	177	11	132	11	345
IOWA, Ames	174	10	253	7	326	5	403	5	480	5	541	5	436	4	460	6	367	6	274	7	187	6	143	7	345
KANS., Dodge City	255	5	316	7	418	7	528	7	568	7	650	9	642	8	592	8	493	9	380	9	285	9	234	10	447
Manhattan	192	3	264	3	345	3	433	3	527	4	551	4	531	4	526	4	410	4	292	4	156	4	156	4	371
KY., Lexington	172	9	263	9	357	10	480	11	581	11	508	14	617	10	563	10	494	13	357	13	245	14	174	11	411
LA., Lake Charles	245	11	306	11	397	11	481	11	535	11	591	11	526	11	511	11	449	11	402	11	300	10	230	10	418
New Orleans	214	14	259	14	335	15	412	16	483	14	443	13	417	15	416	15	383	15	357	13	278	14	198	14	347
Shreveport	232	10	292	10	384	10	446	10	523	10	557	10	573	11	528	11	414	11	354	10	254	11	205	11	400
MAINE, Caribou	133	8	231	9	364	8	400	10	476	10	535	9	537	9	486	8	366	10	212	9	111	9	107	8	316
Portland	152	7	235	7	352	7	400	11	514	9	539	9	540	10	522	10	453	10	237	5	157	8	137	9	350
MASS., Amherst	116	2	*		300	2	*		431	2	514	2	541	2	509	6	417	6	324	5	152	2	132	2	345
Blue Hill	153	27	228	27	319	26	389	26	469	27	510	27	502	27	449	27	354	28	266	28	162	28	135	28	328
Boston	129	16	194	17	290	17	528	17	445	16	483	16	411	16	460	16	334	17	235	16	136	16	115	15	301
Cambridge	153	4	235	3	323	3	400	3	420	3	476	3	482	4	464	4	367	4	253	4	164	4	124	4	322
East Wareham	140	12	218	13	305	12	385	14	452	14	508	14	495	14	436	13	365	13	258	14	163	14	140	13	322
Lynn	118	2	209	2	300	2	394	2	454	2	549	2	528	4	432	3	341	2	241	2	135	3	107	3	317
MICH., East Lansing	121	10	210	10	309	15	359	11	483	10	547	10	547	10	466	11	373	11	255	11	136	11	108	11	311
Sault Ste. Marie	130	10	225	9	356	9	416	10	523	10	557	10	573	11	472	10	322	10	216	9	105	9	96	9	333
MINN., St. Cloud	168	8	260	8	368	8	426	8	496	8	535	8	537	8	486	8	366	8	237	7	146	8	124	8	348
MO., Columbia (C. O.)	173	10	251	10	340	10	434	11	530	11	574	11	574	11	522	10	453	10	332	10	225	10	158	10	380
University of Missouri	166	5	248	6	322	6	429	6	501	6	560	6	583	6	509	6	417	6	324	5	177	5	146	5	365
MONT., Glasgow	154	8	258	8	385	7	434	8	568	8	605	8	643	8	531	8	410	8	267	8	154	8	116	7	388
Great Falls	140	8	232	8	366	9	434	9	528	8	583	8	639	9	532	9	411	9	264	9	154	10	112	10	366
Summit	122	3	162	2	268	3	414	3	462	3	493	3	560	3	354	2	407	2	264	2	102	2	102	2	312
NEBR., Lincoln	188	39	259	39	350	39	416	39	494	40	544	38	568	38	510	38	396	38	296	36	199	40	159	39	363
North Omaha	193	3	299	3	365	3	463	3	516	3	546	3	568	4	519	3	410	4	298	3	204	3	170	4	379

STATES AND STATIONS	JAN	YRS	FEB	YRS	MAR	YRS	APR	YRS	MAY	YRS	JUNE	YRS	JULY	YRS	AUG	YRS3	SEPT	YRS	OCT	YRS	NOV	YRS	DEC	YRS	ANNUAL
NEV.																									
Ely	236	7	339	9	468	9	563	9	625	10	712	10	647	11	618	11	518	11	394	10	289	10	218	10	469
Las Vegas	277	11	384	11	519	11	621	11	702	11	748	10	675	11	627	11	551	11	429	11	318	11	258	11	509
N.J.																									
Seabrook	157	8	227	8	338	8	403	8	482	9	527	8	509	8	455	8	385	9	278	7	192	8	140	8	339
N.H.																									
Mt.Washington	117	2	238	2	238	2	*		*		*		*		*		*		*		*		96	2	---
N.Mex.																									
Albuquerque	303	13	386	13	511	13	618	13	686	13	726	13	683	12	626	13	554	14	438	15	334	15	276	14	512
N.Y.																									
Ithaca	116	22	194	21	272	23	334	23	440	24	501	23	515	23	453	23	346	21	231	22	120	23	96	23	302
N.Y. Central Park	130	34	199	34	290	33	369	35	432	35	470	34	459	35	470	35	331	36	247	36	214	36	115	35	298
Sayville	160	11	249	11	335	10	415	10	494	10	565	10	543	10	462	10	385	10	289	10	186	10	142	11	352
Schenectady	130	8	200	9	273	9	413	9	413	9	448	9	448	8	397	8	299	8	218	8	128	8	104	8	282
Upton	155	8	232	8	339	8	428	8	502	8	573	8	543	8	475	7	391	7	293	6	182	7	146	7	355
N.C.																									
Greensboro	200	7	276	9	354	9	469	9	531	10	564	10	544	10	485	10	406	10	322	10	243	10	197	8	383
Hatteras	238	10	317	9	426	9	569	9	635	10	652	10	625	10	562	11	471	10	358	11	282	11	214	11	443
Raleigh	235	3	302	3	*	8	466	3	494	2	564	2	535	3	476	3	379	3	307	3	235	3	199	3	---
N.D.																									
Bisarck	157	7	250	8	356	6	447	6	550	9	590	9	617	10	516	10	390	11	272	10	161	10	124	10	369
OHIO.																									
Cleveland	125	6	183	8	303	7	286	8	502	8	562	8	562	8	494	8	278	8	289	9	141	9	115	7	335
Columbus	128	7	200	9	297	10	391	11	471	6	562	4	542	5	477	4	422	4	286	4	176	4	129	5	340
Put-in-Bay	126	10	204	9	302	10	386	11	468	11	544	11	561	10	487	10	382	11	275	11	144	11	109	11	332
OKLA.																									
Oklahoma City	251	10	319	10	409	9	494	10	536	10	615	10	610	8	593	8	487	9	377	9	291	9	240	9	436
Stillwater	205	8	289	8	390	9	454	9	504	9	600	9	596	10	545	10	455	11	354	10	269	8	209	8	405
OREG.																									
Astoria	90	7	162	8	270	8	375	8	492	8	469	8	539	8	461	7	354	7	209	8	111	8	79	8	301
Corvallis	89	2	*		287	3	406	3	517	3	570	3	676	3	558	4	397	4	235	4	144	4	80	4	---
Medford	116	6	215	11	336	11	482	11	592	11	652	11	698	11	605	11	447	11	279	11	149	11	93	11	389
PA.																									
Pittsburgh	94	6	169	5	216	6	317	6	429	6	491	6	497	7	409	6	339	6	207	5	118	6	77	5	280
State College	133	20	201	19	295	20	380	20	456	20	518	20	511	20	444	20	358	20	256	20	149	20	118	20	318
R.I.																									
Newport	155	23	232	22	334	23	405	23	477	23	527	24	513	24	455	24	377	24	271	24	176	24	139	24	338
S.C.																									
Charleston	252	11	314	11	388	11	512	11	551	11	564	11	520	11	501	11	404	11	338	11	286	11	225	11	404
S.D.																									
Rapid City	183	11	277	11	400	11	482	11	532	11	585	11	590	11	541	11	435	11	315	10	204	11	158	10	392
TENN.																									
Nashville	149	18	228	19	322	19	432	19	503	18	551	18	530	17	473	17	403	17	308	19	208	18	150	19	355
Oak Ridge	161	11	239	11	331	11	450	11	518	11	551	11	526	11	478	11	416	11	318	11	213	11	163	11	364
TEXAS																									
Brownsville	297	10	341	10	402	10	456	11	564	11	610	9	627	8	568	8	475	11	411	11	296	11	263	10	442
El Paso	333	11	430	11	547	11	654	11	714	11	729	11	666	11	603	11	576	11	460	11	372	11	313	11	536
Ft.Worth	250	11	320	11	427	11	488	11	562	11	651	11	613	11	593	11	503	11	403	11	306	11	245	9	445
Midland	283	9	358	8	476	9	550	8	611	8	617	8	608	7	574	7	522	9	396	9	325	8	275	8	466
San Antonio	279	9	347	9	417	9	445	9	541	9	612	9	639	9	585	9	493	10	398	10	295	10	256	8	442
UTAH																									
Flaming Gorge	238	2	298	2	443	2	522	2	565	2	650	2	599	3	538	3	425	3	352	3	262	3	215	3	426
Salt Lake City	163	8	256	8	354	8	479	8	570	8	621	7	620	6	551	6	446	8	316	8	204	8	146	8	394
VA.																									
Mt.Weather	172	2	274	2	338	2	414	2	508	3	525	3	510	3	430	3	375	3	281	2	202	2	168	3	350
WASH.																									
North Head	*		167	2	257	2	432	2	509	3	487	3	516	3	436	3	321	3	205	2	122	3	77	3	---
Friday Harbor	87	8	157	7	274	8	418	8	504	8	578	10	586	10	507	10	351	8	194	8	102	8	75	8	320
Prosser	117	3	222	4	351	4	521	5	616	4	680	4	707	4	604	4	458	4	274	4	136	4	100	4	399
Pullman	121	4	205	2	304	2	462	2	558	5	653	5	699	5	562	5	410	4	245	5	146	5	96	5	372
University of Washington	67	9	126	9	245	10	364	9	445	10	461	10	496	10	435	10	299	9	170	9	93	9	59	9	272
Seattle-Tacoma	75	9	139	9	265	9	403	9	503	9	511	9	566	9	452	10	324	10	188	9	104	9	64	10	300
Spokane	119	8	204	8	321	8	474	8	563	9	596	9	665	9	556	9	404	10	225	9	131	9	75	7	361
WIS.																									
Madison +	148	46	220	46	313	45	394	47	466	47	514	47	531	47	452	47	348	47	241	47	145	44	115	46	324
WYO.																									
Lander	226	3	324	3	452	3	548	3	587	3	678	3	651	3	586	3	472	3	354	3	239	3	196	3	443
Laramie	216	3	295	3	424	3	521	3	554	3	643	3	606	3	536	3	458	3	274	3	229	3	186	3	408
ISLAND STATIONS																									
Canton Island	588	9	626	7	634	7	604	7	561	9	549	8	550	9	597	9	640	9	651	9	600	8	572	8	597
San Juan, P.R.	404	4	481	4	580	4	622	4	519	9	536	6	639	5	549	6	531	6	460	6	411	6	411	6	512
Swan Island	442	6	496	6	615	6	646	6	625	6	544	6	588	8	591	7	535	8	457	7	394	8	382	8	526
Wake Island	438	7	518	7	577	7	627	7	642	8	656	6	629	7	623	7	587	6	525	7	482	7	421	7	560

NOTES:

*   Denotes only one year of data for the month -- no means computed.
--- No data for the month (or incomplete data for the year).
Barrow is in darkness during the winter months.
+   Madison data after 1957 not used due to exposure influences.
‡   Riverside data prior to March 1952 not used-instrumental discrepancies.

Langley is the unit used to denote one gram calorie per square centimeter.

these expanded units would only be used during limited periods, the extra cost is not justified by the fuel savings.

In general, the total cost of a solar heating system depends upon the quality and size of the system, the warranty offered, installation costs, and other variables. Because of fuel savings, the system can return the original costs within a reasonable period of time. The payback period depends upon the cost of conventional heating, the amount of water used by your household, the isolation rate, and the efficiency of the system. It is possible for these systems to save 50 to 80% of water heating costs depending upon the system used and its location. Table 6 gives a comparison of installed cost of a solar system in various parts of the United States and the length of payback period as a function of fuel costs.

A summary of the pros and cons of a solar system using air or water as media for heat transfer are:

Air Advantages

- Moderate cost
- No freezing problems
- Minor leaks of little consequence
- As air is used directly to heat the house, no temperature losses occur in heat exchangers
- No boiling or pressure problems

Air Disadvantages

- Can only be used to heat homes, not presently economical for cooling
- Large air ducts needed
- Larger storage space needed for rocks
- Heat exchangers needed if system is to used to heat water

Water or Liquid Advantages

- Holds and transfers heat well
- Water can be used as heat storage

- Can be used both to heat and cool homes
- Compact storage and small conduits

Water or Liquid Disadvantages

- Leaking, freezing, and corrosion can be problems
- Corrosion inhibitors needed with water when using steel or aluminum. There are liquids which are noncorrosive. However, they are toxic and some flammable
- A separate collector loop using a nonfreezing fluid and a heat exchanger or, alternatively, draining water or a treated water system are required to prevent freezing. In warm regions where freezing is infrequent, electric warmers can be used.

Solar collectors must be exposed to the sun in the first place for the solar energy system to be evaluated. The amount to be invested in such a system is a function of the maximum amount of solar energy that can impinge upon a surface offset by the thermal losses of the system. Only then can the net heating (or cooling) be evaluated against the cost of conventional energy systems and fuel costs.

Solar energy system designers use solar radiation charts to determine how much radiation falls upon different areas each month or annually. Units of radiation are called langleys, where one langley equals one gram calorie per square centimeter. Table 7 are isothermal maps of the United States by month, and annually giving usable solar radiation, direct and diffuse in langleys that fall upon a horizontal surface. Table 8 gives the mean daily solar radiation by month on various United State's cities and the number of years that data have been taken. All information has been gathered by the U.S. Department of Commerce.

SECTION II

HVAC Components

1

AIR CONDITIONING APPARATUS

Air conditioning is a term covering a wide range of apparatus which, when assembled perform such functions as heating, cooling, dehumidifying, filtering, and ventilating to suit a particular application. Cooling comes about by use of an integral refrigerating unit comprised of a condenser, evaporator, fan, and compressor. Heating is done by use of hot water or steam coils or electric resistance coils. Ventilation comes about by blowers which circulate the air through filters that remove particulates from the airstream.

Air conditioning units come in many forms and combinations. It is up to the engineer and designer to select the proper package that is most economical and functional for the project involved. This Chapter will discuss the features of central station air conditioners, unitary systems, room air conditioners, and air conditioners for computer and data processing applications. These descriptions are followed by estimated labor costs to install various systems and combinations.

1.1. CENTRAL STATION

A central station air conditioning or air handling unit is essentially an all-air system which originated with the forced warm air heating system. The forced warm air system had centrally located heating equipment that warmed the air and distributed it through ducts. Adding cooling and dehumidification components to this system was found to improve the control of temperature and humidity of the space to be conditioned. Today this system is used in a multitude of applications in low-, medium-, and high-pressure distribution systems.

The central station is normally located outside the air conditioned area in a mechanical room usually in the core of a building, or on a roof or in a basement. Depending on the application, the central system includes some, or all, of the following items:

- water and condensate piping
- cooling tower
- humidifiers
- control valves and controls
- water heaters
- pumps
- ventilation air heaters
- chillers
- auxiliary heat exchangers

Figure 1 shows outline drawings for three types of central station air conditioning units: a vertical type, single zone, where floor space is limited; a horizontal single-zone type, where headroom is limited; and a third type, called a multizone unit. Figures 2 and 3 show perspective views of commercially available central station units.

1.2. UNITARY SYSTEMS

Unitary systems are complete HVAC, factory assembled units. They are generally classified as a split system or as a self-contained system. A

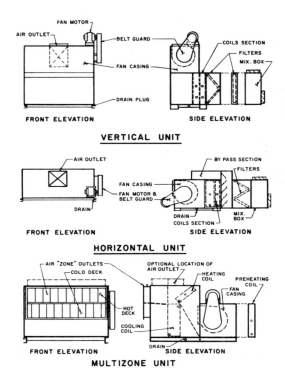

FIGURE 1. Central station air conditioning units.

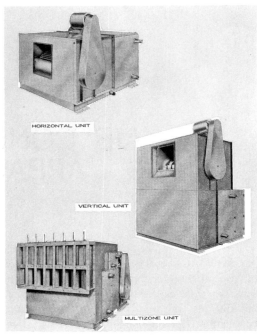

FIGURE 2. Single-zone and multizone central station air conditioning units. (Manufactured by York, Division of Borg-Warner Corp.)

split system is further classified according to its assembly into the following:

- a fan unit with a remote water chiller
- a fan unit with a remote condenser
- a fan unit with a remote condensing unit

The advantages and the flexibility of unitary systems make them competitive with central systems. Some of their unique advantages are:

- The installation and removal of unitary systems is simple.

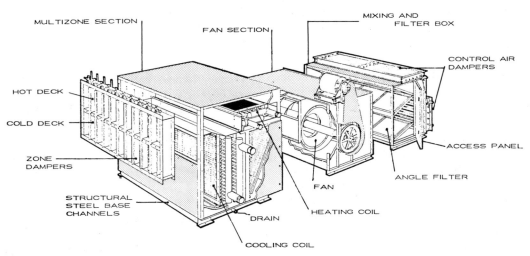

FIGURE 3. Multizone central station air conditioning unit. (Manufactured by York, Division of Borg-Warner Corp.)

- The cost of installation is low.
- They can be used with or without duct-work.
- Splitting the system allows the equipment to be installed outside the air conditioned

space with a suitable duct connecting the fan discharge outlet to the space.

Figure 4 is a view of a split system air conditioner with a remote condensing unit. The over-

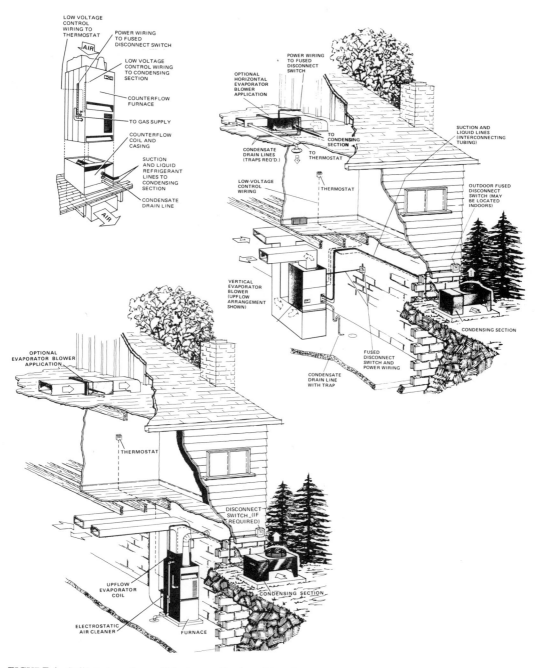

FIGURE 4. Split system air conditioning application. (Manufactured by York, Division of Borg-Warner Corp.)

all picture shows the unit installed in the basement of the building, where air enters the back of the unit near its bottom and leaves at the top. The enlarged inset view shows a first floor location. For this, air enters at the top and leaves at the bottom. Figure 5 shows gas heating unitary units, one for installation on a rooftop, the other at ground level. Figures 6 and 7 are artist's sketches of rooftop installations of self-contained unitary air conditioning units. Both types are similar except for their size and capacity. The latter unit is designed for large scale commercial or industrial application.

1.3. ROOM AIR CONDITIONERS

Room air conditioners are small, completely self-contained units designed primarily to cool a single room or space. They vary in capacity from $\frac{1}{2}$ to $3\frac{1}{2}$ tons. Most are air cooled. They are designed to be installed on a window sill or to protrude through a hole cut in a wall. Floor-mounted models are called consoles.

The basic function of a room air conditioner is to provide comfort by cooling, dehumidifying, filtering, and circulating the room air. It may also provide ventilation by introducing

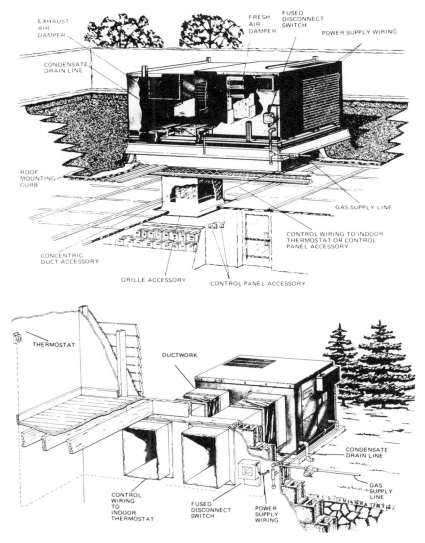

FIGURE 5. Unitary system gas heating applications. (Manufactured by York, Division of Borg-Warner Corp.)

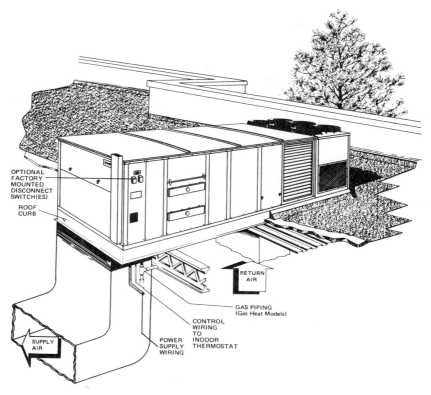

FIGURE 6. Self-contained unitary air conditioner for rooftop installation. (Manufactured by York, Division of Borg-Warner Corp.)

outdoor air into the room, and/or by exhausting room air to the outside. The conditioner can also be designed to provide heating, by reverse cycle (heat pump) operation, by steam or hot water coils, or by electric resistance elements.

Figure 8 is a cross-sectional view of a room air conditioner showing locations of components within the conditioner and airflow. Commercially available window units are represented in Figure 9. The selection for these installations is based upon such factors as:

- lower operating costs (highest Btu/watt)
- low-sound level
- small physical size
- lowest initial cost (no ductwork)
- electric service limitations

Room air conditioners are usually installed in:

- schools
- hospitals
- nursing homes
- high-rise motels and hotels
- office buildings
- residential homes

1.4. COMPUTER AND DATA PROCESSING AREAS

Computers and electronic data processing (EDP) systems must be operated in a closely controlled environment. Three closely related conditions which must be maintained: temperature, relative humidity, and filtration. Operating limits for these conditions vary with each computer application, and are usually furnished by the computer manufacturer.

The importance of a well-designed HVAC application for a computer room cannot be overstressed. The investment for electronic equipment is far greater, usually 100 times over, then the HVAC apparatus required for the

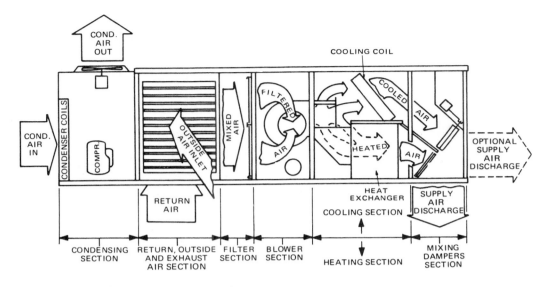

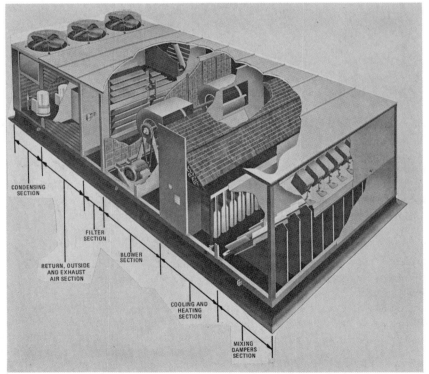

FIGURE 7. Self-contained unitary air conditioner for rooftop installation, high capacity. (Manufactured by York Division of Borg & Wagner Corp.)

environment. A bad design can create a malfunction, with continuous breakdowns that no client will excuse. It is the designer who has the responsibility to create the right environment for optimum equipment performance.

Computer rooms are usually fitted with special raised floors in order to give a space or plenum between the permenant floor and the platform upon which the equipment rests. This plenum serves two purposes: it is first a conduit

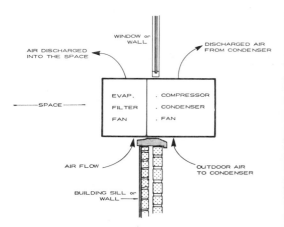

FIGURE 8. Detail of window air conditioning unit.

for cables and wiring; secondly, it is a channel for cooling air. The flooring is designed so that the air can enter the electronic equipment directly without need of any additional ducting.

Figure 10 shows a commercial process cooling unit that is placed directly on the raised

FIGURE 10. Electronic data process cooling. (Manufactured by AC Manufacturing Co.)

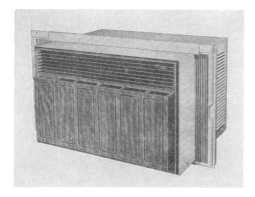

FIGURE 9. Window air conditioning units. (Manufactured by York, Division of Borg-Warner Corp.)

floor, sometimes referred to as the computer floor. Cool air is discharged into the plenum and can escape only through openings under the equipment to be cooled. Figure 11 shows section drawings for two EDP installations. The first is a unit that uses chilled water as the refrigerant, the second a split system with an air-cooled condenser located on the building roof. In both cases, cooled air is discharged into the underfloor plenum, enters the bottom of the computer equipment where it becomes warm, and is discharged to the room near the ceiling. The warmed room air is then drawn into the top of the cooling unit. This airflow cycle is largely confined to spaces internal to equipment and the underfloor space, thus reducing air movement and turbulence in the room, where the occupants are virtually unaware of the air movement.

Some designers accept the fact that cool air for comfort air conditioning is best distributed from the ceiling where it mixes with room air and gradually drifts downward, since it is heavier than the warmer room air. This is not satisfactory in computer rooms. If cool air is

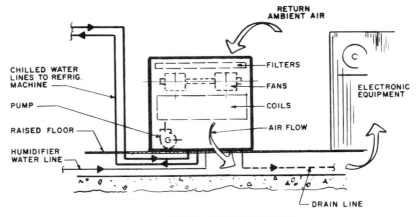

CHILLED WATER SYSTEM

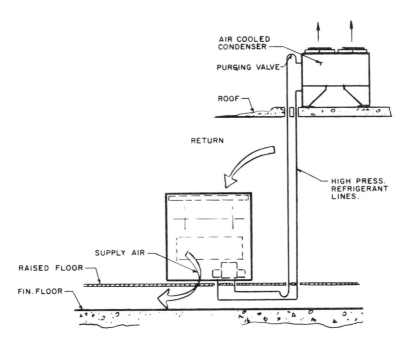

AIR-COOLED DIRECT EXPANSION
SYSTEM-W/REMOTE AIR COOLED CONDENSER

FIGURE 11. Computer and EDP cooling systems.

introduced at the ceiling it has to find its way to the floor past the upflow of hot air from the computers. The downflow of cool air mixing with the upflow of hot air causes a turbulence which creates uncomfortable working conditions. Also, this mixed air enters the computer at a higher temperature than the cooled air from the cooling system. This results in a higher computer operating temperature and, therefore, increased failure rates can be expected.

Although cooling air introduced under the floor will tend to reduce room temperatures somewhat, this is more than offset by the radiated heat from the computers. One of the reasons computer rooms must be maintained at a lower temperature for comfort (70° to 72°F) than normal office space (78° to 80°F) is because of this radiated heat. Experience indicates that floor-fed systems give superior performance both for human comfort and efficient equipment cooling and is therefore highly recommended.

1.5. ESTIMATED LABOR TO INSTALL AIR CONDITIONING APPARATUS

The design for building construction must include several cost estimating steps to assure the building owner and others that the project can be constructed within budget. The building must be divided into its major systems and components, and each of these divisions analyzed. The analysis, while not complicated, contains many variables which must be taken into account. The total cost of any installation is made up of: materials, labor, and equipment costs; the cost of doing business (overhead); and finally the contractor's profit for risking his time and money.

Material costs vary with time and location. Overhead and profit also vary according to market conditions. Labor costs are often a major component in the total cost. These come about by knowing how many man-hours are required to do a specific job and multiplying these by a rate of dollars per hour. The rate fluctuates widely according to location, whether union or nonunion labor is used, and the prospect of near future raises in pay. One of the most stable elements in the cost analysis is the number of man-hours required. This is a function of productivity. Assuming proper supervision and slow advances in technology, this factor tends to remain fairly constant.

Man-hour requirements come about after averaging the productivity of the workcrews over many installations. This section and others that follow will pass on this man-hour informa-tion, but will leave it to the reader to provide the materials, equipment, and labor costs to suit his location, time, and economic circumstances.

Description	cfm	gpm	Labor (man-hours)
Air washer type	5,500	45	48
(with flat filter,	14,000	110	64
without	24,100	190	96
circulating pump)	37,500	290	96
	48,500	370	110
	61,000	475	124
	72,800	550	124
	81,000	650	124
	90,000	700	148
	100,000	780	148

Description	cfm	Labor (man-hours) Floor Mounted	Labor (man-hours) Ceiling Mounted
Heating and	1200	24	32
ventilating type	2500	24	32
(with heating coil	3500	24	32
and flat filter)	4500	28	36
	5500	30	36
	7000	32	42
	8500	36	48
	10,000	48	54
	15,000	64	72
	20,000	72	96
	25,000	72	120
	34,700	96	136
	48,000	96	136

Description	cfm	Weight (tons)	Labor (man-hours)
Self-contained	1200	3	12
type	2000	5	12
(vertical, with	3000	7.5	12
water-cooled	4000	10	16
condenser)	6000	15	16
	8000	20	24
	10,000	25	24
	12,250	30	24
	16,000	40	24
	20,000	50	30
Self-contained	2000	5	16
type (horizontal,	3000	7.5	16
rooftop with heat-	4000	10	24
ing coil and air-	5000	12.5	24
cooled condenser)	6000	15	30

Description (continued)	Labor (man-hours)
Inclement self-contained air handling units with DX[a] coil	
9000 Btu cooling with up to 12,500 Btu hot water heating	18
11,700 Btu cooling with up to 13,300 Btu hot water heating	18
14,100 Btu cooling with up to 15,500 Btu hot water heating	24
9000 Btu cooling with up to 15,800 Btu steam heating	24
14,100 Btu cooling with up to 16,500 Btu steam heating	24
9000 Btu cooling with up to 11,300 Btu electric heating	12
11,700 Btu cooling with up to 11,400 Btu electric heating	12
11,700 Btu cooling with up to 15,370 Btu electric heating	14

[a]For a description of direct expansion coils (DX), see Chapter 8.

Description (cont)	cfm	Labor (man-hours)
Multizone air handling unit with DX coil, heating coil and roll filter	1750–2750	48
	1850–3750	48
	2250–6600	64
	2700–8300	64
	3350–8700	72
	4250–12,500	72
	5050–16,000	98
	6500–19,000	98
	7700–23,500	126
	10,200–27,000	126
	13,280–35,000	146
	18,600–47,000	160
	55,000	160

Description	tons	Labor (man-hours)
Split system air conditioners with outdoor condensing unit	$1\frac{1}{2}$	12
	2	12
	$2\frac{1}{2}$	18
	3	18
	$3\frac{1}{2}$	18
	4	24
	5	24
	$7\frac{1}{2}$	24
	10	32
	15	32
	20	32

2
AIR CURTAINS

The air curtain unit is literally a curtain of air which prevents outside air from entering open doors in buildings. The system uses air at ambient temperature and operates only when doors are open. It is controlled by the door-operating equipment. The air curtain is used for:

- warehouses
- coolers and freezers
- receiving areas
- loading docks
- meat packing plants
- food processing buildings
- bakeries
- hospitals
- laboratories
- lobbies

This system, by its isolation of incoming air, will reduce heat loss, and the effects of wind, dust, etc. There are a number of advantages such as reduced pollution and contamination for better manufacturing controls and employee comfort, and reduced refrigeration losses and door maintenance.

Manufacturers offer air curtains with and without a recirculating air barrier. The recirculating air curtain is a modified ventilating system which discharges high-volume, low-velocity air from the ceiling to a floor intake where it is drawn in by blowers, filtered, heated, and recirculated back to the discharge. This type of system is used for commercial installations where it is desired to have low velocities and a

minimum amount of air to enter or leave the building. This sytem is used especially in banks, supermarkets, and department stores.

A nonrecirculating air curtain is manufactured in various capacities. It is compact, has a high velocity airstream, and is economical. The major difference between the two systems, besides the velocity, is that there is no return grating in the floor and the curtain air is allowed to dissipate into the ambient air. These systems are used for industrial loading dock doors, where temperature separation of industrial areas is required, and for maintaining positive air pressure in certain building areas. There are numerous other applications.

When designing an air curtain we can follow a simple principle in physics. Air has weight. When air is moved, it has energy because of its weight and velocity (force × distance). Thus, the movement of every cubic foot of air results in a form of energy. The energy can be related to a specific pressure or force. A wind of a certain velocity results in such a force being applied to a building wall or roof.

An air curtain projects a spray of air downward and outward so that when a building door is opened, the air would move outside the building if there were no wind blowing toward the door. You can better visualize the manner in which the system functions by noting that in Figure 12:

1. The energy of the wind per square foot can be summarized as a single energy vector A.

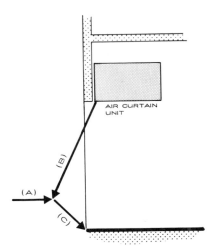

FIGURE 12. Schematic of the air curtain function.

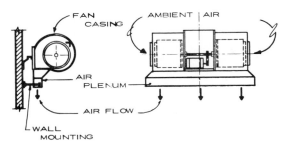

FIGURE 13. Wall-mounted air curtain unit.

2. The energy of the air curtain can be summarized in the same way as an outwardly inclined vector *B*.
3. When two vectors meet, their final path is called the resultant, *C*.
4. Therefore, the forces of the air curtain and the wind neutralize each other. Whatever air is left moving due to differences in energy could be projected back into the plant. Obviously, if the air curtain is stronger than the resultant, the air will flow towards the outside. When the wind is much stronger than the air curtain, the cold air blows in, bending the air curtain at its base.

The system is very simple in application because it maintains an invisible barrier that keeps the warm air in and the cold air out. It is inexpensive to run because the system uses the indoor ambient air.

The energy of the air curtain and the wind can be calculated from the formula

$$E = MV^2$$

where

E = energy (in ft lb/min.)
M = mass flow (in cfm)
V = velocity (in ft/min.)

It is assumed that there is no difference in density between inside and outside air.

Figure 13 shows side and end views of a wall-mounted air curtain unit. The air plenum is wide enough to at least cover the door opening it protects. Figure 14 illustrates how an air curtain could be installed in conjunction with an overhead door (top), and wall-mounted vertical installations for upflow and downflow applications. Actual installations of commercially available air curtain units are shown in Figure 15. These are wall-mounted vertical units (downflow) in industrial plants. Note that in the lower left hand view, the installation is over a doorway leading to a loading dock and not over the outside door.

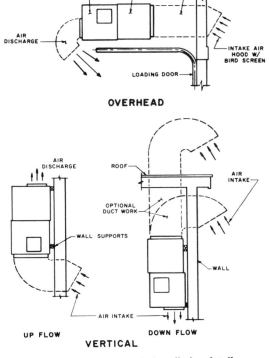

OVERHEAD

VERTICAL

FIGURE 14. Air curtain installation details.

FIGURE 15. Air curtains for industry: (Manufactured by Dynaforce Corporation.)

3

MAKE-UP AIR UNITS

Most manufacturing facilities are equipped with some type of exhaust system to remove fumes and contaminants from the working area. While in operation, an exhaust system creates a negative pressure within the building. If the means of maintaining equilibrium within the building is inadequate, infiltration of cold, unfiltered air through cracks and crevices will occur. Drafts will develop near the points of infiltration. Fumes and dirt will be dispersed through the plant. Hot and cold spots are likely to develop since the airflow is not uniform throughout the building. This infiltration of cold air also reduces the efficiency of the primary heating plant.

Make-up air systems for environmental comfort is essential. The exhaust systems required in most facilities cannot work efficiently without a full supply of clean, fresh air. When these systems do not operate efficiently, drafts, fumes, and odors take their toll on the workers' health and efficiency. Even with a sufficient supply of make-up air for exhaust requirements, if the type of unit and the introduction of air to the work areas is not properly designed, drafts and cold spots will continue to be a problem.

Not all make-up air problems are alike; each calls for a specific solution. For this reason, the following considerations should be given to the design:

1. The volume of make-up air should be 5 to 10% greater than the total volume exhausted to maintain the space under a slight positive pressure.

2. Normally, make-up air is conditioned to approximately the same temperature as that desired in the room.

3. Locate the make-up air intake at a remote point from the exhaust or areas of contamination.

4. Make-up air should be introduced into the working zone of the building at the 8 to 10 ft level. Air discharged in this manner provides both effective general ventilation for the people, and air for the exhaust systems.

5. Adjustable grilles should be used to provide control of the outlet air pattern for maximum comfort in summer and winter without changing the volume of air being supplied.

6. Fresh air cooling or spot cooling can be accomplished by increasing the air motion at the work station. Acceptable velocities for continuous exposure range from 75 to 125 fpm for people seated to 100 to 200 fpm for people standing. As the air motion and velocity increases, the effective air temperature decreases.

7. Make-up air can be used to provide supplemental space heating. Subtract the total capacity of space heating equipment from the building heat loss and add the difference in Btu/hr to the capacity of make-up air.

FIGURE 16. Direct gas-fired make-up air unit. (Manufactured by American Air Filter Company.)

A gas-fired unit is shown in Figure 16.

3.1 ESTIMATED LABOR TO INSTALL MAKE-UP AIR UNITS

Description	cfm (nominal)	Weight (lb)	Labor (man-hours)
Direct gas-	5000	1500	48
fired type	7500	1575	48
	10,000	1900	48
	15,000	2250	48
	20,000	2375	48
	25,000	2750	72
	35,000	3000	72
	50,000	4250	96
	60,000	5550	96
	75,000	7000	96

4

BOILERS

A *boiler* is defined as a closed vessel containing water which by thermodynamic process absorbs the heat generated from the combustion of such fuels as oil, gas, charcoal, and electricity and transfers it to the water. The hot water or steam produced is used in heating systems. The expanding industrial, institutional, and commercial need for heat has provided a large demand for low- and medium-pressure process and space heating. The result is that a wide selection of boilers is available to meet differing requirements.

Boilers may be classified in two general groups: *steel firebox* and *cast iron*. Cast iron boilers are usually shipped in sections and assembled at the job site. However, smaller cast iron boilers are factory assembled. Steel firebox boilers are divided into three general types: *water tube*, *fire tube*, and *tubeless*. The construction of boilers is covered by the ASME code for low-pressure heating boilers. Boilers are rated on the basis of square feet of steam radiation (240 Btu/sq ft/hr) or square feet of hot water radiation (150 Btu/sq ft/hr).

The choice of steel vs. cast iron firebox boilers depends upon such factors as pressure fuel, cost of installation, and others. Since the capacities of cast iron and steel boilers overlap, the selection of a boiler might be based on the following considerations:

1. In the larger sizes, steel boilers are generally more efficient than cast iron.
2. The cost of steel boilers in the smaller

sizes is greater than that of cast iron boilers of the same capacity.

3. When future boiler capacity is expected to increase in the extension of a system, additional sections may be added to a cast iron boiler, whereas an additional or re-

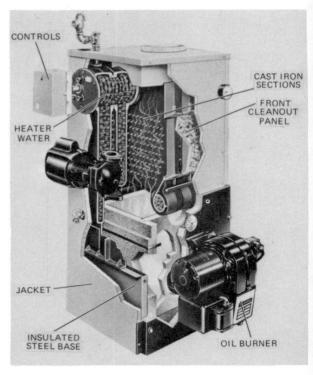

FIGURE 17. Cast iron packaged type oil-fired boiler, hot water.

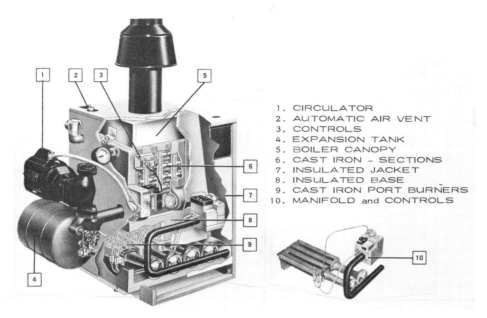

1. CIRCULATOR
2. AUTOMATIC AIR VENT
3. CONTROLS
4. EXPANSION TANK
5. BOILER CANOPY
6. CAST IRON – SECTIONS
7. INSULATED JACKET
8. INSULATED BASE
9. CAST IRON PORT BURNERS
10. MANIFOLD and CONTROLS

FIGURE 18. Packaged gas-fired boiler, hot water.

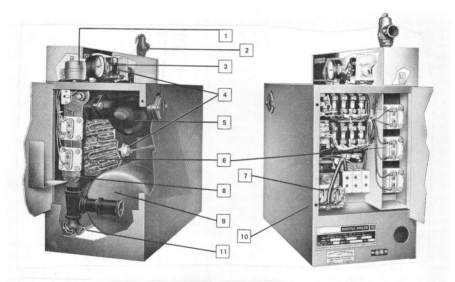

1. AUTOMATIC AIR VENT
2. WATER PRESSURE RELIEF VALVE
3. LIMIT CONTROL
4. PRESSURE CONTROL
5. CIRCULATOR
6. ELECTRIC HEATING ELEMENT
7. SEQUENCING RELAY SWITCH
8. CAST IRON BOILER SHELL
9. EXPANSION TANK
10. BOILER JACKET
11. DRAIN VALVE

FIGURE 19. Packaged electric boiler or hot water.

placement steel boiler of a larger size would not be practical or feasible.

4. Steam fitters are required to assemble the heating sections of a cast iron boiler. The steel boiler has only to be placed into position. However, space restrictions may favor that sections of a cast iron boiler be readily carried through doors or windows.

"Packaged boiler" is a term for a boiler with all components such as the burner, boiler, and controls assembled as a unit. Three such packaged boilers are shown in Figures 17, 18 and 19. All are cast iron and generate hot water, but differ in the fuel or energy used to heat the water. The most commonly used boilers are described below.

4.1. WATER TUBE BOILERS

In this type boiler, the water is inside the tubes, and the hot combustion gases pass around the outside of the tubes. There is less volume of water contained in water tube boilers than in fire tube boilers of the same rating. These boilers are fast to steam because of the large amount of heating surface. Water tube boilers

can differ in design; the tubes may be straight, bent, horizontal, or vertical.

4.2. FIRE TUBE BOILERS

In this type boiler, the hot gases of combustion pass through the vertical tubes. The tubes are partially submerged in the water, which is around the outside of the tubes. Heat from the fuel is transferred to the water through that portion of the tubes in contact with the water. Fire tubes boilers have straight tubes. Of the fire tube boilers, the three most popular types are: horizontal return tubular, firebox water leg fire tube, and scotch marine fire tube.

4.2.1. Horizontal Return Tubular Boilers

These boilers have a firebox or combustion chamber in which the fuel is burned. The pressure vessel is horizontal and is set above the firebox. The firebox is built separately from the pressure vessel and is either steel with a refractory lining or built of regular brick, refractory lined. The heat from the fuel contacts the lower outside shell (belly) of the boiler, then passes

FIGURE 20. Typical firetube boilers. Heavy oil or gas-fired boiler. Light gas or oil fired boiler. (Manufactured by Cleaver-Brooks, Division of Agua-Chem, Inc.)

FIGURE 20. (*Continued*)

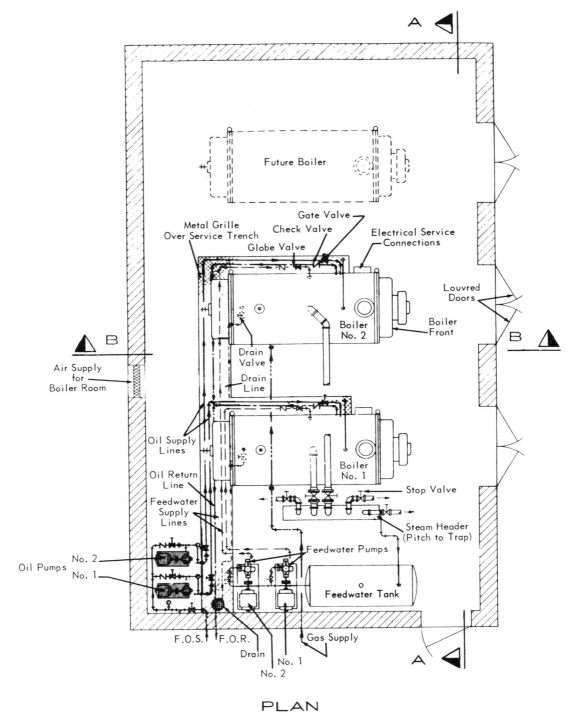

PLAN

FIGURE 21. Boiler room layout for two low-pressure boilers (gas or oil fired). (*Courtesy* of Cleaver-Brooks, Division of Agua-Chem, Inc.)

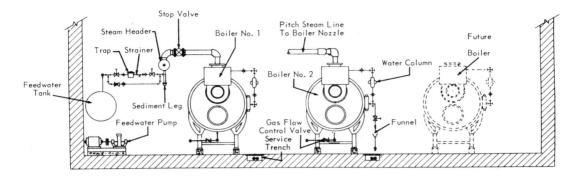

SECTION " A-A "

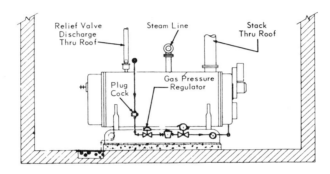

SECTION " B-B "
FIGURE 21. (*Continued*)

upward at the rear and returns through the fire tubes.

4.2.2. Firebox Water Leg Fire Tube Boilers

This type of boiler has water walls extending down the sides and front of the combustion chamber to make the combustion chamber an integral part of the horizontal pressure vessel. Water is contained in this pressure vessel as well as in the walls of the combustion chamber. The fire tubes run horizontally in the pressure drum. Thus, the heat from the burning fuel contacts the walls of the combustion chamber, then strikes the lower outside shell (belly) of the horizontal shell passing to the rear, then upward and returns through the fire tubes.

4.2.3. Scotch Marine Fire Tube Boilers

Scotch marine boilers are horizontal and have a circular furnace which runs through the pressure drum. Fire tubes are above or around the circular furnace. Combustion occurs at the front of the circular furnace and combustion gases move through the furnace to the rear and return through the fire tubes. There can be two or more passes of heat through the boiler. The furnace as well as the fire tubes are surrounded by the water.

Figure 20 shows photographs and a cutaway view of commercially available fire tube boilers that use gas or oil as a fuel. The cutaway view illustrates a scotch marine type boiler. Note the circular furnace. Arrows show how the combustion gases move through the tubes in two passes to heat the surrounding water.

4.3. TUBELESS BOILERS

The name tubeless comes from the fact that these boilers have neither water tubes nor fire tubes. They are designed usually as a round, vertical shell containing water to a certain level, and with the upper portion forming the steam chamber. The heat from combustion is directed under and around the outside shell. Refinements may include fins or other extended surfaces welded to the outside of the shell to help transfer the heat from the burning fuel to the water for steam generation.

4.4. BOILER INSTALLATION

A typical boiler room layout for two low-pressure steam boilers fired with oil or gas is shown in Figure 21. Note such features as:

1. Ample room provided around the components for maintenance.
2. Room for a future boiler.
3. Combustion air supply from windows and louvered doors.
4. Service trench with cover so that oil and feedwater piping will not interfere with the movement of service personnel.
5. Double door positioned in direct line of the boiler for space to pull out the boiler tubes.

The Hartford loop or return connection is used to connect the condensate return to the boiler in such a way that a water hammer will occur to give warning of low-water level in the system. A trap is formed by raising the condensate return line to a point only several inches below the normal water line in the boiler, then, dropping to the level of the boiler return line. This technique is illustrated in Figure 22.

4.5. ESTIMATED INSTALLATION COSTS (per boiler)

Description	Size (Bhp or MBH)	Labor (man-hours)	Average Material Cost ($)
Marine-type packaged boilers (with burners and basic controls)			
#2 Oil-fired	40 Bhp	80	4000
	60	80	
	80	80	
	100	80	
	125	85	10,000
	150	90	
	200	90	17,800
#2 Oil- or gas-fired	250 Bhp	110	19,500
	300	120	
	350	120	
	400	120	
	500	130	
	600	130	
	700	140	32,000

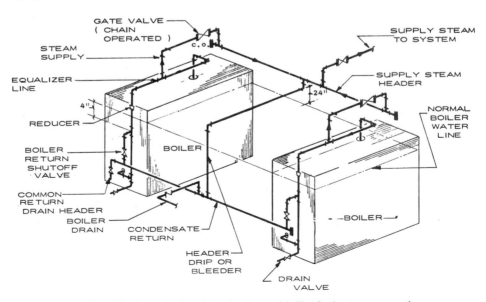

FIGURE 22. Steam boiler piping hookups with Hartford return connection.

Description	Size (Bhp or MBH)	Labor (man-hours)	Average Material Cost ($)
Heavy duty #6 oil-fired	80 Bhp	60	12,200
	100	60	
	125	60	
	150	84	
	200	84	20,000
	300	84	
	400	120	
	500	120	
	600	120	
	700 ↓	124	35,400 ↓
Steel fire box boilers low-water line, package (with burners and basic controls) #2 Gas- or oil-fired	50 Bhp	60	6500
	60	80	
	80	80	
	110	80	10,450
	170	80	
	200	120	22,100
	270	120	
	330	120	
	400	124	
	600 ↓	124	28,650
#4 Heavy duty oil	90 Bhp	80	12,400
	110	80	
	130	80	
	170	80	
	200	80	20,150
	270	120	
	330	120	
	400	120	
	460 ↓	120	29,300
Cast iron sectional boilers (with burners and basic controls) #2 Oil fired	40 Bhp	80	7450
	60	80	
	80 ↓	100	↓

Description	Size (Bhp or MBH)	Labor (man-hours)	Average Material Cost ($)
Cast iron sectional boilers (with burners and basic controls) #2 Oil fired	100	115	10,000
	120	120	13,000
	140	120	↓
	160 ↓	140	17,000
Cast iron sectional boilers (with burners and basic controls) Gas fired	40 Bhp	80	8650
	60	80	
	80	100	↓
	100	115	13,000
	120	120	15,700
	140	125	↓
	160 ↓	130	17,100
Cast iron residential hot boiler (with burner controls, circulator, and tankless heater type) Gas fired	73 MBH	24	700
	97	24	
	121	24	
	146	24	
	170	24	
	194	24	↓
	250	24	1000
	375	24	1500
	500	32	↓
	700 ↓	32	2575
#2 Oil fired boilers	90 MBH	24	700
	150	24	
	200	24	↓
	250	24	1300
	375	24	2000
	500	32	↓
	700 ↓	32	3000

5
BURNERS

Burners are an integral part of a heating plant. This is where the oil or gas fuel is mixed with air and burned to release the energy required for the heating system. The efficiency of the heating system must begin with an efficient burner that produces a clean flame with only a small excess of air. Burners can be classified in a number of ways: the type of fuel, domestic vs. commercial or industrial; the rate of burning fuel; the type of atomizer; etc. The American National Standards Institute has developed standards* for proper use of domestic and industrial burners and continuous research is being undertaken to develop newer and more efficient types of fuel burner equipment because of the need for energy conservation.

5.1. BURNER TYPES

Burners are classified by type: atmospheric and power.

5.1.1. Atmospheric Type

The atmospheric type burner is an assembly of multiple tubes or burners, each with venturi, gas orifice, and burner heads. These burners are of inshot and upshot design. Inshot burners are horizontal, making them particularly adapt-

---

*Z21.8-1971 Installation of Domestic Gas Conversion Burners
Z21.17-1974 Domestic Gas Conversion Burners
Z96.2-1974 Safety Standard for Oil Burners

able for firing scotch type boilers. Upshot burners are arranged vertically, making them more adaptable in firebox type boilers.

5.1.2. Power Type

Power type burners employ a fan to supply and control combustion air. These burners can be of either natural or forced draft design. In natural draft installations, a chimney is required to draw the products of combustion through the boiler or furnace. The burner fan supplies only enough power to move the air through the burner. Many natural draft power burners have a configuration similar to that of an inshot atmospheric burner to which a fan and windbox have been added. More complex gas–air mixing patterns are possible, and combustion capabilities are thereby improved.

The size and speed of power burner fans have been gradually increased and the combustion process modified so that the fan not only moves air through the burner, but also forces it through the boiler. Only a vent is required to release the flue gas products at a height so that the vent products will not be a nuisance. Combustion occurs under pressure. These burners are classified as forced draft burners.

In a forced draft power burner, gas input is usually controlled by throttling valves or other means external to the burner. The gas is introduced into a controlled airstream designed to produce thorough gas-air mixing but still capable of maintaining a stable flame front. In a

FIGURE 23. Forced draft pressure atomizing burner. (Manufactured by Dunham-Bush, Inc.)

ring burner, the gas is introduced into the combustion airstream through a gas-filled ring just ahead of the combustion zone. In a premix burner, gas and primary air are mixed together, and the mixture is then introduced into secondary air in the combustion zone.

The power burner affords superior control of combustion, particularly in restricted furnaces and under forced draft. In Figure 23, a forced draft, pressure atomizing burner is shown. Oil and air are precisely mixed. Fuel is burned in units ranging in capacities of from 7.5 to 30 gal./hr. This type of burner can be used in high- or low-pressure boilers, including the Scotch marine type.

5.2. ESTIMATED LABOR TO INSTALL BURNERS

Description	Size (MBH or gph)	Labor (man-hours)
Gas type	300 MBH	5
	500	9
	800	14
	1000	18
Oil type	4–6 gph	8
	7–12	14
	11–16	25
	10–20	35
	25–50	55[a]

[a]Rotary type only

6

CHILLERS

Chillers are refrigeration machines for cooling liquids, usually water. They are used in air conditioning applications and to cool process water in industrial plants. Many factors enter into their selection which include: whether they will be used for air conditioning or process water; inlet and outlet temperatures of the water; load variations; limits of space; and first costs and maintenance costs.

Refrigeration comes about as the result of a number of processes. The two of concern in this Chapter are 1) by absorption of heat by a gas that has been cooled by expansion, and 2) by evaporating liquids with low boiling points. The *centrifugal machine*, discussed below, uses a centrifugal compressor to compress the refrigerant gas that absorbed the heat of the water as it evaporated. The *reciprocating machine* uses the same thermodynamic process but employs a reciprocating instead of a centrifugal compressor. The *absorption machine* also uses evaporation as the cooling process. In this case, the evaporation is controlled by control of the vacuum pressure.

6.1. THE CENTRIFUGAL MACHINE

Centrifugal refrigeration equipment is built for dependable, heavy-duty, continuous operation for all types of commercial and industrial applications. The machine consists of a centrifugal compressor, a cooler, and a condenser. Water in the tubes of the cooler is chilled by transferring its heat to the cold liquid refrigerant causing the refrigerant to boil and vaporize. This refrigerant vapor is drawn into the suction side of the compressor where it is compressed then discharged into a condenser. Since the gas temperature is higher than the water in the condenser tubes, the vapor gives up its heat to the water, changes to a liquid, and drains into a float chamber. Here, a float valve automatically maintains a liquid seal to prevent gas from passing into the cooler. As the refrigerant level in the float chamber rises, the float valve opens and allows the liquid refrigerant to pass into a cooler where it is subjected to a lower pressure. Because of this lower pressure, a small amount of the refrigerant flashes off. This evaporation cools the remaining portion of the liquid to a temperature which corresponds to the pressure. The liquid refrigerant is distributed evenly along the entire length of the cooler. Figure 24 shows a centrifugal liquid chilling unit.

The prime mover or driver as it is called in Figure 24, may be a gas engine, steam or gas turbine or an electric motor. Nominally, it requires about 1 hp/ton of refrigeration which, for an electric motor drive is about 0.85 kW. If a gas engine is used, approximately 10 cu ft of gas is required. For a steam turbine, approximately 15 lb of steam is needed, depending on the steam pressure and the condensing pressure. The heat, extracted from the building, is rejected usually to a cooling tower. If cooling towers are used, the condenser water flow rate is about 3 gpm/ton.

A commercially available turbine–centrifugal

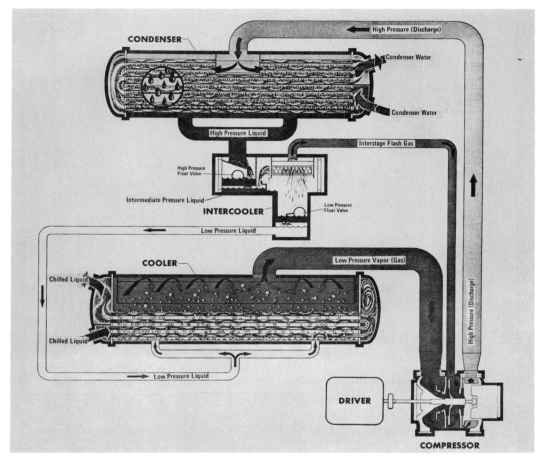

FIGURE 24. Schematic refrigerant flow diagram of a centrifugal liquid chilling unit. (*Courtesy* of York, Division of Borg-Warner Corp.)

EVAPORATOR   Refrigerant enters the top of the lower shell and is sprayed over the evaporator tube bundle. Heat from the liquid being chilled evaporates the refrigerant.

ABSORBER   The refrigerant vapor then migrates to the bottom half of the lower shell. Here the vapor is absorbed by a lithium bromide solution. Lithium bromide is, basically, nothing more than salt water. But lithium bromide is a salt with an especially strong attraction for water. With the lithium bromide spray, it is as if hundreds of little sponges are sucking up the refrigerant vapor. The mixture of lithium bromide and the refrigerant vapor—called the "dilute solution"—now collects in the bottom of the lower shell.

GENERATOR   The dilute solution is then pumped through the heat exchanger where it is preheated by hot concentrated solution from the generator. The heat exchanger improves the efficiency of the cycle by reducing the amount of steam or hot water required to heat the dilute solution in the generator.

   The dilute solution then continues to the upper shell containing the Generator and Condenser, where the pressure is approximately one-tenth that of the outside atmosphere, or seventy millimeters of mercury. The dilute solution flows over the generator tubes and is heated by steam or hot water. The amount of heat input from the steam or hot water is controlled by a valve and is in response to the required cooling load. The hot generator tubes boil the dilute solution releasing refrigerant vapor.

CONDENSER   The refrigerant vapor rises to the condenser and is condensed. The liquid refrigerant flows back to the lower shell, and is once again sprayed over the evaporator. The refrigerant cycle has been completed. Now the concentrated lithium bromide solution flows from the generator back to the absorber in the lower shell ready to absorb more refrigerant. Its cycle has also been completed.

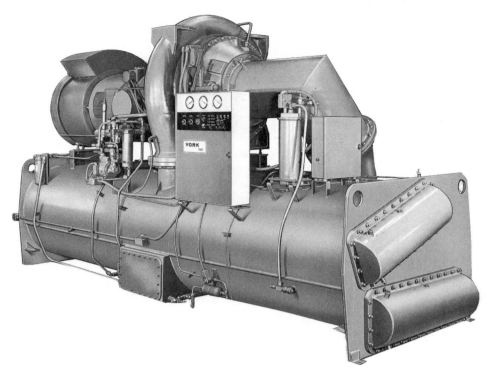

FIGURE 25. Hermatic turbine-centrifugal liquid chiller. (Manufactured by York, Division of Borg-Warner Corp.)

LEGEND

1   Chiller
2   Air Unit
3   Condenser Water Pump
4   Chilled Liquid Pump
5   Fused Disconnect Switch
6   Starter

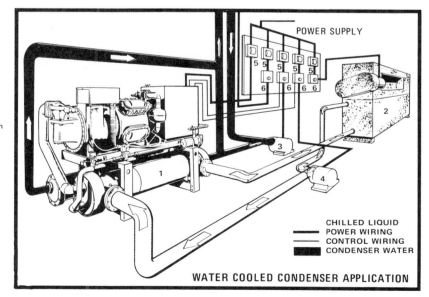

FIGURE 26. Piping and wiring arrangement. (*Courtesy* of York, Division of Borg-Warner Corp.)

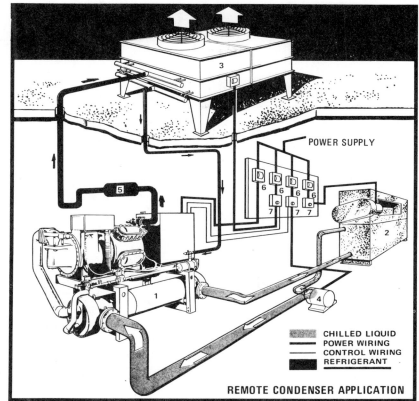

LEGEND

1   Chiller

2   Air Unit

3   Air
Cooled Condenser

4   Chilled Liquid Pump

5   Discharge Line Muffler

6   Fused Disconnect Switch

7   Starter

POWER SUPPLY

CHILLED LIQUID
POWER WIRING
CONTROL WIRING
REFRIGERANT

REMOTE CONDENSER APPLICATION

FIGURE 26. (*Continued*)

liquid chiller is shown in Figure 25. It is called a hermetic unit because the motor and compressor are hermetically sealed within the envelope. Lubricating oil and refrigerant gas circulate in and around the motor.

Condensers may be water-cooled and located near the chiller or be an integral part of the package. Condensers can also be air-cooled and located remotely from the cooler. These two concepts are shown in Figure 26.

Condenser water can be cooled by use of a water tower. This is illustrated in Figure 27 in a perspective sketch of a typical chiller unit installation. Major piping, electrical components, and wiring connections are also shown.

6.2. THE ABSORPTION MACHINE

The absorption type chiller operates on the principle that certain chemicals have a strong affinity for water. For example, common table salt absorbs water vapor from the water

and by evaporation, cools the remaining water. The temperature of evaporation is a function of the pressure (vacuum) in the machine. The temperature of the coolant is regulated by controlling the vacuum; hence, chilled water is available. The diluted solution is reconcentrated by adding heat from steam or hot water. It requires approximately 18.5 lb of steam at 12 psi pressure to produce a ton of cooling. Hot water at about 250°F may be substituted for the steam.

Piping layouts for absorption liquid chiller installations using steam or hot water are shown in Figure 28. A photograph of an absorption chiller manufactured by York, a Division of Borg-Warner Corp. is shown in Figure 29.

6.3 THE RECIPROCATING MACHINE

This reciprocating type of chiller is ideally suited to smaller jobs. Basically, it is the same as the centrifugal except for the method of

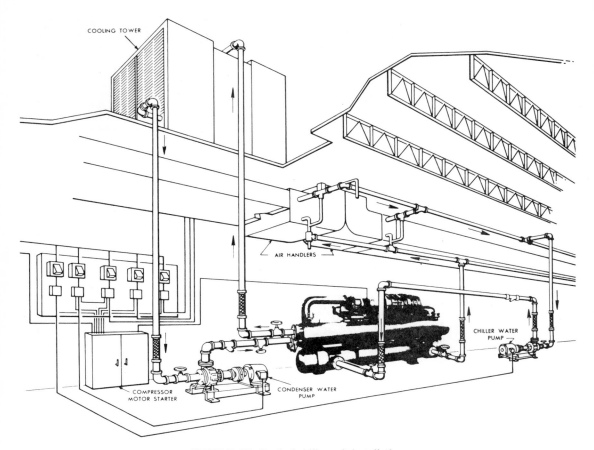

FIGURE 27. Typical chiller unit installation.

compression. Reciprocating systems lend themselves nicely to an air-cooled heat rejection device.

Figure 30 shows a commercially available reciprocating compressor unit, both alone (upper part of Figure) and in conjunction with a condenser.

Piping designs for connecting condensers and chillers are given schematically in Figures 31 and 32.

A commercially available hermetic package chiller is shown in Figure 33. This Dunham-Bush unit is a self-contained assembly of chiller, condenser, compressor, and controls.

6.4. ESTIMATED LABOR TO INSTALL CHILLERS AND COMPRESSORS

Description	Size (tons or hp)	Labor[a] (man-hours)
Water chiller-	3 tons	14
absorption type	4	14
gas-fired	7	26

Description (cont.)		Size (tons or hp)	Labor[a] (man-hours)
↓		10	26
		15	30
Water chiller-	(s)[b]	100–200 tons	75
absorption type	(s)	200–500	100
steam or hot water	(m)[c]	525–700	125
	(m)	725–850	150
	(m)	850–1250	150
	(m)	1250–1500	200
	(m)	1500–1750	225
Water chiller-	(s)	100–200	55
centrifugal type	(s)	250–400	80
	(s)	400–700	150
	(m)	750–1000	200
	(m)	1500	425
	(m)	2000	600
Water chiller type	(m)	2000–3000	1800
turbine drive	(m)	3000–5500	3000
↓	(m)	5500–8500	4200
Water chiller-		10–50	16
reciprocating type		60–100	24
air-cooled			

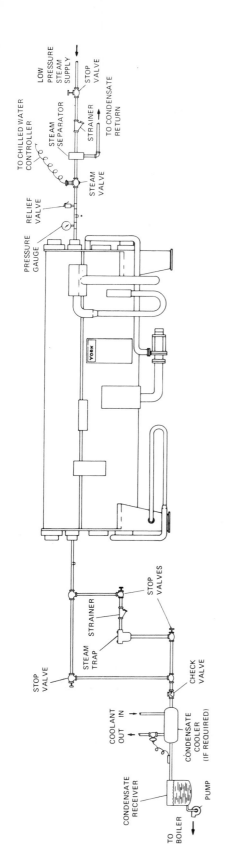
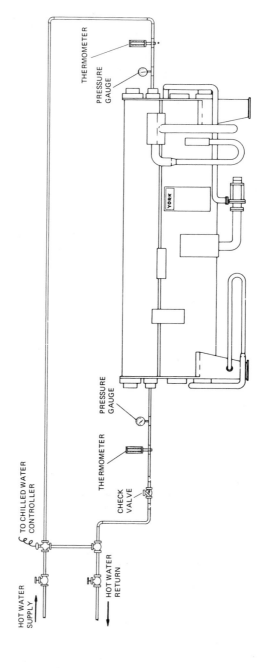

FIGURE 28. Absorption liquid chiller piping hookups.

FIGURE 29. Absorption liquid chiller. (Manufactured by York, Division of Borg-Warner Corp.)

Description		Size (tons or hp)	Labor[a] (man-hours)	Description (cont.)	Size (tons or hp)	Labor[a] (man-hours)
Water chiller-		10–50	12		15	12
reciprocating type		50–100	16		25	16
water-cooled		100–175	24		30	16
					40	16
Water chiller-	(s)	100–300	45		50	16
hermetic centrifugal	(s)	325–400	85		60	32
type	(s)	400–450	150		70	32
	(s)	500–600	185		100	40
	(m)	600–1200	300		120	40
	(m)	1100–1500	375		125	40
	(m)	2000	450			
Compressor		5	12	Compressor type	40 hp	24
reciprocating type		7.5	12	direct drive for	50	24
for HVAC		10	12	HVAC refrigeration	60	24

FIGURE 30. Hermetic reciprocating compressor and condensing units. (Manufactured by York, Division of Borg-Warner Corp.)

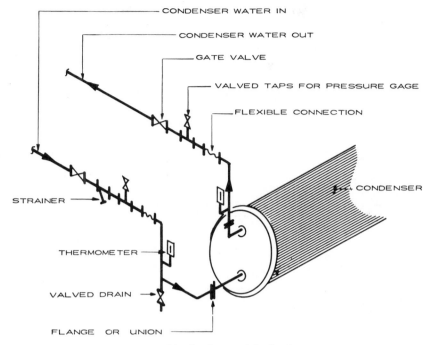

CONDENSER WATER IN

CONDENSER WATER OUT

GATE VALVE

VALVED TAPS FOR PRESSURE GAGE

FLEXIBLE CONNECTION

CONDENSER

STRAINER

THERMOMETER

VALVED DRAIN

FLANGE   OR   UNION

FIGURE 31. Condenser piping hookups.

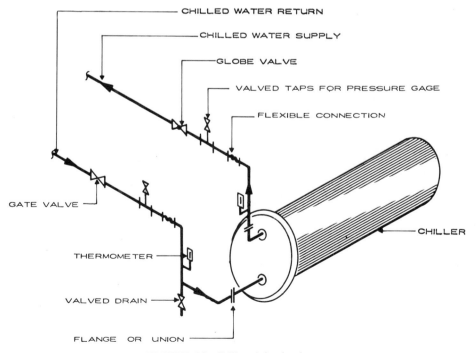

CHILLED WATER RETURN

CHILLED WATER SUPPLY

GLOBE VALVE

VALVED TAPS FOR PRESSURE GAGE

FLEXIBLE CONNECTION

CHILLER

GATE VALVE

THERMOMETER

VALVED DRAIN

FLANGE   OR   UNION

FIGURE 32. Chiller piping hookups.

FIGURE 33. Hermetic package chiller. (Manufactured by Dunham-Bush, Inc.)

Description (cont. p. 142)	Size (tons or hp)	Labor[a] (man-hours)
Belt drive	75	24
	100	32
	50	32
	60	32
	75	32
	100	36
	125	36

Description	Size (hp)	Receiver Capacity (cu ft)	Labor (man-hours)
Air compressor (with receiver, accessories, piping, control, etc.)	1	8	8
	1.5	8	8
	2	8	8
	3	10	8

Description	Size (hp)	Receiver Capacity (cu ft)	Labor (man-hours)
	5	10	12
	7.5	15	12
	10	15	12
	15	15	14
Air compressor (with mounted receiver tank)	1		12
	1.5		12
	2		12
	3		14
	5		14
	7.5		14
	10		14
	15		16

[a] Labor given for equipment (s) installed in place only.
[b] (s) = single piece of equipment
[c] (m) = multiple pieces of equipment

7

CHIMNEYS OR STACKS*

The function of a stack or chimney is to produce a draft so that the air and combustion gases can be drawn through the furnace, fuel, and boiler. The air, which carries the oxygen necessary for proper combustion, is thereby furnished to the fuel. The stack also carries the products of combustion to such a height before discharge that they will be dispersed and diluted, and, therefore, cause no discomfort or damage.

The basic principle of a natural or chimney draft is as follows: when a lighter gas is submerged in a heavier one, the lighter gas is forced upward by the heavier. A hot air balloon ascends in the cooler, denser atmosphere, similar to a cork submerged in water. For the same fundamental reason, a chimney produces the pressure difference which forces the gas through the boiler furnace.

If a stack is required to provide draft for the fuel burning equipment, follow carefully the recommendations of the manufacturer on the diameter and height of the stack. The stack should be higher than any adjacent buildings and should extend at least 4 ft above a roof, parapet, or bulkhead, or as required by local building codes. If a cap is placed on the top of the chimney to prevent entry of rain, it should

be mounted at a height above the stack equal to its diameter.

If a breeching is used from the boiler to a chimney it should be as short and with as few bends as possible. The breeching should slant upward toward the chimney. If it is necessary to run the breeching more than 6 ft, add 1 ft of chimney height for every 2 ft of breeching. If more than two, 90° ells are used, add 1 ft of height for every extra ell.

It is far better to have more rather than less stack height than required, as excess draft can be controlled by the use of a *barometric draft regulator*. This can be set to give the correct draft for proper combustion of the fuel being burned.

On boilers requiring a stack, it is possible to eliminate a high stack by installing an *induced draft blower unit*. Such units consist of a blower operated by an electric motor. With this device it is only necessary to run the breeching from the boiler to the draft unit and a stack from the draft unit to the outside to vent the products of combustion.

Induced draft units save the cost of high stacks of large diameter, plus erection and maintenance costs. Such units maintain a proper draft uniformly, regardless of weather, wind velocity, or barometric conditions. They provide instant and constant draft to the fuel at the required rate.

Induced draft units are installed at or near the stack connection on the boiler. Their blower

---

*A chimney is a vertical conduit or duct of masonry construction for the conveyance of flue gases. A stack is used for the same purpose but is made of steel or cast refractory material.

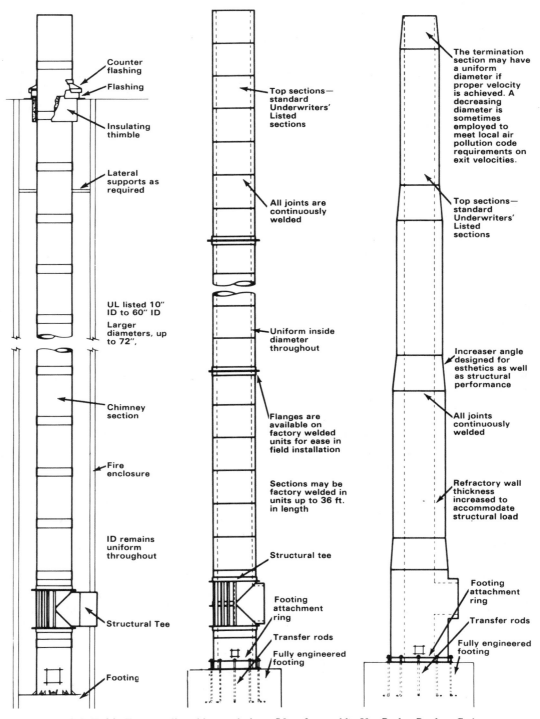

INTERIOR FREE STANDING CHIMNEY

- Counter flashing
- Flashing
- Insulating thimble
- Lateral supports as required
- UL listed 10" ID to 60" ID
- Larger diameters, up to 72",
- Chimney section
- Fire enclosure
- ID remains uniform throughout
- Structural Tee
- Footing

EXTERIOR FREE STANDING CHIMNEY

- Top sections—standard Underwriters' Listed sections
- All joints are continuously welded
- Uniform inside diameter throughout
- Flanges are available on factory welded units for ease in field installation
- Sections may be factory welded in units up to 36 ft. in length
- Structural tee
- Footing attachment ring
- Transfer rods
- Fully engineered footing

EXTERIOR FREE STANDING ARCHITECTURAL CHIMNEY

- The termination section may have a uniform diameter if proper velocity is achieved. A decreasing diameter is sometimes employed to meet local air pollution code requirements on exit velocities.
- Top sections—standard Underwriters' Listed sections
- Increaser angle designed for esthetics as well as structural performance
- All joints continuously welded
- Refractory wall thickness increased to accommodate structural load
- Footing attachment ring
- Transfer rods
- Fully engineered footing

FIGURE 34. Free standing chimney designs. (Manufactured by Van-Packer Product Co.)

draws the air from the boiler room into the boiler combustion chamber, through the boiler, and discharge the products of combustion out of the vent stack.

Figure 34 gives design details for three chimney types: an interior free standing, an exterior free standing, and an exterior free standing architectural. The term free standing means that no additional supports are used. Figure 35 shows a layout of a boiler chimney. Note the guy band near its top for the attachment of guy wires to support the stack from wind forces. This is not a free standing design. A breech connection between boilers and chimney is illustrated in Figure 36.

A commercially available incinerator chimney and adjacent charging chute are shown in Figures 37 and 38. Both use guy wires for support and are made of 28 ga. aluminized steel with a jacket of 11 ga. galvanized steel.

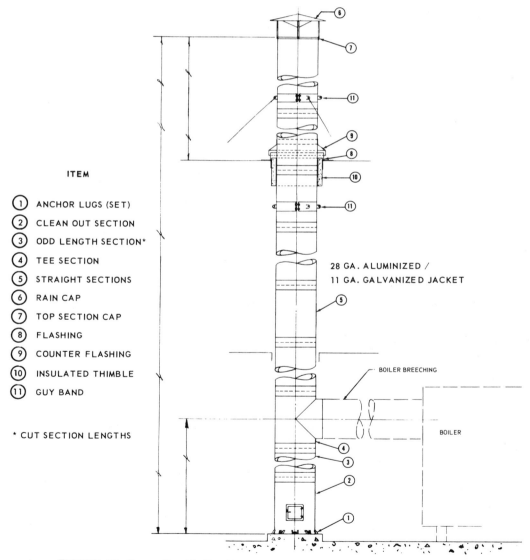

ITEM

1. ANCHOR LUGS (SET)
2. CLEAN OUT SECTION
3. ODD LENGTH SECTION*
4. TEE SECTION
5. STRAIGHT SECTIONS
6. RAIN CAP
7. TOP SECTION CAP
8. FLASHING
9. COUNTER FLASHING
10. INSULATED THIMBLE
11. GUY BAND

* CUT SECTION LENGTHS

28 GA. ALUMINIZED / 11 GA. GALVANIZED JACKET

BOILER BREECHING

BOILER

FIGURE 35. Conventional boiler chimney. (Manufactured by Van-Packer Products Co.)

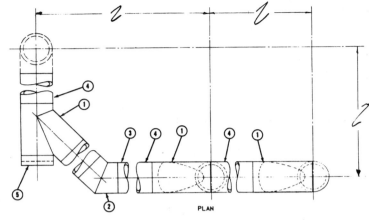

ITEM

① 45°/90° TEE SECTION
② 45° ELBOW
③ STRAIGHT SECTIONS
④ ODD LENGTH SECTION
⑤ END CAP (SPECIAL)

PLAN

200° TO 300°F. FLUE GAS

11 GA. GALV. JACKETS, WELDED INSTALLATION.

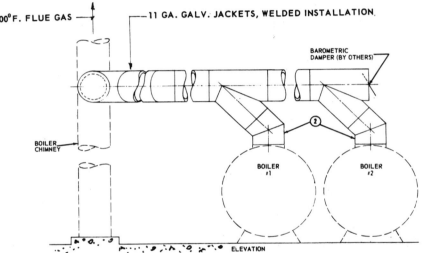

ELEVATION

FIGURE 36. Conventional boiler breeching. (Manufactured by Van-Packer Products Co.)

7.1. ESTIMATED LABOR TO INSTALL CHIMNEYS

Prefabricated refractory chimneys

Diameter, Inside (in.)	Height (ft)	Approximate Free Standing weight (lb)	Labor (man-hours)
18	30	5,500	16
24	40	10,000	25
30	50	16,000	35
36	60	28,000	42

Steel Stacks

Diameter, Inside (in.)	Height (ft)	Approximate Free Standing weight (lb)	Labor (man-hours)
48	80	15,000	25
60	100	26,000	35
72	120	48,000	42
84	150	70,000	62
96	160	95,000	78

ITEM

1. ANCHOR LUGS (SET)
2. CLEANOUT SECTION
3. ODD LENGTH SECTION*
4. TEE SECTION
5. COMPANION SECTION**
6. FLOOR SUPPORT SECTION
7. STRAIGHT SECTION
8. TWO PIECE BAND
9. 45° TEE SECTION
10. INSULATED THIMBLE
11. FLASHING
12. COUNTER FLASHING
13. POSIDRAFT, SIZE
14. GUY BAND
15. SPARK SCREEN

*CUT SECTION LENGTHS

**COMPANION SECTION LENGTHS

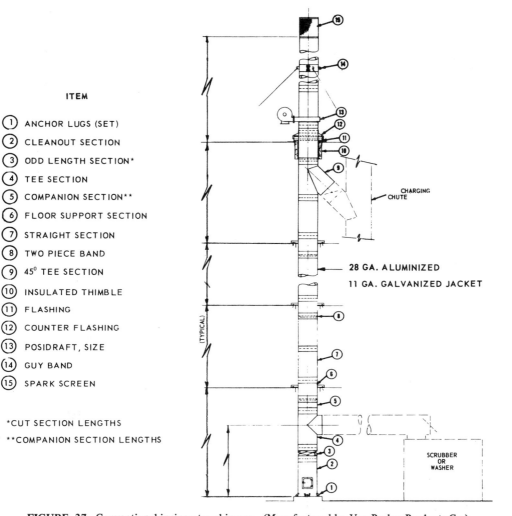

FIGURE 37. Conventional incinerator chimney. (Manufactured by Van-Packer Products Co.)

ITEM

1 COMPANION SECTION*

2 FLOOR SUPPORT SECTION

3 CHARGING CHUTE SECTION

4 TWO PIECE BAND

5 45° TEE SECTION (SPECIAL)

6 INSULATED THIMBLE

7 FLASHING

8 COUNTER FLASHING

9 ODD LENGTH SECTION**

10 GUY BAND

11 STRAIGHT SECTIONS

12 SPARK SCREEN

*COMPANION SECTION LENGTHS

**CUT SECTION LENGTHS

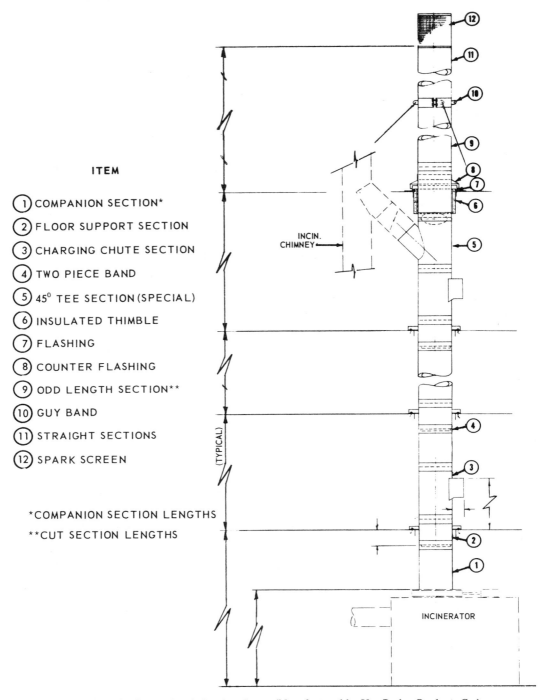

FIGURE 38. Conventional charging chute. (Manufactured by Van-Packer Products Co.)

8

COILS

Coils are designed mainly for the heating or cooling of air or other gases. Certain types of coils, however, are built for heat recovery, i.e., air or another gas is tempered with air as a heating medium. For the sake of simplicity, only the heating and cooling of air is dealt with here.

A coil is composed of a number of parallel tubes in one or more rows in the direction of the airflow. The coil depth is determined by the number of tube rows. The heating or cooling medium (hot or cold water or steam) flows through the tubes, and the air passes around them. In order to obtain a sufficiently large heating or cooling surface, thereby, making up for the lower heat transfer coefficient on the air side, the tubes are usually fitted with fins or gills.

The tubes can be connected in such a way that the heating or cooling medium flows crosswise, parallel to, or against the airflow (see Figure 39). In the first case, the rows are connected in parallel and in the latter two cases, in series. A cross-flow connection is used for condensing steam, and to a certain extent for hot water. The counter-flow connection is used for heating or cooling water. This type of connection gives the highest mean temperature difference and hence the greatest heat transfer capacity. Parallel-flow is sometimes used for heating water. It has a lower thermal capacity than the other types of connections, but the difference is insignificant with only a small number of tube rows and with large differences

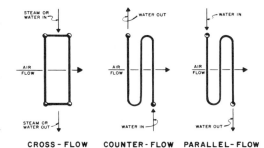

CROSS - FLOW   COUNTER - FLOW   PARALLEL - FLOW

FIGURE 39. Flow arrangements.

in temperature between the air and the heating medium. For those cases where the airflow direction or the direction of the heating or cooling medium is decisive to the capacity of coil, the direction is marked on the coil.

An array of commercially available coils is shown in Figure 40. An electric coil is shown at the upper left. The enlarged inset illustrates the resistance wire heating element imbedded in the finned tube. Direct expansion (DX) coils are used for volatile refrigerants. A thermal expansion valve feeds only enough refrigerant so that all of it will evaporate as it enters the coil. This protects the compressor in the system from unevaporated refrigerant.

8.1. ESTIMATED LABOR TO INSTALL COILS

Figures 41 through 45 show piping connections to various types of coils or coil assemblies

ELECTRIC COIL

STANDARD
STEAM COIL

WATER COIL

(DX) DIRECT
EXPANSION COIL

CLEANABLE WATER
TUBE COIL

STEAM COIL

FIGURE 40. Heating and cooling coils. (Manufactured by American Air Filter Company.)

along with the estimated man-hours required for their installation. Figure 41 illustrates a single coil hookup for a low-pressure steam preheat coil, with labor estimates for double and triple coils. Figures 42 and 43 give piping and installation data for single and double hot water reheat coils. Similarly, Figures 44 and 45 show connections and man-hour requirements for a single chilled water coil and a double chilled water coil. The Figures are followed by the installation time for various sized DX type water or steam coils.

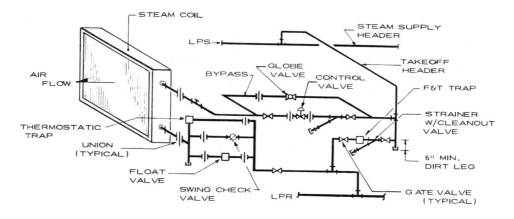

INLET SIZE CONNECTION	3/4"	1"	1-1/4"	1-1/2"	2"	2-1/2"	4"
LABOR HOOKUP (MH) SINGLE COIL	12	14	16	22	24	32	–
DOUBLE COIL (not shown)	16	20	24	26	30	48	–
TRIPLE COIL (not shown)	–	–	48	64	68	72	82

FIGURE 41. Low-pressure steam preheat coil.

DESIGN NOTES:

1. Locate pipe unions and arrange piping to facilitate coil removal.

2. Provide similar piping arrangement on each side for two sections wide coil.

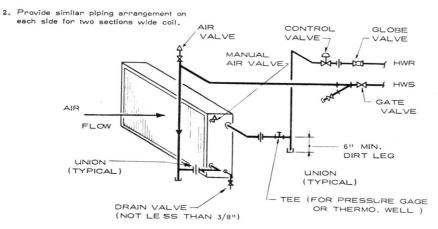

INLET SIZE CONNECTION	1/2"	3/4"	1"	1-1/4"	1-1/2"	2"
LABOR HOOKUP TIME (MH)	12	14	16	18	18	24

FIGURE 42. Hot water reheat coil.

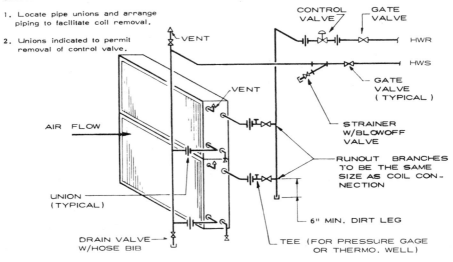

DESIGN NOTES :

1. Locate pipe unions and arrange piping to facilitate coil removal.

2. Unions indicated to permit removal of control valve.

INLET COIL CONNECTION	3/4"	1"	1-1/4"	1-1/2"	2"
LABOR HOOKUP TIME (MH.)	12	16	18	20	24

FIGURE 43. Hot water reheat coil (double coil).

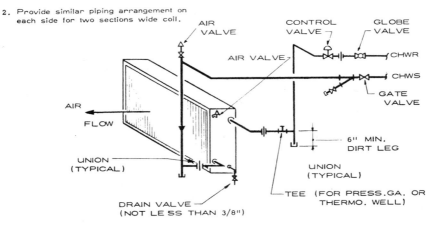

DESIGN NOTES:

1. Locate pipe unions and arrange piping to facilitate coil removal.

2. Provide similar piping arrangement on each side for two sections wide coil.

INLET SIZE CONNECTION	3/4"	1"	1-1/4"	1-1/2"	2"	2-1/2"	4"	6"
LABOR HOOKUP TIME (MH)	12	14	16	20	24	36	72	112

FIGURE 44. Chilled water coil.

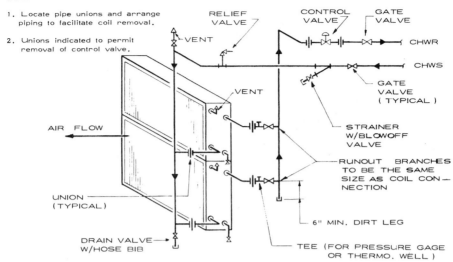

DESIGN NOTES :

1. Locate pipe unions and arrange piping to facilitate coil removal.

2. Unions indicated to permit removal of control valve.

INLET SIZE CONNECTION	3/4"	1"	1-1/4"	1-1/2"	2"	2-1/2"
LABOR HOOKUP TIME (MH)	–	–	32	36	38	42

FIGURE 45. Chilled water coil (double coil).

Description	Coil Area (sq ft)	Labor (man-hours)	Description	Coil Area (sq ft)	Labor (man-hours)
DX-Type Water or Steam Coils	2	4		25	12
	4	4		30	12
	6	4		35	18
	8	7.5		40	24
	10	7.5		42	24
	12	7.5		50	30
	15	10		60	36
	20	10		72	48

9

CONDENSERS AND CONDENSING UNITS

As previously described in the Chapter on the refrigeration cycle, in Section I, a condenser removes the heat from the high-pressure gas causing the gas to condense into a liquid. There are three types of condensers which have been developed for the HVAC industry: water-cooled, evaporative, and air-cooled.

9.1. AIR-COOLED CONDENSING UNITS AND CONDENSERS

Condensing units are normally designed for outdoor installation, but some models are used for indoor or through-wall applications. Air-cooled condensing units contain a compressor, condenser, fan, and electrical controls. These units are usually part of a prematched system of a condenser working in tandem with an air handling unit containing a fan blower, an evaporator, a filter, and a heating coil where needed. These prematched components, acting in unison, are also called a "split system," if they are not separated by more than 50 ft. An exception is where refrigerant lines to and from the condenser, located more than 50 ft from the air handler, are sized for the distance and approved by the manufacturer. Thus, the warranty for the performance is not violated.

The air-cooled condenser performs the same function as the condensing units. The only difference is that it does not house a compressor which is instead remotely located. Both types have their pros and cons for particular applica-

tions but the main disadvantage of the air-cooled condenser is that a fan and motor is required to expell the heat on hotter days, when the condenser capacity is lowest and most needed. However, their popularity is wide spread for residential and commercial applications.

Design applications of air-cooled condensing units are shown in Figure 46. Such locations as wall and roof installations as well as ground level on concrete and wooden supports are illustrated. Typical clearance dimensions of a commercial condensing section is found in Figure 47.

9.2. EVAPORATIVE CONDENSERS

An evaporative condenser is an apparatus which combines the properties of a forced draft cooling tower with a water- or air-cooled condenser unit. It is usually installed outdoors except for small capacity units. Evaporative condensers operate by drawing in the air at the bottom, forcing it upward across refrigerant coil tubes, then through a series of water sprays and plate eliminators, and into a fan mounted at the top of the unit.

Air and water are used as cooling media. In the process about 3 to 7% of the water evaporates. A continuous make up of water is therefore needed. These units are manufactured as a complete package ranging from 10 to 100 tons.

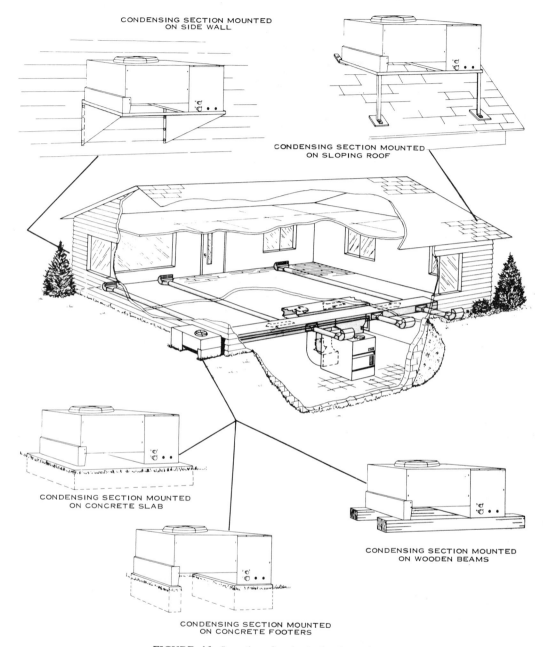

FIGURE 46. Locations for air condensing units.

9.3. WATER-COOLED CONDENSERS

Generally this type of condenser is of the *shell-and-tube* type, where the water (usually city water) is circulated through the tube and the condensing refrigerant gases are in the shell.

Both shell and tube (also called a tube in a tube) are commonly sealed at their ends where the water and refrigerant lines are connected.

Water-cooled condensers come in many diameters and lengths to suit the required thermal conditions. However, corrosion, high

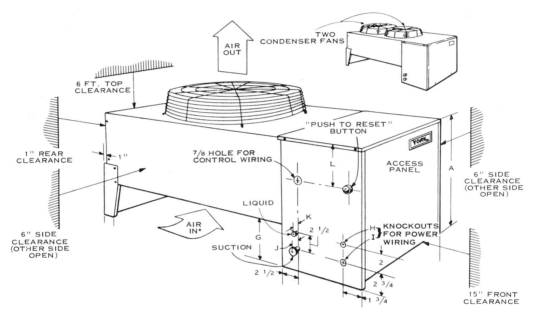

FIGURE 47. Air condensing section. (Manufactured by York, Division of Borg-Warner Corp.)

cost, and the need for water disposal tends to limit their use.

9.4. ESTIMATED LABOR TO INSTALL CONDENSERS

Description	Size (tons)	(gpm)	Weight (lb)	Labor (man-hours)
Shell and tube size, water-cooled				
	15	51	–	12
	25	85	–	12
	30	102	–	14
	60	204	–	24
	100	340	–	24
	150	510	–	40
	175	595	–	40

Description (cont.)	Size (tons)	(gpm)	Weight (lb)	Labor (man-hours)
	200	680	–	40
	300	1010	–	48
	400	1360	–	48
	500	1700	–	48
	600	2000	–	52
	700	2360	–	52
air-cooled type	2.5	–	70	8
	5	–	175	8
	7.5	–	200	8
	10	–	375	8
	15	–	400	8
	20	–	485	8
	30	–	575	8
	40	–	1275	12
	50	–	1500	12

Description	Size (tons)	(gpm)	Weight (lb)	Labor (man-hours)
	65	–	2000	14
	70	–	2500	16
	75	–	3000	16
	100	–	3500	16
Reciprocating, air-cooled type				
(s)[a]	10	–	850	18
(s)	15	–	950	18
(s)	20	–	1725	24
(s)	25	–	1950	24
(s)	30	–	2450	24
(s)	40	–	2700	24
(m)[b]	50	–	3150	30
(m)	60	–	3800	30
(m)	80	–	4950	30
water-cooled type				
	120	–	4575	36
	150	–	5175	36

Description (cont.)	Size (tons)	(gpm)	Weight (lb)	Labor (man-hours)
	200	–	7175	48
	300	–	7775	48
	400	–	11,500	60
	500	–	11,550	60
Evaporative condensers type				
	15	–	1275	12
	25	–	1450	12
	30	–	1550	12
	40	–	1850	12
	50	–	2250	14
	60	–	2750	14
	70	–	3150	18
	80	–	3450	18
	100	–	3750	20

[a](s) = single unit assembly.
[b](m) = multiple unit assembly.

10
CONTROLS

10.1. CONTROL TERMINOLOGY

The definitions of control terms can be some-what confusing. Several different expressions have been used to convey a single idea or concept. This section will define the most commonly used terms in the HVAC industry. The definitions will conform, as nearly as possible, to the automatic control terminology used in other fields of control.

The control system in Figure 48 consists of a controller and a controlled device which measures, by means of a sensing element, *a controlled variable*, such as temperature, humidity, or pressure. The controlled variable is compared with an imput signal to produce a suitable action or impulse for transmission to the controlled devices. Thermostats, humidistats, and pressure controllers are examples of *controllers*. The controlled device reacts to the signal received from a controller and varies the flow of the control agent. The *controlled device* may be a valve, damper, electric relay, or a motor driving a pump or fan.

The control agent is the medium manipulated by the controlled device. It may be air or gas flowing through a damper, gas, steam, or water flowing through a valve. The process plant is the air conditioning apparatus which is being controlled. It reacts to the output of the control agent and effects the change in the controlled variable. It may include a coil, duct, fan, or the occupied space of the building. The con-

trolled variable is the condition, such as temperature, humidity, or pressure being controlled.

10.2. CONTROL SYSTEMS

The functional requirement of control in an air conditioning system is to alter the system variables in some prescribed manner, such that equipment capacity is changed to meet the load. This is accomplished by adding one or more components which are capable of sensing changes in variables and altering the position of flow elements to respond to prescribed conditions. This combination is defined as a control system or control loop. There are two types of control loops for modern HVAC systems: *open loop* and *closed loop*.

An open loop control system usually takes corrective action to offset the effects of external disturbance on the variable of interest. Sometimes this action is referred to as feed-forward control, because it anticipates the effect of an external variable on the system.

An outdoor thermostat arranged to control the flow of heat to a building based on changes in outdoor temperature is an example. In essence, the designer of this system presupposes a fixed relationship between outside air temperature and the heat requirement of the building and takes the control action based only on outdoor air. The actual temperature inside the building has no effect on the controller. Complete control of the space temperature is not achieved with this system.

An improvement over the open loop system is the closed loop control system. A typical example of the closed loop system is shown in Figure 48, where the discharge temperature of the air is being controlled. The controller in this system, a thermostat, measures the change in the variable and actuates a controlled device, the valve, to bring about an opposite change which is again measured by the controller. The corrective action is a continuous process until the variable is brought to a desired value within the design limits of the controller. This system of transmitting information about the results of an action or operation back to its origin is known as feedback; this makes true automatic control possible.

Control systems are divided into four main groups according to the primary source of energy involved:

1. *Pneumatic systems.* Compressed air, usually at a pressure of 15 to 35 psig is the source of energy. The air is supplied to the controller, which in turn regulates the pressure in the controlled device.
2. *Electric/electronic systems.* These systems utilize electricity, either low or line voltage, as the energy source. The electrical energy supplied to the controlled device is regulated by the controller, either directly or through relays or pneumatic-electric transducers. Systems which include electronic sensing and amplification devices are generally classed as electronic.

Controlled devices for these systems include relays, contactors, electromechanical and hydraulic actuators, and solid state regulating devices.

3. *Self-powered systems.* Here, the change in properties of the environment is used to initiate the necessary corrective action. The measuring system derives its energy from the process under control, or from an auxiliary source of energy. Temperature changes at the sensor result in pressure or volume changes of the enclosed media, which are transmitted directly to the operating device of the valve or damper. A system using a thermopile in a pilot flame for the generation of electrical energy is also self-powered.
4. *Hybrid systems.* Hybrid systems are combinations of the above systems which utilize such multiple power sources as electrical and pneumatic. Transducers convert the signals so that either type of source can actuate a controlled device using a different power source.

For convenience, the controls for heating, ventilating, and air conditioning systems are subdivided into four groups:

* sensing elements
* controllers
* controlled devices
* auxiliary devices such as switches and relays, timers, thermometers, gauges, pilot

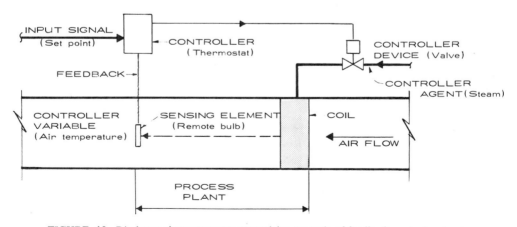

FIGURE 48. Discharge air temperature control (an example of feedback control system).

lights, and other indicators for observing the operation of the system.

10.2.1. Sensing Elements

A *sensing element* is a device which measures changes in the controlled variable and produces a proportional signal for use by the controller. Pressure, temperature, humidity, and water flow are some of the more commonly monitored variables.

Sensing elements for other purposes, such as flame detection, smoke density, specific gravity, current, CO_2, CO, etc., are often necessary for the complete control of a heating, ventilating, or air conditioning system.

Sensing elements may be mounted in the following ways to monitor different environments:

1. Room-mounted on a wall for response to room temperature.
2. Mounted on a duct, with its sensing element extending into the duct.
3. Tank or pipe-mounted, with a fluid-tight connection to allow the measuring element to extent into the fluid. A separable socket well is often used with immersion thermostats to avoid draining the system when removing the thermostat. Since the separable socket will reduce the response rate of the thermostat, the socket should be no heavier than necessary and should fit snugly around the element.
4. Remote bulb, for applications where temperature measurement is at some distance from the desired thermostat location, such as the central panel mounting of the controller. The remote bulb element may be of either the insertion or immersion type.
5. Surface-mounted to sense the temperature of a pipe or other surface.

Temperature Sensing: Temperature sensing elements include:
- bimetal strips
- rod and tube of dissimilar metals
- a sealed bellows, with or without a remote bulb
- electrical elements

A bimetal element is composed of two thin strips of dissimilar metals fused together. Because the two metals have different coefficients of thermal expansion, the element bends as the temperature varies to produce a change in position. Depending on the available space and the movement required, elements can be in the form of a straight strip, u-shaped, or wound in a spiral. Elements are most commonly used for room thermostats, but are also used for insertion and immersion thermostats.

A rod and tube element is a metal tube having a high coefficient of expansion, with an inner, low-expansion rod attached to one end of the tube. The tube changes length with changes in temperature, causing the free end of the rod to move. The rod and tube element is commonly employed on certain types of insertion and immersion thermostats.

A sealed bellows element is either vapor-filled, gas-filled, or liquid-filled, after being evacuated of air. Changes of temperature cause changes in pressure and volume of the gas or liquid, resulting in a change movement. This element is often used in room thermostats.

A remote bulb element is a sealed bellows or diaphragm to which a bulb or capsule is attached by means of a capillary tube. The entire system is filled with vapor, gas, or liquid. Changes of temperature at the bulb result in changes of pressure and volume. The pressure change is communicated to the bellows or diaphragm through the capillary tube. The remote bulb element is useful where the temperature to be measured is remote from the desired thermostat location. It usually is provided with fittings suitable for insertion into a duct, pipe, or tank.

A resistance element is made of wire whose electrical resistance changes with temperature. A thermistor is a special kind of semiconductor in which electrical resistance changes with temperature. A thermocouple is a union of two dissimilar metals joined at their ends which generates a voltage that varies as a function of temperature change. These various elements are available in forms suitable for measuring room temperature or for insertion into a duct, pipe, or tank.

Dial thermometers, as shown in Figure 48a, are used for direct reading of control-agent tem-

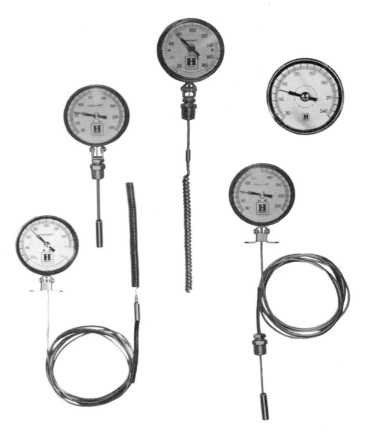

FIGURE 48a. Dial thermometers. (*Courtesy Honeywell, Inc.*)

peratures, such as air in ducts and liquid in storage tanks or pipes.

The liquid-filled bulbs and Bourdon tube sensing elements have rigid, bimetal sensing elements.

Pressure Sensing. Pressure sensing elements can be divided into two general classes, depending upon pressure range.

1. For pressures or vacuums measured in pounds per square inch or inches of mercury, the element is usually a bellows, diaphragm, or Bourdon tube. One side of the element may be open to the atmosphere, in which case the element responds to pressures above or below atmospheric. A differential pressure element has connections to both sides so that it will respond to the difference between two pressures.

2. For low ranges of pressure or vacuum measured in inches of water, such as the static pressure in an air duct, the measuring element may be an inverted bell immersed in oil, a large slack diaphragm, or a large flexible metal bellows. The element usually is of the differential type when orifices are used. Pitot tubes and similar accessories measure flow, velocity, or liquid level as well as static pressure.

Humidity Sensing. There are two types of humidity sensing elements. The first type uses hygroscopic materials such as hair, wood, paper, animal membranes, and nylon. These materials readily absorb moisture and change size or form in doing so to cause a mechanical deflection which is proportional to the absorbed moisture. The second type is electrical. Here the element changes its electrical characteristics (resistance or capacitance) as a function of its hygroscopic properties.

Water Flow Sensing.   Water flow sensing elements use a variety of basic sensing devices such as an orifice plate, pitot tube, venturi, flow nozzles, turbine meter, magnetic flow meter, and vortex shedding flow meter.

Each of these has characteristics of range, accuracy, complexity, and cost, that make them suitable for different situations. In general, the pressure differential type (orifice plates, pitot tubes, venturis, and flow nozzles) are simple and inexpensive but have limited range, so their accuracy depends upon how they are used. The more sophisticated flow sensing devices such as turbines, magnetic, or vortex-shedding flow meters, are more expensive but have better accuracy over a wider range of flow.

10.2.2. Controllers

Controllers take the sensor information, compare it with the desired control condition, and regulate an output signal to cause various types of control action. Controllers are classified into a number of categories; electric/electronic, non-indicating, indicating or recording, transducers and pneumatic. The configuration of the controller is another method of classification. Thermostats and humidistats are controllers which have the sensing elements and controller functions in one device. Sensor (or transmitter)-receiver controller systems use a sensing element remote from the controller function.

Electric/Electronic Controllers.   For two-position control, the controller output may be: a simple electrical contact, or that which starts a burner or pump; or that which actuates a spring-return valve or damper operator. Single pole, double throw (SPDT) switching circuits are used to control a three-wire unidirectional motor operator. Single pole double throw circuits also are used for heating-cooling applications. Either SPST (single pole, single throw) or SPDT circuits may be modified to obtain timed two-position action.

For floating control, the controller output is a SPDT switching circuit with a neutral zone where no contact is made. This type of control is used with reversible motor operators. Proportional control provides continuous or incremental changes in output signal to position an electric actuator or control device.

An alternative control configuration is to have two devices, one for sensing and one for control. Included in this category are the sensor-receiver controller systems. The sensor is sometimes referred to as a transmitter in this arrangement.

Nonindicating Controller.   The nonindicating controller is most common for heating, ventilating, and air conditioning work. It includes all types in which the sensing element does not provide a visual indication of the controlled variable. If an indication is desired, a separate thermometer, relative humidity indicator, pressure gauge, etc., is required. With room thermostats, for example, a separate thermometer often is attached to the cover.

Indicating or Recording Controller.   Controllers may be of the indicating or recording type. An indicating controller has a pointer added to the sensing element or attached to it by a linkage, so that the controlled variable is indicated on a suitable scale. A recording controller is similar to an indicating controller, except that the pointer is replaced by a recording pen which provides a permanent record.

Proportional band is a term used with indicating and recording controllers and means the same as throttling range. It is usually expressed in percent of the scale or chart range of the controller.

Pneumatic Controllers.   Pneumatic controllers are normally combined with sensing elements having a force or position output to obtain a variable air pressure output. The control mode is usually proportional but other modes may be used. They are generally classified as a *nonrelay* type or a *relay* type.

The nonrelay type is the simplest form of pneumatic controller. It uses a restrictor in the air supply and a bleed nozzle. The sensing element positions a flapper which varies the nozzle opening, resulting in a variable air pressure output applied to the controlled device, usually a small pneumatic operator. The nonrelay type is limited to applications requiring small volumes of air or where long response times can be tolerated.

In a relay type, the variable pressure from the sensing element, either directly or indirectly (through a restrictor, nozzle, and flapper arrangement) actuates a relay device which increases the air volume available for control.

Pneumatic controllers are further classified by construction as *direct* or *reverse-acting*. Direct-acting controllers increase the output air pressure as the controlled variable increases. For example, a direct-acting thremostat increases output pressure when the sensing element detects a temperature rise. Reverse-acting controllers increase the output air pressure as the controlled variable decreases. A reverse-acting thermostat will increase output pressure when the temperature drops.

Transducers. Hybrid systems consisting of combinations of electric or pneumatic control devices may be required. For these applications, devices commonly called *transducers* are used to convert electrical signals into pneumatic output or vice versa. Transducers may convert proportional input to either proportional or two-position output.

Thermostats. Many thermostats are of the simple, single-purpose type, but there are others for various special purposes. These include day-night, heating-cooling, multistate, submaster, wet-bulb, and dew point thermostats, which are described below.

The day-night or dual room thermostat controls at a reduced night temperature. It may be indexed (changed from day to night operation) individually or in groups from a remote point by a manual or time switch. Some electric types have an individual clock and switch built into the thermostat.

The pneumatic day-night thermostat uses a two-pressure air supply of 13 and 17 psig, or 15 and 20 psig. Changing the pressure actuates switching devices in the thermostat and indexes them from day to night or vice versa. Supply air mains are often divided into two or more cir-

cuits so that switching can be accomplished in various areas of the building at different times. For example, a school building may have separate circuits for classrooms, offices and administrative areas, the auditorium, and other areas.

The heating-cooling or summer-winter thermostat is reversible and, if desired, can have its set point changed by indexing. It is used to actuate valves or dampers that regulate a heating source at one time and a cooling source at another. The thermostat is often manually indexed in groups by a switch, or automatically, by a thermostat sensing the temperature of the control agent, the outdoor temperature, or another suitable variable. The pneumatic heating-cooling thermostat uses a two-pressure air supply, as described for day-night thermostats.

Multistate thermostats are arranged to operate two or more succesive steps in sequence.

A *submaster thermostat* has its set point raised or lowered over a predetermined range, by a master controller. The master controller can be a thermostat, manual switch, pressure controller, or similar device. For example, a master thermostat measuring outdoor air temperature can be used to adjust a submaster thermostat to control water temperature in a heating system. Master-submaster combinations are sometimes designated as single-cascade action. When such action is accomplished by a single thermostat having more than one measuring element, it is known as compensated control.

A *wet-bulb thermostat* is often used for humidity control in conjunction with proper control of the dry-bulb temperature. A wick, or other means for keeping the bulb wet, and rapid air motion to assure a true wet-bulb measurement, are essential.

A *dew point thermostat* is a control device which reacts to changes in dew point temperatures.

A number of thermostats, a humidistat, a pressure controller, and ancillary equipment are shown in the following illustrations. The upper part of Figure 49 is a liquid filled insertion thermostat, a direct-acting, one-pipe, bleed-type, remote-bulb controller for air or water systems. The averaging element model is used as a discharge controller in central fan systems. The lower portion of Figure 49 shows a pneu-

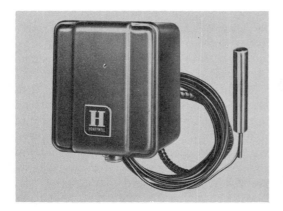

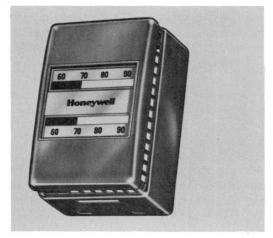

FIGURE 49. (a) Liquid filled insertion thermostat. (b) Pneumatic dehumidifiers. (*Courtesy Honeywell, Inc.*)

matic thermostat. Pneumatic two-pipe thermostats give proportional control of pneumatic devices such as valves and damper operators in heating, ventilating, and air conditioning systems.

The pressure controller at the top of Figure 50 is a one-pipe, bleed type used for proportional control of pneumatic motors and valves, or for control of the pressure of steam, air, or noncorrosive liquids and gases. A pneumatic humidistat is shown in the lower part of Figure 50. This is a two-pipe proportional device for controlling air motors of valves and dampers in air conditioning systems, requiring control for humidification and dehumidification.

Figure 51 (top) shows a duct sampling chamber used with humidistats or thermostats. It

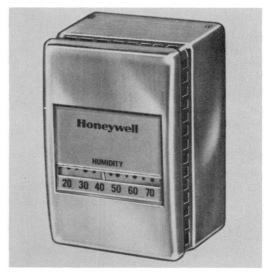

FIGURE 50. (a) Pressure controller. (b) Pneumatic thermostat. (*Courtesy Honeywell, Inc.*)

provides an easy means for sampling air within a duct to provide humidity or temperature control. Air compressors are used to provide and maintain an air pressure source for pneumatic temperature control systems. Such a compression is shown at the bottom of Figure 51.

FIGURE 51. Air compressor. (*Courtesy Honeywell, Inc.*)

Figure 52 shows a static pressure regulator designed for use with pneumatic control devices to control the static head for differential pressure in a central fan installation.

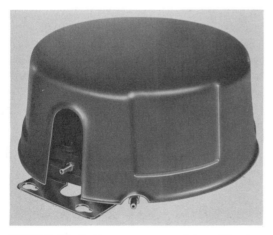

FIGURE 52. Static pressure regulator. (*Courtesy Honeywell, Inc.*)

10.2.3. Controlled Devices

Controlled devices may be automatic valves, valve operators, automatic dampers, or damper operators. Other controlled devices include electric heaters, relays, and motors on such equipment as fans, pumps, burners, refrigeration compressors, and similar apparatus.

Automatic Valves. An automatic valve is designed to control the flow of steam, water, gas, and other fluids, and may be considered as a variable orifice which is positioned by an electric or pneumatic operator in response to impulses from the controller. It may be equipped with a throttling plug or v-port of special design to provide the desired flow characteristics.

Renewable composition disks on the valve stem are common. They are made of materials best suited to the media handled by the valve, the operating temperature, and the pressure. For high pressures or superheated steam, metal disks are often employed. Internal parts of valves, such as the seat ring, throttling plug, v-port skirt, disk holder, and stem, are made of stainless steel or other hard and corrosion-resistant metals for use in severe service. Automatic valves should be properly selected and sized for specific applications.

The functions of various types of automatic valves are:

- A *single-seated* valve (Figure 53a) is designed for tight shut-off. Appropriate disk materials for various pressure ranges and media are used.
- A *pilot piston* valve uses the pressure of the control agent to operate the valve. It usually is single-seated and used where large forces are required for valve operation.
- A *double-seated* or balanced valve (Figure 53b) is designed so that the pressure acting against the valve disk is essentially balanced, thereby reducing the force required of the operator. It is widely used where the fluid pressure is too high to permit a single-seated valve to close. It cannot be used where tight shutoff is required.
- A *three-way mixing valve* (Figure 53c) has two inlet connections, one outlet connection, and a double-faced disk

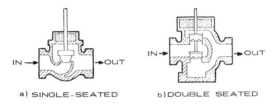

a) SINGLE-SEATED   b)DOUBLE SEATED

SINGLE AND DOUBLE-SEATED VALVES

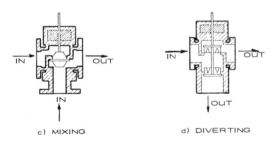

c) MIXING   d) DIVERTING

THREE-WAY MIXING AND DIVERTING VALVES

FIGURE 53. Typical automatic valves for HVAC control.

operating between two seats. It is used to mix the two fluids that enter through the inlets and leave through the common outlet.

- A *three-way diverting valve* (Figure 53d) has one inlet, two outlets and two disks and seats. It is used to divert the flow to either of the outlets or to proportion the flow to both outlets.
- A *butterfly valve* (not shown) has a body with a heavy ring seat enclosing a disk which rotates on an axis at or near its center. It is similar to a round single-blade damper. The disk seats against the the ring or a resilient liner in the body.

Valve Operators. Valve operators are of three general types: solenoid-operated, electric motor-operated, and pneumatically operated.

1. A solenoid consists of a magnetic coil that operates a movable plunger. It is used for two-position operation. The most common use of a solenoid-operated valve is to control the flow of gas, water, air and refrigerants, and it is generally limited to valves up to 3-in. diameter.

2. An electric motor operates the valve stem through a gear train and linkage. Electric motor

operators are classified in three types: unidirectional, spring-return, or reversible.

  a. Unidirectional operators are for two-position operation. The valve opens during one-half of the revolution of the output shaft and closes during the other half revolution. Once started, it continues until the half revolution is completed, regardless of action by the controller. Limit switches built into the operator stop the motor at the end of each stroke. If the controller has been satisfied during this interval, the operator will continue to the other position.

  b. Spring-return operators are used for two-position operation. An electric-powered motor drives the valve to one position and holds it there. When the circuit is broken the spring returns the valve to its normal position.

  c. Reversible operatores are installed for floating and proportional operation. The motor can run in either direction and stop at any position. It is sometimes equipped with a return spring. For proportional control applications, a potentiometer for rebalancing the control circuit is also driven by the motor.

3. A pneumatic operator consists of a spring-opposed, flexible diaphragm or bellows attached to the valve stem in such a way that an increase in air pressure moves the valve stem and simultaneously compresses the spring. Sequence operation of two or more devices can be achieved by proper selection or adjustment of the springs. Springless pneumatic operators, using two opposed diaphragms or two sides of a single diaphragm are also used, but are generally limited to special applications involving large valves or high pressures. Pneumatic operators are designed primarily for proportional control. Two-position control is accomplished by use of a two-position controller, or a two-position pneumatic relay to apply either full air pressure or no pressure to the valve operator.

A pneumatic operator is shown in Figure 54. It is a sturdy, compact device used primarily on unit ventilator control systems and can be adapted to other ventilator systems as required.

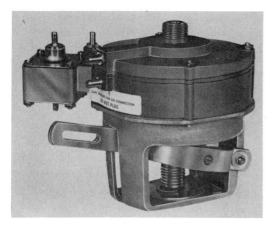

FIGURE 54. Pneumatic operator. (*Courtesy Honeywell, Inc.*)

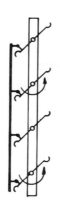 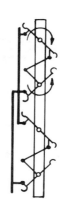

PARALLEL
OPERATION

OPPOSED
OPERATION

FIGURE 55. Multiblade dampers.

Pneumatic valves and valves with spring-return electric operators can be classified as normally open or normally closed. A normally open valve will assume an open position when all operating force is removed. A normally closed valve will close when all operating force is removed.

An electric-hydraulic actuator is similar to a pneumatic one, except it uses an incompressible fluid which is circulated by an associated electric pump.

Automatic Dampers. Automatic dampers are designed to control the flow of air or gases and function like valves in this respect. Steel blades and frames are the most common materials of construction, but other materials are used for special applications, such as those involving corrosive fumes. Large dampers are often made in two or more sections for strength and convenience in handling, the sections being interconnected so as to operate as a unit. The damper types commonly used are:

- *Single-blade*, which is generally restricted to small sizes because of the difficulty of securing proper operation with high-velocity air.
- *Multiblade or louver damper* (Figure 55) which has two or more blades linked together. It may be arranged for parallel operation whereby all blades rotate in the same direction, or opposed operation in

which adjacent blades rotate in opposite directions. Figure 56 illustrates an opposed louver damper.

- *Mixing dampers*, composed of two sections interlinked so that one section opens as the other closes.

Electric damper operators may be unidirectional, spring-return, or reversible, as are electric

FIGURE 56. Opposed louver damper. (*Courtesy Honeywell, Inc.*)

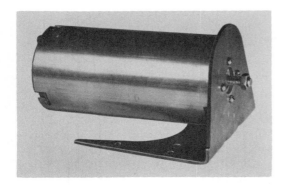

FIGURE 57. Pneumatic damper operator. (*Courtesy Honeywell, Inc.*)

motor-valve operators. Pneumatic damper operators are similar to pneumatic valve operators except that they usually have a longer stroke, or the stroke is increased by means of a multiplying lever. Damper operators are mounted on the damper frame and are connected by linkage directly to a damper blade. They are also mounted outside the duct and connected to a crank arm attached to a shaft extension of one of the blades. On large dampers, two or more operators may be required, and are usually applied at separate points on the damper. Normally open or normally closed operation is obtained according to the method of mounting the operator and connecting the linkage. A pneumatic damper operator is shown in Figure 57. In addition to dampers they can be used for air valves and shutters in high-velocity mixing boxes.

10.2.4. Auxiliary Control Equipment

In addition to the conventional controllers and controlled devices described, many control systems require a auxiliary devices to perform various functions. Auxiliary controls for electric control systems include:

- *Transformers* to provide current at the required voltage.
- *Electric relays* for control of electric heaters or to start and stop motors on oil burners, refrigeration compressors, fans, pumps, or other apparatus for which the electrical load is too large to be handled directly by the controller. Other uses

include time-delays; circuit-interlocking safety applications, etc.
- *Potentiometers* for manual positioning of proportional control devices, or for remote set point adjustment of electronic controllers.
- *Manual switches* for two-position or multiple position type of operation with single or multiple poles.
- *Auxiliary switches* on valve and damper operators for providing a selected sequence of operations.

Auxiliary control equipment for pneumatic systems includes:

- *Air compressors* and accessories, including driers and filters, to provide a source of clean, dry air at the required pressure. An air compressor was shown in Figure 51.
- *Electro-pneumatic relays*, which are electrically actuated air valves for operating pneumatic equipment.
- *Pneumatic-electric relays*, which are actuated by air pressure to make or break an electrical circuit.
- *Pneumatic relays*, which are actuated by the pressure from a controller to perform numerous functions. They may be divided into two groups:
  1. Two-position relays, which permit a controller actuating a proportional device to also actuate one or more two-position devices. They are also used for automatic switching operations.
  2. Proportional relays, which are used to reverse a proportional controller, select the higher or lower of two pressures, average two or more pressures, respond to the difference between two pressures, add or subtract pressures, amplify or retard pressure changes, or perform other similar functions.

- *Positioning relays*, which are devices for accurately positioning a valve or damper operator in response to changes in controller pressure. They monitor the position of the operator and the controller pressure. Whenever the two are out of balance, these relays will change the pressure to the operator until balance is restored.

- *Switching relays*, which are pneumatically operated air valves for diverting air from one circuit to another, or for opening and closing air circuits. Figure 58 is a relay that can be used for either positioning or switching.
- *Pneumatic switches*, which are manually operated devices for diverting air from one circuit to another, or for opening and closing air circuits. They may be of the two-position or multiple-position type.
- *Gradual switches*, which are proportional devices for manually varying the air pressure in a circuit.

Auxiliary control devices common to both electric and pneumatic systems include:

- *Step controllers*, for operating a number of electric switches in sequence by means of a proportional electric or pneumatic operator. They are commonly used to control the steps to increase refrigeration capacity. Controllers may be arranged to prevent simultaneous starting of compressors or to alternate their sequence to

equalize wear. They may also be used for sequence operation of electric heating elements and other equipment, in response to the demands of a proportional controller.

- *Power controllers*, for controlling electrical power input to resistance-type electric heating elements. The controlled device may be a variable autotransformer, a saturable core reactor, or a solid-state power controller. They are available with ratings for either single- or three-phase heater loads. Normally, they are arranged to regulate power input to a heater in response to demands of proportional electronic or pneumatic controllers. However, solid-state controllers may also be used in two-position control modes.
- *Clocks or timers* are used for turning apparatus on and off at predetermined times, for switching control systems from day to night operation, and for other time sequence functions.

10.3. MECHANICAL SYSTEMS CLASSIFICATION

In this section we will deal with general classifications of control systems. Specific systems or pieces of equipment will be discussed elsewhere. Control applications in this section are on a functional basis and apply to both on-site built-up systems and to unitary systems that are factory-built. These functions, seen in Figure 59, are separated into three divisions: primary source apparatus, distribution systems, and terminal equipment.

10.3.1. Central Fan Systems

Since there are numerous control combinations for central fan systems, the following sections will deal with only a few of the more typical ones such as for controlling outdoor and return

FIGURE 58. Pneumatic snap-acting relay. For use in HVAC systems to transform a gradual air pressure change from a controller to a positive (two-position) pressure change at a pneumatic valve or damper operator. It can also be used to divert a supply line to one of two branches by snap action. (*Courtesy Honeywell, Inc.*)

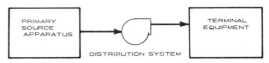

FIGURE 59. Three general divisions of control systems.

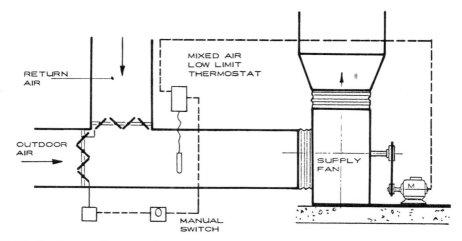

FIGURE 60. Manual adjustment of outdoor air quantity with low limit control of mixture temperature.

air dampers, preheat coils, heating coils, cooling coils, and humidity control. The complete control system is obtained by combining the control of each component into an integrated set of controls. Proper consideration must be given to the interrelation of the several parts the sequence of operation, and the design parameters of the whole system.

Outdoor Air Damper Control. Outdoor air for ventilation or natural cooling should be available whenever the fan is running, and is usually provided by means of an outdoor air damper which opens to a minimum position when the fan is started. The damper may be further opened by automatic control (see Figure 60). Another ventilation method is by the use of a minimum outdoor air damper which opens fully when the fan is started. A maximum outdoor air damper may also be provided for natural cooling, as seen in Figure 61. Recirculated air and exhaust air dampers must be synchronized to operate with the outdoor dampers. When the system geometry permits, mechanical linkages

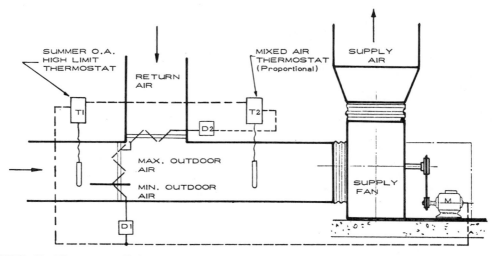

FIGURE 61. All season ventilation control with fixed minimum outdoor air plus additional outdoor air when practicable.

should be used. Outdoor air and exhaust air dampers should close when the fan stops.

In Figure 60, the outdoor air damper is controlled by a manual switch. A low-limit mixed air thermostat is desirable for protection against freezing air. Figure 61 shows an all-season system with a fixed minimum outdoor air damper which opens whenever the fan is running. A maximum outdoor air damper is automatically controlled by a mixed air thermostat, and a high limit outdoor air thermostat returns the system to a minimum outdoor air condition. Whenever the outdoor air is above the setting of the mixed air thermostat and below the setting of the high limit, 100% outdoor air will be used. Since this provides essentially free cooling, it has been called the economy cycle.

For improved economy, the high limit may be a wet-bulb or enthalpy thermostat to provide a better indication of total heat in outdoor air. Also, the use of cool summer night air can store cool air in the system to be used as outdoor air temperature rises. This system is shown in Figure 62. Also shown is mixed air

reset, whenever possible, from zone demand for maximum energy conservation.

10.3.2. Control Interlocks for Fire Safety

Control interlocks are especially important in large fan systems for high-rise buildings. Design concepts on this subject are in a state of change and vary from city to city. The information given here covers both the old static approach and the new dynamic approach to fan system interlock for fire and smoke control. Any system used must be approved by the local authorities. It may be that some jobs will be done best by use of the dynamic approach in the first phases of a fire. This would be followed by manual control by fire department personnel who would return the system to the static approach as they see fit.

The static approach requires smoke or thermostatic detectors in supply and exhaust ducts that will shut off the fan system and close the smoke dampers to restrict the circulation of smoke when activated. If the fan system is greater than 15,000 cfm, smoke detectors are

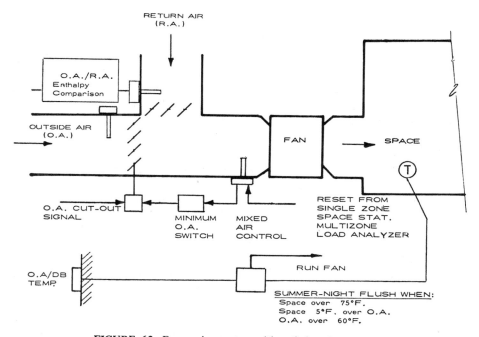

FIGURE 62. Economizer system with enthalpy changeover.

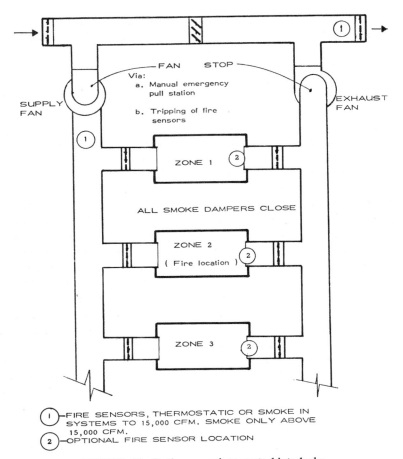

FIGURE 63. Static approach to control interlocks.

more commonly used than thermostatic detectors. Figure 63 is a schematic of the static approach to control interlocks.

A new dynamic approach to smoke controls deals with buildings of multiple zones where evacuation is impractical (such as hospitals) or excessively time-consuming (such as high-rise buildings). In these cases, the fire zone air supply is cut off, but its exhaust continues either through the central system which goes 100% supply and exhaust, or by a separate exhaust system. Figure 64 illustrates the dynamic approach to control interlocks.

10.3.3. Preheater Coil Control

The function of a preheater is to temper the outdoor air to prevent freezing beyond the coil. Care must also be taken to avoid freezing in the coil itself. The proper selection, installation, and sizing of steam traps are therefore, very important to assure rapid elimination of condensate to prevent freezing. Other important features of a preheater are: vacuum breakers to assure condensate removal; provision for rapid air elimination from the coil; and the use of vertical rather than horizontal tubes wherever possible, to prevent nonflowing water.

Figure 65 shows an outdoor thermostat, usually set a 35°F controlling a two-position preheat coil valve. Since this arrangement provides full flow of the heating medium (steam or hot water), the possibility of coil freezing is minimized. However, there is no control of the downstream temperature, and, under certain conditions, this may become excessive.

Control of an electric preheat coil is similar

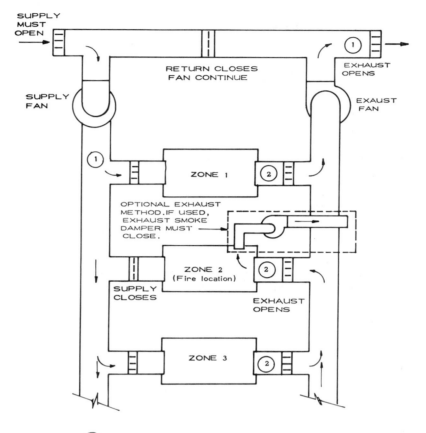

FIGURE 64. Dynamic approach to control interlocks.

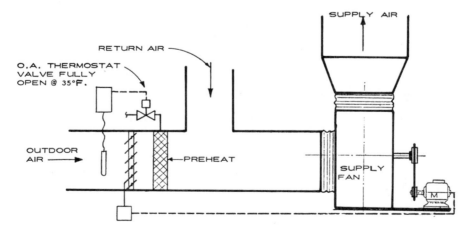

FIGURE 65. Control of preheat coil from outdoor air.

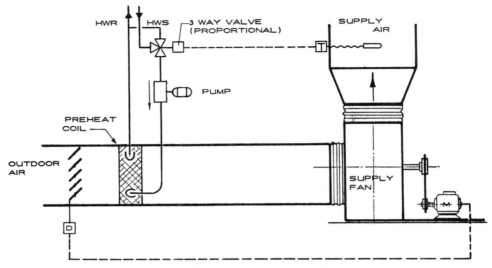

FIGURE 66. Control of hot water preheat coil to prevent freezing.

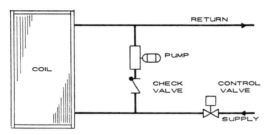

FIGURE 67. Pump and control valve arrangement for a preheat coil.

to that illustrated in Figure 65, except that time proportioning or proportional action control with a wide throttling range is recommended over two-position action. This limits energy consumption to that based on outdoor conditions and maintains better downstream temperature control. The electric preheat coil would normally be proportioned from full off at 35° or 40°F to full on at design conditions.

Figure 66 is a hot water preheat control

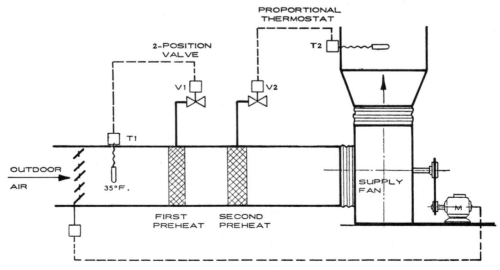

FIGURE 68. Two preheat coils in series.

system providing both protection against freezing and accurate control of down-stream temperatures. The circulating pump provides a constant water flow through the coil, at a velocity sufficient to prevent freezing (2 to 3 ft/sec, depending on the approach air temperature). The three-way valve admits enough hot water to provide downstream air temperature control. This system is especially useful with 100% outdoor air, or in systems where stratification of mixed air cannot be avoided. A flow switch should be added to stop the fan and close the outdoor air damper in case of pump failure. The pump may be started and stopped by an outdoor air thermostat and should automatically restart after power failure.

An alternate pump arrangement for a preheat coil using primary-secondary pumping and a three-way valve is shown in Figure 67. The check valve will permit hot water circulation if the pump fails during a call for heating from the discharge thermostat.

Preheat control of a 100% outdoor air system using steam coils is shown in Figure 68. The first coil is provided with a two-position control to prevent freezing; the second is provided with proportional control for downstream temperature control. The first coil must be selected for a very low-temperature rise to prevent overheating.

10.3.4. Heating Coil Control

All of the control systems shown in Figures 66, 67, and 68 can be used for heating and reheating coil control. In particular, where very close control of downstream temperature is required, circulating pump systems should be used.

Heating coils, when used for tempering air for ventilation, can be controlled by a proportional type of insertion thermostat, preferably located in the fan discharge where the air is least stratified. Averaging type sensing elements are recommended for mixed air or discharge control unless a single-point control is specified. This thermostat operates the heating coil valve(s), or face and bypass dampers, to maintain a constant discharge air temperature. When face and bypass dampers are used, it is good practice to have a valve on the coil which closes as the face damper closes. The valve closing will prevent overheating due to damper linkage or heat picked up by air sweeping over the exposed coil face.

Figure 69 shows a typical control arrangement for a heating coil when the fan system is used for heating and ventilation. Here, a room or a return-air thermostat controls the heating coil valve until the temperature of the conditioned space is satisfied. A discharge duct thermostat then assumes control of the valve to

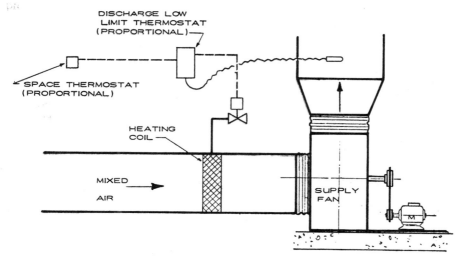

FIGURE 69. Control of heating coil from both the conditioned space and discharge air using a valve.

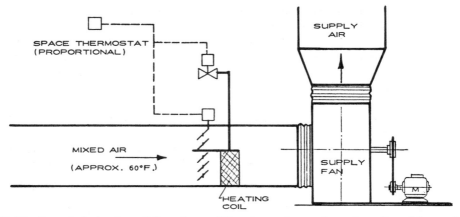

FIGURE 70. Control of a heating coil from the conditioned space temperature using valve and bypass dampers.

maintain the desired minimum air discharge temperature. With this system, no portion of the air entering the coil should be less than 32°F.

An optional plan for this type of control is the master–submaster arrangement. With this arrangement, the space thermostat, acting as a master controller, resets the temperature sensed by the submaster element in the discharge air. This arrangement will result in more stable room temperature control for applications where a large variation in load occurs, or for systems with large coil capacities. It is also

used to control gas-fired heaters or electric heating coils in a two-position or on–off manner. For light load conditions, on demand from the space thermostat (master), the discharge element (submaster) cycles the heating element and maintains a closer control of the discharge temperature than a system with only space temperature control and a two-position high limit. With any type of master–submaster control, the maximum amount of reset should be limited if such unsafe conditions (excessively low- or high-discharge temperatures) could re-

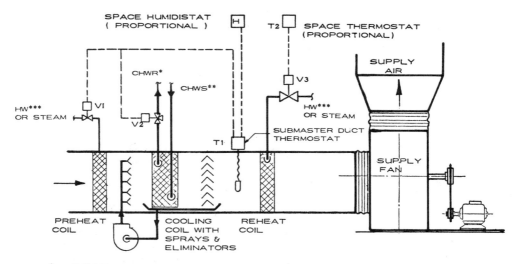

*   CHILLED WATER RETURN

**   CHILLED WATER SUPPLY

***   HOT WATER

FIGURE 71. Dew point control of supply air.

sult from extreme load conditions, either during start-up or manual space element if it is the type with adjustable set point.

Figure 70 shows an arrangement using face and bypass dampers on the heating coil, and two-position control of the valve. It should be noted that without a low-limit controller, the entering air temperature must be the desired minimum discharge air temperature. If it is possible that entering air could be below the minimum desired discharge temperature, a low-limit discharge control similar to that shown in Figure 69 should be used. With face and bypass application, it is particularly important that the low-limit controller sense the temperature of the mixture of air passing through the coil and the bypass section. This is best sensed downstream of the fan. Averaging type sensing elements should be used if there is any possibility that thorough mixing will not take place at the low-limit sensing element. If the air temperature entering the coil could drop below $32°F$, a controller located before the coil should open the coil valve wide.

Where an electric heating coil is used, a power controller or step controller may be substituted for the valve shown in Figures 69 and 71. The face and bypass method (Figure 70) isn't recommended for electric heat. On ventilating or makeup air applications, the discharge temperature is reset as a function of the outdoor temperature. For larger coils, time-proportional or solid-state control is preferred over sequence control because of difficulty in preventing air stratification unless the coil is specially constructed. Where air volume varies, such as in the hot deck of a multizone system, proportional or solid-state control and special coil designs are necessary to prevent overheating of elements at low air volumes. A differential pressure switch (as well as fan interlock) is desirable as a safety interlock.

10.3.5. Cooling and Humidity Controls

The simplest cooling control system involves a coil with a two-way valve controlling the flow of chilled water or a solenoid valve controlling the flow of refrigerant in response to a room or duct thermostat. Figure 72 shows a three-way valve system which gives the best response of the three alternatives. A circulating pump may be used to further improve control response. See Figures 66 and 77. This control system does not provide humidity control, especially at light sensible loads.

A simple system which allows improved response with a two-way valve is shown in Figure 73. The face and bypass dampers provide close control of supply air temperature. When this system is used with a direct expansion coil, it is necessary to keep the suction pressure above

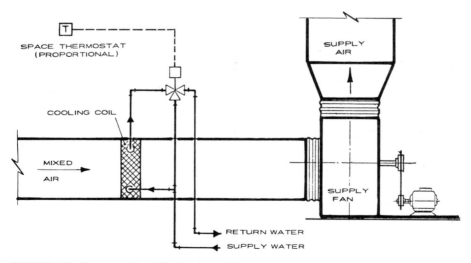

FIGURE 72. Control of the chilled water coil from the space temperature using a mixing valve.

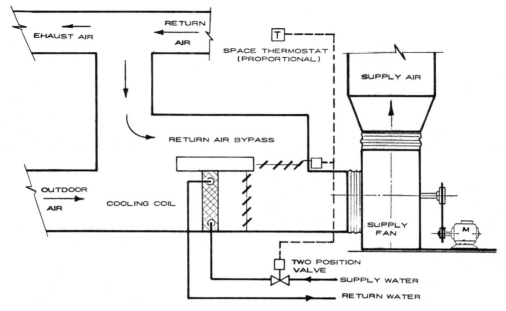

FIGURE 73. Control of chilled water coil from space temperature using return bypass damper and valve.

freezing to prevent ice formation on the coil. This may be done by means of a back pressure regulating valve or pressure-operated unloaders (a proportional relay) on the compressor.

Close control of the dew point may be obtained by spraying the cooling coil with water or brine. Figure 71 shows such a system. Since the air leaves the coil in essentially a saturated

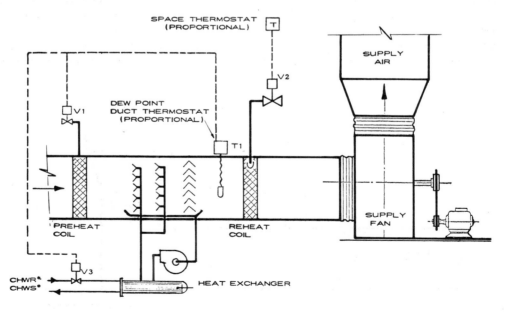

* CHILLED WATER SUPPLY
** CHILLED WATER RETURN

FIGURE 74. Air washer control of space temperature and humidity.

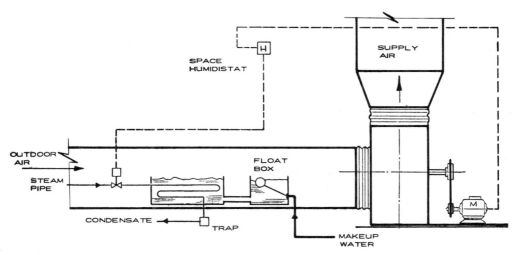

FIGURE 75. Evaporative pan humidifier.

condition, the downstream thermostat, frequently called a dew point controller, can be a dry-bulb instrument. This is more reliable and easier to maintain than a wet-bulb instrument. The thermostat may be a submaster controller reset by a humidistat in the conditioned space to provide better control of space humidity. A preheat coil may be needed to bring the mixed air to a wet-bulb temperature corresponding to the dew point of the desired space conditions.

A reheat coil is necessary to provide space temperature control. This will provide accurate relative humidity control, provided space dry-bulb temperature is maintained. A direct expansion coil is usually limited to small systems. To vary relative humidity, the setting of the duct thermostat must be changed.

A similar system uses an air washer instead of a coil (see Figure 74). Preheat is provided or spray water is cooled as required by dew point

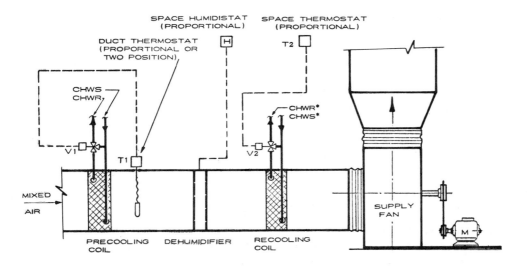

* CHILLED WATER SUPPLY
** CHILLED WATER RETURN

FIGURE 76. Chemical dehumidification.

duct thermostat T 1 to maintain a constant leaving dew point temperature. Space thermostat T 2 provides reheat control to maintain space conditions by dry-bulb temperature. Sprayed coil systems, while providing very accurate control, are somewhat expensive to operate and cannot provide relative humidities above 60%. For more economical operation or for very high relative humidities, other humidiification methods are required.

Figure 75 shows an evaporative pan humidifier with a steam heater. Steam supply is controlled by a space or duct humidistat. An interlock is provided to shut off the heater when the fan is stopped. Hot water or electrical heating systems may also be employed to provide heat to the humidifier. Steam jet humidifiers are controlled in a similar manner. Atomizing water spray humidifiers use a two-position valve controlled by the space humidistat. For slinger atomizers, which are spray nozzles, the motor is started and stopped by the humidistat.

Figure 76 shows a chemical dehumidifier, used for obtaining low humidities. The control of the dehumidifier will vary with the manufacturer. The recooling coil is necessary because all chemical dehumidification processes add heat.

10.3.6. Static Pressure Control

Since fans are selected to produce desired results for a certain system characteristic, any change in the system characteristic will change the output of the fan. Such changes occur, for example, on single-duct systems in which the volume of air delivered is varied, and in double-duct systems in which quantities of air are shifted from one duct to the other to obtain final space control.

10.3.7. Variable Air Volume Control Systems

In variable volume systems, as terminal units reduce the air flow, static pressure in the duct rises if no corrective action is taken at the fan. At lower air flow, the increased static pressure developed by the fan is wasted energy and results in increased noise at the points of pressure reduction. Inlet vanes, variable speed drives, or variable pitch blades are generally used to reduce the static pressure to an acceptable level and conserve energy. These devices are generally controlled by a static pressure controller located near the end of the duct system or at a low-pressure point in the ductwork. Such a controller is shown in Figure 77. Separate controllers should be used if major variations in flow and pressure occur in different zones with corresponding changes in load. High-static limit controls are recommended at the fan discharge to override the remote sensor in the event of a fire, or for a smoke damper closing between the fan and the remote sensor. Static pressure control for a variable volume system is shown in Figure 77.

Return Fan Control for Variable Air Volume Systems. Building pressure will vary if return fan volume does not vary with supply fan volume. Three common methods have been used to prevent this from occurring.

1. The Building Static Control Method senses differential pressure between a typical room and the outdoors, and increases the volume of air handled by the return exhaust fan as the building pressure increases. Difficulty in determining which space can be regarded as a typical room and obtaining a stable outdoor reference, limits the application of this control method to a special application where the following two methods may not be practical.

2. The Open-Loop Control Method, shown in Figure 78, utilizes pilot positioners on the supply and return fan controls to sequence the operation of the return fan with the supply fan. This system requires close coordination with the balancing contractor to properly adjust the pilots. If system loads vary significantly between major zones served by the supply fan, it is possible that the resistance in the return system will not vary proportionally with the resistance in the supply system. The open-loop method of control does not sense the resistance variance between supply and return systems, and building pressure may vary when major load variation occurs.

3. The Closed-Loop Control Method, shown in Figure 79, senses changes in the volume of air delivered by the supply fan and resets the re-

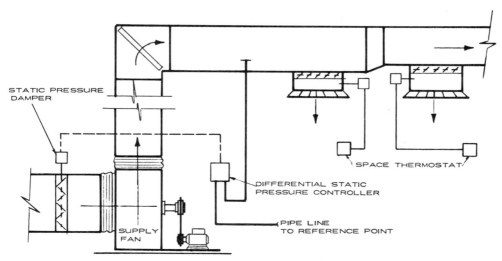

FIGURE 77. Static pressure control in a variable volume system.

turn fan by means of a receiver controller that has a submaster input proportional to return fan flow. By controlling flow, the effects of different fan or vane characteristics are eliminated. The controller can be set to the difference in cfm required between supply and return fan to maintain building pressurization and accommodate auxiliary exhaust systems. If this flow difference between supply and return is also the minimum amount of outside air required by the system, the fan controllers will maintain the minimum rate regardless of variations in outside and return damper settings. Multiple-point pitot tubes or flow measuring stations should be used for sensing velocity pressure at the fans on variable volume systems since the velocity profile will vary with flow, and single-point pitot tubes are inaccurate for total flow.

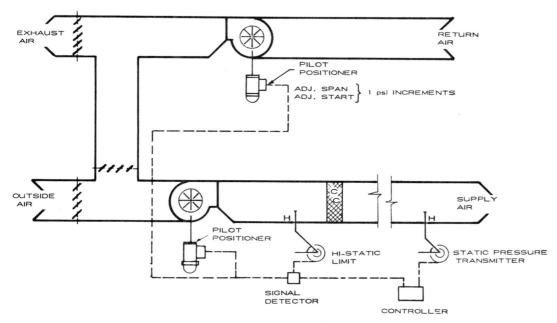

FIGURE 78. Open-loop fan volume control.

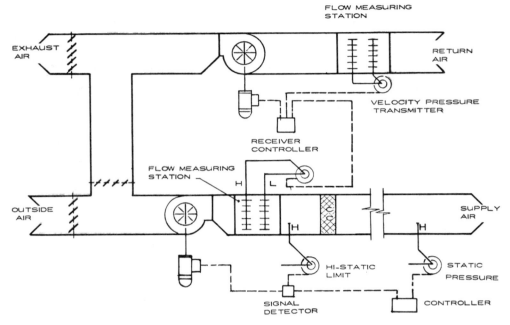

FIGURE 79. Close-loop fan volume control.

Outdoor Air Volume. On large, variable air volume systems where the flow rate of the return fan is reset based on supply fan volume, and the minimum temperature of outdoor air is 40°F or greater, successful control of mixed air temperature is obtained by modulating only the return and relief damper. The outdoor air damper for this type of system can be operated in a two-position manner from a supply fan interlock.

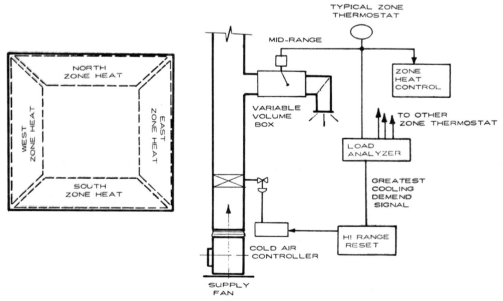

FIGURE 80. Mixed loads with demand reset.

Zone Heating–Cooling Coordination. Compromised zones, i.e., areas where heating and cooling are separately controlled, occur frequently in variable-volume cooling systems where perimeter radiation is the source of heating. Coordinated controls measure the heating and cooling requirements, then reset supply temperatures of both heating and cooling sources. Figure 80 shows an example of mixed zones of variable volume and heating with load demand reset to both supply temperatures.

10.3.8. Multizone Control

Variations of multizone control, Figure 81, occur in three system areas; room control, hot and cold deck control, and mixed air control. These are described below.

Room Control Variations. In the Zone Direct Control, the room thermostat directly controls the zone dampers that proportion the hot and cold air. Usually, this is sufficient for satisfactory control.

For the Zone Discharge Control, the space thermostat resets a discharge controller in each zone. This system will overcome instability control problems due to high-temperature differ-

ence, ΔT, between decks and/or long runs of ductwork.

Hot and Cold Deck Temperature Control Variations.

Deck Discharge Control. This is the simplest form of control where the maximum hot and cold deck temperatures are maintained at all times. The chance of room control being unstable is highest with this system due to the high ΔT between decks. The economy of operation is poor since the system is running at highest energy losses due to mixing of hot and cold air.

Reset of Deck Discharge Control. As with discharge control of room temperature, a discharge controller maintains the deck temperature. Reset is by a signal from a source that will predict load. This can be an outside air sensor, manual reset, deck air volume, return air temperature, or the room thermostat calling for the most heating or cooling. Most installations will have reset of the hot deck, although added economy will result from resetting the cold deck as well.

Control from the zone calling for the most heating or cooling is more economical since it allows the deck to reduce the energy delivered to less than the most severe load conditions.

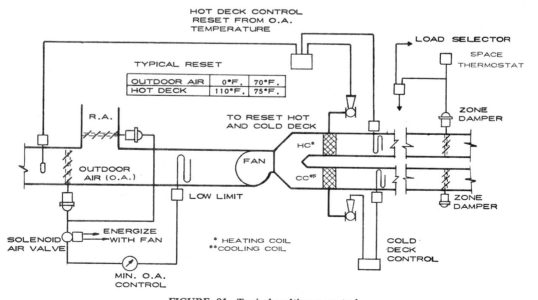

FIGURE 81. Typical multizone control.

The ΔT between decks is reduced. One caution on room reset of discharge control is that if the system designer undersizes the air quantity to one zone, it will be at full demand for all or most of the time. This will result in the deck temperature being held at an extreme value, and the economy of operation will be reduced.

Direct Deck Control from Greatest Demand Space Thermostat. In this system, the space thermostat that calls for the most heating directly controls the heating valve or relay. The space thermostat that requires the most cooling directly controls the cooling valve or relay. This system also maintains minimum ΔT between hot and cold decks, thereby improving economy of operation, similar to the greatest demand zone thermostat resetting a deck discharge control. This system, which also uses direct control of all the zone dampers, is suitable where there is no instability caused by high ΔT between decks or long runs of ductwork. In some cases, deck discharge control may be required because of the high gains of heating coils or cooling coils, especially at low-deck airflow conditions.

Mixed Air Control Variation. The basic system uses a thermostat that controls the mixed air at a constant temperature. This system will not reduce outdoor air to a minimum during the cooling season, so it is only practical for systems with no mechanical cooling.

The Economizer System is a mixed air control with an outside air cutout that reduces this air to a minimum when it is not a source of cooling. For some jobs, the mixed air temperature is the final low limit; on other jobs, the minimum outside air is the final limit.

An Economizer with Reset System is basically the same as the economizer system except that the mixed air is reset from outside air or from maximum space cooling demand. This allows the system to provide a higher cold deck temperature when cooling loads are at a minimum.

The Economizer with Direct Mixed Air Control from Space System allows the space thermostat which calls for the most cooling, to directly control the outside air dampers. A low-temperature limit is used to prevent the

cold deck from becoming too low, and a minimum position relay ensures that a minimum outside air quantity is provided regardless of the call for cooling. During the summer, an outside air changeover thermostat will switch the unit to the fixed minimum outside air setting for ventilation and minimum load to the mechanical cooling system. This system is generally used with reset of cold deck from the space thermostat, in which the outside air damper is sequenced with the cooling media so that all the available "free" cooling is used prior to the introduction of any mechanical cooling.

10.3.9. Temperature Control in Hot Water Systems

Hot water heating systems permit the use of relatively simple automatic temperature control which reduce operating costs while providing comfortable temperatures. For installations where different supply water temperatures are not required simultaneously, and where the boiler is not used to heat domestic hot water, outdoor reset control of the supply water temperature may be obtained by direct control of the burner. This control is from a thermostat, located in the boiler water, which is reset in accordance with outdoor temperatures. Separate high and low-limit controls should be provided. The water temperature controller should not be set below 140°F in order to prevent condensation of flue gases. See Figure 82 for a system layout.

In systems using converters (heat exchangers) for hot water supply, a controlling thermostat actuated by the temperature of outlet water modulates the heat source to the converter. The set point of this supply water thermostat can be readjusted in accordance with outdoor temperature.

In some systems, a boiler or converter supplies water at constant temperature, and three-way mixing valves are used to supply water at different temperatures for various uses, or to multiple zones. For these cases, a three-way mixing valve, under control of a thermostat, located in the valve discharge, blends boiler water with return water to provide supply water at the desired temperature. The control setting of this thermostat can be reset in accordance with

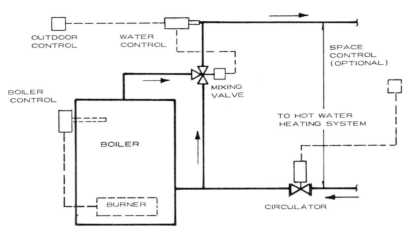

FIGURE 82. Control of hot water heating without domestic hot water service.

outdoor temperatures. Since low flow through the boiler can cause damage, it is recommended that a minimum flow by means of pumping system, be used.

On systems using outdoor-reset supply water and lowered night or weekend temperatures, provision should be made for warm-up on restoration to day or normal operation. This can be accomplished by a programmed control, a warm-up space thermostat, or some similar means of providing higher than normal water temperatures during the warm-up period.

Precautions should be taken to prevent the freezing of water in a hot water heating system whenever design outdoor temperatures are lower than 20°F. This is particularly true if the system is to be operated on an intermittent basis, as in schools where normal operation is not required on weekends or during vacation periods. In selecting a means of freeze protection, consideration should be given to the type of systems as well as outdoor temperature. Water freezing in the mains can be minimized by using an outdoor thermostat to provide continuous circulation of the water whenever the outdoor temperature drops below 30°F. Another arrangement uses an interval timer and outdoor thermostat for intermittent cycling of the circulating pump whenever the outdoor temperature drops below some predetermined temperature. A thermostat in the return water at the boiler or other heat source can be used to start the burner if the temperature of the return water drops to about 40°F.

Hot Water From Refrigeration Condensers. Split or double-bundle condensers may be used to capture heat rejected from refrigeration systems to heat water. Such a system must control the heating supply and assure heat rejection capability. Normally, other mechanical equipment must be available to supply heat when rejected heat is not available. Figure 83 shows control of a double-bundle condenser in sequence with an auxiliary heat source and a conventional condenser and evaporative cooling tower. Water–air or air–air heat exchangers can extract heat from exhaust air to preheat incoming outside air. The primary control requirement here is to shut off the extra heat when it is not needed. Figure 83 also shows a schematic diagram of multiple chiller plants automatically put on line by load as measured by chilled water differential temperature.

Automatic Plant Capacity Lineup. The method of selecting the most efficient source of cold water at any particular time can be applied to large plants with groupings of similar equipment. Various combinations of equipment will best meet different load requirements and by using a straightforward schedule to match components to the load. Figure 84 shows multiple chillers automatically put on line by load as measured by the chiller water differential temperature. For 30% of the load, one chiller is operated. Between 30 and 60%, two chillers carry the load. All three chillers are used above 60% of the load.

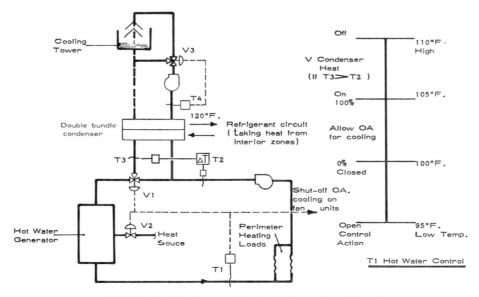

FIGURE 83. Heating supplemented with condensed heat.

10.3.10. Zone Control of Steam and Water Systems

Most zone control systems are designed to supply heat to a zone at a rate equal to the heat loss. This rate of heat input may be established by room thermostats, by controllers responsive to outdoor conditions, or by a combination of both. The outdoor controller can respond to temperature, or to solar radiation, and wind velocity and direction. Features which may be added to the basic zone control system include:

1. Means for maintaining a lowered temperature for economy, or completely shutting off the heat at night or at other times when the zone is unoccupied. This may be accomplished manually or with a clock.
2. An arrangement providing rapid warm-up following a period at the nonoccupancy temperature. During this warm-up period, full heat may be admitted to the system or the heat input may be controlled at a higher than normal rate. The warm-up

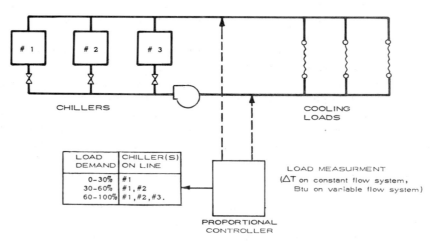

FIGURE 84. Optimizing of chiller start-up.

may be accomplished by a manual switch or be part of the clock program. A thermostat located within the zone may be arranged to terminate the warm-up period independently of the switch at the desired temperature.

3. A manual switch to provide either full heat or no heat independently of other controls and to permit manual selection of the various clock functions.
4. Means of shutting off the heat completely when the outdoor temperature reaches a point where heat is no longer needed.
5. A high-limit thermostat in the zone to shut off the heat completely when the zone becomes overheated.

Steam: Zone Distribution.   In continuous flow steam systems, the quantity of steam supplied to the zone is varied in accordance with the demands of the control system. To obtain equalized distribution of the steam, metering orifices are generally required on the inlets to all radiators or convectors. Supplementary controls for low heat output are often desirable. The following two arrangements are common:

1. Varying the difference in pressure between the supply and return results in the partial filling of heating units in accordance with demand. This method can be used on atmospheric or vacuum return systems.
2. Varying the pressure in the system while maintaining a constant differential pressure between the supply and return results in the heating units being filled with steam at a temperature proportional to demand. A vacuum pump maintains a vacuum slightly below the supply pressure. Vacuums as high as 20 in. Hg during mild heating weather are common.

In intermittent flow systems, steam at full pressure is supplied intermittently to the zone. The length of the on and off periods are varied in accordance with the demands of the control system. Two common arrangements are:

1. The use of a timing device controlled by an outdoor thermostatic element. It can vary the on-off periods of the boiler operation or the opening and closing of a steam valve as a function of outdoor temperature.
2. A control that responds to an outdoor element and one attached to a radiator or convector. This type of control varies the length and frequency of the on-off intervals in such a way that the radiator or convector temperature is varied in accordance with outdoor temperature.

For proper operation of zone control equipment, the heating system must be carefully designed, installed, and maintained. Vents, traps, vacuum pumps, and valves must be inspected, repaired, or replaced when required. Piping must be of adequate size and must be graded and dripped. Return piping must be vented; any pockets or lifts must be removed.

Water: Zone Distribution.   The usual method of obtaining zone control of hot water systems is by varying the temperature of water supplied to the zone inversely with the outdoor temperature. This is accomplished by a submaster thermostat in the zone supply line, which is readjusted by either an outdoor master thermostat, or by a single controller having one measuring element in the supply line and another outdoors. On single-zone installations, the controller may actuate the burner directly, as shown in Figure 85. If the boiler is also used for domestic hot water, the controls may be arranged as in Figure 82. In this case, to obtain the desired zone water temperature, the boiler is held at a fixed temperature, and the zone controller operates a valve to proportion hot water from the boiler with cooler return water. On multiple-zone installations, the boiler is held at a fixed temperature, and each zone is controlled by a separate valve and outdoor-compensated controller. Each zone should be equipped with a separate pump because the return water from some zones is warmer than the supply water for other zones. By using individual zone pumps, the designer can separate the returns and thus prevent overheating.

To compensate for variations in internal load, the zone control can be modified by a

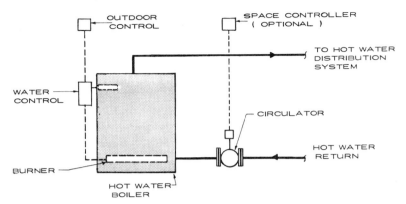

FIGURE 85. Zone control of hot water system by varying the water temperature inversely with the outdoor temperature.

room thermostat in each zone, which either starts and stops the zone pump, modifies the relationship between outdoor temperature and water temperature, or operates an automatic valve in the zone supply or return line. Each zone may be further subdivided by providing a separate thermostat and valve for each.

Another zone control method for hot water systems consists of maintaining a constant water temperature and varying the volume of water supplied to the zone by means of a room thermostat. The thermostat either starts and stops the zone pump or operates a two-position valve in the zone supply or return line.

Whenever flow control valves are used in water circuits, it is necessary to employ a relief valve or pressure regulating valve in a bypass around the pump. Alternately, the flow control valves may be three-way, with a connection to overcome head pressure at the pump.

10.3.11. Control of Electric Heat

With the exception of variable autotransformers and saturable core reactors, practically all methods to control the output of electric heating equipment are some form of on–off control. The speed of response and the cycling rate, usually dictated by the system requirements, dictate which control mode will be used. The two basic control modes are two-position action, incremental and/or proportional action.

Two-Position Action.   Two-position action may be direct, indirect (pilot), or timed.

1. Direct mode on–off control systems provide an economical means of controlling temperatures where the load is small enough to be switched directly by the room or remote bulb thermostat. Mild temperature fluctuations and varying degrees of offset can be expected, depending on sensing element location, system time constant, and operating characteristics of the thermostat.

2. Indirect or pilot mode systems are similar to direct on–off control, except that a separate contact device is added to handle the load current. Three systems are commonly used:

   a. A pilot-duty electric thermostat controlling a contactor.
   b. A pneumatic thermostat piloting a pressure-electric switch.
   c. An electronic thermostat piloting a relay or a solid-state switch to control heat input in proportion to load.

The pressure-electric switch may control load current directly, or may pilot a separate contactor to increase load current capability. Control is generally more accurate with piloted systems because of better thermostat operating characteristics.

3. Timed two-position action systems are similar to indirect ones, except that they include heaters or other means of introducing

controlled heat anticipation. This increases cycling rate and reduces temperature overshoot.

Incremental Action and/or Proportional Action. Incremental action may be sequence controlled or time-proportional (vernier control mode).

1. Sequence control utilizes motor-driven switches or several pressure-electric switches for incremental control of electric heaters, until the demand is satisfied. The number of steps depends on such factors as allowable temperature rise per step (a function of accuracy desired), the number of circuits available in the heater, and the load ratings of switch contacts. Contactors (indirect mode) may be used to increase load rating per step of the switching device.

2. The time-proportional mode approaches full proportional control of heater output by establishing a fixed time base, and allowing current to flow into the heater for a percentage of this time base. The time on varies with heating demand as sensed by a proportional output thermostat.

    a. Electromechanical controllers are time-proportional devices that use a mechanical or electronic timer to provide a fixed time base somewhere between 0.5 and 5 min. Load current may be handled directly by built in switching, or through an auxiliary relay. Results approaching full proportional control may be expected with electromechanical controllers on medium to high-mass heaters.

    b. Solid-state power controllers utilize electronic timers to provide an extremely short time base. Load current is handled directly by solid-state semiconductor switching devices. The percentage of on time is varied in response to demand by a proportional electronic thermostat, or a pneumatic thermostat and pressure-to-resistance transducer.

Power regulation of the load may be accomplished in one of two ways. The first, known as phase control, allows the solid-state switch to operate for part of each cycle. The percentage of the ac wave form applied to the load is varied in response to demand.

The second method (burst control) switches short bursts of full cycles on and off. The number of "on" cycles applied to the load is varied from 0 to 100% in response to heating demand. This method is preferred over the first because it is free from radio interference and other transient noise generation.

Solid-state power control permits increased sensitivity of operation, and extremely fast cycling rates which approach true modulation of heater output. Power losses in the controller are minimal, varying from approximately 1% at 120 V to 0.25% at 480 V ac.

The vernier control mode combines sequence control with solid-state power control to provide economical full modulation of larger heaters. The heating coil is divided into several sections, with one section modulated by a solid-state controller, and the rest stepped in by the sequence controller. On a demand for heat, control is placed on the modulated section until it reaches 100% output, at which time the next stage is stepped in. As the increased heat supply is sensed by the discharge thermostat, power to the modulated section is cut back as required. This arrangement allows the modulated section to act as a vernier on each of the other sections as they are stepped in or out. The modulated section capacity should be 25 to 50% larger than the stepped sections to prevent cycling between stages.

10.3.12. Control of Gas-Fired Air Handling Units

Gas-fired air handling units are used for makeup air, heating, heating–ventilating, and heating–ventilating–cooling applications where gas is readily available. Custom-built or packaged units are available for either indoor or outdoor installations. The trend has been to locate this equipment on the roof to achieve lower installed costs and smaller distribution systems. For these installations, care in the selection and location of controls is important because of the adverse effects of wide ambient temperature ranges on both the controls and the devices controlled. Weatherproofed and heated compartments or walk-in areas will minimize these temperature problems.

The gas-fired air handling unit may be of the direct- or indirect-fired type.

Direct-fired units have no combustion chamber. The products of combustion are burned and mixed directly with the makeup air and discharged into the building space. Applications are limited to makeup air heaters to replace exhaust air with tempered 100% outdoor air, alleviating negative pressure problems. Indirect-fired heaters vent products of combustion to the outside atmosphere. The arrangement requires a combustion chamber and heat exchanger. Applications are generally unlimited for heating, ventilating, and cooling units, including gas-fired, single-zone and multi-zone unit applications.

Direct-fired (makeup air) units require a modulating gas valve with a minimum 15:1 turndown ratio. Turndown ratios of 25:1 or higher are recommended. Indirect-fired units may be controlled by on–off, off–low–high, or modulating gas valve (with low-fire setting. Gas-fired units are controlled by control of burner discharge air reset by outdoor air temperature, space temperature, or a manual switch. Other controls are similar to central fan systems discussed previously.

Airflow proving devices (fan motor and 100% outdoor air damper interlock), flame-failure supervision, and high-temperature limit switches are required to shut down or prevent burner operation in the event of abnormal equipment operation. Makeup air units may also require proof of exhaust systems operation. Local codes and regulating bodies will dictate the required safety controls and interlocks.

Centralized control is recommended for remote (rooftop) and multiple units.

10.3.13. Terminal Equipment

Terminal equipment are individual units that are located in or near the conditioned space. They provide the final conditioning required by the space controller. There are five basic control sequences associated with terminal equipment:

1. A single sequence incorporates either a heating or a cooling cycle. With single sequences, there is no changeover, and therefore, heating or cooling modes must be selected initially. Single sequences are referred to as heating only or cooling only.

2. A heating and cooling sequence incorporates both a heating and cooling cycle. A changeover control selects either mode. Each cycle operates independently but they cannot work simultaneously.

3. A heating and ventilating sequence incorporates heating and ventilating cycles. Fresh air cooling is obtained on the ventilating cycle, but is not considered a true cooling mode. Ventilation and heating are simultaneously provided.

4. The heating and cooling arrangement incorporates sequential heating and cooling modes. Selection of cycles is unnecessary. No automatic changeover is required because both cycles are available to satisfy both extremes of the controlled environment.

5. Heating, ventilating, and cooling control incorporates three cycles, all working sequentially as required to satisfy demand conditions.

10.3.14. Radiator and Convector Systems

When individual room control is used, each radiator and convector is equipped with an automatic control valve. Depending upon the size of the room, one thermostat may control one valve or several.

Electric baseboard or convectors may be used either as the primary room heating system, or for supplemental heating at the perimeter to offset building transmission losses. The control scheme used will depend on which of the two functions the radiation is to perform.

Several methods are used for incremental room control:

- A built-in line voltage thermostat in return air.
- A remote wall-mounted line or low voltage, timed two-position thermostat.
- A time-proportional power controller.

Return air control is generally the least accurate, and results in wide space temperature fluctuations. Where space is to be controlled for the comfort of seated occupants, indirect wall-mounted thermostats with timed two-position action or time-proportional power controllers will give the best results.

Where perimeter radiation is being used only to offset skin heat losses (with zone or space load handled separately), outdoor control should be considered. Radiation may be zoned by exposure, as in hot water systems. An outdoor thermostat is located to sense temperature, solar load, or both. Time-proportional power controllers may be piloted from the outdoor thermostat (wide throttling range) to modulate power input to the radiation, in accordance with a preselected schedule.

10.3.15. Design Considerations

Size of Controlled Area.   No individually controlled area should be too large, because the difficulties in obtaining good distribution and finding a representative location for controls become greater as the area increases. Each area must have similar load characteristics throughout. To achieve uniform conditions proper distribution must be provided. This comes about by competent engineering design, careful sizing of equipment and proper balancing of the system. A control can measure conditions only at the point at which it is located. It cannot compensate for variable conditions throughout the area caused by improper distribution or inadequate design. Areas or rooms having dissimilar load characteristics, or different conditions to be maintained, should be individually controlled. The smaller the area, the better will be the system approach to optimum performance and flexibility.

Location of Space Controllers.   Space controllers must be located where they will sense the variables they are to control, and where the condition is representative of the whole area or zone served by the controller. In large open areas that have more than one zone, thermostats should be in the middle of their zone to reduce the effects of the surrounding zones. Three different locations for space temperature controllers are in common use:

1. Wall-mounted thermostats are placed on inside walls or columns. Outside wall locations should be avoided. Thermostats should not be mounted where they will be affected by heat from other sources, such as direct sunlight, pipes, ducts in the wall, convectors, air currents from diffusers, etc. The location should provide ample air circulation, unimpeded by furniture or other obstructions, and should be protected from damage. They should never be placed in corridors, lobbies, or foyers, unless for the exclusive control of these areas.

2. Return air thermostats are used to control such floor-mounted unitary conditioners as induction or fan-coil units and unit ventilators. On induction and fan-coil units, the sensing element can be located behind the return air grille. However, for unit ventilators in classrooms which use up to 100% outdoor air for natural cooling, a forced flow air sampling chamber will house the sensing element.

If return air sensing is used with central fan systems, the sensing element should be located as near as possible to the controlled space. This location will eliminate the influence from other spaces and the effects of any heat gain or loss in the duct. Where supply-return air is combined with lighting fixtures in a ceiling plenum, the return air opening in the lighting fixture can be used to locate the return air sensing element.

Precautions should be taken in locating the sensing element to avoid radiant effects and to assure adequate air velocity across the element.

3. Diffuser-mounted thermostats usually have sensing elements mounted on ceiling supply diffusers and depend on aspiration of room air into the supply airstream. They are used only on high-aspiration diffusers that are adjusted for a horizontal air pattern. The diffuser in which the element is mounted should be in the center of the occupied area of the controlled zone.

10.3.16. Control Application Limitations

To use automatic controls most effectively, knowledge of their limitations, and operating characteristics is essential.

Controls for Mobile Units.   Any control that relies on pressure to operate a switch or valve is subject to variations in the operating point as the atmospheric pressure changes. Normal variations in atmospheric pressure do not noticeably change the operating point, but a change in altitude can affect the control point. This characteristic is especially important when controls are being selected for use in planes and vehicles that are subjected to altitude variations.

In addition to pressure changes, vibration is present that will affect the operation of some types of controls. Controls that are unsatisfactory for mobile applications employ mercury tubes, slow-moving contacts, or have parts that must be kept level. Only those controls that have positive contact action and use mechanical or magnetic detents on their contacts will operate properly.

Explosive Atmospheres.   Since sealed-in-glass contacts are not considered explosion-proof, other means must be provided to eliminate any possible spark in an explosive atmosphere. The designer can use an explosion-proof case to isolate the control case and contacts, permitting only the probe capsule and the capilary tubing to extend into the conditioned space. It is also frequently possible to use a long capilary tube and mount the instrument case in an explosive-free atmosphere.

Because a pneumatic control system utilizes compressed air as its source of energy, safety is inherent in what otherwise would be a hazardous location if electrical controls were used.

Limit Controls.   A very important safety factor to consider when selecting automatic controls is the type of operation needed for control of high limits of temperature, pressure, or current. This control may be inoperative for long periods and then be required to operate immediately to prevent serious injury or damage to equipment or property. Separate operating and limit controls are always recommended, even for the same functions.

10.3.17. Design Precautions

Since several specific precautions have already been discussed, this section will deal with general precautions that should be considered when designing a heating or cooling system.

Electric Heating Coils.   Electrical heating devices differ from those using steam or hot water in that the latter tend to be somewhat self-regulating. For example, if the airflow through a steam or hot water coil is stopped, the temperature of the coil surfaces will rise towards, but not exceed the temperature of the entering steam or hot water. The only heat removed from the steam or hot water will be losses by convection and radiation. Generally no damage to the coil results from no airflow.

The same is not true of electric coils or heaters. If the airflow through an electric coil were stopped and power still applied, the heater would continue to generate its rated heat output. High-element temperatures and unsafe conditions will result. Therefore, it is imperative that control and power circuits be interlocked with heat transfer devices so that on shutdown of the device, power to the heater will be shut off. Where air or water flow is provided mechanically by a fan or pump, the use of flow or differential pressure switches for interlocking is recommended (in addition to electrical interlocks through auxiliary contacts in motor starters). Limit thermostats should also deenergize heaters if safe operating temperatures are exceeded.

Safety Considerations for Duct Heaters.   The current in individual elements of duct heaters will normally be limited to a maximum safe value established by the National Electric Code or local codes.

In addition to an airflow interlock device, an automatic reset high-limit thermostat, and a manual reset high-limit device are usually used to control duct heaters, see Figure 86. The automatic reset high limit normally deenergizes the control circuit. However, if the control circuit has an inherent time delay or uses solid-state switching, a separate safety contactor may be desirable. The manual reset backup limit is arranged to interrupt all current to the heater should all other control devices fail.

Cooling and Dehumidification.   Valve control of coils for sensible cooling often results in

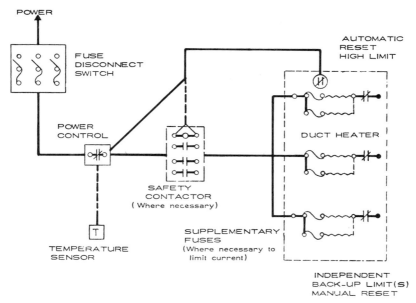

FIGURE 86. Duct heater control.

higher coil temperatures than the dew point temperature, and no dehumidification is accomplished. Bypassing some return air around the coil to control sensible cooling permits the coil to remain at the lowest possible temperature, for maximum dehumidification with higher exit temperatures. A satisfactory relative humidity can be reached without excessively low dry-bulb temperatures, and therefore, less sensible heat will have to be added.

Humidification. Humidifier location should be such that the volume and temperature of the duct air is sufficient to absorb the required amount of moisture discharged by the humidifier. It should be located where air flow patterns are relatively consistent, and normally at least 3 ft away from fans, elbows, splits, transitions, etc. A high-limit humidistat is recommended whenever the maximum design capacity of the humidifier could exceed the ability of duct air to hold this amount of moisture under light load conditions. Humidity controls should be interlocked to prevent humidifier operation on fan shutdown or power failure.

Refrigeration Compressor Operation. On direct-expansion systems, the compressor should not be allowed to operate unless the fan in the conditioning unit is also operating. Likewise, in a chilled water coil installation, the compressor should not operate unless the water circulating pump is functioning. Proper control writing between a fan or pump starter and compressor starter will provide this feature.

Direct Expansion Coils. The control sequence used for direct expansion coils or the manner in which the direct-expansion system is applied should not reduce airflow to the point of frosting the coil.

Outdoor Air Dampers. It is desirable to prevent outdoor air from entering the conditioning unit when the fan is off, particularly during the winter when a freeze-up might occur. This may be accomplished automatically with a motorized damper which closes when the fan is off, or by an electropneumatic relay in the air line to the pneumatic damper operator. In either case, deenergizing the fan closes the outdoor air damper.

Seasonal Changeover. Where changeover from summer to winter is by automatic control, there should be a period of no operation of heating and cooling equipment. Otherwise, serious cycling from one function to the other will occur.

This period of inoperation can be accomplished by a two-position controller with an adjustable temperature differential that senses outdoor temperature, or by a controller sensing the temperature at a representative conditioned space.

Lowered Night Temperature. When temperatures during unoccupied periods are lower than those for periods of occupancy, the proper day and night temperature cycle is established by an automatic timer. Sufficient time should be allowed in the morning to pick up the conditioning load well before there is any heavy load increase in the conditioned spaces.

Leakage Ratings. Dampers are classified by their leakage. Leaking dampers have a number of adverse effects. It may decrease the capacity of a heating or cooling unit, limit the controllability of a unit, or limit the effectiveness of isolating an inoperative air handling unit.

The amount of leakage in a particular application is determined by the type and size of the damper and the differential pressure. The leakage ratings of simple dampers that have no special sealing provisions are in the range of 50 cfm/sq ft of damper when there is a static pressure of 1.5 in. of water across the damper.

If low-leakage dampers are required, particular care should be taken in the damper and duct installation to assure such performance. Low-leakage dampers are usually rated to leak no more than 10 to 20 cfm/sq ft when there is a static pressure of 4 in. of water across the damper.

Multiple Thermostats per Zone. Buildings or zones laid out in a modular concept may be designed for subdividing to meet occupants needs. Until actual subdivision takes place, operating inefficiencies can occur if more than one thermostat is located in a zone. If the system is such that one thermostat can turn on heating while another is calling for cooling, the two zones or terminals should be controlled from a single thermostat until subdivision occurs to prevent such inefficient operation.

10.3.18. Estimated Labor to Install Control Instruments (Mounted in place with connections)

Description	Labor (man-hours)
Flow Instruments	
(panel mounted)	
Electronic flow recorder	25.0
Pneumatic/electric receiver recorder	20.0
Flow counter	3.5
Flow-to-current converter	4.0
Position indicating meter	7.0
Annunciator point	1.2
(local mounted)	
Flow meter	6.5
Flow nozzle	7.5
Electrical flow transmitter	15.0
Pneumatic/differential pressure transmitter	18.0
Electrical/differential flow transmitter	20.0
Pneumatic receiver indicator	12.0
Rotameter	6.5
Indicating and recording rotameter	10.0
Manometer	6.0
Sight glass	3.5
Orifice union	3.5
Orifice flanges	6.0
Flow switch	3.5
Pressure Instruments	
(panel mounted)	
Pneumatic receiver indicator	10.0
Differential pressure switch	3.5
Gas regulator	2.0
Mercoid switch	1.5
Pneumatic recorder	22.0
Electrical indicating recording controller	24.0
Electrical indicating controller	22.0
(local mounted)	
Gauge	1.5
Regulator	2.0
Gauge with limit switch	2.5
Electric operating damper	4.0
Draft gauge	4.0

Description (continued)	Labor (man-hours)
Vacuum regulating valve	5.0
Pressure relief valve, 4 in.	4.5
Electrical transmitter	10.0

Combustion Control
(panel mounted)

Fireye relay control	4.5
Alarm point	1.5
Alarm switch	3.5
Flame safety timer	5.5
Electronic recorder	24.0

(local mounted)

Flame arrester	18.0
Fireye scanning element	3.5
Combustibles analyzer	22.0

Temperature Control
(panel mounted)

Millivolt meter	8.0
Selector switch	3.5
Annunciator or alarm point	2.0
Electronic temperature recorder	22.0
Electronic indicating receiver recorder	30.0
Electronic indicating controller	30.0
Multipoint recorder	38.0
12-Point temperature recorder	35.0
Alarm switch	10.0
Automatic selector with two pneumatic controller	46.0

(local mounted)

Thermometer	2.5
Micro switch	2.5
Thermowell	2.5

Description (continued)	Labor (man-hours)
Temperature	3.5
Thermocouple	3.5
Adjustable air valve	3.5
Solenoid valve	3.5
Thermocouple and protecting tube	5.0
Wall thermostat	5.0
Control valve, 4 in.	5.5
Resistance bulb and thermowell	5.0

Miscellaneous Controls
(local mounted)

Vent or test valve	2.5
Control valve	4.0
Motor operated control valve	9.0
Conductivity cell	9.0
Relay	9.0
Torque alarm station	9.0
Bindicator switch	5.0
Torque switch	4.0
Push button station	3.5
Rotary bindicator controller	26.0

(panel mounted)

Voltmeter	4.5
Manual control station	4.5
Ammeter	4.0
12-Point annunciator alarm panel	30.0
16-Point annunciator alarm panel	30.5
Station potentiometer	8.0
Time totalizer	15.0
Indicating controller	18.0
Operation recorder	20.0

11

CONVECTORS, BASEBOARD, RADIATORS, AND FINNED TUBE UNITS

Convectors, baseboard, radiators, and finned tube units are heating elements used with steam or hot water to supply heat uniformly in a given space. They are designed to produce heat quickly by convection with some small heating effect by radiation and low air motion. These units are strategically located to offset the cold transmission at such locations as: under windows, along perimeter walls, and at entrances.

11.1. CONVECTORS

A convector may be made of cast iron finned sections, or steel, or copper finned tubing that when bonded together create a large amount of secondary surface. This surface has an inlet for air at the bottom and an outlet grill at the top which provides a gravity circulation of air. The air is heated both by convection in a stack effect and by radiation from the heated surfaces. Units are available in different heights, depths, and lengths. They may be installed free standing, wall hung, and semi-or fully recessed into a wall.

Figure 87 shows construction details of a cast iron convector. This unit is semirecessed, as is the unit illustrated in Figure 88, which is a convector with a copper heating element. Piping hook-ups for convectors using low-pressure steam and hot water are given in schematic drawings that appear in Section 11.4, Estimated

Labor to Install Convectors, Baseboard, Radiators, and Finned Tube Units.

11.2. BASEBOARDS

Baseboard units are available in three types: radiant, radiant/convector, and finned tube.

1. The radiant type is generally made of cast iron or steel. Figure 89 shows cast iron baseboards, one with a smooth surface, the other with extended surfaces. Hot water or steam enters the cores of the units producing instant heat which is transmitted to the room by radiation.

2. The radiant/convector type is also made of cast iron or steel. These units are provided with an enclosure which has an opening at the top and bottom, thus, permitting the air to circulate over the wall side of the unit. The wall side of the unit has an extended surface to provide increased heat output. The heat is transferred by convection, with some contribution by radiation.

3. Finned tube or "fin-tube" units are designed for maximum heat output in a minimum of space. Each consists of a finned tube element with an enclosure where most of the heat is transferred by convection. Figure 90 illustrates how baseboard heating panels are installed, both an isometric and sectional view. The principal components are identified and their dimensions are given. Figure 91 is a close-up view

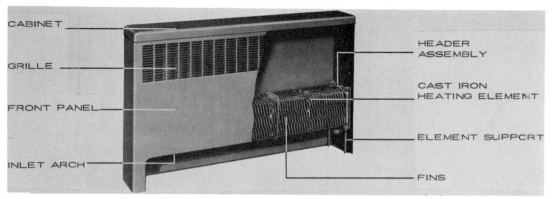

CABINET

GRILLE

FRONT PANEL

INLET ARCH

HEADER ASSEMBLY

CAST IRON HEATING ELEMENT

ELEMENT SUPPCRT

FINS

FIGURE 87. Cast iron convector.

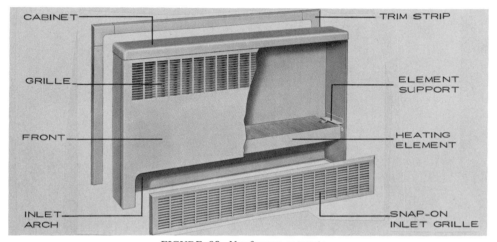

CABINET

GRILLE

FRONT

INLET ARCH

TRIM STRIP

ELEMENT SUPPORT

HEATING ELEMENT

SNAP-ON INLET GRILLE

FIGURE 88. Nonferrous convector.

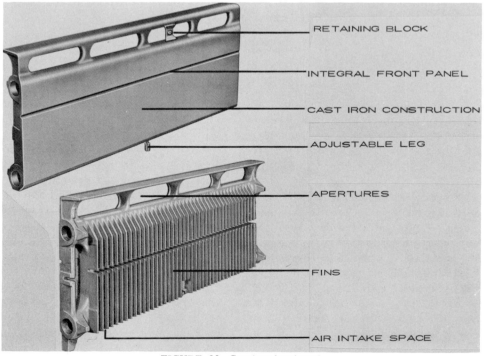

RETAINING BLOCK

INTEGRAL FRONT PANEL

CAST IRON CONSTRUCTION

ADJUSTABLE LEG

APERTURES

FINS

AIR INTAKE SPACE

FIGURE 89. Cast iron baseboard.

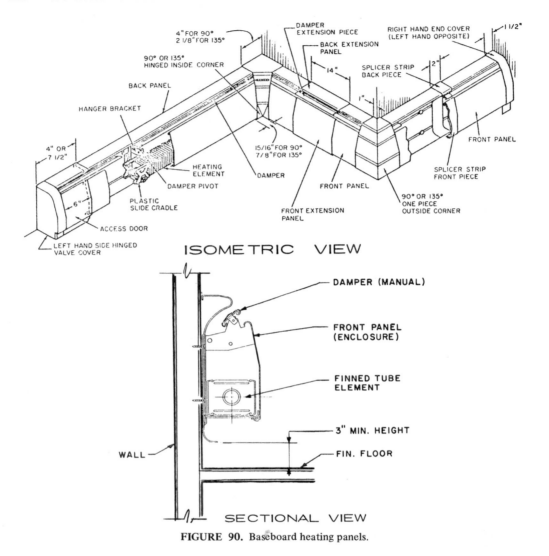

ISOMETRIC   VIEW

SECTIONAL   VIEW

FIGURE 90. Baseboard heating panels.

of the fin-tube in its enclosure. Note the cradle that permits the element to expand and contract as heat is cycled without thermal stress or noise. The upper tube is part of a second fin element.

The basic advantage of baseboard heating units is that they can be placed along a perimeter wall to produce a uniform heat output where the greater cold infiltration occurs. Another advantage is that it is inconspicous and offers minimum interference with the physical room layout while maintaining the heat source near the floor. This application is most often used where cold floors are prevalent such as basementless homes.

11.3. RADIATORS

Today's industry offers a slimmer type of radiator which uses less space than the older bulky models. These units are generally made of cast iron and consist of a series of cored tubes which are bounded together in sections by means of nipples to make the assembly as long as required for specific heat output. Figure 92 shows cast iron radiators for use with hot water or steam. They come in a number of styles and can be free standing or wall-recessed much as are convector units. The diagram near the upper right of the Figure shows how heat is transferred to the room by convection and radiation.

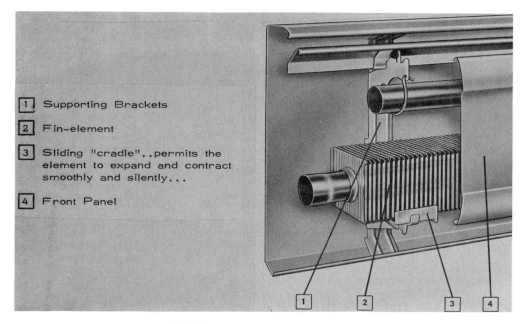

1 Supporting Brackets

2 Fin-element

3 Sliding "cradle"..permits the element to expand and contract smoothly and silently...

4 Front Panel

FIGURE 91. Fin-tube baseboard unit.

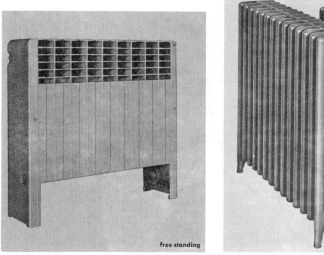

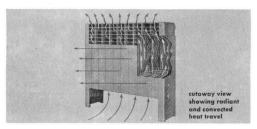

FIGURE 92. Cast iron radiators, hot water or steam.

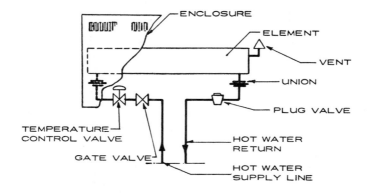

CONVECTOR PIPING HOOKUP

(HOT WATER)

Supply Line Size (Inches)	3/4	1	1-1/4	1-1/2	2
Labor Hookup (Man-hours)	10	12	14	14	17.25

FIGURE 92a. Convector piping hookup, hot water.

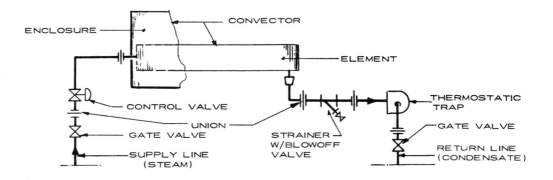

CONVECTOR PIPING HOOKUP

(LOW PRESSURE STEAM)

Supply Line Size (Inches)	3/4	1	1-1/4	1-1/2	2
Labor Hookup (Man-hours)	12	14.50	15.50	16.75	18

FIGURE 92b. Convector piping hookup, low-pressure steam.

11.4. ESTIMATED LABOR TO INSTALL CONVECTORS, BASEBOARDS, RADIATORS, AND FINNED TUBE UNITS

Description	Size (Height in in.)	Labor (man-hours)
Baseboard, radiant cast iron	8	0.50
	10	0.55

Description	Size (in.) Depth × Length		Labor (man-hours)
Freestanding and/or	4	20	2.50
semirecessed convector	4	28	2.50
(with a 24-in. high	4	36	3.00
enclosure)	4	40	3.25
	4	48	3.50
	6	20	2.50
	6	28	2.50
	6	36	3.00
	6	40	3.25
	6	48	3.50
	8	22	2.50
	8	28	2.50
	8	36	3.00
	8	40	3.25
	8	48	3.50
	10	20	2.50
	10	28	3.00
	10	36	3.25
	10	40	3.25
	10	48	4.00

Description	No. of Fins/ft × Enclosure Height, in.			Labor (man-hours)
Fin-tube				
$1\frac{1}{4}$-in. diameter	(s)[a]	32	14	0.75
element, single	(s)	40	14	0.75
row	(c)[b]	48	14	0.80
	(s)	32	20	1.00
	(s)	40	20	1.20
	(c)	48	20	1.50
	(s)	32	24	1.00
	(s)	40	24	1.50
	(c)	48	24	1.50
$1\frac{1}{4}$-in. double	(s)	32	14	1.15
row pipe	(s)	40	14	1.15
element and	(c)	48	14	1.15
enclosure type	(s)	32	20	1.65
	(s)	40	20	1.65
	(c)	48	20	1.65
	(s)	32	24	1.65
	(s)	40	24	1.65
	(c)	48	24	1.65

The labor estimate for the supply line connections to the hot water and low-pressure steam convector units, as a function of the pipe sizes, are given in the following Tables.

Description	No. of Tubes × Height, in.		Labor (man-hours)
Radiators, cast iron	6	32	0.11
	6	25	0.11
	6	19	0.11
	5	25	0.10
	5	22	0.10
	4	25	0.10
	4	19	0.10
	3	25	0.10

Description	Size (in.) Tube Diameter × Enclosure Height		Labor (man-hours)
Baseboard, copper	$\frac{1}{2}$	8	0.25
tube (with	$\frac{3}{4}$	8	0.30
aluminum element)	1	8	0.35
	$\frac{3}{4}$	10	0.30
	1	10	0.35
	$1\frac{1}{4}$	10	0.40

[a] (s) = steel
[b] (c) = copper

12

COOLING TOWERS

In many HVAC applications there is a need to collect and remove the heat generated in the condenser. As previously described, in the condenser there is a thermodynamic exchange of heat, whereby, the refrigerant transfers its heat to the water in the condenser. Where the supply of water is limited or expensive this water must be re-used. For this to occur the water must 'be cooled. Evaporative cooling is the most practical process for this purpose.

Cooling towers, of which several types are available, transfer heat from the water to the atmosphere by exposing as much water surface to the air as is practical in a limited space. The air is moved through the water by natural or forced draft, or the water may be transformed to a spray or otherwise slowed down. All of these techniques are designed for maximum contact between the air and water so that evaporation can take place. The evaporation cools the remaining water which then returns to the condenser to pick up more heat from the refrigerant. Therefore, the cycle continues with essentially the same water except for a small percentage that evaporates in the cooling tower.

Cooling towers are essentially shells that are open at their top and bottom. The heated water from the condenser is pumped to the top of the tower and falls through decks or as a spray to a reservoir below. Air enters the tower at the bottom and moves upward, cooling the falling water. There are two types of cooling towers most widely used today: the mechanical draft and the atmospheric. Depending upon their size they can be located on the ground just outside the building they service, or be mounted on the roof if weight is not excessive.

The selection of each type of tower depends upon many different factors. The designer and manufacturer must be in close consultation to determine the type of tower to be used.

12.1. MECHANICAL DRAFT TOWER

The mechanical draft tower is available in two different configurations: forced draft and the induced draft. In both cases, except for fan location, the structural configuration and operation are the same. The forced draft tower has the fan mounted at the base of the tower and the air is forced .from the bottom up, whereas the induced draft tower has the fan mounted at the top of the tower and the air is pulled upward.

Mechanical towers are more efficient than atmospheric ones and occupy less space for equal capacity. They are also independent of the reduced efficiency that comes about in

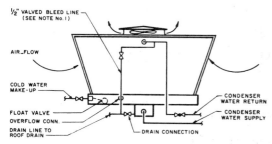

FIGURE 93. Induced draft cooling tower.

atmospheric towers when wind velocities fall below a certain point.

An induced draft tower is shown schematically in Figure 93. The inlet water to the tower, labelled condenser water return, is piped to the top, then falls as a spray through the rising airflow. Note the fan at the top to induce the airflow and the return of cooled water from reservoir at the bottom of the tower.

Commercially available mechanical draft towers are shown in Figure 94. An induced draft unit appears at the top, while a forced draft (fan at bottom) is in the lower portion of the Figure.

12.2. ATMOSPHERIC DRAFT TOWER

The atmospheric cooling tower is an enclosed spray system where the water is released at the top and collected at the bottom. In the flow process the water is in contact with a series of decks or plates which are cooled by the air as it passes horizontally through the tower. The process depends entirely upon the wind for airflow, although the spray nozzles have the tendency of inducing vertical air movement. Figure 95 illustrates the major components of an atmospheric cooling tower, which is sometimes called a natural draft tower. Here, a roof unit receives the warm condenser return water and discharges it as a spray at the top of the tower. The bottom of the tower is a tank that collects the cooled water for return to the HVAC system. A photograph of an actual natural draft tower is shown in Figure 94(b).

(a)

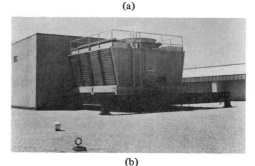

(b)

(c)

FIGURE 94. Commercially available cooling towers. (Manufactured by © Baltimore Aircoil Company, Inc. 1977)

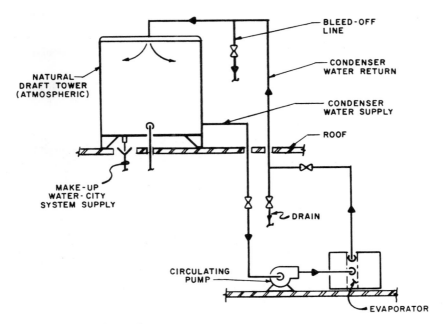

FIGURE 95. Atmospheric (natural draft) cooling tower.

12.3. ESTIMATED LABOR TO INSTALL COOLING TOWERS

Description	Capacity (tons of refrigeration)	Weight (ton)	Labor (man-hours)	Description	Capacity (tons of refrigeration)	Weight (tons)	Labor (man-hours)
Steel package	10	700	12		325	10,200	72
unit, single	15	750	12		350	10,500	72
unit	20	850	12		375	12,000	72
	25	1000	18		400	12,750	72
	30	1300	18		425	13,400	72
	35	1375	18		450	14,500	72
	40	1400	18		475	14,750	72
	45	1450	18		500	15,000	96
	50	1550	18		525	15,750	96
	60	1850	18		550	16,750	140
	70	2000	24		600	18,750	140
	75	2150	24		650	19,750	140
	80	2450	24		700	23,000	200
	90	2600	24		750	23,750	200
	100	2850	24		800	25,000	200
	110	3000	32		850	27,000	220
	125	3500	32		900	28,500	220
	150	4250	48		950	29,000	220
	160	5000	48		1000	30,000	220
	175	5500	48		1050	32,500	220
	200	6050	48		1100	36,500	220
	250	7000	48		1150	38,500	220
	275	8500	48		1200	40,000	220
	300	9500	72				

13

DAMPERS

Dampers are devices that vary the air going through a duct by closing or opening their cross-sectional area. They are part of an air distribution system and are vital to its function and performance. Poor distribution of air will result in an uncomfortable condition even though the correct volume of air at a proper temperature is being brought to the conditioned space.

Dampers are "valves" in the air stream which control airflow. This regulation can be minor or can actually shut off the airstream. In this respect they function much like valves in piping systems and must be selected with the same care. By this it is meant that dampers should not introduce undue air resistance, for this will result in pressure drops, uneven air distribution, and noise.

Ducts are generally constructed of a galvanized steel frame with connecting flanges at both ends. A number of contrarotating blades are joined by means of coupling rods outside the frame. One blade has an extended shaft at both ends for connection to a linkage device.

Dampers are basically manufactured in two different styles: parallel blades and opposed blades. It is important to analyze the flow characteristics of each and the recommendations of the manufacturers before a selection is made. Each type of damper will control air volume, but with different leakages and pressure drops. Opposed blade designs tend to give a more uniform airflow around the centerline of the duct.

Dampers are classified according to their use as follows:

- outdoor air damper
- return air damper
- exhaust air damper
- fire damper
- volume air damper
- relief air damper

These uses are shown in Figure 96.

13.1. OUTDOOR AIR DAMPER (O.A.D.)

This damper controls the amount of outside air introduced into the air system. It is usually sized to match the size of the outdoor stationary louver. It is of an opposed blade type with a capability to modulate by manual or motorized linkage from full open to 25% of its open area. This range will allow outdoor air to enter the louver and pass through the damper without a pressure drop or a velocity change when the damper is full open and to regulate design conditions to some minimum volume without completely shutting off outside air, see Figure 97.

13.2. RETURN AIR DAMPER (R.A.D.)

This damper, also called a mixing air damper, maintains constant pressure at the inlet side of the fan. In selecting return air dampers it is important that the pressure drop is the same as the combined pressure drop across the louver

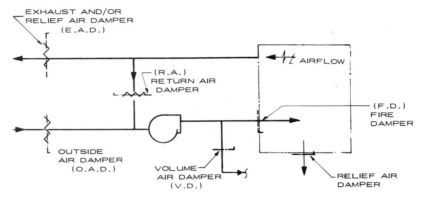

FIGURE 96. Dampers classified by use.

and damper at the intake air side, thus, producing a constant air volume in the system without the problem of air stratification. This damper, shown in Figure 97 is usually of the opposed blade type and can be interlocked with the outside air damper.

13.3. EXHAUST AIR DAMPER (E.A.D.)

This damper functions in conjunction with the return air damper. Its selection is based upon the combined pressure drops of the ductwork

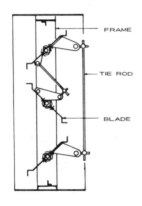

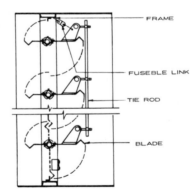

FIGURE 97. Air volume control damper. (Manufactured by Dowco Corp.)

FIGURE 98. Fire damper. (Manufactured by Dowco Corp.)

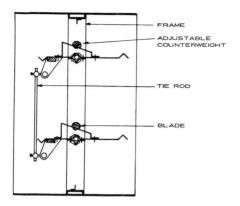

FRAME

ADJUSTABLE
COUNTERWEIGHT

TIE ROD

BLADE

FIGURE 99. Relief air damper. (Manufactured by Dowco Corp.)

system and that of the louvered exhaust with a screen. This damper which works as the outdoor air damper will be sized to allow passage of 100% air exhaust and be adjustable down to a minumum of 25%. It is usually of the opposed

blade type with controlling linkage. This damper is also shown in Figure 97.

13.4. FIRE DAMPERS (F.D.)

Fire dampers are used to close off ventilation ducts in order to reduce the spread of fire. Their construction and application are detailed in building codes. Fire dampers are obtainable for mounting in walls and ceilings and are square or round in shape.

A fire damper is shown in Figure 98. Note the fusible link which holds the damper in the open position. An elevated airstream temperature that would result from a fire, melts the fusible link, which separates and the damper shuts due to gravity.

13.5. VOLUME AIR CONTROL DAMPER

Volume control dampers are used to control airflow to individual spaces or zones. They are of the opposed blade type as shown in Figure 97.

13.6. RELIEF AIR DAMPERS

These dampers are installed in an enclosed area to provide control of pressure by a continuous introduction of air to the space. They can be counterweighted, adjustable, fixed, or motorized, When motorized, the controls are energized by a pressure sensor located within the area. Figure 99 illustrates the construction of a relief air damper (top), and shows a commercial unit at the bottom of the figure.

14

DUCTWORK

Ductwork is used to convey air, much as a wire is a conduit for electricity and piping for water. The air is conveyed through ducts as part of a HVAC distribution system, to remove contaminants or pollutants from industrial processes, and in other applications. Ductwork must be designed so that the air is conveyed at proper velocity and flow rate and be distributed evenly to all spaces. Noise rates must be within tolerable levels, heat losses must be controlled, and space requirements must be taken into account since ducts take space that might otherwise be put to good use. In addition to all of these design requirements, the cost of ducts must be commensurate with their use.

From a mechanical point of view, ducts must be well supported so that buckling cannot occur. Joints must be tight and permanent. Flexible connections should be provided between fan or furnace plenum boxes and ductwork to isolate noises or allow for thermal expansion.

14.1 MATERIALS

Air ducts are predominantly built of galvanized steel, and depending upon the design application, they are also fabricated of aluminum, stainless steel, asbestos, plastic, and fiberglass.

Galvanized Steel. Because of its lower cost, good workability, and structural strength, galvanized steel is used for most air conditioning ducts not subject to acid or humid conditions,

volume dampers, duct supports, louvers, and hoods for general ventilation.

Aluminum. The use of aluminum is usually reserved for conveying air containing humidity, some acid gases, and where outdoor exposure is required. Aluminum is usually left unpainted since it withstands corrosion in salt-air free atmospheres.

Stainless Steel. Stainless steel is expensive and is reserved for special applications such as kitchen hoods where a bright finish is desired, or in corrosive laboratory areas where aluminum may not be suitable for fume hoods.

Asbestos. Asbestos is used mainly for high-temperature exhaust flue stacks, or for general ventilation where no moisture is present, such as gas ventilation. Because of its carcinogenic properties, asbestos must be used with great care in ventilation systems.

Plastic. The use of plastic for ducts is usually guided by the manufacturers claims of physical properties and limitations. An economic evaluation should be based on certified data from manufacturers.

Fiberglass. Fiberglass ducts are becoming more popular for ventilation because it incorporates acoustic and thermal insulation properties with its mechanical properties.

14.2. DUCTWORK SHAPES

Ductwork is made in round, rectangular, and oval shapes. Round ductwork is the cheapest to install and fabricate. Rectangular ductwork is more expensive than round. However, its application is more flexible then the round and can fit in areas where most rounds do not. The sizes of round and rectangular ducts are given later in the section on labor estimates. The flat oval duct combines the advantages of round and rectangular duct. It can be joined with a simple slip-type coupling and can be furnished in varying aspect ratios. The flat oval duct has considerably less flat surface and is less susceptible to vibration. This duct requires less reinforcement than a corresponding size of rectangular duct because of its inherently stronger cross section. The term aspect ratio is commonly used in duct design, it is simply the ratio of the duct width to its depth.

14.3. DUCT FITTINGS

A multitude of duct fittings exist in a ductwork system. The following illustrations include take-offs, bands, elbows, turning vanes, reducers, transformations, flexible connections, dampers, offsets, stack fittings, register heads, hanger supports, and several combinations. The intent of this section is to familiarize the reader with the most common ductwork practices in use today. However, for more accurate approach to ductwork design and application, the guidelines issued by the SMACNA Manual* and by ASHRAE Handbooks[†] are recommended.

Bends, turns, reductions, transformations, offsets, and other fittings that change the shape,

---

*Sheet Metal and Air Conditioning Contractors' National Association, Inc.
[†]American Society of Heating, Refrigerating and Air Conditioning Engineers.

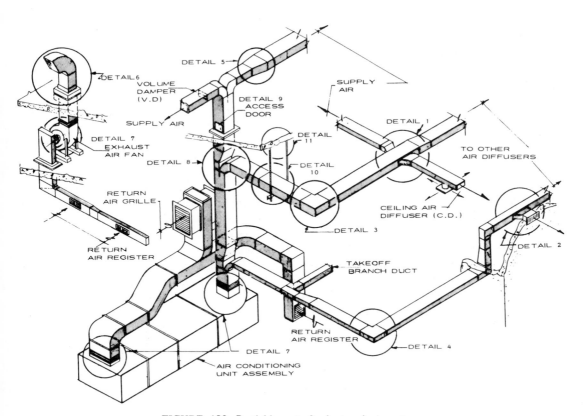

FIGURE 100. Partial layout of a ductwork air system.

size or direction of the ductwork, should be kept to a minimum. This does not mean that essential fittings should be omitted. It does mean that serious thought should be given during design to keeping the plenum, branches, and stacks as simple as possible and still provide satisfactory air quantities and distribution. Unnecessarily complicated combinations of fittings, and poor planning in locating the plenum, branch runs, and stacks results in an increase in materials, installation time, and system friction.

Figure 100 is a ductwork layout of an air system showing a number of duct fittings. These are encircled and noted as follows:

Detail 1, Tee Connections
Detail 2, Register and Grille Connections
Detail 3, Elbow Details
Detail 4, Elbows with Vanes
Detail 5, Tapers and Offsets
Detail 6, Rectangular Gooseneck
Detail 7, Flexible Duct Connections
Detail 8, Branch Duct Takeoff
Detail 9, Access Doors in the Duct
Detail 10, Hangers
Detail 11, Concrete Inserts

14.3.1. Tee Connections

Figure 101 shows eight types of tees commonly found in duct systems. They are noted by letters A through F on Figure 101 and are described as follows:

A—*Tee tap-in at 90°* shows a branch tee cut into a decreasing taper in the main, using a clinch lock connection. The decreasing taper forces the air into the branch.

B—*Straight tap-in with vanes* shows a branch tee cut into the straight side of the main, using a clinch lock connection. An adjustable set of vanes, to turn the required amount of air into the branch, is installed at the joint.

C-1—*Radius tap-in* shows a branch tee cut into the straight side of the main, with a radius in the branch throat and larger radius in the branch back, clinch locked to the main.

C-2—*45° Take-off* shows the branch connected on a 45° angle. This construction

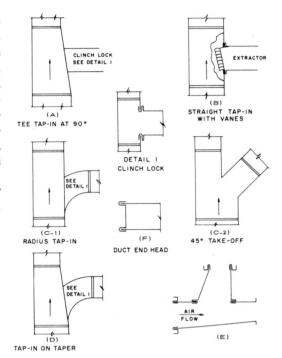

FIGURE 101. Tee connections, datails. (Reprinted with the permission of the Sheet Metal and Air Conditioning Contractors' National Association, Inc.)

proportions the areas to the amount of air required in the branch and continuing main. The top and bottom may be one piece with the sides installed with a Pittsburgh lock, described below and shown in Figure 114.

D—*Tap-in on taper* shows the same assembly as C-1 with the branch installed in a taper in the main as in A.

E—*Duct tap-in* shows method of connecting branch tee into a taper in the main using a clinch lock to main and pocket lock for continuation of fitting and main. Clinch lock is shown in Figure 101, pocket lock in Figure 113.

F—*Head in end of run* shows a method of installing a head or end in a duct. The lock is secured with button punch, rivets or metal screws through the flanges.

14.3.2. Register and Grille Connections

Figure 102 illustrates the methods used to connect a register and grille to the ducts. The

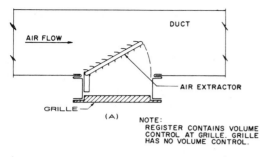

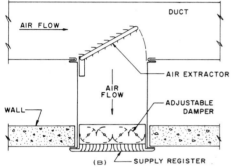

FIGURE 102. Register and grille connections. (Reprinted with the permission of the Sheet Metal and Air Conditioning Contractors' National Association, Inc.)

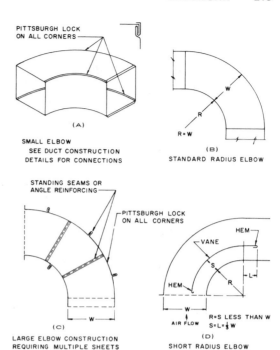

FIGURE 103. Elbow details. (Reprinted with the permission of the Sheet Metal and Air Conditioning Contractors' National Association, Inc.)

method of fastening must be in accordance with the manufacturers' recommendations. The same mounting details for registers and grilles apply to ceiling diffusers. A grille and register connection are shown in Figure 102. They are noted by letters A and B and are described as follows:

A—*Grille* is shown installed in the side of a duct. A short tee is clinch locked into the duct with the end flanged to receive the grille. An air extractor is used to produce an even distribution of air over the grille area.

B—*Register* is shown installed in an extended tee in conjunction with a multiple blade type of volume damper for air adjustment. The extractor is in a fixed position to supply an even distribution of air over the grille.

14.3.3. Elbow Details

Figure 103 illustrates four types of elbows without vanes; small, large, standard radius, and short radius. They are noted by letters A through D and are described as follows:

A—*Small elbows* are constructed with Pittsburgh locks at all corners. The lock is fabricated as part of the sides of the duct; the top and bottom portions are flanged to fit the lock. The gauges of metal for elbows are the same as shown in the section on labor estimates, Section 14.

B—*Standard radius elbows* are designed to eliminate air turbulence and unnecessary static pressure loss. For this to occur, the radius of the throat should be equal to the width of the elbow. In Figure 103, R is the throat radius and W is the width; therefore, $R = W$.

C—*Large elbows* use Pittsburgh locks at all corners. The gauge of metal and joints are usually specified in the duct specifications. When end joints other than the standing steam are used, angles are used for stiffening. When standing steams are used, the joints are spaced so that they reinforce the sides. Standard seams should be riveted.

D—*Short radius elbow* is used when space will not allow the use of a standard radius elbow. A vane is placed in the elbow and spaced according to the formula on Figure 103. The vane spacing (*L*) is equal to $\frac{1}{3}$ of the duct width. The vane is made of the same gauge as the elbow and is riveted or spot welded securely to the duct sides to prevent vibration. Ends should be hemmed for stiffening.

14.3.4. Elbows with Vanes

Design details for single and double vaned elbows are given in Figure 104. They are noted by letters A, B, and C and are described as follows:

A—*Single vane elbow* spacing is given for a 90° elbow. The opening between the vanes should be maintained as shown to insure an even airflow around the turn without causing turbulence.

Detail 1 indicates the radius and width

of each vane when spaced as shown in Figure 104A.

Detail 2 is a section of the runner that holds the vanes in place. The runner is riveted, spot welded, or screwed to the duct sides. (The runners, Details 2 and 5, are commercially available or may be made by the contractor.)

B—*Small double vane elbow* spacing is given for 90° elbows up to 36 in. in width. In ducts of greater width, two vanes of equal length should be installed. An intermediate runner should be securely fastened together by riveting or by other means to assure rigidity. The opening between vanes should be maintained as shown in Figure 104.

Detail 3 indicates the radius and width of both pieces of a small double vane and the method of assembly.

C—*Large double vane elbow* spacing is given for a 90° elbow. The vane openings should be maintained at $2\frac{1}{4}$ in.

Detail 4 shows the radius and width of both pieces of a large double vane and their assembly method.

Detail 5 is the runner that holds the large vanes in place. The runner is riveted, spot welded, or screwed to the duct sides. All vanes and runners are made of the same thickness of metal that is used for ducts and elbows.

14.3.5. Tapers and Offsets

Figure 105 gives design data for equal tapers in ducts where flow is diverging or converging. Offsets to avoid obstructions or change directions for other purposes are noted by letters A through E described as follows:

A—*Equal taper* design is shown with end joints at the top and bottom. Pocket locks, described in Section 14.3, are used on all sides. The longitudinal joints of tapers are made with Pittsburgh locks or button punch snap locks, both described in Section 14.3, on all corners. Metal gauge and reinforcing are described in Section 14.3.

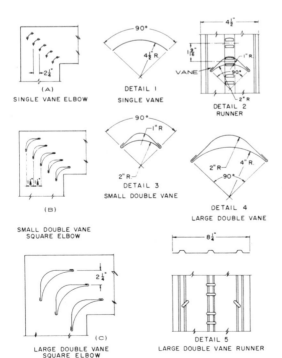

FIGURE 104. Vaned elbow details. (Reprinted with the permission of the Sheet Metal and Air Conditioning Contractors' National Association, Inc.)

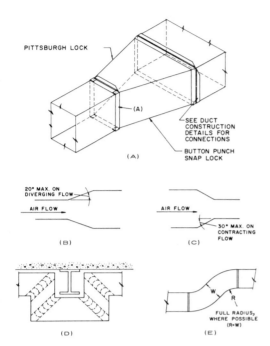

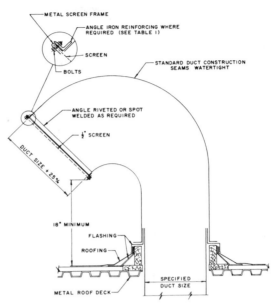

FIGURE 105a. Rectangular gooseneck detail. (Reprinted with the permission of the Sheet Metal and Air Conditioning Contractors' National Association, Inc.)

FIGURE 105. Tapers and offsets. (Reprinted with the permission of the Sheet Metal and Air Conditioning Contractors' National Association, Inc.)

B—*Equal taper for diverging airflow* requires that the sides be pitched to a maximum of 20° to prevent turbulence or an additional increase of static pressure.

C—*Equal taper for contracting airflow* requires that the sides be pitched to a maximum of 30° to prevent turbulence or an additional increase of static pressure.

D—*Offsets* used to avoid an obstruction, as shown, where space limitation will not allow a full radius turn. Square elbows with turning vanes should be installed to eliminate turbulence and to keep the static pressure increase as low as possible.

E—*Offsets* used to change the horizontal or vertical direction of the duct. The minimum radius of the turn (R) is to be equal to the duct width (W) as shown.

14.3.6. Rectangular Gooseneck

As shown in Figure 100, a gooseneck fitting is the portion of a duct that penetrates a roof or deck and almost reverses its direction to prevent rain from entering the duct system. Construction details are given in Figure 105a.

The gooseneck should be made using standard duct construction with Pittsburgh locks at the corners, standing seams, or reinforcing on the sides when required, and all seams watertight. The duct walls at the outer end of the gooseneck should be flanged outward or an angle installed for reinforcing and for fastening the screen. The screen should be installed in a metal frame which is bolted to the flange or angle. The area of the outer opening should be equal to the duct size plus 25% to compensate for the screen. If the gooseneck extends to an appreciable height above the roof line, angle iron support legs can be added. Metal flashing may be made up separately.

14.3.7. Flexible Duct Connections

Flexible connections should be installed between a fan and metal ducts or casings to prevent transfer of fan or motor vibration. Flexible materials are used for this purpose. They are installed loosely, in folds, so that they can flex easily as the equipment vibrates. If canvas is used, it should not be painted.

The following is quoted from Standard No. 90–A of the National Fire Protection Association, paragraph 113(b).

> Vibration connectors in duct systems, other then as covered by paragraph 113(c), shall be made of woven asbestos or approved flame proofed fabric or shall consist of sleeve joints with packing of rope asbestos or other approved noncombustible material. Vibration connectors of fabric shall not exceed 10 in. in length.

Figure 106 shows methods for making fan connections and for connecting ducts. At the top of the Figure is an elevation and a side view of the connections at the fan inlet and discharge. To avoid turbulence and air pulsation, the width of the duct leading to the fan inlet should be equal to or wider than the fan wheel diameter. The side view shows the position and rotation of the fan in relation to the duct elbow above. The rise of the duct should be equal to or greater than the fan wheel diameter.

Details 1 and 2, in the middle of Figure 106, show how flexible material is installed between sections of round duct or the fan intake collar using draw bands. Detail 2 is an enlarged end view of the drawbands with the bolt and nut for clamping the flexible material to the fan collar. Detail 3 shows how a flexible connector is designed for a rectangular duct with the method of locking the fabric to the duct.

14.3.8. Branch Duct Take-Off

A branch, which is tapped into a main duct at the transverse joint, is generally used to proportionally divert the airflow. Such a branch take-off is shown in Figure 107. The illustration shows that the branch is made of three parts: main duct ahead of the branch; main duct after the branch; and the branch duct. The enlarged insert shows a pocket lock joint, but any of the joints shown in Section 14.3.11 can be used.

The side of Figure 107B (the main duct after the branch) is hooked over the back edge of part C (the branch) to make the intersection air tight and to reinforce the edge. For branch elbows construction see Figure 103.

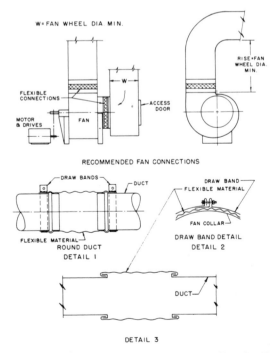

FIGURE 106. Flexible duct connections. (Reprinted with the permission of the Sheet Metal and Air Conditioning Contractors' National Association, Inc.)

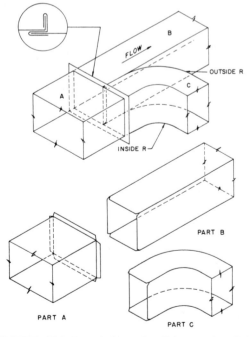

FIGURE 107. Branch duct takeoff from main. (Reprinted with the permission of the Sheet Metal and Air Conditioning Contractors' National Association, Inc.)

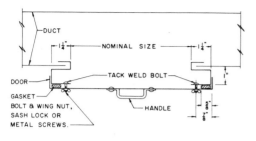

REMOVABLE DOOR

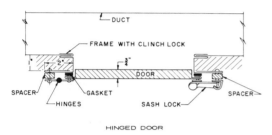

HINGED DOOR

FIGURE 108. Access doors in ducts. (Reprinted with the permission of the Sheet Metal and Air Conditioning Contractors' National Association, Inc.)

14.3.9. Access Doors in the Duct

Two types of access doors are shown in Figure 108. The first is removable (top of figure), while the second operates on hinges (bottom).

In the removable access door, the frame is clinch locked or tee locked to the duct with the outer edge flanged inward to receive the door. The pan type door is fastened to the flange with bolts and wing nuts, metal screws, or sash locks. If bolts are used, the bolt should be welded to the flange. A gasket is used to make the door air tight.

The hinged door may be insulated or uninsulated. The frame is clinch locked to the duct with the outer edge flanged outward and hemmed back for reinforcement. The flange is wide enough to receive the hinges and latches. The hinges and sash locks are installed with spacers on the duct side, bolted or riveted to duct and door. The door is formed and locked together as shown with the outer flange lapped over the frame flange. The gasket is applied to either frame or door with adhesive. Sash locks are used to tighten the door on the gasket. When required, insulation is installed in the door. If insulation is installed on the outside

of the duct, the flange depth of the door frame should be equal to the thickness of the insulation.

14.3.10. Hangers

Ducts must be properly supported in order to prevent sagging, which could open joints in the duct, or bring the weight of the duct to bear upon a part of the building that might be damaged, or collapse the duct itself. There are many ways of supporting a duct, depending upon its size, whether it is horizontal or vertical, and the structure from which the support will be taken. Many parts of a building, such as heating, plumbing, and sprinkler pipe, electrical conduits, lighting fixtures, and ceilings, require hanging. Consequently, there is a continuing effort to improve hanging systems.

Figure 109 illustrates a band iron strap and a shelf bracket that attach to a vertical wall, and angle supports for a vertical duct. For each type there is an accompanying table of the size of the supporting member and its spacing, where applicable. The band iron strap hanger, shelf bracket, and floor support angles are noted by letters A, B, and C; they are shown in Figure 109 and are described as follows:

A—*Band iron strap hangers* support the horizontal duct on the wall, and are recommended for small size ducts that have widths less than their heights. The band is anchored into the wall at top and bottom with powder-actuated fasteners or concrete anchors.

B—*Shelf brackets* support the horizontal duct on the wall. They are recommended for ducts that have widths greater than their heights. The vertical angle is anchored securely into the wall with similar fasteners to those used for strap hangers. The size of the angles for the bracket will vary with the duct size and spacing as shown in the Figure.

C—*Supports for vertical ducts at floor level* are achieved by extending the angles beyond the opening in the floor. The angles are screwed to the ducts. The duct seam is just above the floor level so it will not

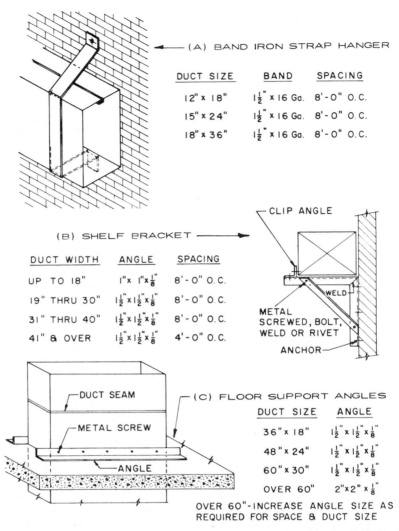

FIGURE 109. Hangers for ducts. (Reprinted with the permission of the Sheet Metal and Air Conditioning Contractors' National Association, Inc.)

interfere with the angle supports. The size of angles will vary with the duct size.

Additional supports are shown for vertical ducts in Figure 110. As with the previous Figure, the size and spacing of supports are tabulated for various duct dimensions. These are noted by letters D and E and are described as follows:

D—*Band iron strap hanger* for support of small vertical ducts from a wall is shown. Power-actuated fasteners or concrete anchors are used to affix the hanger to the wall.

E—*Angle bracket* are used to support large sized ducts. The bolt shown holding the bracket to the wall is screwed into an anchor.

Horizontal ducts, both round and rectangular, are suspended by strap hangers, hanger rods, and trapeze hangers as shown in Figure 111. Strap hangers for rectangular ducts are supported by two straps. The straps are screwed to the sides and bottom of the duct and extend upward to one of a variety of fasteners (clamps that tie into the building structure). These are shown in Figure 112.

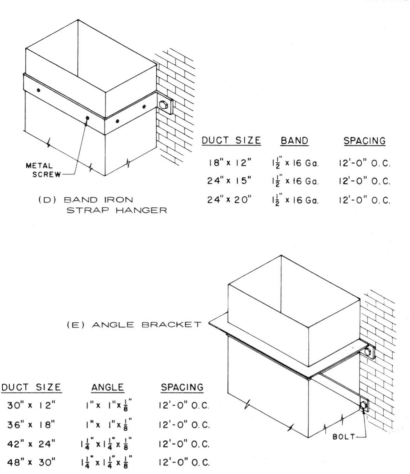

DUCT SIZE	BAND	SPACING
18" x 12"	$1\frac{1}{2}$" x 16 Ga.	12'-0" O.C.
24" x 15"	$1\frac{1}{2}$" x 16 Ga.	12'-0" O.C.
24" x 20"	$1\frac{1}{2}$" x 16 Ga.	12'-0" O.C.

(D) BAND IRON STRAP HANGER

(E) ANGLE BRACKET

DUCT SIZE	ANGLE	SPACING
30" x 12"	1" x 1" x $\frac{1}{8}$"	12'-0" O.C.
36" x 18"	1" x 1" x $\frac{1}{8}$"	12'-0" O.C.
42" x 24"	$1\frac{1}{4}$" x $1\frac{1}{4}$" x $\frac{1}{8}$"	12'-0" O.C.
48" x 30"	$1\frac{1}{4}$" x $1\frac{1}{4}$" x $\frac{1}{8}$"	12'-0" O.C.

FIGURE 110. Hangers for vertical ducts. (Reprinted with the permission of the Sheet Metal and Air Conditioning Contractors' National Association, Inc.)

Round ducts have metal clamping bands, either one piece that goes around the duct and bolt to a hanger strap, or two pieces that extend over half the diameter and bolt to two hanger rods or straps. Trapeze hangers are recommended for larger ducts. The supporting shelf, or trapeze is made of angle iron and may be attached to the supporting rods, straps, or angles by welding, bolting, or push nuts as illustrated in Figure 111.

Recommended hanger sizes for rectangular and round ducts are given in the following Tables. These are for single ducts. Where two or more ducts are carried by a single trapeze, or where piping is also supported, sizes and spacing of support members must be altered accordingly.

Hanger straps, both round and rectangular in cross section, and angle straps must be securely attached to the building structure. A wide variety of methods is shown in Figure 112. If concrete is the structural material, fastening can be done by means of concrete inserts, powder actuated fasteners, or by anchors.

Concrete inserts must be installed prior to pouring the concrete. They are used primarily where the duct layout is simple and when there is enough lead time for accurate placement. The simplest insert is merely a piece of bent flat bar. Manufactured inserts are available individually or in long lengths. The latter is used where many hangers will be installed in a small area, or where individual inserts cannot be precisely spotted at the time of placing concrete.

Concrete fasteners are installed after the

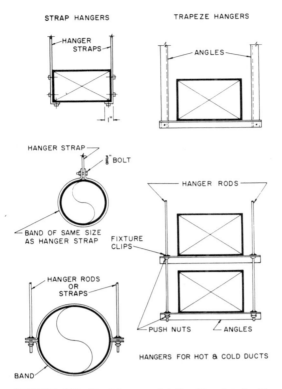

STRAP HANGERS

HANGER STRAPS

TRAPEZE HANGERS

ANGLES

1"

HANGER STRAP

$\frac{3}{8}$" BOLT

BAND OF SAME SIZE
AS HANGER STRAP

FIXTURE
CLIPS

HANGER RODS

HANGER RODS
OR
STRAPS

PUSH NUTS   ANGLES

BAND

HANGERS FOR HOT & COLD DUCTS

FIGURE 111. Duct hangers details. (Reprinted with the permission of the Sheet Metal and Air Conditioning Contractors' National Association, Inc.)

concrete is set. They are more flexible than concrete inserts because their exact location can be determined after all interferences between trades have been coordinated. Powder-actuated fasteners are placed by an explosive charge. These fasteners should not be used in certain lightweight aggregate concretes, nor should they be used in slab sections of less than 4 in. thick.

Expanding concrete anchors should be made of steel. Nonferrous anchors like lead tend to creep with vibration. Holes for expansion anchors are drilled either by a carbide bit or by teeth on the fastener itself. The expansion shield is "set" by driving it into the hole and expanding it with the conical plug. The expansion nail is a light duty fastener used for small duct and flexible tubing.

In all of the above applications, there are possibilities of interference with steel reinforcing in the concrete. The installer must exercise good judgment and have some knowledge of reinforcing patterns.

When the structural support is taken from a steel member, a number of fastening techniques are used. A c-clamp in conjunction with a retaining clip is fastened to the lower flange of an

Recommended Hanger Sizes for Rectangular Duct

Longest Dimension of Duct	Round Hangers	Strap Hangers	Trapeze Shelf Angles	Maximum Spacing
up to 18"	8 ga. wire	1" × 16 ga.	$1'' \times 1'' \times \frac{1}{8}''$	10'0"
19" to 30"	8 ga. wire	1" × 16 ga.	$1'' \times 1'' \times \frac{1}{8}''$	10'0"
31" to 42"	$\frac{1}{4}''$ rod	1" × 16 ga.	$1\frac{1}{2}'' \times 1\frac{1}{2}'' \times \frac{1}{8}''$	10'0"
43" to 60"	$\frac{3}{8}''$ rod	1" × 16 ga.	$1\frac{1}{2}'' \times 1\frac{1}{2}'' \times \frac{1}{8}''$	10'0"
61" to 84"	$\frac{3}{8}''$ rod	$1\frac{1}{2}''$ × 16 ga.	$2'' \times 2'' \times \frac{1}{8}''$	8'0"
85" to 96"	$\frac{3}{8}''$ rod	$1\frac{1}{2}''$ × 16 ga.	$2'' \times 2'' \times \frac{1}{16}''$	8'0"
over 97"	$\frac{3}{8}''$ rod		$2'' \times 2'' \times \frac{1}{4}''$	8'0"

Recommended Hanger Sizes for Round Duct

Duct Diameter	Round Hangers	Strap Hangers	Maximum Spacing	Number of Hangers
up to 18"	8 ga. wire	1" × 16 ga.	10' 0"	1
19" to 36"		1" × 12 ga.	10' 0"	1
37" to 50"		2" × 16 ga.	10' 0"	1
51" to 84"		2" × 16 ga.	10'0"	

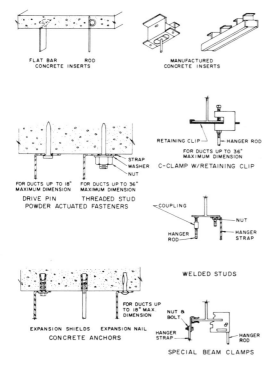

FLAT BAR · ROD
CONCRETE INSERTS

MANUFACTURED
CONCRETE INSERTS

RETAINING CLIP — HANGER ROD
FOR DUCTS UP TO 36"
MAXIMUM DIMENSION
C-CLAMP W/RETAINING CLIP

STRAP
WASHER
NUT

FOR DUCTS UP TO 18" · FOR DUCTS UP TO 36"
MAXIMUM DIMENSION · MAXIMUM DIMENSION
DRIVE PIN · THREADED STUD
POWDER ACTUATED FASTENERS

COUPLING

NUT

HANGER
ROD

HANGER
STRAP

WELDED STUDS

FOR DUCTS UP
TO 18" MAX
DIMENSION

EXPANSION SHIELDS · EXPANSION NAIL
CONCRETE ANCHORS

NUT &
BOLT

HANGER
STRAP

HANGER
ROD

SPECIAL BEAM CLAMPS

FIGURE 112. Hanger strap and rod attachments. (Reprinted with the permission of the Sheet Metal and Air Conditioning Contractors' National Association, Inc.)

H-beam, I-beam, or tee. The retaining clip prevents lateral slippage. As shown in Figure 112, the hanger rod is screwed into the bottom of the clamp and held by a locknut.

Stud welders are able to securely weld threaded studs to metal surfaces. These studs are then used to fasten a strap hanger or are attached directly to a hanger rod by means of a threaded coupling. Still another method of tying in to the flange of a steel member is by means of friction devices which are driven into the edge of the flange and in turn support the hanger rod or strap.

14.3.11. Duct Joints and Seams

Ducts are fabricated and joined to themselves and to fittings by means of joints that are transverse to the length of the duct and longitudinal seams. Transverse, or cross joints are shown in Figure 113. Longitudinal seams are illustrated in Figure 114.

Listed below are descriptions of 12 types of cross joints keyed to Figure 113. Note that the preferred direction of airflow is given for some cases. If this recommendation is followed, the airflow will be smoother since it will not hit the exposed edge of the joint.

A—*Drive slip*—Ends of the ducts are inserted into the cleat. They are used for the narrow sides of the ducts that are 18 in. or less. Drive slips that are 19 to 30 in. must be reinforced with a 1 in. × 1 in. × $\frac{1}{8}$ in. angle. A combination of a drive slip and any S-slip (B, C, E, F, G, H) completes the transverse joint.

B—*Plain S-slip*—Ends of the ducts are inserted into open ends of "S." They are used on wide sides of small ducts. (Use a drive slip on narrow sides.)

C—*Hemmed S-slip*—They are similar to plain S-slip except that edges are hemmed to produce stiffness.

E—*Bar slip*—They are similar to plain S-slip except for the standing edge which is formed to provide reinforcement.

F—*Alternate bar slip* (standing S-slip)—They are the same as the bar slip except the standing leg is folded to three thicknesses for stiffness.

G—*Reinforced bar slip*—They are similar to the bar slip except for the addition of a steel reinforcing bar inserted in the standing edge.

H—*Angle slip*—They are the same as reinforced bar slip except for the use of a reinforcing angle in place of the reinforcing bar. The angle may be inside or fastened to outside of slip.

I—*Standing seam*—Ends of the adjoining ducts are joined as shown with button punch indentations at 6-in. centers, added after assembly.

J—*Angle reinforced standing seam*—They are the same as the standing seam except for the reinforcement with an angle, one leg of which is fastened to the duct and the other leg to the standing edge of the seam.

K—*Pocket lock*—Normally used on four sides of a duct. The pocket section is clip punched or "hickey" punched to the duct near corners and then at every 6 in. The

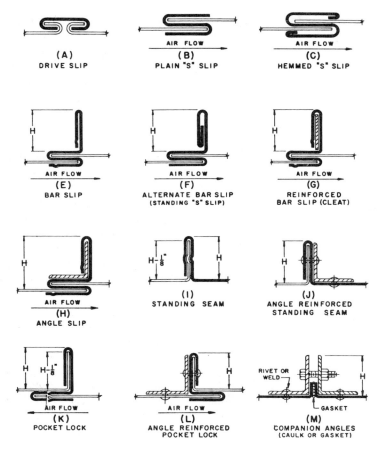

FIGURE 113. Duct connections-cross joints. When the longest side of the duct is up to 42 in., H = 1 in.; when the longest side is between 43 and 96 in., H = 1-1/2 in.; when the lonest side is over 96 in., H = 2 in.

other end is flanged outward to join with second sheet and then hammered down.

L—*Angle reinforced pocket lock*—They are the same as the pocket lock with an added angle stiffener.

M—*Companion angles*—Angle frames are riveted, bolted, or welded to duct ends and are then bolted together with gaskets or caulking to prevent air leakage. Recommended for use where duct sections must be periodically removed.

Longitudinal seams, which are parallel to the length of the duct, are important because these locks must hold the duct pieces securely. They should not leak under pressure and should be readily and swiftly put together on the job or in the shop.

There are four such seams illustrated in Figure 114: the Pittsburgh lock, Acme lock, double seam and button punch snap lock. These are noted by the letters N, O, T, and Z and are described as follows:

N—*The Pittsburgh lock* is the most common longitudinal seam. Originally formed in the brake or press brake, today roll forming machines are used to form the pocket in one piece and the flange in the other piece. After one piece is inserted in the pocket the "tail" is hammered over to close the lock.

O—The *Acme lock* originally called a "lock-grooved seam" was popular because it provided snug nesting and a smooth exterior surface. Today this lock is used to join two flat sheets for increased width.

T—*Standing seams* or double standing seams are used mostly on the inside of ducts and for certain sizes of ducts where their use leads to economical sheet cutting.

Z—*Button punch snap locks* are a recent inovation. Originally the continuous snap lock was used on light gauge stove and furnace pipe to permit nested shipping. The pipe section was then snapped together. Button punch spaces the buttons on approximately 2-in. centers along the flange to be inserted into the pocket. The continuous folded edge in the pocket prevents the button from leaving the pocket. The dimensions of the pocket and the flange are critical in high-pressure duct systems. The pocket and the flange must be fabricated by a machine suited to the gauge of metal being formed. If this is not adhered to, the pocket will be loose, reducing the stiffness and air tightness of the duct.

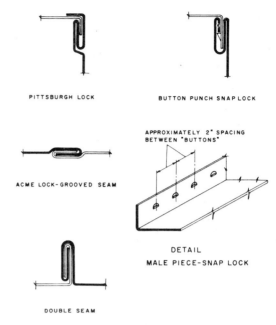

PITTSBURGH LOCK

BUTTON PUNCH SNAP LOCK

ACME LOCK-GROOVED SEAM

APPROXIMATELY 2" SPACING
BETWEEN "BUTTONS"

DETAIL
MALE PIECE-SNAP LOCK

DOUBLE SEAM

FIGURE 114. Longitudinal seams. (Reprinted with the permission of the Sheet Metal and Air Conditioning Contractors' National Association, Inc.)

14.3.12. Estimated Labor to Fabricate and Install Duct and Duct Fitting

Rectangular Sheet Metal and Aluminum Duct—Shop Fabrication

Duct Size (in.)	Thickness (US gauge)	Wyes	Man-hours Each			
			Tees	Ells	Tap-ins	Offset Transition
12 × 12	24	0.08	0.07	0.05	0.05	0.06
24 × 16	24	0.08	0.07	0.05	0.05	0.06
36 × 16	24	0.09	0.08	0.07	0.07	0.06
24 × 24	22	0.09	0.08	0.07	0.07	0.06
36 × 24	22	0.10	0.09	0.07	0.07	0.07
48 × 24	20	0.10	0.10	0.08	0.08	0.07
48 × 30	20	0.12	0.11	0.08	0.08	0.08
48 × 36	20	0.12	0.11	0.08	0.08	0.08
48 × 42	18	0.14	0.12	0.09	0.08	0.08
60 × 60	18	0.14	0.12	0.10	0.10	0.08

Rectangular Sheet Metal and Aluminum Duct—Field Installation

Duct Size (in.)	Thickness (US gauge)	Wyes	Man-hours Each			
			Tees	Ells	Tap-Ins	Offset Transition
12 × 12	24	0.10	0.09	0.06	0.06	0.08
24 × 16	24	0.10	0.09	0.06	0.06	0.08
36 × 16	24	0.12	0.10	0.07	0.08	0.08
24 × 24	22	0.12	0.10	0.08	0.08	0.08

Rectangular Sheet Metal and Aluminum Duct—Field Installation

Duct Size (in.)	Thickness (US gauge)	Wyes	Man-hours Each			
			Tees	Ells	Tap-Ins	Offset Transition
36 × 24	22	0.13	0.10	0.08	0.08	0.10
48 × 24	20	0.14	0.12	0.10	0.10	0.10
48 × 30	20	0.16	0.14	0.10	0.10	0.12
48 × 36	20	0.16	0.14	0.10	0.10	0.12
48 × 42	18	0.18	0.16	0.12	0.12	0.12
60 × 60	18	0.18	0.16	0.14	0.14	0.12

Rectangular Galvanized Sheet Metal Duct[a]

Duct Size (in.)	Thickness (US gauge)	Weight (lb.)	Average cfm[b]	Man-hours/Linear Foot of Duct	
				Shop Fabrication	Field Installation
6 × 6	26	1.80	150	0.11	0.09
12 × 6	26	2.70	350	0.16	0.14
12 × 12	24	4.80	1000	0.34	0.24
18 × 12	24	6.00	1500	0.42	0.30
24 × 12	24	7.20	2000	0.50	0.36
24 × 18	24	8.40	3750	0.60	0.42
24 × 24	24	9.60	5250	0.67	0.48
36 × 18	22	13.06	6350	1.04	0.65
36 × 36	22	16.80	16,000	1.34	0.84
48 × 12	22	14.00	4800	1.12	0.70
48 × 18	22	15.40	8700	1.23	0.77
48 × 24	22	16.80	14,000	1.34	0.84
48 × 36	22	19.60	22,500	1.57	0.98
48 × 48	22	22.40	30,000	1.80	1.12
54 × 24	22	18.20	15,000	1.46	0.91
54 × 36	22	21.00	27,000	1.68	1.05
54 × 48	22	23.80	40,000	1.90	1.19
54 × 54	22	25.20	46,000	2.02	1.26
60 × 24	20	23.80	17,500	2.38	2.38
60 × 36	20	27.20	30,000	2.72	2.72
60 × 48	20	30.60	42,000	3.07	3.06
60 × 54	20	32.30	52,000	3.23	3.23
60 × 60	20	34.00	60,000	3.40	3.40

Round Duct Galvanized Sheet Metal

Duct Diameter (in.)	Thickness (US gauge)	Man hours/Linear Foot of Duct[c]	
		Shop Fabrication	Field Installation
4	22	0.08	0.10
6	22	0.12	0.15
8	22	0.18	0.20
10	22	0.20	0.25
12	20	0.24	0.35
14	20	0.28	0.45
16	20	0.32	0.50
18	20	0.38	0.55
20	20	0.42	0.62

Round Duct Galvanized Sheet Metal

Duct Diameter (in.)	Thickness (US gauge)	Man hours/Linear Foot of Duct[c]	
		Shop Fabrication	Field Installation
22	18	0.50	0.75
24	18	0.60	0.90
26	18	0.68	1.00
28	18	0.72	1.10
30	18	0.80	1.15
32	16	1.00	1.30
36	16	1.12	1.70
40	14	1.25	1.85

Duct Finishes

Type of Finish	Man hours[d] (per sq ft of duct surface)
Paint, two coats	0.022
$\frac{1}{2}$-in finish cement	0.040
Weatherproofing	0.030
Canvas (8 oz)	0.025
Asphalt finish	0.070
Cement, two coats	
Plaster with wire mesh and canvas covering	0.080

[a]Estimate Duct Size with or without fitting and or hanger supports.
[b]cfm—based on 0.1 in. of water static pressure drop/ 100 ft of duct.
[c]Man hours includes hauling to and from fabricating shop. For straight section only.
[d]Man hours are for application only. Scaffolding, preparation etc. not included.

15

DUST CONTROL AND AIR POLLUTION EQUIPMENTS

New air pollution control codes have put a focus on high-efficiency gas cleaning. Many codes now limit stack emissions to a maximum of 0.02 grains/cu ft or even less. Present day equipment can achieve this requirement. Efficiencies of over 99.9% are not uncommon. But, the higher the efficiency requirements, the more critical and exacting the control system becomes.

Before attempting to select a specific cyclone separator, baghouse, scrubber, or precipitator, the designer must first understand the efficiency requirements. He must know the local codes and feel assured that the manufacturer has related the code requirements to his equipment. Selection of equipment comes about by an understanding of the properties of the air or gas to be cleaned and the particulates that must be extracted.

The operation of an air pollution control system is affected by many variables. The most important of these in selecting equipment are as follows:

1. Normal gas flow (in cubic feet per minute) at the inlet conditions is required as well as the maximum flow rate. The flow must be known in order to select the properly sized equipment. The temperature and pressure of the gas are such inlet conditions.

2. The maximum inlet temperature must be known in order to determine the materials of construction for the system. If extreme temperatures are encountered, it may be necessary to design sprays, bleed-ins, or radiant coolers ahead of the system for cooling and conditioning the gases.

3. The density of the gas (in pounds per cubic feet) at the inlet and its moisture content are important properties that affect a systems size.

4. Chemical properties of the gas—The identity and the volume or weight percents of combustible or corrosive gases, such as sulfur dioxide, sulfur trioxide, hydrochloric acid, fluorine, etc., affect safety and the materials of construction.

5. Concentration of particulates—The average weight and peak amount of solid matter should be stated in total pounds per hour or grains per cubic foot (7000 grains/lb). These should be correlated to the inlet volume at operating conditions. The concentration of solids affects the system's size and the sizing of related equipment.

6. Particle size distribution—Since the particulate matter to be collected could be either solids or condensible substances, their size distribution is required. Two forms of particle size determination can be used: *In-situ* sampling and laboratory techniques. Determination in the gas stream (*in-situ*) is most accurate because it eliminates the possibility of agglomeration of sampled particles and the effects from changes in the atmosphere.

7. Chemistry of particulates—A complete chemical analysis of the solids is important to avoid chemical attack on the construction materials. Even small traces of such halogens

as fluorine, chlorine, and bromine can cause serious problems. Solubility of the solids should be also known.

8. Physical size—If a process can be controlled by several different methods of collection, then size, weight, and general layout become major factors in the equipment selection. The total size includes the necessary auxiliaries and the access space needed for periodic maintenance. If possible, equipment should be located within the building. If this is done, weight becomes important to size the supporting steel. As a general rule, a bag filter will occupy the largest amount of square footage followed by a precipitator, a scrubber, and a cyclonic collector.

15.1. AIR POLLUTION CONTROL EQUIPMENT

15.1.1. The Cyclonic Collector

The basic principle of a cyclonic collector is that when the dirty gas is whirled around in a cylindrical chamber there is a force, called a centrifugal force, which tends to push the suspended particles to the outer wall of the chamber, where they then fall by gravity through a part in the bottom of the chamber. The force at which these particles are thrown out depends upon how fast the gas is whirled and on the radius of the chamber. These two important factors influence the efficiency and

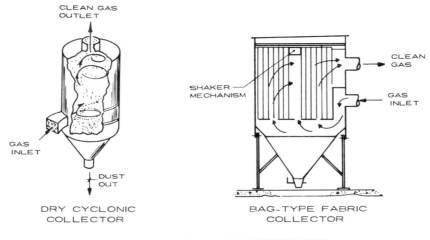

DRY CYCLONIC COLLECTOR

BAG-TYPE FABRIC COLLECTOR

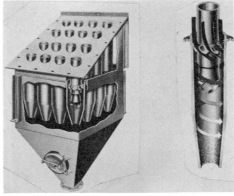

MULTICLONE DUST COLLECTOR

FIGURE 115. Dust collectors. Multiclone is the trademark of Western Precipitation Division of Joy Manufacturing Co.

cleansing action of the collector. A sketch of a cyclonic collector is shown in Figure 115 (upper left). It is sometimes called a dry collector since water is not used to cleanse the gas.

15.1.2. The Multiclone Collector®

As with other types of cyclone separators, the multiclone collector creates a column of whirling gas. But how this whirling action is created is one of the major differences between conventional cyclonic equipment and the multi-clone collector. In conventional cyclones, the whirling action is set up by bringing the gas tangentially through a side inlet, into the collection chamber. From this inlet, the gas follows the curved inside surface of the chamber, setting the required whirling action. The gas follows an involute path and gradually moves into the final separating vortex. The construction of individual separators prevents the use of a compact assembly of multiple units.

From years of development and research, the multiclone collector has been produced to overcome this disadvantage in conventional collectors. Multiclone collectors differ from conventional cyclones in that instead of bringing the gas in at the side to set up the whirling action, gas is brought in at the top of the chamber and is made to whirl by a vane that deflects the gas in a circular path. The vane has the added advantage of uniformly distributing the gas around the circumference of the chamber, by causing the gas to flow through a number of evenly spaced small entrances. The unique multicollector design is more efficient and takes less space than the conventional side-entry models. This is illustrated in Figure 115 (bottom). Multiclone collectors are often used as precleaners to lighten the load on a secondary collector. This application results in a more efficient system and a lower total system cost.

15.1.3. Baghouses

All baghouses operate in basically the same way. Dirty gas is ducted to the unit where it is filtered by cloth tubes or bags. The filtering is extremely efficient and normally results in removal of better than 99% of the entrained particles. The bags must be periodically purged of the collected material, and the method and frequency of cleaning characterize one type of baghouse from another. Cleaning can be automatic on an intermittent or continuous basis. Intermittent baghouses are cleaned after the unit has ceased filtering at the end of a work day. They cannot be cleaned while on-line, and so are limited to low dust loadings or infrequent operation. Continuous automatic baghouses, on the other hand, can operate 24 hr a day without rest and can handle high dust loadings. Baghouses are also characterized according to the method used to remove collected material from the bags. These include shaking the bags, reversing the direction of airflow through the bags, or rapidly expanding the bags by a pulse of compressed air.

Shaker Type. The bags in shaker-type baghouses are supported by a structural framework that is free to oscillate when driven by a small electric motor. When operated as an intermittent filter, the unit is shut down and the bags are shaken for about one min. to free them of dust. To operate a shaker filter as an automatic continuous unit, a timed damper isolates a compartment of the baghouse so that no air flows through. The bags in the compartment are then shaken for approximately one min. at which time the collected dust cakes are dislodged from the bags and fall into a hopper below for removal. The dampers then open, allowing the section to go back on line. A shaker baghouse is shown in Figure 115 (upper right).

Reverse Flow. The reverse flow-type baghouse is sectionalized as in the automatic continuous filter by time dampers. At the same time an auxiliary fan damper is opened forcing air through the bags in the direction opposite to flow during filtration. This backwash action collapses the bag and fractures the dust cake allowing it to drop into the hopper. When the bag is brought back on-line and reinflated, more of the fractured dust cake is dislodged into the hopper. This procedure may be repeated several times during the two or three min. cleaning cycle.

Reverse Pulse.   Reverse pulse baghouses have been used increasingly in recent years. This design utilizes a short pulse of compressed air through a venturi that is directed from the top to the bottom of the bag. The air pulse aspirates secondary air as it passes through the venturi. The resulting pressure front violently expands the bag to cast off the dust cake. No sectionalizing is required as the pulse of compressed air actually stops the flow of the air that would otherwise enter the bag being cleaned.

15.1.4. Wet Scrubber

A wet scrubber mixes the gas stream with a liquid, entrapping the material to be collected in the liquid and draining it away. Water is normally used, but other liquids are used for special applications. The methods of mixing the air and liquid vary in different types of scrubbers, but the basic operating principle is the same.

In the process of mixing, the gas is both cooled and saturated adiabatically. As you might recall, an adiabatic process is one in which no heat is added or removed. Psychrometric curves and tables are used to determine the outlet volume of the wet scrubber. All wet scrubbers are sized on the saturated outlet gas volume. Therefore, it is necessary to know the exact inlet conditions so that properly sized equipment can be selected.

Wet scrubbers lend themselves to controlling corrosive gases that are hot and difficult to

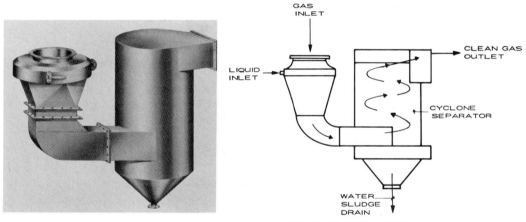

VARIABLE THROAT VENTURI SCRUBBER

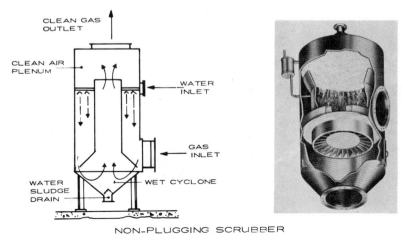

NON-PLUGGING SCRUBBER

FIGURE 116. Wet scrubbers. (Manufactured by Joy Manufacturing Co.)

handle. But because the collected material is mixed with the liquid, the resulting slurry can create a handling or disposal problem. It is sometimes possible to return the slurry to the process. However, a wet scrubber improperly applied can turn an air pollution problem into a water pollution problem. Properly applied, it can be a highly efficient, relatively inexpensive solution.

Figure 116 shows a variable throat venturi scrubber and a nonplugging type scrubber. A variable throat venturi scrubber lets you adjust the unit for changing volumes or process conditions. Incoming gases are directed to the flooded venturi by an inlet thimble that is not subject to any moisture; therefore, there's no wet–dry line to cause build-up or plugging.

The nonplugging scrubber is a high efficiency, low energy nonplugging device for large volumes. The vertical flow design takes a minimum of floor space, yet provides high efficiency.

15.1.5. Electrostatic Precipitator

Electrostatic precipitators clean industrial gases by electrically charging the particles in the gas stream, causing them to be attracted to and "precipitate" upon collector plates. High voltage is applied to a discharge electrode system within the precipitator. The electrodes are rows of vertical wires suspended between grounded parallel plates about 8 to 10 in. apart. Applying 35,000 or more volts causes a corona discharge

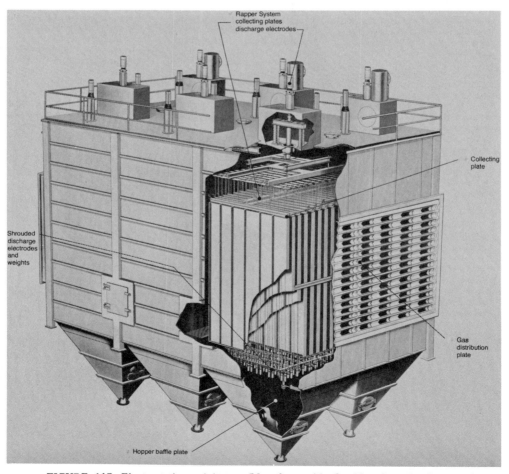

FIGURE 117. Electrostatic precipitator. (Manufactured by Joy Manufacturing Company.)

which charges particles in the gas stream and sends them to the grounded plates. At suitable time intervals, the collecting surfaces are rapped with electromagnetic or pneumatic hammers to transfer the dust to hoppers below.

Precipitators are particularly well suited to large volume applications and where ultra-high efficiencies in the 99% plus range are required. For these applications there are a number of distinct advantages to contemporary precipitators. They are exceptionally durable with low maintenance. There are virtually no moving parts, only the rappers, and these are easily accessible for inspection and repair. While exceptionally high voltages are used, current is low and power requirements are low. The essentially open flow design results in a low draft loss, usually less than one-half in. of water. Also, precipitators can handle high-temperature gas of up to 800°F, or more. On the negative side, precipitators require ample space. And, because most precipitators must be custom-engineered and constructed to fit each specific application, initial costs are relatively high. A cutaway view of a commercially available electrostatic precipitator is shown in Figure 117.

15.2. COST OF AIR POLLUTION EQUIPMENT

Unit Types	Weight (lbs/cfm)	Cost ($/cfm)	Pressure Loss (in. water)	Space Reqd. (sq ft)	Efficiency (%)
Cyclone	0.22	0.32	3.2 to 6	66	90 to 92
Multiclone	0.17 to 0.44	0.22 to 0.48	3.3 to 4	66 to 112	85 to 90
Baghouse	0.42 to 0.70	0.69 to 1.26	4 to 6	350 to 450	99+
Wet scrubber	0.16 to 0.23	0.51 to 0.95	5 to 15	115 to 125	99+
Electrostatic precipitator	2.2	2.7	0.5	484	99+

[a]The cost shown is for equipment cost. Labor cost and the escalation of equipment cost should be verified at the time the estimate is made.

Notes: Most air pollution control systems will require the installation of an additional fan. The power to run the fan can be an important operating cost of the system. It is fairly easy to calculate the total cost of operating air pollution equipment by use of the following:

1. Fan Brake Horsepower = Actual Air Volume × Total Static Pressure
2. Pump Brake Horsepower = GMP × Total Head × Specific Gravity × 3900 × Pump Efficiency
3. Total Brake Horsepower × 0.7457 = Kilowatts
4. Kilowatts × Cost/Kilowatt-hour = Hourly Operating Cost

16

FAN COIL UNITS

Fan coil units are prefabricated air conditioners, each containing a fan and motor, air filter, coil bypass damper, thermal and acoustic insulation, and a coil or coils for chilled or hot water. These components are enclosed in a metal cabinet. Fan coil units are used in chilled or hot water systems (or both), for air conditioning a wide variety of spaces.

Fan coil units can be used in lieu of a fan room, since units are used with or without ductwork. There are outlets at the top, front, and ends of the cabinet for use singly or in multiples with ductwork for zone air distribution, or for free air delivery without ducts. When the air conditioner is installed outside the space being served, ductwork between the space and the cooling coil bypass inlet is needed to ensure that the return air is bypassed and not used for make-up air. The centrifugal fan in the fan coil unit is direct- or belt-driven. The belt drive offers flexibility in varying fan capacity for specific applications involving the use and extent of ductwork.

The coil has copper tubes, fins, and headers, and is provided with vent and drain cocks. The vent and drain are accessible through a removable access panel. The unit assembly is provided with a fixed bypass, manual locking type damper mounted on a removable end panel.

All working mechanisms for the damper are enclosed inside the cabinet to prevent adjustment of the damper without the panel being removed.

Provision is made for access to and removal of the fan, fan drive, motor, and cooling coil through removable end panels. These panels are interchangeable on each assembly. A front access opening permits servicing, adjustment, and removal of the motor, fan bearings, pulleys, and belts. The removable panels and access openings are gasketed and airtight under normal operating conditions. For servicing the air filters, access openings are provided on each end of the cabinet.

The unit is designed predominantly for use in offices, hotels, restaurants and schools, where the heating and cooling requirements may be high in relation to the ventilation requirements. Significant advantages of the fan coils are:

• Low noise level.
• The room temperature can be rapidly adjusted, with forced air circulation.
• Unrestricted location in the room.
• Control of the air and water flows provides a wide range of temperature control.
• Cooling and heating of the air independent of the ventilation flow is possible.
• Energy saving can be achieved by disconnecting the unit in premises not used.

Figure 118 shows two applications of fan coil units. The upper part of the Figure illustrates a ceiling-mounted unit, the lower, a floor-mounted unit. Figure 119 shows photographs of four types of commercially available fan coil units.

The fan coil air conditioner in Figure 119a is designed for use in concealed locations, such as above ceilings or closets in hotels, motels, and apartment buildings. This unit has a supply air

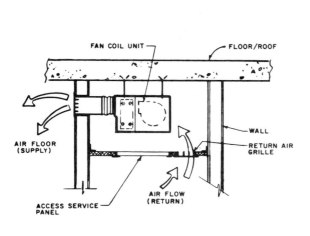

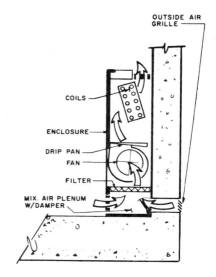

CEILING MOUNTED   FLOOR MOUNTED

FIGURE 118. Fan coil unit applications.

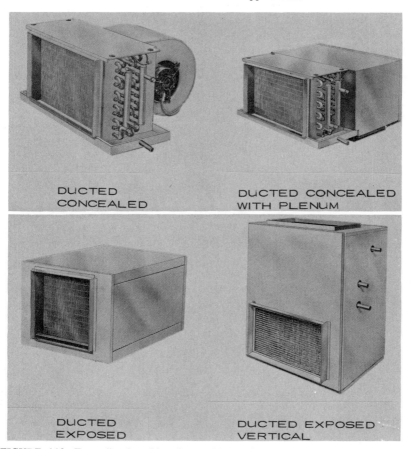

FIGURE 119. Fan coil units. (Manufactured by York, Division of Borg-Warner Corp.)

duct collar to facilitate field connection of the ductwork. The basic cabinet has four bolt holes with rubber grommets for hanging the unit. Servicing and cleaning the basic unit is simple because the motor-blower assembly is easily removable, exposing the entire face of the coil for cleaning. A drain pan is provided to collect the condensate.

The fan coil unit in Figure 119b is similar to that in Figure 119a except for the addition of a plenum section. The enclosed fan section serves as a return air plenum, making the unit compatible with the return air ductwork. The return air duct collar can be fitted with a filter and filter rack. The standard arrangement has the return air at the bottom of the unit and a removable rear access panel. The return air can easily be rerouted in the field to the rear of the unit.

The fan coil unit in Figure 119c is designed for installations where the unit and its ductwork are visible to the occupants, such as in work areas and commercial areas. This unit has a galvanized steel enclosure with a removable bottom panel for access to the valves and piping. The front panel has a supply air duct collar, and the rear panel has a return air duct collar with an integral filter rack.

The fan coil in Figure 119d is an exposed vertical unit designed for installations where vertical discharge is needed, such as in a utility closet. It is completely enclosed in a steel cabinet with a top panel that has a supply air duct collar. The front panel is easily removed to provide complete access to the coil and motor-blower assembly. Filters are removed by sliding them vertically out of the filter rack located in the return air duct collar on the front panel.

16.1. ESTIMATED LABOR TO INSTALL FAN COIL UNITS

cfm	Average Weight (lb)	Floor Mounted (man-hours)	Ceiling Mounted (man-hours)
210	80	3.00	5.25
330	100	3.00	5.25
430	115	3.00	5.25
600	130	3.00	5.25

17
FANS

A fan is designed to produce a flow of air. For such flow to take place a pressure increase in the air is required. Energy is added to the mass of air by the rotation of a fan wheel or blade. As the air passes through the fan both the dynamic and static pressure are increased.

There are two types of fans commonly used: centrifugal and axial. A centrifugal fan is comprised of a fan rotor, and is often called an impeller. The impeller rotater within a casing which collects the air from the impeller and directs it in a spiral or scroll-shaped path. An axial fan is essentially a propeller-like blade within a mounting ring or cylinder. Centrifugal fans increase pressure from the centrifugal force on the confined air and from the kinetic energy of the air leaving the impeller. The axial-type fan increases the air pressure entirely from the increased velocity air that is converted to increased static pressure.

An HVAC air distribution system is made up of ductwork and fittings plus a number of components such as dampers, heating coils, filters, etc. All of these offer a resistance to airflow. A fan must overcome this friction in order to distribute the required volume of air to the conditioned space. Therefore, the designer must add the frictional resistance of all components. The pressure needed to move air through a particular resistance depends upon the volume of flow. Knowing the volume of flow required, it is possible to determine a set of pressure-volume characteristics for a duct system. Each fan has a particular set of characteristics of pressure and

volume at a certain speed. Graphs of system characteristics and fan characteristics are superimposed. If the curves intersect at a point where there is adequate pressure and flow volume, the fan will be suitable for the application, providing the point of intersection corresponds with a reasonable efficiency of the fan. Fan curves are available from manufacturers.

Both centrifugal and axial fans can be used for an exhaust system. Wall fans which operate against little or no resistance are usually of the propeller type. Propeller fans are sometimes incorporated into factory-built penthouses or roof caps. Hooded exhaust fans and central station exhaust fans are typically of the centrifugal type. Axial fans are suitable for exhaust applications, particularly in factory installations.

17.1. AXIAL FANS

In an axial flow fan the air flows axially through the impeller. Axial flow fans are classified as propeller, tubeaxial, and vaneaxial. The tubeaxial fan has its axial flow blade within a cylindrical housing, as does a vaneaxial fan. In the latter case, air guide vanes direct either incoming or outgoing air for even distribution.

The impeller of an axial fan is shown diagrammatically in Figure 120a. An impeller blade can have a fixed angle, which results in a simple, compact fan unit.

A fan with a variable pitch control impeller has small dimensions and the airflow can be varied to maintain high efficiency. They have small

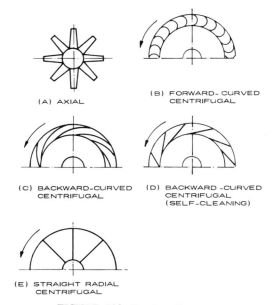

(A) AXIAL

(B) FORWARD- CURVED
CENTRIFUGAL

(C) BACKWARD-CURVED
CENTRIFUGAL

(D) BACKWARD -CURVED
CENTRIFUGAL
(SELF-CLEANING)

(E) STRAIGHT RADIAL
CENTRIFUGAL

FIGURE 120. Fan impeller types.

dimensions, and can easily be adjusted to the required airflow.

17.2. CENTRIFUGAL FANS

Centrifugal fans have impeller wheels that rotate within a scrolled housing. The air moves radially through the impeller. These fans are classified as follows according to the shape of their impellers:

1. Forward-curved bladed impellers, Figure 120b, are small and have low peripheral speed at a given pressure and flow. A change in the system characteristic will involve only modest variations in the total pressure generated.

2. Backward-curved blades, Figure 120c, have high efficiency and should be selected where operating costs are important. A change in the system characteristic will result in only a modest change in the airflow. These fans provide stable conditions for parallel operation and have low noise levels.

3. Backward-curved self-cleaning blades, Figure 120d, combine high efficiency with a high degree of self-cleaning. When using these fans, the airflow will not be significantly affected by variations in flow resistance. These fans have lower peripheral velocity for a given pressure increase than nonself-cleaning backward-curved blades.

4. Straight radial blades, Figure 120e, without an impeller shroud have the highest self-cleaning capacity, and are suitable for pneumatic transport of materials. Straight radial blades with an impeller shroud are suitable for a high pressure increase, and have a good self-cleaning capacity.

There are additional variations of impellers which relate to their width, materials of construction, and method of manufacture. In Figure 121a is shown a backward-inclined, single width impeller. The blades are welded to heavy steel rims and riveted to a hub.

The double width, backward-inclined impeller, Figure 121b, can be of the nonsparking type for hazardous locations. These can be made of aluminum wheels with steel or monel shafts and cast iron hubs. For high-temperture service, monel wheels and shafts are used with cast iron hubs. The forward-inclined blade, Figure 121c has die-forward blades that are welded or riveted to a heavy steel rim and backplate. They have a large capacity, operate at low speed, and operate quietly.

Aside from the width and construction of the impellers, centrifugal fans are further classified by manufacturers into a number of other categories.

1. *Direction of rotation and position of discharge port*, see Figure 122. The direction of rotation is determined from the drive side of the fan. For single inlet fans, the drive side is always opposite the fan inlet. For double inlet, double drive fans, the drive side is that with the higher powered motor. The direction of discharge is found on the manufacturer's diagrams. The angle of discharge is referred taken in degrees above or below the horizontal axis of the fan. For inverted or wall-mounted fans the direction of rotation and discharge is taken as though the fan were resting on the floor.

2. *Location of motor*, see Figure 123 for description. The method shown in the Figure has been adopted for designating the motor on belt driven fans and blowers by fan manufacturers.

3. *Arrangement of drive components.* This method of identifying arrangements of bearing locations is used as shorthand nomenclature in the industry. The Table is shown on page 239.

FIGURE 121. Additional impeller types.

Arrangement Number	Inlet and Width (see legend) FIGURE 124	Description
1	SWSI	For belt drive or direct connection. Wheel overhung. Two bearings on base.
2	SWSI	For belt drive or direct connection. Wheel overhung. Bearings in bracket supported by fan housing.
3	SWSI	For belt drive or direct connection. One bearing on each side of casing and supported by fan housing. Not recommended for 27-in. diameter wheel or smaller.
3	DWDI	For belt drive or direct connection. One bearing on each side and supported by fan housing.
4	SWSI	For direct drive. Impeller wheel overhung on motor shaft. No bearings on fan. Motor is base mounted.
7	SWSI	For belt drive or direct connection. Same as Arrangement 3, plus base for motor. Not recommended for 27-in. diameter wheels and smaller.
7	DWDI	For belt drive or direct connection. Same as Arrangement 3 plus motor base.
8	SWSI	For belt drive or direct connection. Same as Arrangement 1, plus extended base for motor.
9	SWSI	For belt drive. Wheel overhung, two bearings, with motor outside base.
10	SWSI	For belt drive, wheel overhung, two bearings, with motor inside base.

Direction of Rotation is determined from drive side for either single or double inlet fans.
(The driving side of a single inlet fan is considered to be the side opposite the inlet regardless of
actual location of the drive.) For fan inverted for ceiling suspension, Direction of Rotation and Discharge
is determined when fan is resting on floor.

FIGURE 122. Designations for rotation and discharge of centrifugal fans.

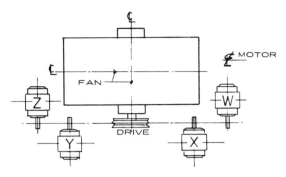

FIGURE 123. Classification of motor locations. The location of the motor is determined by facing the drive side of the fan or blower and designating the motor position by letters *W*, *X*, *Y*, or *Z* as shown.

A number of commercially available fans and blowers are shown in Figure 125.

Utility Fans. Utility fans provide a versatile and compact unit suitable for a wide variety of ventilating applications in commercial, industrial, and institutional fields. Useful for both supply and exhaust service, they are ideally suited for ventilating small or medium sized areas where it is necessary to operate against the resistance of ductwork.

As supply fans, units find wide use for such applications as ventilating storage rooms, locker rooms, auditoriums and recreation centers. As

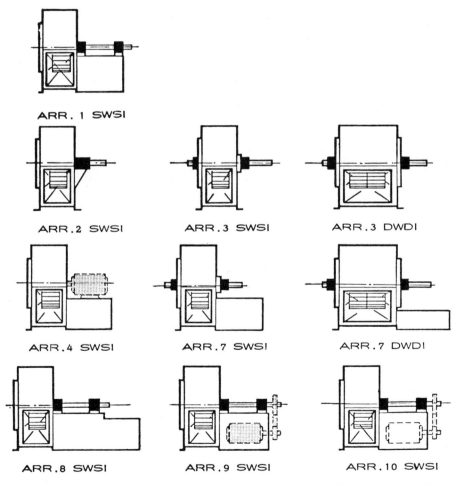

FIGURE 124. Drive arrangements for centrifugal fans. ARP = arrangement; SW = single width; SI = single inlet; DW = double inlet. (See Table on p. 237 for description.)

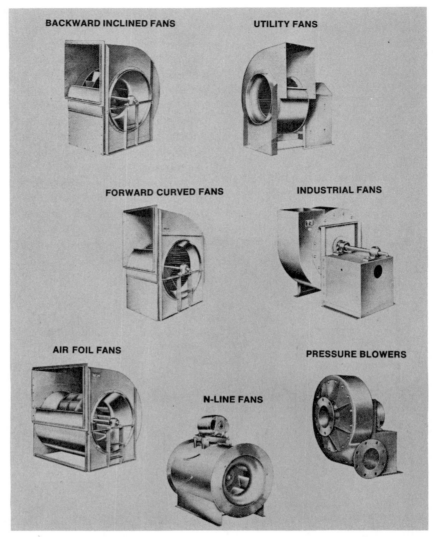

FIGURE 125. Commercially available fans and blowers. (Manufactured by Aladdin Heating Corporation.)

exhaust units, they are particularly well suited for systems required to exhaust air from kitchens, rest rooms, laboratories, and laundry rooms.

Industrial Fans. Industrial fans are manufactured in a wide choice of sizes, arrangements, and special duty wheel types to meet the air volume, pressure, and temperature requirements for all types of industrial process applications. They are ruggedly built, and are designed for trouble-free continuous duty when operating under the most severe conditions required for material handling or for exhausting hot, dirty fumes or gases.

In-Line Fans. An In-line fan is a completely new design concept. This new fan combines the reliable performance of scroll-type centrifugal fans with the spacesaving advantages of axial type fans.

Due to the straight-line airflow through the

tubular, compact housing, the need for inlet and outlet elbows, and transformation connections are eliminated. Thus, for many cases, the in-line fan can be installed in half the space required for conventional scroll-type fans. Both inlet and outlet connections are identical, making it possible to install the unit using one size duct. Units may be suspended from the ceiling, mounted on the floor, wall, or roof to take advantage of otherwise wasted space.

Inherent in the design of the in-line fan are time-proven characteristics which have become accepted standards of performance for general ventilation and industrial process application. The integration of the centrifugal airfoil wheel and multiple airfoil conversion vanes into the straight-line tubular housing has resulted in a highly efficient, quiet, and dependable unit with a stable pressure curve and a nonoverloading horsepower characteristic.

Pressure Blowers. Pressure blowers are designed for applications that require low or moderate air volumes at pressures ranging up to 30 oz/sq in. These fans are manufactured in a compact arrangement with fan rotors mounted directly to the shafts of 3600 rpm motors. The rotor is mounted on a separate shaft held in place by ball bearing pillow blocks. The fan shaft is connected to the motor by means of a flexible coupling.

Three basic sizes of blowers are manufactured, each of which can be furnished with a variety of wheel sizes to cover a wide range of performances up to 50 in. S.P. Wheels and housings can be rotated and the angle of the discharge can be changed in the field as required.

17.3. ESTIMATED LABOR TO INSTALL FANS AND BLOWERS

Aluminum roof-mounted fans (axial blade power exhauster with automatic back draft damper, bird screen and prefab curb and disconnect switch.)

nominal cfm	Motor (hp)	Average Weight (lb)	Labor (man-hours)
200	$\frac{1}{20}$	20	2.5
450	$\frac{1}{20}$	30	2.5
1200	$\frac{1}{4}$	50	2.5
2000	$\frac{1}{2}$	110	3.0
3600	$\frac{3}{4}$	180	3.5
6000	1	300	4.0
12,500	2	400	4.5
14,500	3	480	6.0
18,000	3	800	6.5
20,000	5	1200	8.0

Utility sets (centrifugal blower)

nominal cfm	Motor (hp)	Average Weight (lb)	Labor (man-hours)
250	$\frac{1}{4}$	30	2.0
500	$\frac{1}{4}$	45	2.1
750	$\frac{1}{2}$	65	2.4
1250	$\frac{1}{2}$	75	2.8
1500	$\frac{1}{2}$	120	3.4
2250	$\frac{3}{4}$	150	3.6
2750	1	200	4.1
3200	$1\frac{1}{2}$	250	4.8
4500	$1\frac{1}{2}$	300	5.0
8500	3	425	6.5
9500	5	500	7.0
12,000	$7\frac{1}{2}$	650	8.5

18
FILTERS

Air filter applications vary according to the different degrees of air cleaning required. Atmospheric contaminants are a complex mixture of products of combustion, fumes, granular particles, fibers, aerosols, and other constituants. This variety can make it somewhat difficult to select the best filter for a specific application. Government guidelines and independent institutions such as ASHRAE, in conjunction with manufacturers have established ratings and characteristics for air cleaning devices. This information enables the HVAC designer to specify the proper filter for a specific application.

All of the common types of air cleaners fall into three broad categories:

1. Fibrous media filters, in which accumulating dust increases the resistance to airflow up to some maximum permissible value. During this period, efficiency also increases. However, at high dust loads, dust may adhere poorly to the filter fiber, and efficiency will drop. Filters in such condition should be replaced, or reconditioned, as should filters which have reached their maximum permissible pressure drop. This category includes both viscous impingement and dry type air filters.

2. Renewable media filters, in which fresh media is introduced into the airstream, as needed, to maintain essentially constant resistance. The efficiency of renewable media filters stays approximately constant.

3. Electronic air cleaners, maintain a constant pressure drop and efficiency, unless their precipitating elements become severely dust laden.

Various combinations of the above types of air cleaners are used. For example, an electronic air cleaner may be used as an agglomerator with a fibrous media filter downstream to catch the agglomerated particles blown off the plates. Also, a renewable media filter may be used upstream of a high-efficiency filter to extend its life. There is also a charged media filter in which particle deposition on the fibers is increased by the use of an electrostatic field. In this case, the resistance to airflow with time increases like a fibrous media filter.

Figure 126 shows an analysis of a typical atmospheric dust sample collected at the University of Minnesota. The division of particles on the basis of weight makes particle count an interesting comparison. Atmospheric dust is affected by several variables. The components of contamination and their respective particle sizes vary with location, season of the year, direction of the wind, proximity to manufacturing areas, temperature, and weather conditions.

In Figure 126, the box is assumed to be 48 in. long × 6 in. wide × 12 in. deep. This represents a volume of 2 cu ft, which, when filled with atmospheric dust, would weigh approximately 85 lb. This is roughly the amount of dirt to which a filter would be subjected when handling 100,000 cfm for 12 hours per day for 80 days. The weight is based upon a typical atmospheric dust concentration of 0.1 grains/1000 cu ft. Re-

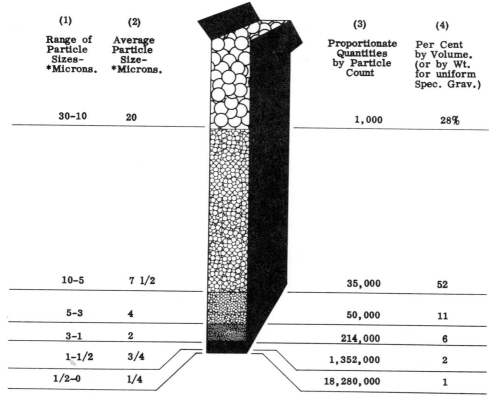

(1) Range of Particle Sizes— *Microns.	(2) Average Particle Size— *Microns.		(3) Proportionate Quantities by Particle Count	(4) Per Cent by Volume. (or by Wt. for uniform Spec. Grav.)
30–10	20		1,000	28%
10–5	7 1/2		35,000	52
5–3	4		50,000	11
3–1	2		214,000	6
1–1/2	3/4		1,352,000	2
1/2–0	1/4		18,280,000	1

FIGURE 126. Size distribution of a typical atmospheric dust sample. (*Courtesy of American Air Filter Company*.)

membering the 48 × 6 × 12-in. box, it may be of interest to mention that those particles having an average diameter of 20 microns* would appear in a full-sized diagram as circles 2 in. in diameter. This represented an actual particle size magnification of 2500 to 1.

Note that in the analysis a large percentage of the dust by weight is equivalent to a small percentage by particle count. Thus, a filter may have a relatively high weight efficiency and be ineffective for the smaller size particles. This fact is encountered every day in the application of air filters. There are many jobs where the moderate cleaning efficiency provided by viscous impingement filters is sufficient. For other applications more effective types of filters with such media as spun glass or synthetic fiber are

*A micron is a unit of linear measurement and by definition there are one million microns in a meter (39.37″). This reduces to 25,400 microns per inch.

required. This is especially true for the removal of small particles of carbon and soot in department store installations. Where the offender is tobacco smoke or dust particles in the submicron range, it may be necessary to install high-density dry-media or electrostatic filters and precipitators.

18.1. FIBROUS MEDIA UNIT FILTERS

The filter materials used for viscous impingement filters include hair, vegetable fibers, fiberglass, metal screens, or wool. Filter materials used for dry type filters include metal screens, cellulose fiber, glass wool, felts, etc.

18.1. Viscous Impingement Filters

Viscous impingement filters are flat panel filters made of coarse fibers and have high porosity. The fibers are coated with a viscous substance, such as oil, which acts as an adhesive for particles that impinge upon the filters; hence, the name viscous impingement. Design air velocity, through the media, is usually in the range of 250 to 700 fpm. These filters are characterized by low pressure drop, low cost, and good efficiency for lint but low efficiency for normal atmospheric dust. They are commonly made $\frac{1}{2}$ to 4 in. thick, with 1 and 2 in. nominal thicknesses being the most popular. In the thicker configurations, the dust-holding capacity is substantial. Unit panels are available in standard and special sizes, up to about 24 × 24 in. This type of filter is often used as a prefilter to higher efficiency filters.

The viscous impingement filter shown in

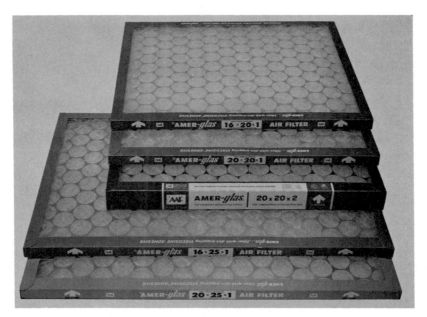

FIGURE 127. Viscous impingement and disposal fiberglass filters. (Manufactured by American Air Filter Company.)

Figure 127 is a washable, high-capacity, low-resistance permanent unit, designed to operate up to 500 fpm. The filtering media is a galvanized screen set on edge to the air stream and crimped to provide pyramid-shaped pockets.

18.1.2. Formed Screen Filters

Here, filter media made of screens or expanded metal are crimped to produce a high-porosity filter that will not collapse. Air flows through the media, and dust impinges on the wires. The relatively open structure allows the filter to store substantial quantities of dust and lint without plugging.

18.1.3. Random Fiber Media

Fibers with or without bonding material, are formed into a matrix of high porosity. Media of this type are often fabricated with fibers packed more densely on the leaving air side than on the entering air side. This arrangement permits the accumulation of larger particles and lint near the entering face of the filter, and the filtration of finer particles at the more closely packed exit face. Fiber diameters may also be graded from coarse at the entry face to fine at the exit face. The fibers used and the methods of manufacture are often cheap enough to allow clogged filters to be thrown away rather

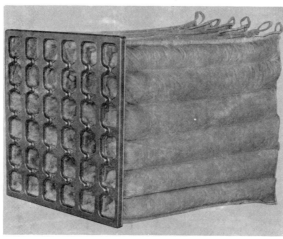

FIGURE 128. Extended surface, medium to high efficiency filters. (Manufactured by American Air Filter Company.)

than being cleaned and reused, as is commonly done with the other two types.

Figure 127 (bottom) is an example of a replaceable random fiber filter. The medium is a mat of interlaced glass filaments held together by a thermosetting bond and held in a fiberboard casing between perforated metal grilles.

18.1.4. Extended Surface, Medium to High Efficiency Filters

Filters in this category are dry type panel or cartridge units that are discarded when loaded. They work on the principle of an extended surface of the filter material in contact with the airstream. The media can be made of cellulose or other fibers, that is made into a mat that has deep pleats or otherwise presents a large surface to the incoming air. Figure 128 (top) shows a medium to high-efficiency disposable filter, ideal not only for new installations but for upgrading existing air filter systems. Incorporated in the filter design are such features as low-initial resistance, with high-dust holding capacity.

The filter at the bottom of Figure 128 is an extended surface cartridge filter which offers high-efficiency, low-resistance, compactness and unusual dust retention. When placed in the ventilating system, the filtering media inflates for maximum efficiency and dust-holding capacity. Available in efficiency ranges from 30 to 94%.

18.2. RENEWABLE OR AUTOMATIC AIR FILTERS

The automatic air filter is a self-cleaning, dry air filter designed specifically to provide for efficient, economical collection and disposal of lint, dust, etc. This filter is made in vertical and horizontal units as shown in Figure 129. The filter is a paper sheet installed on a roller at the top of the casing, and it is transported in a continuous screen down the face of the filter and is rerolled automatically at the bottom. The contaminants collected during the time the paper is exposed to the airstream are rolled up with the filter. Disposal of collected dust is simply a matter of removing the bottom roll.

The filter screen is moved by a fractional hp motor, operating through a reduction gear and chain drive. The rate of feed will vary depending upon the amount of contaminant in the air, and the air resistance of the filter. The frequency at which the curtain moves and the distance it travels is controlled by a pressure switch in the drive motor circuit. The switch is actuated by the air pressure differential across the filter

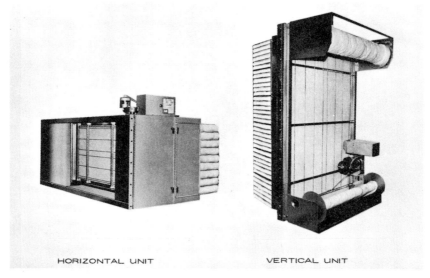

HORIZONTAL UNIT   VERTICAL UNIT

FIGURE 129. Automatic Air Filters. (Manufactured by American Air Filter Company.)

curtain and may be set for any desired resistance. When the dust load on the filter curtain raises the resistance to airflow to this predetermined point, the switch will start the drive motor. After a sufficient amount of clean filter surface has been introduced to lower the resistance of the filter (approximately 0.1 in. wg.), the pressure switch will open the motor circuit and stop the curtain travel. This method of controlling the operation of the automatic filter provides a comparatively uniform operating resistance and assures the most economical use of media. It is sensitive to any changes in dust concentrations and will feed new media into the filter curtain only as it is needed to lower the resistance below the established maximum.

18.3. ELECTRONIC AIR CLEANERS

Electronic air cleaners use electrostatic precipitators to clean the air. These are described in Section 15.1.5. However, precipitators used for cleaning ventilating air are smaller, lighter duty units than those used for cleaning stack gases. Only 12,000 V are required to ionize the dust particles for this application as opposed to the more than 35,000 V for the industrial dust control.

Electronic air cleaners offer very little resistance to airflow and can remove fine dust and aerosols from the airstream.

18.4. FILTER SELECTION AND APPLICATION

Major considerations go into the selection of a filter. These include filter efficiency, resistance, maintenance, and space accessibility.

1. *Efficiency*—A primary consideration is the air cleanliness required in the air conditioned space. This dictates the efficiency requirements of the filter. The trend in recent years has been toward higher levels of filter efficiency. Higher efficiencies produce benefits that can readily justify their higher initial cost.

2. *Resistance*—Resistance to airflow affects the horsepower required to force the air through the filter. The resistance values are usually given for a clean filter at its rated air volume flow. Also given is the recommended resistance when the filter should either be reconditioned or replaced. If a specified minimum airflow is required in a system, the resistance of the air filter just prior to its replacement should be used. If a constant volume of air delivery is desired, it will be necessary to maintain a uniform system resistance by means of dampers, or by choosing a constant resistance filter.

3. *Maintenance*—The problem of removing dust from an air-stream is twofold. The first concern is the effectiveness of the filter in removing the dust. The second concern is the means of removing the collected dust from the filter. The question of maintenance has taken on special significance in recent years because of increased labor rates. There are four methods of maintaining filters: replace the complete filter, replace the filter media only, recondition by washing or cleaning, and automatic self-cleaning.

4. *Accessibility and Space*—There are many instances when the choice of an air filter is limited to restraints of the air handling equipment. Fan coil units and self-contained air conditioners limit selection to a simple panel filter. However, central station air handling equipment is not limited to unit or panel filters as was the former practice. Today, side-access housings can house a variety of air filters.

19

FURNACES

A *furnace* can be defined as a self-enclosed apparatus that heats air which is circulated through the enclosure. The heated air is discharged directly into the space being heated or is conveyed through ducts. A warm air furnace is comprised of a fan, including motor and belt drives, a combustion chamber, with heat exchanger burner; a flue gas fan; an airflow casing; and a control station. There are three types of air furnaces: oil-fired, gas-fired, and electrical.

Furnaces intended for commercial and industrial uses are also known as heavy duty heaters, heavy duty fan furnaces, nonresidential warm air heater, and by a variety of other terms.

19.1. GAS FURNACE

In gas-burning furnaces, the combustion takes place within a metal-walled heat exchanger. The circulating air passes over the outside surfaces of the heat exchanger. Heat transfer takes place through the heat exchanger wall, and the circulating air does not come in contact with the fuel or the products of combustion. The products of combustion are conveyed to the outside atmosphere through a flue or vent.

Gas furnaces may burn natural gas, or manufactured, mixed, liquified petroleum, and liquified petroleum-air gases. Most residential gas furnaces are equipped with one of two types of heat exchangers. The first is a single combustion chamber, usually cylindrical in shape, with extended flue-gas passageways which are either adjacent to the combustion chamber, or encir-

cle it. In the second case, the heat exchanger is made of a series of individual sections connected to each other near the bottom and to a common flue-gas breeching at the top.

The American National Standards Institute's Requirements for Gas Fired Gravity and Forced Air Central Furnaces (ANSI Z21.47), is universally used in testing and rating. All gravity gas furnaces certified by the American Gas Association under these requirements are assigned a rating based on no less than 75% efficiency. Forced air gas furnaces are rated on a minimum of 80% efficiency.

19.2 OIL FURNACE

Oil furnaces are available as complete packages including the burner and controls. Pressure-type or rotary burners are included as a part of the package. Oil-fired furnaces equipped with pressure-atomizing or rotary burners are rated in accordance with commercial standard test methods. This requires a minimum efficiency of 80% for forced air furnaces and 75% for gravity furnaces.

Furnaces are available in a variety of shapes as shown in Figure 130. These include upflow, counterflow, low-boy, and horizontal. A description of the geometry of these is found in the section on Electric Furnaces, 19.3. One such type, an upflow oil-fired furnace, is illustrated as a sectional view in Figure 131; its principal parts are identified and their function described.

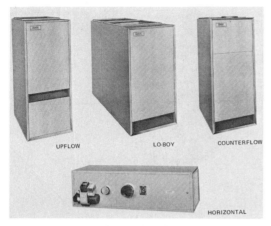

FIGURE 130. Oil-fired furnace types.

Oil-fired furnaces can be used alone for warm air heating or combined with an air cooling system. Figure 132 shows an upflow oil-fired furnace in combination with a split cooling system. The condensing section is located outdoors, and the evaporator coil is placed at the outlet of the furnace, which is in the basement of the residence.

19.3. ELECTRIC FURNACE

In an electric furnace a resistance heating element either heats the circulating air directly or through a metal sheath enclosing the element. Furnaces are manufactured in a variety of ways to best suit the particular design need. The most common types of furnaces are described below.

1. *Horizontal Furnace* (Figure 133, top). This type of furnace has a blower located behind the heat exchanger. The air enters at one end and travels horizontally through the blower and over the heat exchanger; it is discharged at the opposite end. These furnaces are used for locations with limited head room, such as attics, crawl spaces below floors, or suspended under a ceiling. They are often designed for the components to be rearranged so that the airflow may be in the reverse direction from that shown in the Figure.

1. High-pressure gun-type oil burner atomizes either No. 1 or No. 2 fuel oil into super-fine droplets for complete combustion.
2. Casing thoroughly insulated to keep heat inside and the exterior casing at room temperature.
3. Drawer-type assembly (in air tube of burner) groups oil lines, fittings and other components into one assembly for easy inspection and flame adjustment.
4. Flame detector cuts off burner motor if oil fails to ignite.
5. Centrifugal type blower with rubber isolators.
6. Burner encasement, protects burner and other parts from dust and dirt and presents a neat appearance.
7. Heat exchanger—This stainless steel combustion chamber receives flame from the oil burner, isolates it to prevent corrosion and scaling of the heat exchanger and distributes intense heat over the entire surface. Because of this hot surface, air from the furnace blower is heated quickly and evenly.

FIGURE 131. Sectional view of an oil-fired furnace. (Manufactured by York, Division of Borg-Warner Corp.)

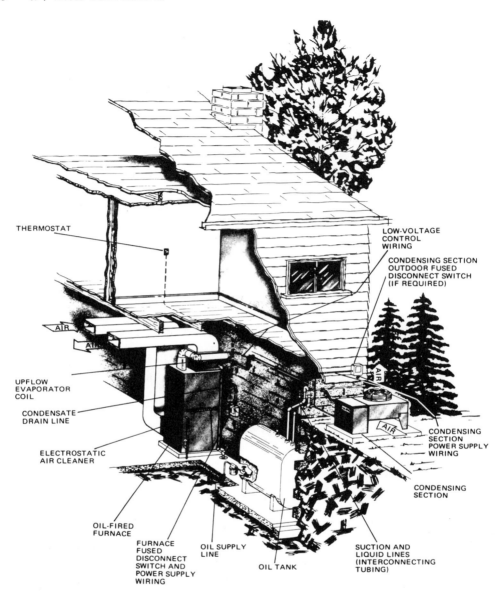

FIGURE 132. Oil-fired furnace applied with split system air conditioner. (Manufactured by York, Division of Borg-Warner Corp.)

2. *Upflow or High-Boy* (Figure 133, bottom). This furnace has a blower beneath the heat exchanger that discharges vertically upward. Air enters through the bottom or the side of the blower compartment and leaves at the top. These furnaces may be used in utility rooms on the first floor of homes without basements, or in basements with the return air ducted down to the blower compartment entrance.

3. *Low-Boy* (Figure 134, top). The low-boy is a variation of the upflow furnace and requires less head room. The blower is located alongside the heat exchanger at the bottom. The air enters the top of the cabinet, is drawn down past the filters through the blower, and is discharged over the heat exchanger. The warm air leaves vertically at the top.

4. *Downflow or Counterflow* (Figure 134,

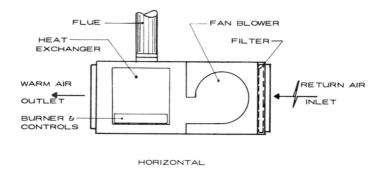

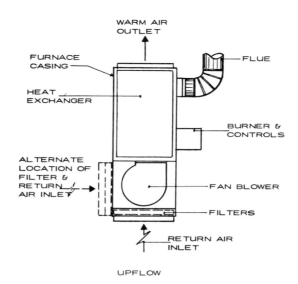

FIGURE 133. Horizontal and up-flow forced warm air furnaces.

bottom). This furnace has the blower located above the heat exchanger and is positioned to discharge downward. The air enters at the top and is discharged vertically at the bottom. This style furnace is used with a perimeter heating system in a house without a basement.

19.4. FORCED AIR FURNACES FOR OUTDOOR INSTALLATION

Forced air furnaces for outdoor installation are similar in many respects to standard central heating furnaces, but they must be designed to withstand the effects of weather. Cabinets and components must be corrosion resistant,

and pilots and controls cannot be affected by wind or a wide range of temperatures.

The majority of outdoor units are roof-mounted, although some are mounted through the walls or on slabs at ground level close to one wall of the structure. In residential applications, the units are connected to conventional duct systems. Commercial installations frequently utilize only one supply outlet and one return inlet. The return air may be taken completely from the conditioned space, or completely from the outside, or from both locations.

These outdoor units are frequently combinations of heating and self-contained cooling units. Various combinations of gas, oil, and electricity are used as the source of energy.

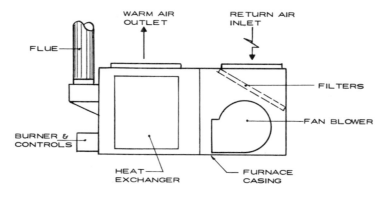

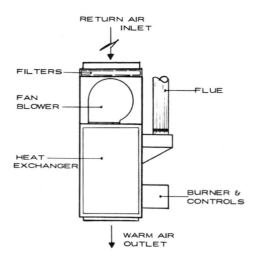

FIGURE 134. Low-boy and downflow forced warm air furnaces.

19.5. ESTIMATED LABOR TO INSTALL FURNACES

Blower unit, gas-fired type

Nominal cfm	Average Weights (lb)	Labor (man-hours)
1600	250	8
2000	300	8
3200	375	8
3600	450	12
4000	650	12

Duct, gas-fired type

Nominal cfm	Average Weights (lbs)	Labor (man-hours)
600	100	4
900	150	4
1250	200	6
1600	225	6
2000	275	6
2250	300	8
2500	325	8
3000	375	8
4000	425	10
5000	550	10

20
GRILLES, REGISTERS, AND DIFFUSERS

The objective of selecting the proper supply outlet is to provide a uniform and desirable level of air motion within the conditioned space. If this is achieved, temperature within the space will be uniform and air motion will be such that the occupants will not experience a sensation of stagnant air. At the same time, the air velocities will be low enough for the occupants to not experience drafts.

The type and style of a supply outlet or return intake for a specific application is usually determined jointly by the architect and the mechanical engineer. Their decisions are influenced by a number of factors, including the dimensions and use of each space, desired appearance, duct system location, etc. After making the preliminary decisions on style, type, and general location, the specific selection can be made by the design engineer.

Like every other step in the design of mechanical systems, the selection of supply outlets and return intakes is a mixture of art and science. Although specific guidance is contained in the selection tables of manufacturers' catalogs, an understanding of the principles of air diffusion and judgement in the use of the data are essential to proper selection.

The design engineer needs accurate, reliable product data upon which to base his decisions, providing this is the responsibility of the manufacturer. The data must be conveniently presented. The tabulations must make sense not only to the experienced designer, but to the novice as well. Selection data must be compre-

hensive enough for the engineer to select registers, grilles, and diffusers within the limitations imposed by all of the system design conditions.

20.1. GRILLES AND REGISTERS

A *grille* is a louvered covering placed over an opening through which air is introduced into or drawn out of the room. A *register* is an assembly of a grille plus a damper for controlling the volume of air which passes through the grille. A supply grille or register is placed over the opening through which air is introduced into the room. Usually the louvers of a supply grille are adjustable so that the pattern of the supply air can be controlled. A return grille or register is placed over the opening through which air is withdrawn from the room. Often, the louvers of a return grille are not adjustable. Their angle and design are determined by permissible pressure loss and sound level, and by the amount of "see through" which is permitted from the room into the return opening.

A single deflection grille or register has one bank of louvers. The louvers may be either horizontally or vertically oriented to control the air pattern in one plane. A double deflection grille or register contains two banks of louvers, one horizontal and the other vertical. Each bank of louvers controls the air delivery pattern in one plane.

Vertical louvers control the spread angle of the airstream. Although the vertical louvers in a grille may be set to provide any specified spread

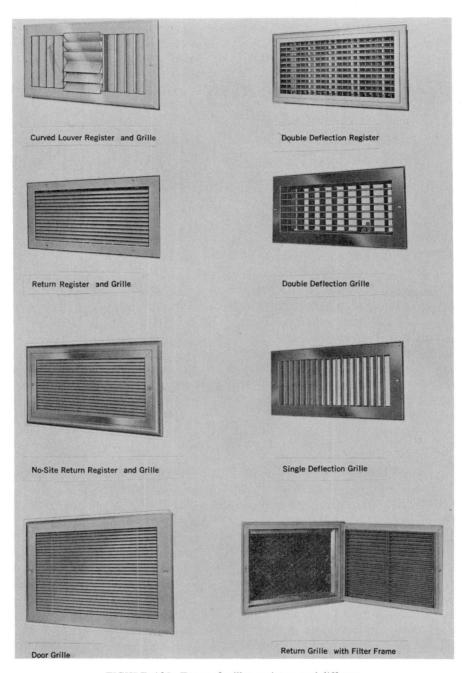

Curved Louver Register and Grille

Double Deflection Register

Return Register and Grille

Double Deflection Grille

No-Site Return Register and Grille

Single Deflection Grille

Door Grille

Return Grille with Filter Frame

FIGURE 135. Types of grilles, registers, and diffusers.

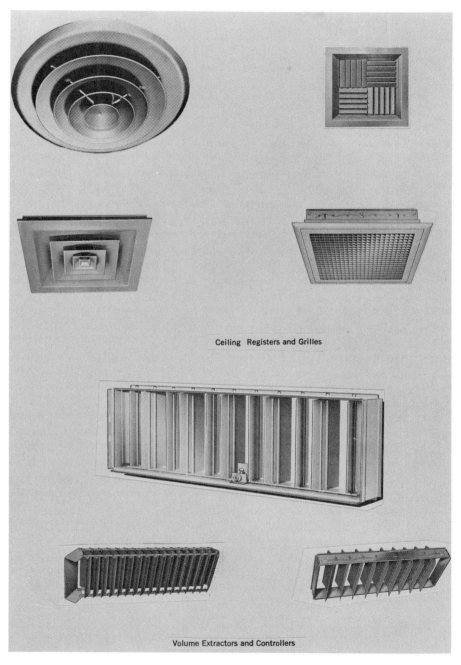

Ceiling Registers and Grilles

Volume Extractors and Controllers

FIGURE 135. (*Continued*)

angle, three commonly used angles are $0°$, $22\frac{1}{2}°$, and $45°$. The spread angle is measured between a line perpendicular to the grille face (airflow direction) and a line paralled with the louver setting.

Horizontal louvers control the vertical deflection of the airstream. Deflection angles commonly used are from $0°$ to $20°$, upwards. Downward deflection may be specified for some applications. Single deflection grilles and registers can control either horizontal spread or vertical deflection, whereas double deflection grilles and registers can control both. Dampers control only the volume of air supplied through the grille. They do not significantly affect the shape of the delivery pattern. A number of grilles and registers are shown in Figure 135.

Many different grille areas can be defined. The core is that portion of the grille contained within the frame. The core area is the largest of the areas used to describe a grille or register, since it does not take into account the portion of the core occupied by louvers.

Free area is the minimum area in the grille face through which air can pass. This is sometimes called the "open area" and is, in fact, the open space between the vanes. Although the concept of free area is useful, its use has two significant drawbacks. Determining free area requires that physical measurements be taken at right angles to louvers or vanes, which may be deflected, offset, or curved. Second, not all of the free area will be used if the air is flowing at fairly high velocity and is being deflected either horizontally or vertically by the louvers. Hence, free area is of limited use in grille selection.

A much more useful concept is that of area factor, symbolized by Ak. The area factor is determined in the laboratory by measuring the average air velocity at the grille face when a known air volume is delivered through the grille. The area factor is defined by the equation:

$$(Ak)\ \text{Area Factor} = \frac{\text{Air Volume}}{\text{Air Velocity}}$$

where the air volume is expressed in cubic feet per minute (cfm), and the air velocity is expressed in feet per minute (fpm).

Even though Ak is a calculated number, it does express the area characteristic of a grille or register in a meaningful way. Its units are square feet.

20.2. DIFFUSERS

A supply outlet which is capable of delivering air in a pattern generally parallel to the surface in which the outlet is located is called a diffuser. Although diffusers can be located in walls or floors, they are most commonly located in the ceiling. Square or rectangular diffusers are used to deliver air in almost any direction and pattern desired. Several types of diffusers are shown in Figure 135. The delivery pattern of adjustable diffusers can be changed so that air is delivered more nearly perpendicular to the ceiling. The adjustment is made at the diffuser face and is made for seasonal changes in the air pattern.

Many of the terms commonly used in diffuser selection are identical to those used for register and grilles. Such terms as face velocity, static pressure, and total pressure have the same meaning with directional diffusers as with registers and grilles. Similarly, area factor (Ak) is determined with the formula in the first column.

Diffuser size is usually described by the dimensions of the neck of the diffuser—that part of the diffuser which is attached to the supply duct. The average velocity of the air as it passes through the diffuser neck is called the neck velocity.

The air velocity at the face of a directional diffuser, the face velocity, is often quite variable. When balancing the system, face velocity must be determined in strict accord with balancing data and instructions published by the manufacturer.

20.3. ESTIMATED LABOR AND MATERIAL COSTS FOR GRILLES, REGISTERS, AND DIFFUSERS

Round diffusers, steel primed

Diameter (in.)	Unit EA	Labor (man-hours)	Average Material Cost ($)
6	↓	1.52	16
8		1.52	16

Round diffusers, steel primed

Diameter (in.)	Unit EA	Labor (man-hours)	Average Material Cost ($)
10		1.60	16
12		2.34	35
14		2.34	42
16		2.34	42
18		3.14	42
20		3.14	82
24		5.47	140
30		5.47	240
36	↓	5.47	300

Registers, face adjusted slots, steel primed

Size (in.)	Unit EA	Labor (man-hours)	Average Material Cost ($)
8 × 4		0.95	9.00
8 × 6		0.95	
10 × 4		1.04	
10 × 6		1.04	
10 × 8		1.04	
12 × 6		1.18	
12 × 8		1.18	
12 × 10		1.18	
14 × 6		1.20	
14 × 8		1.20	
14 × 10		1.20	15.00
14 × 12		1.20	
16 × 6		1.32	
16 × 8		1.32	
16 × 10		1.32	
16 × 12		1.32	
18 × 6		1.32	
18 × 8		1.32	
18 × 10		1.32	
18 × 12		1.32	20.00
20 × 6	↓	1.55	16.00

Registers, face adjusted slots, steel primed

Size (in.)	Unit EA	Labor (man-hours)	Average Material Cost ($)
20 × 8		1.55	
20 × 10		1.55	
20 × 12		1.55	
24 × 6		2.05	20
24 × 8		2.05	20
24 × 10		2.05	20
24 × 12		2.05	20
30 × 6		2.64	20
30 × 8		2.64	20
30 × 10		2.64	20
30 × 12		2.64	30
30 × 14		2.64	30
30 × 16		2.64	30
30 × 18		2.64	30
36 × 6		3.27	30
36 × 8		3.27	30
36 × 10		3.27	30
36 × 12		3.27	40
36 × 14		3.27	50
36 × 16	↓	3.27	60

Grilles steel, air transfer type

Size (in.)	Unit	Labor (man-hours)	Average Material Cost ($)
6 × 6	EA	0.95	6.50
8 × 6		0.95	6.50
10 × 6		1.04	6.50
14 × 6		1.20	6.50
16 × 8		1.32	6.50
18 × 6		1.32	12.50
22 × 10		2.05	14.00
24 × 12		2.05	14.00
30 × 8		2.64	14.00
36 × 12		3.27	30.00
36 × 24	↓	3.27	45.00

21

HEAT EXCHANGERS

Heat exchangers are chambers designed to allow the transfer of heat from one fluid (liquid or gas) to another. Boilers, distilling plants, and deaerating feed tanks are primary examples of heat exchangers. However, these large and complex pieces of equipment are seldom referred to as heat exchangers, but by their specific names.

As mentioned in Section I, for heat to be transferred from one substance to another, a temperature difference is required. Heat flow can occur only from a substance that is at a higher temperature to a one at a lower temperature. If two objects at different temperatures are placed in contact or near each other, heat will flow from the warmer object to the cooler one until both are at the same temperature. Heat transfer occurs at a faster rate when there is a large temperature difference. As the temperature difference approaches zero, the rate of heat flow also approaches zero.

Some heat exchangers raise the temperature of one fluid. Fuel oil heaters, combustion air preheaters, lube oil heaters, and many other heat exchangers used on HVAC systems, serve this function. Other heat exchangers lower the temperature of one fluid. Lube oil coolers, boiler water sample coolers, and desuperheaters are examples of this type of heat exchanger. In condensers, latent heat is removed from a fluid to change it from a gas to a liquid. Very often, we want to remove the latent heat without removing any sensible heat, that is, we want to change the state of the fluid but do not want to lower its temperature. For example, the purpose

of the main condenser is to remove the latent heat from turbine exhaust steam so that the steam will condense. In this process we do not want to lower the temperature of the condensate. Since any heat removed from the condensate must be replaced in the deaerating feed tank or in the boiler, lowering the temperature of the condensate is wasteful of heat and therefore wasteful of fuel.

In still other heat exchangers, we want to add latent heat to a fluid in order to make it change from a liquid to a gas. The generating part of a boiler is a good example of this type of heat exchanger. Since it is impossible to raise the temperature of the steam as long as it is in contact with the water from which it is being generated, the steam does not increase in temperature until it has been drawn off into another heat exchanger, the superheater.

Heat exchangers may be classified according to the path of heat flow, the relative direction of the flow of the fluids, the number of times that either fluid passes the other, and such general construction features as the type of heat transfer surface and the arrangement of component parts. The types of heat exchangers in common use in HVAC applications are described in terms of these basic classifications.

21.1. PATH OF HEAT FLOW

When classified according to the *path* of heat flow, heat exchangers are of two basic types, *indirect* or surface type and *direct* contact type. In the *indirect or surface type*, the heat flows from one fluid to the other through some kind of tube, plate, or other surface that separates the two fluids; consequently, there is no mixing of the fluids. In the *direct contact type* of heat exchanger, the heat is transferred directly from one fluid to another as the two fluids mix. In surface heat exchangers, the fluids may flow parallel to each other, counter to each other, or at right angles to each other. The latter case is called *crossflow*.

In parallel flow, both fluids move in the same direction. If a parallel flow heat exchanger has a long enough heat transfer surface, the temperatures of the two fluids will be practically equal as the fluids leave the heat exchanger. *In counterflow*, the two fluids flow in opposite directions. Counterflow heat exchangers are used in many applications where it is necessary to obtain a large temperature change in the cooled or heated fluid. In crossflow, one fluid flows at right angles to the other. Crossflow is particularly useful for removing latent heat and thus condensing a vapor to a liquid.

Surface heat exchangers may be also classified as single-pass if one fluid passes another only once; or as multipass, if one fluid passes the other more than once. Multipass flow may be obtained either by the arrangement of the tubes and the fluid inlets and outlets or by using baffles to guide a fluid so that it passes the other fluid more than once before it leaves the heat exchanger.

Surface heat exchangers are known as plain surface units, if the surface is relatively smooth, or as extended surface units, if the surface is fitted with rings, fins, studs, or some other kind of extension. The main advantage of the extended surface lies in the fact that the extensions increase the heat transfer area without requiring any substantial increase in the overall size and weight of the heat exchanger.

Surface heat exchangers are often called by names that indicate general features of design and construction. Basically, all surface heat exchangers are of shell-and-tube construction. However, the shell-and-tube arrangement is modified in various ways and in some cases it is not easy to recognize the basic design. Shell-

and-tube heat exchangers include such types as:

- straight tube,
- U-tube,
- helical or spiral tube,
- double-tube, and
- plate tube heat exchangers.

In straight tube heat exchangers, the tubes are usually arranged in a bundle and enclosed in a cylindrical shell. The ends of the tubes are expanded into a tube sheet at each end of the bundle, or they may be expanded into one tube sheet and packed and ferruled into the other. The use of ferrules allow the tubes to expand and contract slightly with temperature changes, thus avoiding the stress if they were rigidly joined at both tube sheets. U-bend heat exchangers, sometimes called return bend heat exchangers, consist of a bundle of U-shaped tubes inside a shell. Since the tubes are U-shaped, there is only one tube sheet. The shape of the tubes provides a sufficient allowance for expansion and contraction.

Helical tube or spiral tube heat exchangers have one or more coils of tubing installed inside a shell. The tubes may communicate with headers at each end of the shell, or in the case of relatively simple units such as boiler water sample coolers, the ends of the tubing may pass through the shell and serve as the inlet and the outlet for the fluid in the tubing.

Double-tube heat exchangers have one tube inside another. One fluid flows through the inner tube and the other flows in the annulus between the outer and inner tubes. The outer tube may thus be regarded as the shell for the

FIGURE 136. Steam heat exchanger (converter). (Manufactured by Dunham-Bush, Inc.)

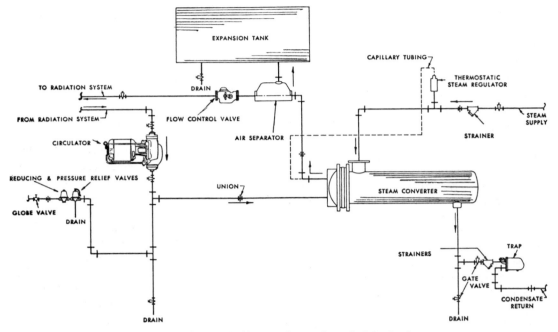

FIGURE 137. Typical heat exchanger (steam) piping hookup.

inner tube. The shells or outer tubes are usually arranged in groups, called banks and are connected at on end by a common tube sheet and a partitioned cover that serves to direct the flow. Many double-tube heat exchangers are of U-bend construction to allow for expansion and contraction.

A steam converter is shown in Figure 136. This surface type of heat exchanger is used primarily for heating radiation water with steam.

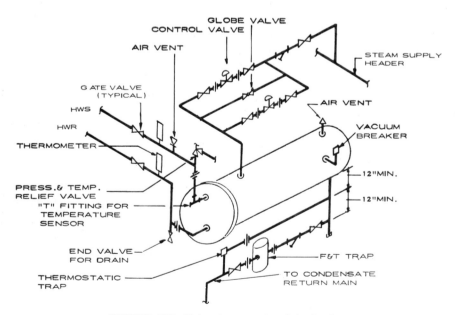

FIGURE 138. Hot water converter piping hookup.

Definite advantages are to be found for large installations when zoning is desirable. Steam as the source of heat may simplify the piping design required for the exclusive use of water. Converters often provide heating water for office space and are used for heating in high buildings to keep static heads at a reasonable level and simplify piping design.

If steam is used for process work, a converter is a logical selection for providing heating water for the occupied space.

Figures 137 and 138 show piping arrangements for a steam converter for use with water for radiators and for hot water for domestic or process applications.

21.2. ESTIMATED LABOR TO INSTALL HEAT EXCHANGERS

Shell and coil type (installed in place)

Nominal Dimensions (Diameter × Length, in in.)	Average Weight (lb)	Labor (man-hours)
8 × 24	40	4
8 × 36	50	4
12 × 24	100	8
12 × 30	100	8
12 × 42	120	8

Shell and tube, liquid chiller type

Nominal Diameter (in.)	Average Weight (lb)	Labor (man-hours)
8	700	12
8	900	12
8	1200	12
10	925	12
10	1200	12
10	1750	12
12	1300	12
12	1700	12
12	2200	14
14	1500	12
14	2000	12
14	2600	14
16	1800	12
16	2500	14
16	3250	14
20	2500	14
20	4000	14

Shell and tube, liquid chiller type

Nominal Diameter (in.)	Average Weight (lb)	Labor (man-hours)
20	5500	20
24	3650	14
24	5500	20
24	7000	20
30	5500	20
30	8250	20
30	10000	24

Shell and tube converter type

Nominal dimensions (Diameter × Length in in.)	Average Weight (lb)	Labor (man-hours)
4.5 × 32	50	4
4.5 × 48	80	6
4.5 × 84	130	6
6.5 × 32	110	6
6.5 × 48	170	6
6.5 × 84	230	6
6.5 × 102	290	7.5
8.5 × 30	170	6
8.5 × 60	300	6
8.5 × 84	370	6
8.5 × 120	500	6
10 × 40	270	6
10 × 60	400	6
10 × 84	550	6
10 × 100	690	12
10 × 132	830	12
12 × 48	470	6
12 × 60	650	12
12 × 100	850	12
12 × 120	1150	12
14 × 48	570	12
14 × 66	790	12
14 × 96	1020	12
14 × 124	1300	12
14 × 130	1450	12
16 × 48	700	12
16 × 72	1000	12
16 × 96	1300	12
16 × 124	1650	12
16 × 132	1750	12
20 × 48	1120	12
20 × 72	1600	12
20 × 98	2000	12
20 × 124	2450	14
20 × 136	2650	14

22

HEAT PUMPS

Heat pumps are heat transfer devices. They have the same elements as a compression refrigerating plant, namely:

1. an outdoor element that extracts heat from the outside or rejects heat to the outside air. This can be called a heat absorber, evaporator, or refrigerant-to-air heat exchanger.
2. an indoor element that adds or extracts heat from conditioned space. This can be called a condenser or refrigerant-to-water heat exchanger.
3. a compressor, to elevate the heat in the heat in the "refrigerant" so it may be used for heating, or rejected to the outside air.
4. a reversing valve.
5. controls.

Heat pumps are reversible and can move heat in either direction. In the winter, small quantities of heat are extracted from outside air, from water or another heat source, and are used for heating. Conversely, during the summer, heat is extracted from the air in occupied areas by a refrigerant, then released to the outside air or water depending upon the system used. Heating and cooling cycles for a heat pump are shown in Figure 139.

The heat pump has the capacity to transfer latent heat as well as sensible heat. They are available in the following configurations:

- air-to-air
- water-to-air
- water-to-water
- air-to-water

Heat pump sizes range from $\frac{1}{2}$ to 25 tons, as commercially available package units. Larger sizes can be obtained by simple modifications to refrigeration machines. And, they come as horizontal or vertical packages. The horizontal model is used where it is necessary to conserve floor space, and floor mounting is not practical. Designed for concealed installations in overhead locations, the units can be suspended in drop or plenum ceilings above the areas to be served. Access to the units is required on two sides only. All ceiling mounted units must be provided with a hinged ceiling access panel or removable lay-in tile ceiling. Vertical units are designed for for floor mounted, vertical discharge, hideway applications. They are normally furred into partition walls or located in closets, utility rooms, or other concealed locations. Access to the unit components is required on two sides only. It is recommended that unit access for servicing be given careful consideration for any type of unit installation, be it a removable wall panel or framed door. Commercially available horizontal and vertical heat pump units are shown in Figure 140.

Heat pumps as heat exchange units, when in the heating mode, have a high efficiency which extracts more heat than any other heat exchange device. In the heating mode, the power input is approximately $\frac{1}{3}$ to $\frac{1}{5}$ of the heat output. Heat pumps are more favorable than electric resistance

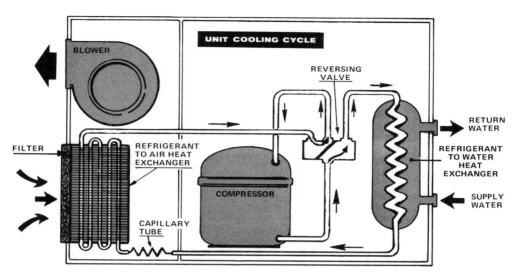

COOLING CYCLE—Upon a demand for cooling, the low voltage thermostat activates the centrifugal blower and de-energizes the reversing valve. If all safeties are made the compressor will start pumping high temperature refrigerant vapor to the refrigerant-to-water coaxial heat exchanger. The refrigerant vapor is condensed to a liquid, giving up its heat to the condenser water loop.

Liquid refrigerant then passes through the capillary tube, into the evaporator (air side) coil, where it expands into a low temperature vapor. Here it absorbs heat from the air passing over the air side coil. To complete the cycle, the refrigerant then flows through the reversing valve back to the suction side of the hermetic compressor.

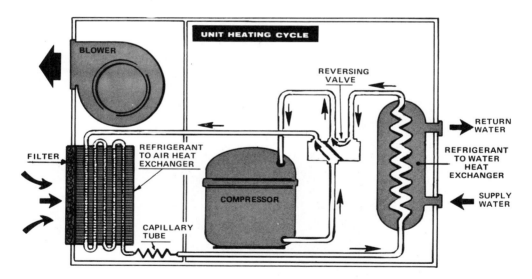

HEATING CYCLE—Upon a demand for heating, the low voltage thermostat activates the centrifugal blower and energizes the reversing valve. The hot refrigerant discharge gas is directed to the air side coil where it gives off its heat to the air passing over it. Condensing

into a liquid, the refrigerant then passes through the capillary tube to the refrigerant-to-water coaxial heat exchanger. The refrigerant absorbs heat from the condenser water loop and then to complete the cycle, returns to the suction side of the hermetic compressor.

FIGURE 139. Heat pump, heating and cooling cycles. (*Courtesy of Dunham-Bush, Inc.*)

FIGURE 140. Water cooled heat pumps. (Manufactured by Dunham-Bush, Inc.)

heating, and they compare well with oil- or gas-fired heating systems. When used in the cooling mode, their performance decreases, and they should be considered exclusively for this purpose.

Heat pumps make heat available from low-temperature sources. For example, heat can be extracted from cold drain water at approximately 50°F with rejection at temperatures high enough for use by domestic hot water systems or for preheating of the air.

22.1. ESTIMATED LABOR TO INSTALL HEAT PUMP UNITS

Single packaged, air-to-air unit type

Size (tons)	Labor (man-hours)
2	6
3	9

Single packaged, air-to-air unit type

Size (tons)	Labor (man-hours)
4	12
5	12

Split system, air-cooled with outdoor section unit

Size (tons)	Labor (man-hours)
3	8
4	10
5	12

Split system, air-cooled unit

Size (tons)	Labor (man-hours)
3	8
4	10
5	12

23

HEAT RECOVERY UNITS
(Thermal Wheels)

Gas shortages and the ever-rising cost of electrical power and fuel oil have created serious problems for consumers, power utility companies, and the engineers. Exhaust heat recovery for industrial and HVAC applications significantly reduced fuel and power consumption, and owning and operating costs. A means of recovering exhaust heat is found in a device called the heat exchanger wheel, making it of great interest to those affected by high fuel costs.

A thermal or heat wheel is a rotating heat exchanger, driven by an electric motor, with a high thermal inertia core. They are capable of transferring energy from one airstream to another and, in very large boiler plants, from flue gas to air.

Many authorities have predicted that by the end of the next decade virtually all new buildings will be completely environmentally controlled. Competition dictates that the most successful planners, builders, and vendors will be those who provide the highest quality at a reasonable cost. The thermal wheel offers great promise of contributing to environmental control with a high degree of cost-effectiveness. Of greatest importance is that this heat recovery system uses all, or a large portion of the outside air at similar or less cost than systems that recirculate the same stale air. Ventilation with outside air improves the controlled environment by helping to achieve safe, healthful conditions by dilution. This reduces ordors, toxic or hazardous dusts or vapors, bacteria, and other microorganisms. Our safety and health authorities recognize this, and make ventilation with outside air compulsory for many critical areas.

Reduction in the size of the heating and cooling plant, results initially in a smaller capital investment, and later in reduced fuel, electricity, maintenance, and other operating costs. The use of a high grade energy such as electricity, which allows a minimum initial investment is virtually 100% efficient in its conversion to heat. Another advantage is its ease of control. Electricity becomes economical with heat recovery, due to lower demand, even though electrical energy costs are high.

It is not conjecture that one day we will find that fossil fuels are of too great a value to burn as a source of heat. Their most valuable use will be for production of petro-chemicals and other materials. Heat recovery must take its place in the conservation of our resources.

23.1 SYSTEM DESCRIPTION

It was noted above that the rotary heat exchange is a driven wheel containing material with thermal heat capacity. Part of the wheel resides in the hot airstream, part in the cold.

The hot and cold airstreams must be immediately adjacent and parallel to permit installation of the heat wheel. Duct modifications may be necessary for existing systems. Two types of thermal wheels are available: a. one which transfers only sensible heat, b. one which transfers both sensible and latent heat.

Exhaust air (from the building or space) and make-up air (from outside), flowing in opposite directions, are introduced to each half of the thermal wheel through separate adjacent dusts. A thin cylinder containing a heat transfer media slowly rotates between the two airstreams. Energy absorbed by the media is transferred from one airstream to the other. The media can be metal, organic fibers, or synthetic organic polymers with random or directional air passages. For this application the media is made of foam, felt, or wire mesh.

A heat recovery package unit is shown in Figure 141. At the upper right is an enlarged view of the rotary heat exchanges, which is also shown assembled in the unit.

During a heating cycle, incoming low-temperature fresh air is heated and humidified by heat transferred from the warm, moist exhaust air. During summer months, cooling and dehumidifying of higher temperature moist, outside air are accomplished in a similar manner by giving up heat to the cooler, exhaust airstream.

1. Twin rotary–Heat exchanger.
2. Exhaust fan.
3. Supply fan.
4. Exhaust back draft damper.
5. Louvers with bird screen.
6. Supply air filters.
7. Exhaust air filters.
9. Heat recovery unit cabinet.
10. Exhaust and supply fan duct connectors.
11. Exhaust and supply fan vibration isolators.
12. Electrical control center and master disconnect switch.
13. Motorized louvers.
14. Motorized damper.
15. Exhaust air bypass motorized damper.
16. Outside air intake filters and motorized louvers.
17. Heating coil.
18. Reheating coil.
19. Cooling coil.

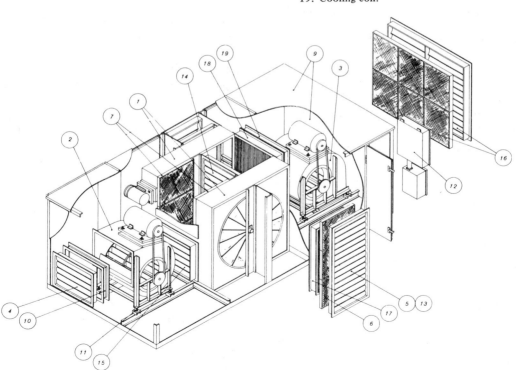

FIGURE 141. Heat recovery package unit. (Manufactured by the Heat Recovery Corp.)

The total system efficiency is equal to the heat transfer efficiency plus the heat of the driving motors in the heating mode. During the cooling cycle, the energy required to rotate the wheel, plus the additional energy to overcome the air resistance, must be deducted from the gross load reduction due to heat transfer.

Thermal wheels typically have an efficiency of 60 to 80% for sensible heat transfer and 20 to 60% for latent heat transfer. These efficiencies should be applied to the theoretical maximum heat transfers, in order to obtain the actual heat transfer rate in Btuh's, when selecting a thermal wheel.

When installing thermal wheels into ventilation systems use a roughing filter and/or insect screen to protect the core. Icing, which can occur under some outdoor conditions with enthalpy wheels, can be eliminated by preheating the air up 10°F. When used to transfer heat from flue gases, the construction of the thermal wheel must be suitable for the temperatures encountered.

Thermal wheels are commercially available in sizes ranging from 3000 to 60,000 cfm. They can be provided with variable speed controls.

23.2 SYSTEM APPLICATIONS*

Heat recovery units are used in schools, hospitals, apartments, lobbies, theaters, and commercial buildings for make-up air heating and air conditioning systems. Five applications are described and illustrated:

1. Heating cycle recovery with 100% outside air.
2. Cooling cycle recovery with 100% outside air.
3. Heating cycle recovery using 100% outside air with additional outside air blending.
4. Recirculating cycle with no outside air.
5. Ventilating cycle.

23.2.1. Heating Cycle Recovery with 100% Outside Air (Figure 142-a)

When supply air temperature discharge from the heat exchanger is below the thermostat

*Design data courtesy of the Heat Recovery Corp.

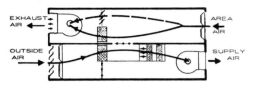

(a) HEATING CYCLE RECOVERY 100% OUTSIDE AIR

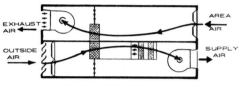

(b) COOLING CYCLE RECOVERY 100% OUTSIDE AIR

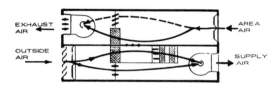

(c) HEATING CYCLE RECOVERY 100% OUTSIDE AIR WITH ADDITIONAL OUTSIDE AIR BLENDING

FIGURE 142. Heat recovery system applications.

set point, the exhaust air will flow as indicated by the solid lines. Auxiliary heating coils may be activated by a thermostat in order to deliver outdoor air to the environment at a desired temperature level. If due to rising outside air temperature, the supply air tends to exceed the thermostat set point, a portion of the exhaust air will flow the route of the dotted lines and will bypass the heat exchanger through a motorized damper thereby controlling Btu exchange.

23.2.2 Cooling Cycle Recovery with 100% Outside Air (Figure 142-b)

During this cycle the exhaust air bypass damper remains closed at all times since over cooling the environment without additional refrigeration is not possible. Auxiliary cooling coils can be incorporated through the heat exchanger to further reduce the supply air temperature to a desired level. The changeover from cooling to heating cycle is effected by heating/cooling thermostats which are incorporated in the heat exchanger unit.

23.2.3. Heating Cycle Recovery with 100% Outside Air with Additional Outside Air Blending (Figure 142-c)

This cycle is applicable for industrial heat recovery applications where recovery from an elevated temperature exhaust is effected and where the discharge temperature of the supply air is in excess of the desired comfort level. This arrangement makes it possible to deliver large quantities of reheated outside air from a relatively small exhaust air quantity and help to alleviate negative plant pressure.

23.2.4. Recirculting Cycle with No Outside Air (Figure 143-a)

During nonoccupancy periods or whenever exhausting of air is not warranted, outside air is not introduced, and the heat exchanger is by-passed. The heat exchanger wheel assumes the roll of an air handling unit with auxiliary heating/cooling coils.

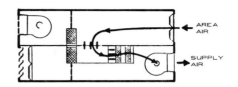

(a) RECIRCULATING CYCLE, NO OUTSIDE AIR

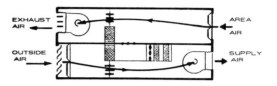

(b) VENTILATING CYCLE

FIGURE 143. Heat recovery system applications.

23.25. Ventilating Cycle (Figure 143-b)

When air must be exhausted and raw outside make-up air can be introduced, the heat exchanger wheel is bypassed to prevent dirt loading.

24

HEATERS, UNIT

Unit heaters can warm relatively large spaces and require a minimum of piping. They occupy a comparatively small amount of space, keep the floor areas free from obstruction, and because of their light weight, can be hung from almost any convenient portion of the building structure.

Unit heaters contain a fan and motor, a heating element, and louvers to direct the heated air, all housed in a cabinet or enclosure. The fan can be either a propeller type or centrifrigal. The heating can be achieved by hot water, steam, gas or oil-burners, and electricity. There is another way of classifying unit heaters, that is by the direction of the discharged air.

A horizontal discharge unit heater has outlet velocities that range from about 500 to 1000 fpm. It is best suited for commercial and industrial buildings. It is used most effectively in rooms or spaces under 16 ft in height, with no major obstructions to the horizontal flow of air. Heaters must be installed to blow air along the outside wall, or toward an exposure, creating a gentle, circular motion of air, see Figure 144.

Horizontal discharge heaters have louvers which permit individual adjustment of the flow of warm air in five directions. Heaters should be located to offset down drafts from roofs of relatively large areas. Low outlet temperature and relatively high outlet velocities promote better heat distribution.

Vertical discharge unit heaters are used primarily in spaces with large roof areas, especially where there are obstructions to horizontal distribution of heat and for the installation of piping. They have outlet velocities from about 1000 to 2000 fpm. A propeller-driven vertical unit heater is shown in Figure 145. Heated air has a tendency to rise. Since it is projected downward by the vertical discharge heater, relatively high outlet velocities and low outlet temperatures are necessary for maximum efficiency.

It is important to consider the type of building in which the unit heaters are to be installed and the occupancy. For example, only unit heaters operating at speeds with low sound ratings should be used in theaters, auditoriums, schools, hospitals, or other places where quiet is necessary.

Figures 146, 147, and 148 show piping hookups for a steam heated unit heater, either horizontal or vertical discharge. Figure 146 is the hookup for high-pressure steam (60 psi and over) with an overhead return. Figure 147 is for low-pressure steam (less than 60 psi), also with an overhead return. Figure 148 illustrates the pipe and fittings for a vacuum return steam unit heater.

FIGURE 144. Horizontal discharge unit heater. (Manufactured by Dunham-Bush, Inc.)

FIGURE 145. Vertical discharge unit heater. (Manufactured by Dunham-Bush, Inc.)

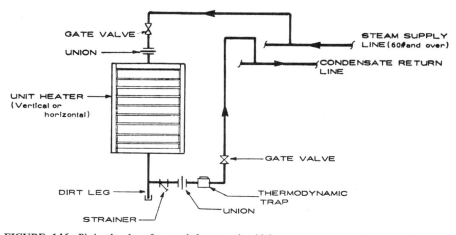

FIGURE 146. Piping hookup for a unit heater using high-pressure steam with overhead return.

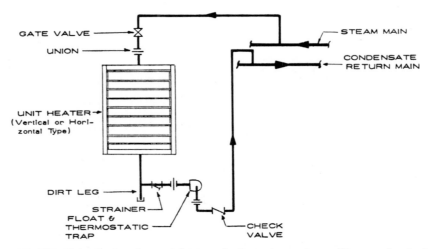

FIGURE 147. Piping hookup for a unit heater using low-pressure steam with an overhead return.

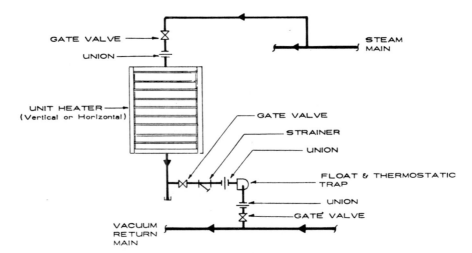

FIGURE 148. Piping hookup to a vacuum return steam unit heater.

24.1. ESTIMATED LABOR TO INSTALL UNIT HEATERS

Propeller unit, gas-fired

Nominal cfm	Average Weight (lb)	Labor (man-hours)
300	60	4
600	85	4
900	115	4
1200	130	4
1500	150	6
1800	175	6
2000	225	8
2500	250	8
3000	275	8
3600	375	12
4500	500	12

Propeller unit, steam or hot water

Nominal cfm	Average Weight (lb)	Labor (man-hours)
250	30	4
300	40	4
500	45	4
600	45	4
850	55	4
1200	60	4
1500	65	6
1750	75	6
2000	100	6
2500	125	6
3000	140	6
3500	165	8
4000	175	8
4500	200	10
5000	225	10

25

HUMIDIFIERS

Humidifiers are devices used in HVAC systems to maintain comfort conditions in the winter when heating systems tend to dry out the air in a home or other conditioned space. The required capacity of a humidifier is determined from the volume of the conditioned space and the number of air changes per hour. This gives the throughput of air. Knowing the desired indoor temperature and humidity and the outdoor conditions, the pounds of moisture per unit time can be determined.

The most commonly used types of humidifiers are: wetted elements, pan, atomizing, steam grid, electrical or steam heated pan, and dry steam types. These are described below.

25.1. WETTED TYPE HUMIDIFIER

This unit makes use of an open textured wetted media through and over which the air is circulated by means of a fan. Evaporation adds moisture to the air. Figure 149 shows a portable wetted type humidifier used mostly for residences.

25.2 PAN HUMIDIFIER

The basic unit, illustrated in Figure 150, is normally installed in the furnace plenum. The water is maintained at a constant level in a shallow pan. Humidification is added to the flow of air by evaporation of the water in the pan. The humidification rate of this apparatus is slow.

The heated pan type uses steam or electricity to heat the water. It is basically the same as the pan type except that the heating process increases evaporation of water and therefore the humidification rate is higher than for the unheated water application.

25.3. ATOMIZING TYPE HUMIDIFIER

This type of humidifier introduces small particles of water into the air by means of a spray of water, or by spinning a disk or cone. The humidification rate of this apparatus depends on the condition of the surrounding air. This condition includes the air velocity, temperature, and moisture content, all of which affect its ability to absorb the atomized water particles. Figure 151 illustrates this type of humidifier, mounted in the ductwork. Plate eliminators are provided to remove droplets of water that may form under some operating conditions.

25.4. STEAM GRID TYPE HUMIDIFIER

This type uses steam injected directly into the air through small holes in the feedpipe. The steam is controlled by a modulating valve which responds to a humidity controller to maintain a constant air–moisture combination in the airstream. In Figure 152, a steam grid humidifier is shown mounted in the ductwork. Note that the pan in the bottom of the duct collects the condensate from the steam that does not get into the airstream.

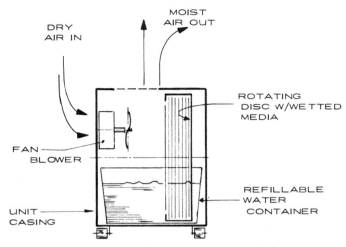

FIGURE 149. Portable, wetted type humidifier.

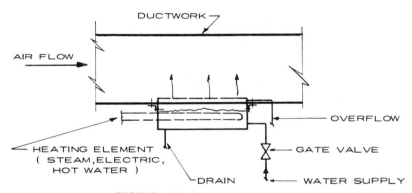

FIGURE 150. Heated pan humidifier.

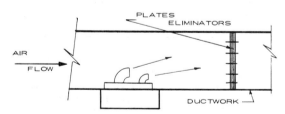

FIGURE 151. Atomizing humidifier with eliminators.

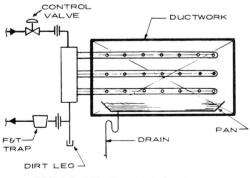

FIGURE 152. Steam grid humidifier.

25.5. DRY STEAM HUMIDIFIER

This unit is similar to the grid type except that it uses a steam-jacketed manifold and a condensate separator which eliminates the eventual condensate vapor into the airstream. This unit is designed to deliver a precise and controlled amount of dry steam for most humidification systems, see Figure 153.

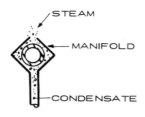

SECTION A-A

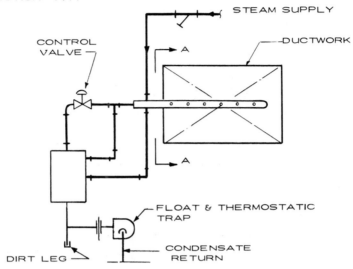

FIGURE 153. Jacketed dry steam humidifier.

25.6 ESTIMATED LABOR TO INSTALL HUMIDIFIERS

Description	Output (lb/hr)	Electrical Coil (watts)	Average Weight (lb)	Labor (man-hours)
Pan type	6	2000	30	3.0
(with electric	12	4000	35	3.0
heating element)	25	8000	35	4.5

Description	Steam (lb/hr)	Average Weight (lb)	Labor (man-hours)
Steam grid type	15–120	70	4
(with control valve	120–225	120	4
assembly duct	225–425	160	6
mounted)	450–800	225	8

26

INSULATION

The primary purpose of insulation is to retard the transfer of heat from a warm surface to the cooler surrounding atmosphere as would be the case in winter, or to prevent the reverse heat transfer of heat from air to a cooler surface in summer. In addition to maintaining desired temperatures without waste of energy, insulation has two other primary functions. It prevents sweating of piping that carries cool fluids. Insulation also protects maintenance people from being burned by coming in contact with hot surfaces.

Insulation covers a wide spectrum of materials used to conserve heat and attenuate sound. This Chapter will deal with duct and pipe insulation usually found in HVAC systems.

Insulation covers a wide range of temperatures, from extremely low temperatures of refrigerating plants, to high temperatures of boiler installations. No one material can meet all the conditions with the same efficiency. For example, cork or rock wool, is used for low temperatures. Minerals or metals such as asbestos, carbonate of magnesia, diatomaceous earth, aluminum foil, argillaceous (clay-like) limestone, mica, fiberglass, and diatomaceous silica are employed for high temperatures. Because of its excellent refractory properties, diatomaceous silica forms the base of practically every high-temperature insulating material.

The designer must select the type of insulation for a particular application based upon a number of properties inherent to each. In the broadest terms, the following should be considered:

1. The ability of the insulation to withstand the highest or lowest temperature to which it may be subjected without its insulating value being impaired.
2. It must possess sufficient structural strength to withstand handling during its application, and the mechanical shocks and vibrations during service, without disintegration, settling, or deformation.
3. Insulation must be chemically stable and retain its insulating characteristics over time.
4. The material must be easy to apply, maintain, and repair.
5. There must be no fire or smoke hazard from its use.
6. The insulation must have a low-heat capacity when used in boiler walls and furnaces, so that start up time is minimized.
7. The material must be moisture repellant and vermin proof.
8. Moistureproofing is as important for high-temperature insulation as for low-temperature insulation. In the former heat is lost because of evaporation, while in the latter case, condensed moisture may freeze. In either case, insulating efficiency is impaired, and eventually the insulating material disintegrates.

26.1. USES AND FORMS OF INSULATION

Cork.   Cork is available in sheet form and is generally used to insulate refrigerated spaces. Molded cork pipe covering, treated with a fire-retardant compound, is used on refrigerant piping.

Mineral or Rock Wool. This material is usually supplied as wire-reinforced pads. This material is suitable for high temperature use, and is particularly useful for insulating large areas. Molded sheets, pads, blankets, or tapes containing long asbestos fibers are suitable for insulating temperatures up to 850°F. This insulation material is cheaper, lighter, and more durable than diatomaceous earth. The pads or blankets are used for applications where they must be taken down fairly often such as insulating flanges, valves, and turbine casings. The pads are molded to fit any shape, and the outer surfaces are fitted with metal hooks to facilitate installation and removal. The blankets are made in various thicknesses and widths, and are also fitted with hooks. Tapes are used for covering small curved piping. They can be used for temperatures up to 750°F and tend to reduce fire hazards, but have poor insulating quality.

Magnesia-Asbestos Pipe Covering. This insulation is most commonly used as insulation on high-temperature piping. The material is supplied in molded cylindrical sections which are 3-ft long. Each section is split in half lengthwise. Suitable openings are available to fit the various pipe sizes. Magnesia-asbestos pipe covering comes in three grades: grades I, II, III, are suitable for temperatures up to 500°F, 750°F, and 1050°F respectively.

Diatomaceous Earth. Diatomaceous means that the material was formed from skeletons of certain microscopic plants. They are combinations of the earth and magnesium or calcium carbonates, bonded together with small amounts of asbestos fibers. These materials are heavier, more expensive, and less insulating than many other materials but their high heat resistance allows their use for temperature up to 1500°F. When practicable, pipe coverings are made of diatomaceous earth as an inner layer, and with an outer layer of magnesia-asbestos. This reduces the overall weight.

Fibrous Glass Slabs, Batts, and Pipe Insulation. Fiberglass is widely used for insulating duct work and living space. The fibrous glass has a low moisture absorption and offers no attraction to insects, vermin, fungus growth, or fire. The slabs are first cut to shape, then secured by mechanical fasteners such as quilting pins. The slabs are covered with a glass cloth facing and stripping tape that are held in place by fire-resistant adhesive cement. Duct wrap insulation is shown in Figure 154a. The fiber blanket has a vapor barrier facing with one 2-in flange. For ducts that operate between 40 and 250°F.

Figure 154b illustrates preformed semirigid and rigid boards of varying densities used to insulate ducts, ovens, tanks, etc. Resilient boards are placed on irregular surfaces where the exterior is supported by welded studs, pins, or other mechanical fasteners. The semirigid board is used generally on equipment and ductwork. Rigid board has high strength and resistance to abuse. For this reason it is used for chillers, hot and cold equipment, and HVAC ductwork.

Fiberglass pipe insulation, seen in Figure 154c, has a vapor barrier and a pressure sealing lap adhesive that eliminates the need for adhesives or bands. It is recommended for hot and cold piping operating from −60 to 450°F for commercial buildings, hospitals, schools, and where fire safety is important.

Insulating Cements.   There are many types of insulating cements, with different heat conductivities, weights, and other physical characteristics. Typical of these variations are cements containing asbestos, diatomaceous earth, and mineral and slag wool. These cements are less efficient than other high temperature insulating materials, but they are valuable for patchwork emergency repairs, and for covering small irregular surfaces (valves, flanges, joints, etc.) The cements are also used for a surface finish over block or sheet forms of insulation, to seal joints between the blocks, and to provide a smooth finish over which asbestos or glass cloth lagging may be applied.

(a) Duct wrap

(b) Boards

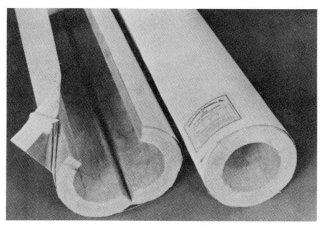

(c) Pipe insulation

FIGURE 154. Fiberglass insulation. (Manufactured by Owens-Corning Fiberglass Corp.)

Asbestos-Free Insulating Block. Asbestos, because it may be a health hazard if ingested, is avoided for some insulation applications. Asbestos-free block insulation is a rigid hydrous calcium silicate heat insulation. It is strong, efficient and highly resistant to abrasion and moisture damage. It is for use on indoor or outdoor equipment operating at temperatures up to 1200°F. The block is suited for use on stainless steel vessels and equipment as it does not contribute to stress corrosion cracking. Typical applications are for boilers, breeching, tanks, and vessels, see Figure 155.

FIGURE 155. Asbestos-free insulating block. (Manufactured by Owens-Corning Fiberglass Corp.)

26.2. LABOR AND MATERIAL COSTS TO INSTALL INSULATION

Description	Size (Diameter, in in.)	Unit	Material Cost ($)	Labor (man-hours)
Piping insulation (with fitting and/ or valving every 10 LF)				
$\frac{3}{4}$-in. Glass fiber insulation	$\frac{1}{2}$	LF	.80	0.15
	$\frac{3}{4}$			0.15
	1			0.18
	$1\frac{1}{4}$			0.20
	$1\frac{1}{2}$			0.20
	2			0.22
	$2\frac{1}{2}$			0.25
	3		2.00	0.29
	4			0.30
	5			0.37
	6			0.45
	8		3.50	0.56
	10			0.75
	12		8.00	0.77
1-in. Glass fiber insulation	$\frac{3}{4}$	LF	1.00	0.15
	1			0.15
	$1\frac{1}{4}$			0.16
	$1\frac{1}{2}$			0.16
	2			0.19
	$2\frac{1}{2}$		1.50	0.20
	3			0.23
	4			0.24
	5		2.60	0.29
	6		2.80	0.32
	8			0.43
	10			0.49
	12		8.20	0.71
$1\frac{1}{2}$-in. Glass fiber insulation	$2\frac{1}{2}$	LF	2.18	0.20
	3			0.23
	4			0.24
	5			0.29

Description (cont.)	Size (Diameter, in in.)	Unit	Material Cost ($)	Labor (man-hours)
↓	6	↓	3.50	0.32
	8		↓	0.43
	10			0.49
↓	12	↓	8.50	0.71

Description	Thickness (in.)	Unit	Material Cost ($)	Labor (man-hours)
Ductwork insulation, Glass fiber	1 Blanket	SF	.40	0.09
	1½ Blanket	↓	↓	0.09
↓	1 Rigid			0.13
	1½ Rigid		↓	0.13
↓	2 Rigid	↓	1.00	0.13

27

PIPING AND FITTINGS

Piping is defined as an assembly of pipe or tubing, with valves, fittings, and related components forming a whole or a part of a system for transferring liquids and gases. Historically, there has not been clear distinction between pipe and tubing, since the designation for each product was established by the manufacturer. If the manufacturer called a product pipe, it was pipe; if he called it tubing, it was tubing. Over the years certain standards were developed which separated the two types. These standards were based upon sizes and how the sizes were measured.

There are three important dimensions of any tubular product: outside diameter (OD), inside diameter (ID), and wall thickness. A product is called tubing if its size is identified by its measured outside diameter (OD) and wall thickness. A product is called pipe if its size is identified by a nominal dimension called iron pipe size (IPS) and by reference to a wall thickness schedule designation.

The size identification of tubing is simple enough, since it consists of actual measured dimensions, but the terms used for identifying pipe sizes may require some explanation. A nominal dimension for an iron pipe size is close to, but not necessarily identical with, the measured dimension. For example, a pipe with an IPS of 3 in. has an actual outside diameter of 3.50 in. One with an IPS of 2 in. has a measured outside diameter of 2.375 in. In the larger sizes (about 12 in.), the nominal pipe size and the

actual measured outside diameter are the same. For example, a 14 in. pipe has an OD of 14 in. Nominal dimensions are used in order to simplify the identity of pipe fittings, pipe taps, and threading dies.

The wall thickness of pipe is identified by schedules established by the American National Standards Association. For example, a 3-in. diameter Sch. 40 steel pipe has a wall thickness of 0.216 in. A Sch. 80 steel pipe of the same size has a wall thickness of 0.300 in. As the schedule number increases, so does the pipe thickness.

A schedule designation does not identify any one particular wall thickness unless the nominal pipe size is also specified. For example, we have seen that a 3-in. Sch. 40 steel pipe has a wall thickness of 0.216 in. But, if we look up the wall thickness of a 4-in. Sch. 40 steel pipe, we find it to be 0.327 in. This concept is understandable if we think of pipe as a pressure vessel. The schedule number is a function of the bursting strength of the pipe. The larger the diameter of the pipe, the thicker it must be for the same strength. In engineering handbooks, tables give schedules, dimensions, and other physical properties.

Pipe is also identified as standard (Std), extra strong (XS), and double extra strong (XXS). These designations, once very common, are still used to some extent. They also refer to wall thickness. However, pipe is manufactured in a number of different wall thicknesses, and some pipe does not fit into the standard, extra strong, or double extra strong classifications. Schedules are being used increasingly to identify the wall thickness of pipe, because they provide for the identification of the widest range of pipe sizes.

The standard ways of identifying size and wall thickness of pipe and tubing have been briefly described. It should be noted, however, that sometimes pipe and tubing are identified in other ways. For example, you may see some tubing specified by ID rather than by OD, others by nominal pipe size, by OD, by ID and by actual wall thickness.

27.1. TYPES OF PIPE AND TUBE

A great many different kinds of pipe and tubing are used in piping systems. Their selection is based on such factors as the temperature and pressure of the gas or liquid contained, its corrosiveness, and the cost of pipe and fittings. Often, building codes require certain choices be made.

Seamless Chromium-Molybdenum Alloy Steel Pipe. This expensive material is used for some high pressure, high-temperature steam systems. The upper limit for the piping is 1500 psig and 1050°F.

Seamless Carbon Steel Tubing. This tubing is used in oil, steam, and feedwater lines operating at 775°F and below. Different wall thicknesses of this tubing are available for each diameter, the type used depends on the working pressure of the system.

Seamless Carbon-Molybdenum Alloy Steel Tubing. This material is used for feedwater discharge piping, boiler superheated steam, and saturated steam lines. Several types of this tubing are available, depending upon the boiler operating pressure and the superheater, outlet temperatures. The upper pressure and temperature limits for any class of this tubing are 1500 psi and 875°F.

Seamless Chromium-Molybdenum Alloy Steel Tubing. As with the pipe of the same material,

tubing is used for high pressure, high-temperature steam service. This type of alloy steel tubing is available with different percentages of chromium and molybdenum with upper limits of 1500 psig and 1050°F.

Welded Carbon Steel Tubing. This tubing is used in some water, steam, and oil lines where the temperature does not exceed 450°F. There are several types of this tubing, each specified for certain services and conditions.

Nonferrous Pipe and Nonferrous Tubing. These are used in many piping systems and for most heat exchangers. Nonferrous materials are used chiefly where their special properties of corrosion resistance and high-heat conductivity are required. Various types of seamless copper tubing are used for refrigeration installations, plumbing and heating, lubrication systems, and elsewhere. Copper-nickel alloy tubing is available as 70% copper and 30% nickel. The 70:30 composition is generally used in piping systems and heat exchangers.

Many additional kinds of pipe and tubing are used in piping systems. It is important to remember that design considerations control their selection. Although pipe and tubing may look almost exactly alike from the outside, they may respond very differently to pressures, temperatures, and other service conditions. Therefore, each kind of pipe and tubing can be used only for specified applications.

27.2. PIPE CONNECTIONS

Pipe sections are connected by threaded, bolted, welded, and silver-brazed fittings, and by expansion joints.

Threaded Joints. Threaded joints are the simplest type of pipe fittings. Threaded fittings are widely used in low-pressure piping systems. Pipe unions are provided in piping systems to allow the piping to be disassembled for repairs and alterations. Other fittings have threaded joints. These include, elbows, tees, crosses, flanges, reducers, among others. Fittings are available in many different materials and designs to withstand a wide range of pressures and temperatures.

The union is used a great deal for joining piping. The pipe ends connected to the union are threaded, silver-brazed, or welded into the tail pieces; then, the two ends are joined by setting up on the union ring. The male and female connecting ends of the tail pieces are carefully ground to make a tight metal-to-metal fit with each other. Welding or silver-brazing the ends to the tail pieces prevents the contact of the carried liquid or gas with the union threading.

Bolted Flange Joints. Bolted flanges are suitable for a wide range of pressures. The flanges are attached to the piping by welding, brazing screw threads (for some low-pressure piping). Some fittings like crosses, tees, and elbows have integral flanges. Flanges are designated by such numbers as 125 lb or 300 lb (or higher) which indicates the pressure, in square inches, that the joint can withstand. A leak-tight joint is made by means of a gasket or seal between the flange faces that are compressed by the bolts. A wide variety of gaskets are available to suit the pressure and temperature conditions and compatibility of materials.

Welded Joints. The majority of joints found in subassemblies of piping systems are welded joints, especially in high-pressure piping. The welding is done according to standard specifications* which define the materials and techniques. There are three general classes of welded joints: butt-weld, fillet-weld, and socket-weld.

Silver-Brazed Joints. Silver-brazed joints are commonly used for joining nonferrous piping in the pressure and temperature range where its use is practicable. The alloy is melted by heating the joint with an oxyacetylene torch. This causes the molten metal to fill the few thousandths of

*The American Welding Society and American Society of Mechanical Engineers publish general welding requirements that are often referred to in specifications.

an inch annular space between the pipe and the fitting.

Expansion Joints. Various types of expansion joints are installed at suitable intervals in long steam lines, because of the pipe expansion and contraction as it is subjected to a wide range of temperature. The types include slip joints, expansion bends, corrugated joints, and bellows joints.

Slip Joints are used for low pressures, such as for auxiliary exhaust piping. A slip joint consists of a stuffing box, packing gland, male sliding tube, female receptacle tube, and stop bolts to prevent separation of male and female sections of the joints.

Expansion Bends are employed for high pressure, high-temperature steam piping in preference to the slip joint. Expansion bends are merely loops of piping to take up the changes in pipe length caused by temperature changes. The expansion bends take many shapes, most common are the U-bend, Z-bend, and L-bend.

Corrugated and Bellows expansion joints are used for both medium and high pressures and temperatures. The principle of these joints is obvious. The expansion-contraction movement is absorbed by the changing curvature of the corrugations or bellows (as with an accordion). Internal sleeves, free to slide axially in these joints, serve to prevent excessive turbulence and erosion of the expansion parts.

27.3. PIPING ESTIMATING

Unlike the other items and equipment covered in this manual, pipe estimating is somewhat difficult. It requires skill and knowledge for the estimator to read the contract drawings, and determine a "realistic" estimate cost. It is not the intent of this section to be the last word on the subject. Instead, it serves as a general guide to the estimator, designer, and engineer for labor and materials costs needed for a given piping system.

The information in the following charts is a result of work in the field where no one condition is like another. While there is no substitute for experience, this information can serve as a first step in estimating a piping layout.

Estimated Labor to Install Piping, Man-hours per Foot
(Based on Standard Pipe Lengths 16 to 22 ft)

Description	Pipe Size (in.)	Schedule					
		10	20	30	40	80	160
Carbon steel, butt-welded pipe	$\frac{3}{4}$				0.08	0.10	0.12
	1				0.12	0.12	0.14
	$1\frac{1}{4}$				0.12	0.16	0.14
	$1\frac{1}{2}$				0.16	0.16	0.18
	2				0.16	0.18	0.25
	$2\frac{1}{2}$				0.22	0.22	0.32
	3				0.25	0.25	0.40
	4				0.30	0.32	0.40
	6				0.42	0.42	0.70
	8		0.50	0.52	0.55	0.55	1.00
	10		0.55	0.54	0.62	0.62	1.20
	12		0.60	0.58	0.75	0.80	1.55
	14	0.70	0.75	0.78	0.85	1.00	1.80
	16	0.75	0.80	0.80	0.95	1.10	2.10
	18	0.80	0.90	0.86	1.10	1.28	2.70
	20	0.90	1.00	0.98	1.20	1.75	3.10
	24	1.00	1.10	1.25	1.50	2.25	4.50
Carbon steel, threaded and coupled (T & C)	$\frac{3}{4}$				0.07	0.09	0.08
	1				0.08	0.09	0.10
	$1\frac{1}{4}$				0.10	0.11	0.13
	$1\frac{1}{2}$				0.10	0.11	0.15
	2				0.13	0.15	0.20
	$2\frac{1}{2}$				0.17	0.20	0.28
	3				0.22	0.30	0.32
	4				0.25	0.30	0.40
	6				0.40	0.50	
	8				0.55		
	10				0.70		
	12				0.80		
Stainless steel pipe, butt-welded	$\frac{1}{2}$	0.08			0.08	0.10	0.12
	$\frac{3}{4}$	0.09			0.12	0.12	0.15
	1	0.12			0.14	0.15	0.18
	$1\frac{1}{4}$	0.14			0.14	0.15	0.18
	$1\frac{1}{2}$	0.15			0.18	0.20	0.25
	2	0.20			0.24	0.28	0.32
	$2\frac{1}{2}$	0.22			0.28	0.36	0.42
	3	0.25			0.32	0.40	0.50
	4	0.30			0.42	0.50	0.62
	6	0.40			0.56	0.65	1.00
	8	0.50			0.70	0.85	1.35
	10	0.60			0.85	1.10	1.70
	12	0.72			1.00	1.30	2.40
Stainless steel pipe, threaded and coupled (T & C)	$\frac{1}{2}$				0.08	0.08	0.08
	$\frac{3}{4}$				0.10	0.10	0.10
	1				0.10	0.10	0.12
	$1\frac{1}{4}$				0.12	0.12	0.14

Estimated Labor to Install Piping, Man-hours per Foot
(Based on Standard Pipe Lengths 16 to 22 ft)

Stainless Steel Pipe (continued)	Pipe Size (in.)	Schedule					
		10	20	30	40	80	160
	$1\frac{1}{2}$				0.12	0.14	0.16
	2				0.14	0.16	0.20
	$2\frac{1}{2}$				0.22	0.24	0.28
	3				0.24	0.28	0.34
	4				0.30	0.35	0.45
Aluminum pipe threaded and coupled (T & C)	$\frac{3}{4}$	0.08			0.08	0.08	
	1	0.10			0.10	0.10	
	$1\frac{1}{4}$	0.12			0.14	0.16	
	$1\frac{1}{2}$	0.16			0.18	0.22	
	2	0.20			0.22	0.25	
	$2\frac{1}{2}$	0.20			0.25	0.25	
	3	0.25			0.30	0.35	
	4	0.30			0.35	0.40	
	6	0.40			0.45	0.58	
	8	0.50			0.60	0.72	
	10	0.62			0.72	0.75	
	12	0.75			0.85	1.25	

Description	Pipe Size (in.)	Pipe Designation	
		Std	XS
Carbon steel pipe, coated and wrapped (beveled end)	$\frac{3}{4}$	0.08	0.10
	1	0.10	0.12
	$1\frac{1}{4}$	0.12	0.12
	$1\frac{1}{2}$	0.14	0.16
	2	0.16	0.20
	$2\frac{1}{2}$	0.20	0.22
	3	0.21	0.25
	4	0.30	0.32
	6	0.35	0.40
	8	0.42	0.50
	10	0.50	0.62
	12	0.60	0.78
	14	0.70	0.90
	16	0.80	1.00
	18	0.95	1.25
	20	1.10	1.50
	24	1.30	1.90

Description	Pipe Size (in.)	Labor (man-hours)
Copper pipe (using 50/50 solder 95/5 solder, or silver solder) types "L," "K," "M"	$\frac{1}{2}$	0.08
	$\frac{3}{4}$	0.08
	1	0.08
	$1\frac{1}{4}$	0.10
	$1\frac{1}{2}$	0.10
	2	0.15
	$2\frac{1}{2}$	0.15
	3	0.20
	4	0.25
	6	0.38
	8	0.48
Copper tubing (using 50/50 solder, 95/5 solder, or silver solder types "L," "K"	$\frac{3}{4}$	0.08
	1	0.10
	$1\frac{1}{4}$	0.10
	$1\frac{1}{2}$	0.14
	2	0.14
	$2\frac{1}{2}$	0.20
	3	0.24
	4	0.30

Description	Pipe Size (in.)	Labor (man-hours)	
Copper pipe, screwed		Std	X-heavy
	$\frac{1}{2}$	0.09	0.09
	$\frac{3}{4}$	0.10	0.10
	$1\frac{1}{4}$	0.14	0.14
	$1\frac{1}{2}$	0.14	0.14
	2	0.18	0.20
	$2\frac{1}{2}$	0.20	0.25
	3	0.30	0.32
	4	0.35	0.38
	6	0.50	0.55
	8	0.62	0.75
	10	0.80	0.92
	12	0.90	
Brass pipe, threaded and coupled (T & C)	$\frac{1}{2}$	0.08	0.08
	$\frac{3}{4}$	0.10	0.10
	$1\frac{1}{4}$	0.10	0.10
	$1\frac{1}{2}$	0.14	0.16
	2	0.18	0.22
	$2\frac{1}{2}$	0.24	0.28
	3	0.26	0.32
	4	0.32	0.40
	6	0.50	0.54
	8	0.65	0.80

Description		Labor (man-hours)					
Polyvinly Chloride (PVC) plastic pipe	Pipe Size (in.)	Threaded & coupled construction Sch.		Welded construction Sch.		Cemented construction Sch.	
		40	80	40	80	40	80
	$\frac{1}{2}$	0.08	0.08	0.08	0.08	0.08	0.08
	$\frac{3}{4}$	0.10	0.10	0.10	0.10	0.10	0.10
	1	0.10	0.10	0.10	0.10	0.10	0.10
	$1\frac{1}{4}$	0.12	0.12	0.10	0.10	0.10	0.10
	$1\frac{1}{2}$	0.12	0.12	0.10	0.10	0.10	0.10
	2	0.12	0.12	0.12	0.12	0.12	0.12
	$2\frac{1}{2}$	0.14	0.15	0.12	0.12	0.12	0.12
	3	0.20	0.20	0.12	0.12	0.14	0.14
	4	0.25	0.30	0.12	0.12	0.14	0.14
	6	0.30	0.35	0.20	0.25	0.20	0.25
	8	0.40	0.55	0.25	0.30	0.20	0.25
	10	0.50	0.60	0.30	0.35	0.30	0.35
	12	0.55	0.75	0.40	0.40	0.40	0.45

28
PUMPS

Pumps are vitally important to the function of a piping system. If they fail, the system they serve fails, and in an emergency, pump failures can prove to be disastrous. Therefore, proper pump selection is essential.

Pumps are used to move any substance which flows or which can be made to flow. Most commonly, pumps are used to move water, oil, and other liquids. Air, steam, and other gases are also considered fluids and can be moved with pumps.

A *pump* is a device which utilizes an external source of power to apply a force to a fluid in order to move that fluid. A pump develops no energy of its own; it merely transforms energy from the external source (steam turbine, electric motor, etc.) into kinetic energy, or motion of the fluid. This kinetic energy is then put to work as in the following examples:

1. To raise a liquid from one level to another, like water in a well.
2. To transport a liquid through a pipe, as oil is carried through a pipeline.
3. To move a liquid against a resistance, as when water is pumped to a boiler under pressure.
4. To force a liquid through a hydraulic system, against an external resistance, for purpose of exerting a load on the object.

Every pump has a "power mover," whether it be a steam turbine, a reciprocating steam engine, a steam jet, or an electric motor. Each pump also has a suction side and a discharge side. The suction side is the low-pressure inlet where the fluid enters the pump. After work is done on the fluid, in the pump, it is discharged at a higher pressure. This increase in pressure is generally referred to as "head" measured in psi or feet of water. There are four types of heads: net positive suction head, suction head, discharge head, and total head. These are described below and are illustrated in Figure 156.

Net Positive Suction Head (NPSH). The net positive suction head combines all of the factors limiting the suction side of a pump: internal pump losses, static suction lift, friction losses, vapor pressure, and atmospheric conditions. It is important to differentiate between required NPSH and available NPSH.

Required NPSH refers to internal pump losses and is determined by laboratory test. It varies with each pump and for each pump capacity and speed. The greater the capacity, the greater the required NPSH. Required NPSH must always be given by the pump manufacturer.

Available NPSH is a characteristic of the system. It can be calculated, or can be determined by field test on an existing installation using vacuum and pressure gauges. By definition, it is the net positive suction head above the vapor pressure available at the suction flange of the pump to maintain a liquid state. Since there are also internal pump losses (required NPSH), the available NPSH in a system must exceed the pump required NPSH—otherwise, reduction in

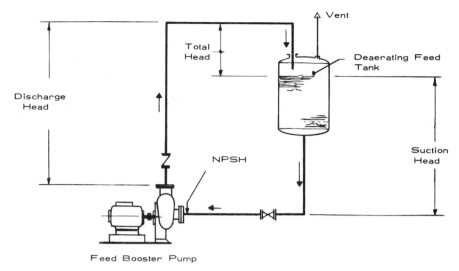

FIGURE 156. Heads (pressure) in pump system.

capacity, loss of efficiency, noise, vibration, and cavitation will result.

Suction Head. The suction head is the vertical distance in feet from the free level of the supply (the deaerating tank in Figure 156), to the horizontal centerline of the pump.

Discharge Head. The discharge head is the pressure of the liquid leaving the pump, or the level of liquid above the level of the pump on the discharge side.

Total Head. The total head is the difference between the suction head and the discharge head. Suction head is usually expressed in feet of water if positive, and in inches of mercury if negative. When a pump operates below the level of a liquid, its suction end receives the liquid under a gravity flow as in Figure 156. When it operates above the level of the liquid, a vacuum must be created to raise the liquid. The liquid may be raised by atmospheric pressure or by another pump. Atmospheric pressure has an important bearing on the suction of the pump.

There are many kinds of pumps being manufactured and used today. For HVAC applications there are three basic types: centrifugal, rotary, reciprocating. Among these, the most common is the centrifugal pump.

28.1. CENTRIFUGAL PUMPS

The centrifugal pump uses the throwing force of a rapidly revolving impeller. The liquid is pulled in at the center or eye of the impeller and is discharged at its outer rim. By the time the liquid reaches the outer rim of the impeller, it has acquired considerable velocity. The liquid is then slowed down by being led through a volute or through a series of diffusing passages. As the velocity of the liquid decreases, its pressure increases. Thus, its kinetic energy is transformed into potential energy.

Centrifugal pumps may be classified in several ways. For example, they may be either *single-stage* or *multistage*. A single-stage pump has only one impeller. A multistage pump has two or more impellers housed in one casing. As a rule, each impeller acts separately, discharging to the suction of the next impeller. Centrifugal pumps are also classified as horizontal or vertical, depending upon the position of the pump shaft.

The impellers used in centrifugal pumps may be classified as single suction or double suction. The single-suction impeller allows liquid to enter the pump from one direction only; the double-suction type allows liquid to enter from two directions. Impellers are also classified as closed or open. Closed impellers have side walls which extend from the eye to the outer edge of

FIGURE 157. Horizontal, single stage centrifugal pump. (Manufactured by Dunham-Bush, Inc.)

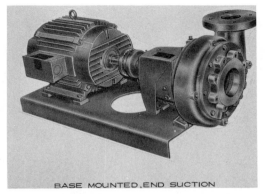

BASE MOUNTED, END SUCTION

HORIZONTAL IN-LINE

the vane. Open impellers do not have these side walls.

As a rule, the casing for the discharge end of a pump with a single suction impeller is made with an end plate which can be removed for inspection and repair. A pump with a double suction impeller is generally made so that one half of the casing may be lifted without disturbing the pump.

FIGURE 158. Centrifugal pumps. (Manufactured by Thrush Products, Inc.)

FIGURE 159. Centrifugal pump, close-coupled volute type. (Manufactured by Dunham-Bush, Inc.)

A number of commercially available centrifugal pumps are illustrated in the following figures. Figure 157 is a horizontal, single-stage pump that is integral to the electric motor drive. An end suction centrifugal pump that is mounted on a base and connected to a driver by means of a flexible coupling is shown in Figure 158, top. At the bottom of the same Figure is a horizontal in-line pump. This pump, and the one shown in Figure 159 can be mounted directly in the pipeline which eliminates the need for foundations and reduces space requirements.

28.2. ROTARY PUMPS

These pumps are assembled with gears, screws, vanes, lobes, or eccentrics, which rotate to trap the liquid at the suction side and force it through to the discharge side of pump. Rotary pumps are referred to as positive displacement pumps since each stroke delivers a specific amount of liquid. They are self-priming and do not require suction or discharge valves.

Rotary pumps are used for high-viscosity liquids, such as pumping oil and grease. Rotary pumps are also used as a source of fluid power in hydraulic systems.

To be efficient and to minimize leakage from discharge back to the suction end, clearances for rotating parts are close. This produces a high vacuum in the suction pipe. Most rotary pumps operate at low speed to maintain clearances and eliminate erosion and excessive wear.

28.3. RECIPROCATING PUMPS

Reciprocating pumps are also positive-displacement pumps. Each stroke displaces a certain definite quantity of liquid, regardless of the resistance against which the pump is operating. Reciprocating pumps are usually classified as;

1. direct acting or indirect acting.
2. simplex (single) or duplex (double).
3. single acting or double acting.
4. high pressure or low pressure.

A direct acting pump has a pump rod that is a direct extension of the piston rod. Therefore, the piston in the power end is directly connected to the plunger in the liquid end. When the piston reaches the end of the first stroke, the cylinder is full of water. As the piston begins its second stroke, the increased pressure in the cylinder forces the liquid into the discharge chamber and into the system.

Single or simplex pumps have only one liquid cylinder. Simplex pumps may be either direct acting or indirect acting. Double or duplex pumps are a single assembly of two cylinders or two single pumps, placed side by side on the same base. The two steam cylinders are cast into a single block, and the two liquid cylinders are cast into another block.

A single-acting pump draws the liquid into the liquid cylinder on the first or suction stroke and is forced out of the cylinder on the return or discharge stroke. In a double acting pump, each stroke serves both to draw in liquid and to discharge liquid. As one end of the cylinder is filled, the other end is emptied. On the return stroke, the end which was just emptied is filled and the end which was just filled is emptied.

A high-pressure pump, has a steam piston that is larger in diameter than the plunger in the liquid cylinder. Since the steam piston area is greater than the area of the plunger, the force exerted by the steam against the piston is concentrated on the smaller area of the plunger in the liquid cylinder. Therefore, the pressure is greater in the liquid cylinder than in the steam cylinder. A high-pressure pump discharges a comparatively small volume of liquid against a high pressure. Low-pressure pumps, on the other hand, have a comparatively low discharge pressure but a larger volume of discharge. In a low-pressure pump, the steam piston is smaller than the plunger in the liquid cylinder.

The standard way of designating the size of a reciprocating pump is by giving three dimensions, in the following order: 1) the diameter of the *steam piston;* 2) the diameter of the *pump plunger;* and 3) the length of the *stroke.* For example, a 12 X 11 X 18 in. reciprocating pump has a 12-in. diameter steam piston, a pump plunger which is 11 in. in diameter, and a stroke of 18 in. The designation enables the reader to tell immediately whether the pump is a high-pressure or low-pressure pump.

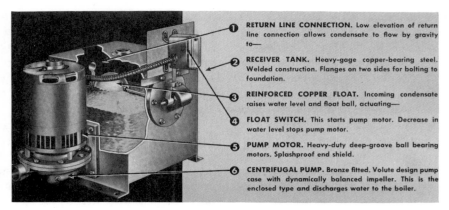

FIGURE 160. Condensate pump. (Manufactured by Dunham-Bush, Inc.)

28.4. PUMP SYSTEMS AND PIPING

Pumps are used principally to circulate hot or chilled water for either heating or cooling systems, to handle hot condensate water from steam heating systems, and to deliver water to a boiler from a deaerating heater.

A condensate pump unit is usually a centrifugal pump with a reservoir and automatic float. Figure 160 shows a condensate pump with its tank and float assembly. Here, the

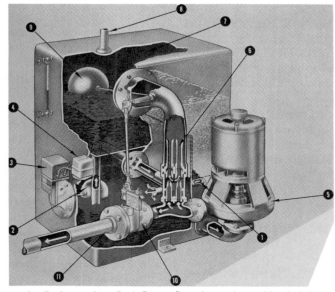

1 *Return Piping Connection To Accumulator Tank*. Return flows by gravity to either built-in accumulator tank with low inlet elevation or to separate tank when pitting is required.
2 *Reinforced Copper Float in Accumulator Tank*. As water returns to accumulator tank the rise of water level raises float ball.
3 *Float Switch*. Rising of float ball actuates the float switch (alternator on duplex) to start pump. Decrease in water level stops pump.
4 *Vacuum Switch*. Pump operation is also controlled by a vacuum switch. It is set as standard to start pump at vacuum below 3" and to stop pump at vacuum over 7".
5 *Centrifugal Pump*. Always-primed centrifugal pump with bronze impeller discharges water under pressure.
6 *Jet Exhauster*. Water from pump discharge passes through 2-stage jet exhauster at high velocity, enveloping air, water vapor and condensate returning from accumulator tank.
7 *Air Separating Tank*. Mixture of air, vapor and condensate is discharged into air separating tank.
8 *Vent*. Air is vented from air separating tank to atmosphere.
9 *Reinforced Copper Float*. As level of water in air separating tank increases . . . float will rise.
10 *Discharge Valve*. Rising float ball opens discharge valve.
11 *Discharge Line*. Excess water passes through discharge valve and is returned under pressure to boiler.

FIGURE 161. Vacuum type condensate pump. (Manufactured by Dunham-Bush, Inc.)

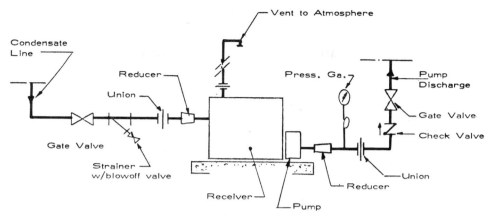

FIGURE 162. Centrifugal (simplex) condensate pump with receiver piping hookup detail.

condensate flows by gravity to the reservoir until it reaches a level to actuate the pump switch. A vacuum-type condensate pump system is also commonly used. In this system, the pump impeller, forces the water through a two-stage jet exhauster at high velocity. This creates a vacuum that serves to drive the steam, air and condensate through the heating system piping. It also creates a positive pressure to return the condensate to a low-pressure boiler. The vacuum-type condensate pump is illustrated in Figure 161.

Figures 162 and 163 show piping details involved with hooking up simplex and duplex condensate pumps and receivers. The schematic in Figure 162 shows the condensate entry into the receiver and the pump discharge which returns the condensate to the boiler. A check valve and gate valve must always be installed on the pump discharge line between pump and boiler to prevent backflow. The duplex unit in Figure 163 has an advantage of the single one in that capacity is increased and a redundancy permits boiler operation in the event of failure of one pump.

The hookup for a double suction centrifugal pump with a horizontally split-case is shown in Figure 164. The rotating assembly can be inspected without disconnecting the piping. Note how flexible pipe connections isolate the piping

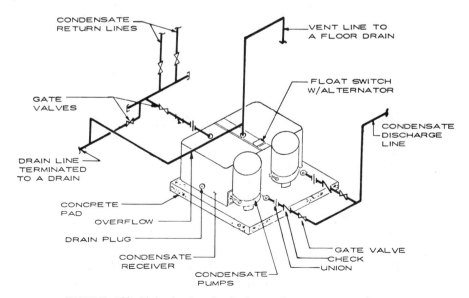

FIGURE 163. Piping hookup for duplex condensate pump receiver set.

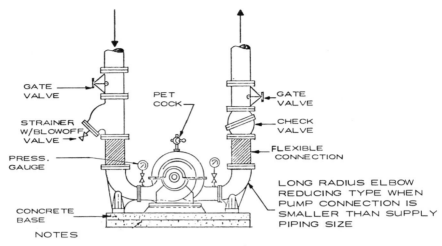

NOTES

Packing gland drips to be piped to a floor drain
Vents, Bleeds, Drains, Blowoff connections to be piped to a drain
Valve strainer and flexible connections to be same size as pipe
connections originated from main line to valves

FIGURE 164. Horizontal split-case pump piping detail.

system from the vibration of the pump. They also take up misalignment in the pipe line components.

The piping hookup for close coupled centrifugal pumps is given in Figure 165. Close coupling reduces misalignment between pump and motor. Note the flexible hose to isolate vibration, and the check valve on the outlet if multiple pumps are used. A system with two centrifugal pumps operating in parallel is shown in Figure 166. It is similar to the installation shown in Figure 165. The check valves prevent backflow from the second pump. Careful attention is paid to vibration isolation. Besides the flexible hose, the pump is mounted on a resilient base. The piping too has vibration dampers in the hanger rod structure and floor mounts.

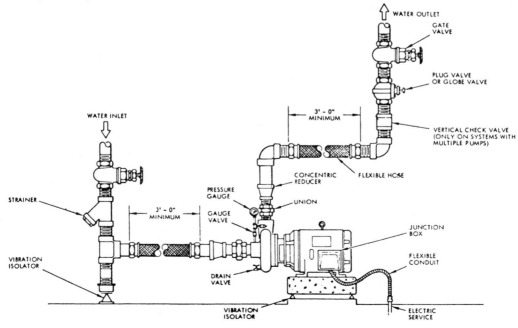

FIGURE 165. Centrifugal closed-coupled end suction pump hookup detail. (*Courtesy of the Singer Co.*)

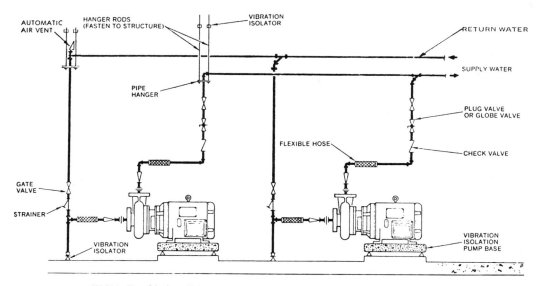

FIGURE 166. Parallel pumps piping hookups. (*Courtesy of the Singer Co.*)

28.5. ESTIMATED LABOR TO INSTALL PUMPS

Estimate Hookup Time (in man-hours)

Inlet Pump Size (in.)	See Figure 162	See Figure 163	See Figure 164	See Figure 165
$\frac{1}{2}$	15.25			
$\frac{3}{4}$	18.00			
1	21.00			
$1\frac{1}{4}$	24.00			
$1\frac{1}{2}$	24.75			
2	27.10	30		
$2\frac{1}{2}$		32		42.5
4		40	72	62.5
6			85	78.0
8			110	92.0
10			130	
12			148	

Description	Suction Sizes (diameter, in in.)	Motor hp	Weight (lb)	Labor (man-hours)
Centrifugal, horizontal split-case, flex coupled	2	10	350	8
	4	30	750	8
↓	4	60	1400	12
	8	125	1650	18
	10	200	2400	18
↓	12	200	3500	24
Centrifugal vertical, split-case	2	10	500	12
	4	30	850	12
↓	4	60	1200	14
	8	125	1900	24
↓	8	125	2400	24

Description	Discharge (diameter in in.)	Motor hp	Weight (lb)	Labor (man-hours)
Centrifugal, vertical, close-coupled	$1\frac{1}{4}$	$\frac{1}{2}$	125	4.5
	$1\frac{1}{2}$	$1\frac{1}{2}$	200	4.5
	2	10	300	8
	$2\frac{1}{2}$	20	400	8
	3	30	500	12
	4	40	600	12
	5	50	700	12
	6	25	550	18
Centrifugal, vertical, flex coupled	2	10	500	16
	$2\frac{1}{2}$	15	700	16
	3	30	900	20
	4	125	2000	32
	5	200	2600	32
	6	200	2800	34
	8	125	2800	34
Turbine, vertical, flex coupled	3	1	350	24
	3	3	400	24
	3	5	450	24
	4	5	450	24
	4	$7\frac{1}{2}$	600	24
	4	10	800	24
	6	$7\frac{1}{2}$	700	24
	6	15	750	32
	6	20	850	32
	8	15	1000	32
	8	25	1050	32
	8	40	1200	48
	10	30	1300	48
	10	40	1250	48
	10	50	1350	48
Centrifugal, close coupled, end suction	$1\frac{1}{4}$	$\frac{1}{2}$	120	4
	$1\frac{1}{2}$	$1\frac{1}{2}$	180	4
	2	10	280	6
	$2\frac{1}{2}$	20	380	12
	3	30	480	12
	4	40	550	14
	5	50	600	14
	6	25	500	14
Gear pumps, base mounted, direct or geared drive type	$\frac{1}{2}$	$\frac{1}{3}$	50	4
	1	$\frac{1}{2}$	100	4
	$1\frac{1}{2}$	$1\frac{1}{2}$	225	8
	$1\frac{1}{2}$	3	300	12
	2	5	450	12
	$2\frac{1}{2}$	10	550	12
	3	15	750	16
	3	20	1250	16
	4	30	2000	24

Description	Inlet (diameter in in.)	Motor hp	Weight (lb)	Labor (man-hours)
Centrifugal, flexible coupled, end suction	$1\frac{1}{4}$	2	250	6
	$1\frac{1}{2}$	5	375	12

Centrifugal, flexible coupled, end suction	Inlet (diameter in in.)	Motor hp	Weight (lb)	Labor (man-hours)
↓	2	10	500	12
	$2\frac{1}{2}$	20	675	12
	3	30	800	14
	4	40	950	14
	5	50	990	14
	6	25	850	14

Description	Water gpm	Air cfm	Motor hp	Weight (lb)	Labor (man-hours)
Centrifugal, duplex, vacuum heating	5	1	1	200	8
↓	10	5	1	210	8
	15	10	$1\frac{1}{2}$	240	12
	25	10	$1\frac{1}{2}$	250	12
	30	15	2	260	16
	40	15	2	280	16
	45	15	2	300	16
	60	15	3	325	16

Description	Suction (diameter, in in.)	Motor hp	Weight (lb)	Labor (man-hours)
In-Line, Circulation:	1	$\frac{1}{4}$	60	3
↓	$1\frac{1}{4}$	$\frac{1}{3}$	70	3
	$1\frac{1}{2}$	$\frac{1}{2}$	80	3
	2	$\frac{3}{4}$	85	3
	2	1	100	4.5
	2	$1\frac{1}{2}$	120	4.5

Description	EDR[a] (sq. ft.)	gpm	Unit	Labor (man-hours)	Average Material Cost ($)
Condensate return, single unit, single phase, 3450 rpm (used for steam boilers, 10 psi)	1000	2	EA	12	350
	2000	3		12	
	4000	6		12	
	6000	9		14	
	8000	12		14	
	10,000	15		14	
	15,000	23		18	
	20,000	30		20	
	25,000	38		20	700
	30,000	45		20	
	40,000	60		20	
	50,000	75		24	
	65,000	98		24	
	75,000	113		24	1400
↓	100,000	150	↓	24	2000
Condensate return, duplex, single phase, 3450 rpm (used for steam boilers, 10 psi)	1000	2	EA	12	750
	2000	3		14	
	4000	6		20	
	6000	9		20	
↓	8000	12		20	
	10,000	15		20	
	15,000	23		20	
↓	20,000	30		24	1000

Condensate pump (continued)	EDR[a] (sq. ft.)	gpm	Unit	Labor (man-hours)	Average Material Cost ($)
	25,000	38		24	
	30,000	45		24	
	40,000	60		24	
	50,000	75		26	
	65,000	98		26	2000
	75,000	113		30	
	100,000	150		32	3000

[a]Equivalent Direct Radiation = 240 Btuh

29

SOUND CONTROL EQUIPMENT
(Noise and Vibration)

Day by day the environmental field is becoming more and more sophisticated. Where once heating, ventilating, and air conditioning were of sole importance, now the environment plays an important part in today's design projects. Preserving or even improvement of the environment takes many forms, from pollution control of air and water to energy conservation. The acoustic environment must also be considered by both designer and manufacturer. It is not only a matter of good design practice to control sound and vibration to a prescribed level. Both federal and local government agencies have issued codes and directives that limit noise pollution.

Machines, in the course of operation, generate vibrations, which if at certain frequencies, will result in unwanted noise. Engineer's are concerned with methods of abating vibration and noise. Excessive vibration from misaligned mechanical components will result in high maintenance costs. Air conditioning ducts serve as a perfect conduit for unwanted noise from HVAC components. A high noise level in the conditioned space can vary from mildly annoying to a threat to the health of its occupants.

Sound is made up of many frequencies. The audible range is between 20 and 100,000 cps second. In the past, engineers and manufactures worked primarily to reduce the amplitude of machinery vibration, by harmlessly dissipating their energy. However, the vibration isolated was only one part of the total noise spectrum. The airborne noise generated by the fan, com-

pressor, or air conditioner had to be controlled as well.

A mechanical system has a tendency to vibrate more easily at some frequencies than at others. If the system is subjected to impact, the natural frequencies are excited and ring out just like a bell excited by a clapper. Whenever the system is excited by a continuous vibration at one of its natural frequencies, a condition of resonance results, in which large amplitudes of vibration are induced by very small amounts of exciting energy.

Damping materials are used to dissipate the vibrational energy of the part to which they are attached, by transforming the energy into heat. These materials are effective in suppressing vibration only at, or near, the natural frequencies, because at these frequencies, they are capable of draining off more energy than it is being supplied. Vibration damping materials function by being deformed so that mechanical energy is dissipated by internal or rubbing friction.

Permissible transmissibilities may range from 50% for small pieces of equipment installed on a rigid floor, to 1% for heavily powered equipment installed on a upper floor of a steel frame building. Put another way, noise and vibration should be attenuated 50% in the first case and 99% in the second. A practical approach to the question of acceptable transmissibility is to consider the hp of the driver, along with the speed of the equipment, as well as the area in which the equipment is located.

29.1. NOISE ISOLATORS

Three types of noise isolators are used to absorb the vibrations produced by HVAC equipment and other types of rotating or reciprocating machinery. They are: steel coil springs, rubber-in-shear, and cork.

1. *Steel springs* are recommended where large deflections are required in order to achieve a high isolation efficiency. The springs should be mounted in a steel housing with neoprene stabilizers for vertical and lateral motion control without binding, and should have a built-in leveling device to eliminate shimming. Neoprene, nonskid, sound deadening pads should be bonded to the underside of the housing to eliminate bolting.

Figure 167 shows examples of spring-type noise isolators. The upper two are called spring-flex mountings. They are free standing, laterally stable without housings, snubbers, or guides. A $\frac{1}{2}$-in. thick ribbed acoustical neoprene pad is cold bonded to the underside of the base plate. All mountings have bolt holes in the base plate and are provided with adjusting bolts for leveling and attachment to the equipment. Horizontal and vertical spring constants are equal to insure the same isolation from horizontal disturbances as from vertical.

2. *Rubber-in-shear* (neoprene) should be used where the deflection will not exceed $\frac{1}{2}$-in. and transmissibility requirements are met. Mountings molded in Neoprene should be used rather than natural rubber because of its oil-resistant qualities. Where equipment does not lend itself to bolting, ribbed neoprene pads may be used either in single or multiple layers with steel shims between each layer. A double defection neoprene-in-shear mounting having a minimum static deflection of 0.40 in. is shown in Figure 168. These mountings consist of a steel top plate and base plate completely imbedded in neoprene. The load of the equipment is taken by the top plate and transferred to the neoprene as a shear force. The mountings are molded with nonskid ribbed construction on the top plate and base plate to eliminate the need for bolting. All mountings, however, are equipped with bolt holes in the base plate and tapped holes in the top plate so that they may

a. Spring-flex mounting

b. Duplex spring mounting

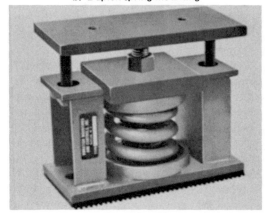

c. Spring mounting with resilient vertical stop

FIGURE 167. Spring noise isolators. (Manufactured by Vibration Mountings and Controls, Inc.)

be bolted to the floor and equipment where required.

3. *Cork*, because of its damping qualities and relatively small deflections is best applied under

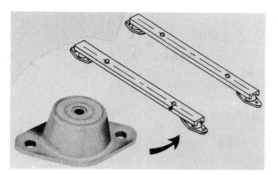

FIGURE 168. Rubber-in-shear mounting.

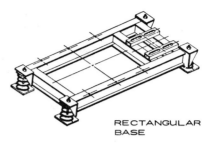

RECTANGULAR BASE

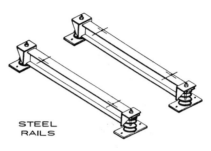

STEEL RAILS

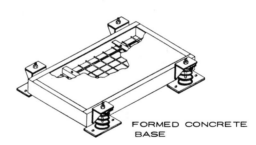

FORMED CONCRETE BASE

FIGURE 169. Steel bases with integral vibration mounts.

concrete foundations for noncritical applications or under equipment to damp high frequencies. Cork is not recommended for upper floor installations or other critical areas.

In addition to the noise attenuators explained above other applications of noise and vibration isolation include:

A *concrete base* under equipment, a misunderstanding often arises as to their need. The additional mass of concrete aids little, if any, towards reducing the transmission of vibration to the building structure. It will, however, reduce the amplitude of oscillation of the equipment, thereby reducing the transmission of vibration through connecting piping. Concrete bases may replace structural steel bases for economic reasons, for better maintenance of alignment of equipment, or when grouting of equipment is required. Concrete foundations are recommended for certain types of equipment, such as: centrifugal pumps with 5 hp motors, centrifugal fans operating at greater than 3 in. static pressure, or centrifugal fans driven by motors 75 hp or larger, and all horizontal and most vertical compressors with unbalanced forces during operation.

Integral structural steel bases incorporating neoprene-in-shear mountings or steel spring mountings should be specified for all belt-driven centrifugal fans not requiring concrete inertia blocks. Figure 169 shows two integral steel bases with spring mountings. The top illustration is that of a rectangular base with adjustable motor slide rails. The middle illustration shows individual steel rails. Both are made of wide flange sections whose depth is no less than $\frac{1}{10}$th the longest span of the base.

All piping connected to resiliently mounted equipment must be made flexible to allow equipment or its foundation to move on its mountings. This can be accomplished by the use of isolation hangers or flexible pipe connections. In some cases, a combination of the two is desirable. The proper treatment of the piping will protect the building structure against noise and vibration that may be transmitted through the pipe lines.

Isolation hangers can incorporate a neoprene-in-shear element, a steel spring, or a combination of both, depending upon the rigidity of the structure and the permissible transmission. Figure 170 shows a number of isolation hangers.

The upper part of the Figure (a) is a steel spring hanger in combination with a neoprene-in-shear element. The total deflection in such a unit is about $1\frac{1}{4}$ in. Part (b) is similar to the one above except that it has the capability of maintaining a fixed elevation of the equipment or piping regardless of load changes. The hanger also incor-

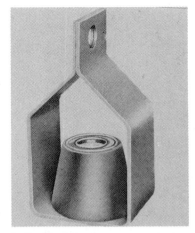

d. Double deflection neoprene-in-shear

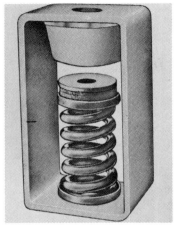

a. Steel spring combined with neoprene-in-shear element

e. Double deflection neoprene-in-shear

FIGURE 170. (*Continued*)

porates an adjusting device to transfer the load to the spring. Part (c) shows a rectangular steel housing incorporating a steel spring in series with a neoprene noise isolation washer. (Design permits installation in hanger rods or at the ceiling.) They are designed to isolate low-frequency mechanical vibration. Part (d) is that of a double deflection neoprene-in-shear hanger with a minimum deflection of 0.40 in. Part (e) is similar to (d) except that it can be installed at the ceiling in addition to hanger rods. They are available in load ranges from 10 to 4000 lb/hanger, with static deflections to $\frac{1}{2}$ in. The elements are molded of color-coded neoprene stock for easy identification of rated load capacity. The hanger configuration is designed for direct attachment to the flat iron duct straps.

Flexible pipe connections at both the suction and discharge ends of a pump are preferred since their longer length provides sufficient flexibility to allow mountings to function. The hose should incorporate a built-in minimum elongation to permit its flexure during operation.

b. Spring and neoprene unit with constant elevation feature

c. Steel spring and neoprene washer

FIGURE 170. Isolation hangers.

Where space is limited, rubber expansion joints can be used: unlike rubber hose, expansion joints require bolts to limit their expansion and contraction. Either type of connection will cushion water hammer and reduce stress on pump flanges. Where temperature, pressure or fluid-carrying requirements cannot be met with rubber connections, metal flexible connections can be used.

Figure 171 shows four types of flexible pipe connections. The top is a flexible connector of duck and butyl construction with integral duck and butyl flanges. Connectors are steel wire re-

a. Steel wire reinforced duct and butyl connector

b. Rubber expansion joint

c. Metal flexible connectors

d. Teflon flexible couplings and expansion joints

FIGURE 171. Flexible pipe connections.

inforced and furnished with steel retaining rings. Lengths are a minimum of six times the diameter through 6-in. I.D. with a maximum length of 36 in. for 8-in. I.D. and larger.

Part (b) is a rubber expansion joint, which has a body and integral flanges made with multiple layers of fabric and rubber, wire reinforced for pressure, vacuum, or both. Expansion joints are helpful in reducing pump noise and vibration transmitted through piping. Their short face-to-face dimensions make them exceptionally useful where space is limited.

Figure 171(c) shows metal flexible connectors, corrugated bronze, stainless steel, or monel flexible hose braided as required to suit working pressures. They are available with a variety of pipe and tube fittings for use with freon, water, steam, air, oil, and exhaust gases.

Teflon flexible couplings and expansion joints in Figure 171(d) shows bellows molded from "FEP" Teflon and assembled to ductile iron flanges. The bellows are reinforced with metal rings. They are furnished with isolated limit rods, preset at the factory, ready for installation. Extremely short face-to-face dimensions make them ideally suited where space is limited.

29.2. NOISE REDUCTION FOR FANS

Centrifugal fans are found throughout air conditioned buildings. The trend is toward high-pressure systems which distribute air through vast networks of ducts with the result that fans, powered by motors 75 hp or more are common. In many cases the installations are adjacent to or directly above occupied spaces. Because of the critical location and the magnitude of forces produced, a high degree of vibration and noise isolation is necessary.

To solve vibration control problems, it is necessary to determine the disturbing frequency as part of the design criteria. In fan and motor systems, the fan speed is usually the lowest disturbing frequency. By providing adequate* deflection for satisfactory isolation of this frequency, we automatically provide protection against transmission of the higher frequencies

*Adequate deflection for satisfactory isolation of fan vibration is given in manufacturers' data for vibration mounts.

caused by electrical hum, motor unbalance, or fan blades.

Selection of the isolation equipment for centrifugal fans is a function of the relationship of fan to motor and the required isolation efficiency. To minimize belt and bearing wear, the fan and motor shafts must be parallel. This cannot be done by individually mounting each component. To insure alignment, the motor must be fixed either to the fan housing, the pedestal, or the fan and motor must be mounted on a rigid common base. If no isolation is required, the floor serves as this base. If isolation is necessary, the base is usually one of the several types described in the following paragraphs:

1. *Structural Steel Base.* Angles or channels are welded or bolted to form a rigid base to which the fan and motor are fastened. This assembly, in turn, is placed upon the vibration isolators. Motor slide rails are usually incorporated with this and other systems to facilitate belt adjustment, see Figure 169.

2. *Framed Concrete Base.* Concrete is poured into a structural steel frame, usually rectangular in shape. Interior reinforcing may be used. Brackets are located on the frame's exterior for the installation of vibration control mountings. Figure 169, bottom, shows a concrete pouring form. Steel members are welded in place to act as a template and are provided with bolt holes for anchoring the supported equipment. Pouring forms for split case pumps are extended in width to provide supports for the suction and discharge elbows. Height saving brackets are located at all mounting positions.

3. *Concrete Slab Base.* The slab is poured and the form is completely removed. All that remains is a slab of concrete (usually 6 in. thick, minimum) with interior reinforcing and foundation bolts for equipment. Isolators are placed directly under the slab.

29.3. NOISE REDUCTION FOR COOLING TOWERS

Each fan gear box and motor is mounted on a rigid integral steel frame that is isolated from

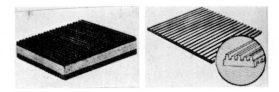

FIGURE 171a. Mounting pads.

the tower steel by means of steel springs selected for proper capacity and isolation efficiency. The outside diameter and wire diameter of the springs are selected to provide spring stability after deflection due to fan thrust. The fan operating speed (rpm) can be considered as the primary disturbing frequency. The vibration mountings should be designed to provide a minimum isolation efficiency of 90% on cooling towers where the motor does not exceed 40 hp. Cooling towers with larger motors should have vibration mountings designed to provide a minimum isolation efficiency of 95%. Each mounting should incorporate a built-in leveling device and a resilient snubber for limiting the vertical rise of the integral base at tower shut down when water is drained from the tower or boiler. Such an isolator is shown in Figure 167c. A minimum clearance of $\frac{1}{2}$ in. is to be maintained between the steel springs and the limit stop housing and around the restraining bolts so as not to interfere with normal spring performance. The mounting shall also be provided with two layers of corrugated neoprene pad separated by a galvanized steel plate in series with the spring. Such a pad is shown in Figure 171a. All anchor bolts shall be isolated from the tower steel by means of neoprene bushings and resilient washers.

On outdoor installations, there should be provision to incorporate a vertical hold-down device with the aforementioned vibration mountings, complete with neoprene bushings and resilient washers. For protection against the weather, all springs and hardware are to be cadmium plated and coated with a neoprene paint, and the housings coated with red lead and a neoprene paint.

by means of steel spring vibration mountings. These consist of a cast, telescoping housing containing one or more steel springs. The mounting can have a built-in leveling device, resilient inserts to act as guide lines for the upper and lower housings, and a corrugated neoprene acoustical pad bonded to the base. One such floor-mounted spring unit is illustrated in Figure 172. It is extremely flexible in that the springs and housings may be varied to suit specific requirements. These mountings may be attached to steel frames or concrete foundations above or below floor level.

Ceiling Suspended Units. Each ceiling suspended unit can be hung from the overhead structure by means of a combination of spring and neoprene-in-shear hangers inserted in the suspension rods. The neoprene-in-shear mounting can be selected in accordance with the load to act as an acoustical barrier, and the spring selected to act as a vibration isolator. Two types of units are shown in Figure 172. The first is a cast-in with hooks to transfer the load to the enclosed steel springs. Mountings may be furnished with a built-in nonrising level with a hydraulic jacking device to facilitate raising and leveling the isolated load. The second type, at the bottom, is similar to the first except that it has a steel housing that is nailed to the concrete form. After the concrete is poured, the housing forms a flush opening to accept the spring. The mounting incorporates a built-in leveling and adjustment bolt as well as neoprene nonskid acoustical pad.

FIGURE 172. Floor and ceiling spring mounts.

29.4. NOISE REDUCTION FOR AIR HANDLING AND HEATING AND VENTILATING UNITS

Floor-Mounted Units. Each floor-mounted unit can be isolated from the building structure

30

STEAM TRAPS

Steam traps are installed in steam lines to drain condensate from the lines without allowing the escape of steam. There are many different designs of steam traps; some are suitable for high pressure use and others for low pressure. In general, traps consist of a container to hold the condensate, an opening to discharge the condensate, and a means to automatically open and close the port. The traps in common use in the HVAC industry include: mechanical, thermostatic, impulse, bimetallic, inverted bucket, and thermodynamic.

30.1. MECHANICAL STEAM TRAPS

There are two types of mechanical traps; the ball float and the open bucket. The operation of the bucket type steam trap is controlled by the condensate level in the trap body. The bucket floats as condensate enters the trap body. The valve is connected to the bucket in such a way that the valve closes as the bucket rises. As condensate continues to flow into the trap body, the valve remains closed until the bucket is full. When the bucket is full, it sinks and thus opens the valve. The valve remains open until enough condensate has blown out to allow the bucket to float, thus closing the valve. The ball float operates on the same principle, except that a closed ball floats on the condensate. Figure 173, bottom shows the open bucket type. At the top of the Figure are several types of commercially available ball float valves.

The term thermostatic valve comes about from the use of flexible bellows that open and close in response to temperature–pressure changes.

30.2. THERMOSTATIC STEAM TRAPS

In general, thermostatic steam traps are more compact and have fewer moving parts than most mechanical steam traps. The operation of a bellows type thermostatic trap is controlled by expansion of the vapor of a volatile liquid, enclosed in a bellows type element. Steam enters the trap body and heats the volatile liquid in the sealed bellows, thus causing expansion of the bellows. The valve is attached to the bellows in such a way that the valve closes when the bellows expand. The valve remains closed, trapping steam in the body. As the steam cools and condenses the bellows cools and contracts, thereby opening the valve and allowing the condensate to drain. The liquid expansion trap works in a similar fashion except that liquid, in a separate container within the trap, expands on contact with the steam moving a piston that closes the valve. Both types are shown in Figure 174(a) and (b).

30.3. IMPULSE STEAM TRAPS (Figure 174c)

This trap uses two orifices in series to provide a pressure impulse to operate the discharge valve. When relatively cool condensate reaches the trap, it passes through the two orifices with-

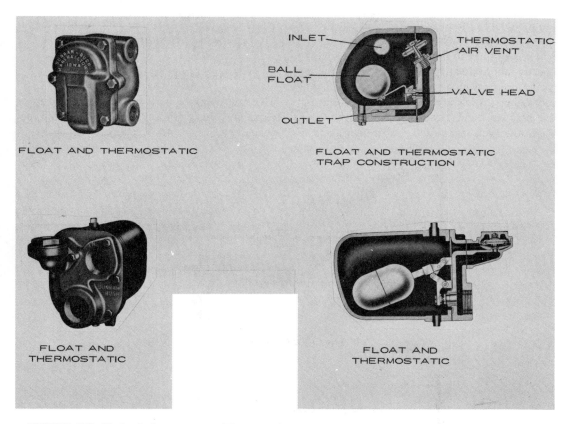

FIGURE 173. Mechanical steam traps. (Float and thermostatic traps manufactured by Dunham-Bush, Inc.)

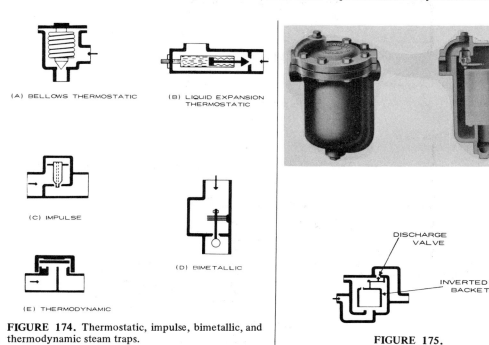

FIGURE 174. Thermostatic, impulse, bimetallic, and thermodynamic steam traps.

FIGURE 175.

out building up sufficient pressure in the control chamber to close the main valve. Condensate continues to flow until it reaches a temperature approximately 30°F below steam temperature, at which point pressure in the control chamber can close the trap. As the condensate is held back and is cooled, the trap again opens, and the cycle is repeated. Under light loads, live steam may reach the trap and be discharged through the bleed orifices.

30.4. BIMETALLIC STEAM TRAPS

The main working parts of this steam trap are a segmented bimetallic element and a ball-type check valve as shown in Figure 174d. One end

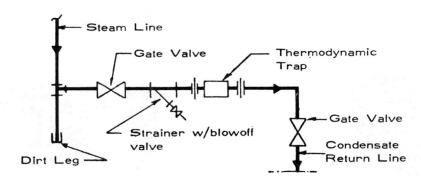

THERMODYNAMIC
TRAP

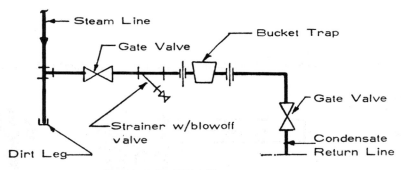

BUCKET TRAP

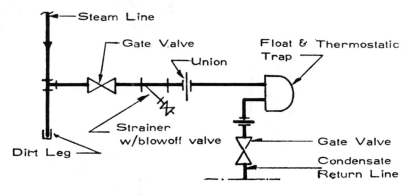

FLOAT & THERMOSTATIC TRAP

FIGURE 176. Hookup details for steam traps.

of the bimetallic element is fastened rigidly to the valve body. The other end, which is free to move, is fastened to the stem of the ball check valve.

Line pressure acting on the check valve tends to keep the valve open. When steam enters the trap body, the bimetallic element expands unequally because of the differential response to temperature of the two metals. The bimetallic element deflects upward at its free end, thus moving the valve stem and closing the valve. As the steam cools and condenses, the bimetallic element moves downward, toward the horizontal position, thus opening the valve and allowing some condensate to flow out through the valve. As the flow of condensate begins, an unbalance of line pressure across the valve is created. Since the line pressure is greater on the upper side of the ball check valve, the valve opens wide and allows a full flow of condensate.

30.5. INVERTED BUCKET TRAPS

The trap body is normally filled with condensate to maintain a seal around the inverted bucket, which serves as a float to operate the discharge valve. Live steam entering the bucket floats it to close the valve. During the closed period, condensate collects in the piping at the inlet side of the valve until steam floating the bucket leaks through a small hole in the top of the bucket and permits the bucket to drop and

open the valve. The condensate is discharged, followed by steam, which is required to actuate the float mechanism. Air can pass through the small hole at the top of the bucket. Some inverted bucket traps are fitted with an auxiliary bimetal air vent. The trap is shown schematically in Figure 175 along with a commercially available bucket trap.

30.6. THERMODYNAMIC TRAPS

In this trap, the steam condensate and air raise the disk and flow freely through the trap. When steam reaches the trap, the velocity under the disk is instantly increased, and recompression above the disk snaps it onto its seat to give a tight shutoff. Heat loss from the small control chamber, which is filled with a steam-condensate mixture, causes the chamber pressure to decrease to a point at which the valve disk opens again to discharge condensate, see Figure 174e. The recompression above the disk (which closes the trap) occurs when the condensate and air in the portion of the chamber above the trap suddenly flashes to steam as it becomes heated from the incoming steam.

30.7. STEAM TRAP HOOKUPS

Several rules should be remembered when installing steam traps. The vertical pipe from the steam line that conducts the condensate back

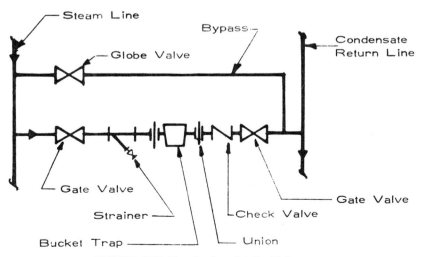

FIGURE 177. Trap hookup detail with bypass.

to the water side of the boiler should be as long as possible and contain a strainer to remove particulate matter such as boiler scale. Gate valves on each side of the trap allow shut off of flow during maintenance. Hookup details for thermodynamic, bucket and float traps are shown in Figure 176. Figure 177 illustrates the piping details for a trap bypass which permits servicing of the trap without interruption of flow. When the condensate must be discharged against a high back pressure, a check valve is put at the discharge side of the trap.

30.8. ESTIMATED LABOR TO INSTALL STEAM TRAPS

Description	Size (in.)	Labor (man-hours)
Bucket traps	$\frac{1}{2}$	1.25
	$\frac{3}{4}$	1.50
	1	1.75
	$1\frac{1}{4}$	2.25
	$1\frac{1}{2}$	2.50
	2	3.15
Float and thermostatic trap	$\frac{1}{2}$	1.25
	$\frac{3}{4}$	1.50
	1	1.50
	$1\frac{1}{4}$	1.75
	$1\frac{1}{2}$	2.00
	2	2.75
Thermodynamic trap (impulse type)	$\frac{1}{2}$	0.95
	$\frac{3}{4}$	1.25
	1	1.55
Thermostatic trap	$\frac{1}{2}$	0.75
	$\frac{3}{4}$	1.25
	1	1.50

Man-hours for piping hookups, refer to Figures 176 and 177

Trap Size (in.)	Thermodynamic	Bucket	Float and Thermostatic	Trap with Bypass
$\frac{1}{2}$	6.00	6.00	6.50	
$\frac{3}{4}$	7.75	8.25	8.50	12.00
1	8.25	9.25	9.50	14.50
$1\frac{1}{4}$		11.00	11.25	15.50
$1\frac{1}{2}$		12.00	12.25	16.25
2		15.25	15.25	18.50

31
STRAINERS

Strainers are placed in pipe lines to prevent the passage of grit, scale, dirt, and other foreign matter. Such matter could obstruct pump or throttle valves, or damage machinery parts. Various types of strainers are used, depending upon the intended service.

Suction Strainer. The suction strainer is located in the pump suction line between suction manifold and pump. Any debris which enters the piping will collect in the strainer basket. The basket is removed for cleaning by loosening the retaining screws, removing the cover, and lifting out the basket. Basket type strainers, provided in most lube oil systems, are fitted with magnets which catch any small metallic particles that may be in the lube oil. When cleaning these strainers, metallic particles that adhere to the magnets should be removed.

Baskets are usually made of perforated metal as shown in Figure 178. They fit into a Y-shaped housing so that the water or oil flow enters the inside of the basket and must exit via the holes. The figure shows the basket being retained by two methods, a flanged cover and by a bushing. A tapped hole in the cover or bushing permits partial cleaning of the basket without its removal.

Duplex Oil Strainer. The duplex oil strainers are generally used in fuel or lube oil lines, where it is important that an uninterrupted flow be maintained. Two strainers are used and the flow may be diverted from one strainer basket to the other, while one is being cleaned. The shutoff device works on the principle of a four-way cock.

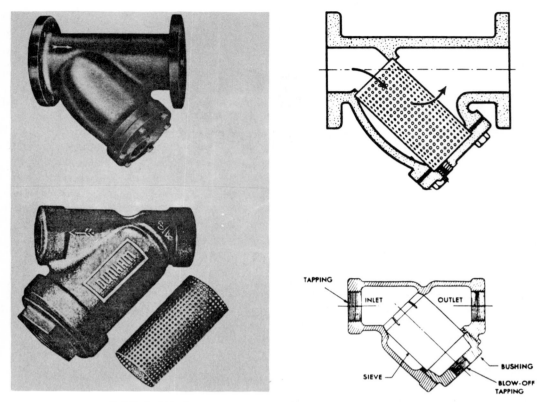

FIGURE 178. Pipe strainers. (Manufactured by Dunham-Bush, Inc.)

30.1. ESTIMATED LABOR TO INSTALL STRAINERS

Description	Size (in.)	Labor (man-hours)	Average Cost of Strainers ($)
Basket, 125# steel, screwed	$\frac{1}{2}$	1.25	
	$\frac{3}{4}$	1.40	
	1	1.75	
	$1\frac{1}{4}$	1.75	
	$1\frac{1}{2}$	2.00	
	2	2.35	
	$2\frac{1}{2}$	3.50	
	3	4.00	
	4	5.95	
	6	9.50	
Basket, 125# cast iron, screwed	$\frac{3}{4}$	2.00	
	1	2.25	
	$1\frac{1}{4}$	2.75	
	$1\frac{1}{2}$	3.00	
	2	3.50	
	$2\frac{1}{2}$	5.00	
Basket, 125# with flanged ends	2	2.00	25
↓	$2\frac{1}{2}$	2.15	60

Description	Size (in.)	Labor (man-hours)	Average Cost of Strainers ($)
Basket 125# (continued)	3	2.25	75
	4	4.00	200
	6	5.50	400
	8	7.50	500
	10	10.25	1000
	12	15.10	1600
Duplex, 125# steel, flanged	1	2.50	
	$1\frac{1}{2}$	2.75	
	2	3.00	
	$2\frac{1}{2}$	3.45	
	3	5.00	
	4	8.00	
	6	12.25	
	8	14.00	
	10	16.15	
	12	18.50	
	14	20.00	
	16	22.00	
	20	30.00	
	24	32.50	

32
TANKS

The most widely used tanks in the HVAC industry are for:

- oil storage
- expansion
- flash tanks
- general storage

Three professional organizations produce guidelines and codes under which tanks are designed, manufactured and tested. These are the American Society of Mechanical Engineers (ASME), who produce standards for unfired pressure vessels and low-pressure heating boilers; the National Fire Protection Association (NFPA); and the American Petroleum Institute (API). The HVAC engineer should specify that these codes, especially ASME and NFPA codes, be the basis for tanks in systems that he is designing.

32.1. OIL STORAGE TANK

These tanks are used to store heating oil. Oil storage tanks should be designed or selected at an ample size to take care of maximum plant requirements and provide a reasonable margin of oil storage over the longest probable period between oil deliveries.

These tanks may be of riveted or welded steel construction. No attempt is made here to state how the tanks should be installed as local ordinances and plant conditions usually dictate this.

Each tank is usually equipped with a vent pipe, with a fire screen in it. Large tanks have a manhole for cleaning, inspection and repairs. A filling connection, a return connection for the excess oil, and finally the main suction connection to the oil pumping system is usually provided.

An oil storage tank is shown in Figure 179. Because it is for Number 6 oil with high viscosity, a hot water or steam coil is wound around the suction and return lines to heat the oil in order to assure flow. This is a buried tank with a concrete vault over it to provide access to the manhole, gauge, and suction and return lines.

32.2. EXPANSION TANK (Compression Tank)

Expansion tanks are used to maintain a limit to pressure in the piping of a cooling or heating system. A cushion of air in the tank compresses as the water expands during heating, or expands as cooled water contracts. This "excess" water is stored in the expansion tank during periods of high operating temperature and is returned to the system when the water temperature is lower. The expansion tank must be sized to store the required volume of water during maximum design temperature without exceeding the maximum allowable operating pressure, and to maintain the required minimum pressure when the system is cold.

Piping systems use open and closed expansion tanks. The open expansion tank has a vented pipe connection to the atmosphere and is generally limited to installations having

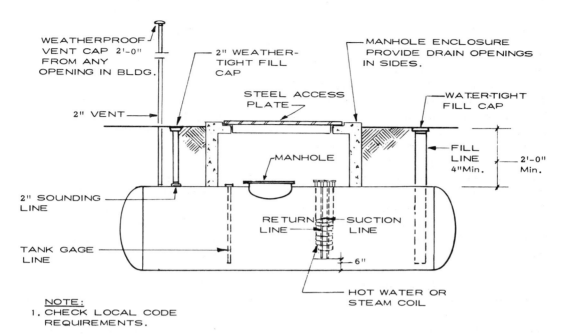

FIGURE 179. Typical oil storage tank (No. 6 oil).

operating temperatures of 180°F or less because of system boiling and water evaporation
problems. The tank should be at least 3 ft
above the high point of the system and be
connected to the suction side of the pump to
prevent subatmospheric system pressures
caused by pump operation.

The closed expansion tank is airtight which
provides a means of pressurizing the system for
operation over a wide range of conditions. As
the excess water due to thermal expansion
moves into the expansion tank, it compresses
the air trapped therein and increases the pressure of the system.

An expansion tank hookup for a closed
system is shown in Figure 180. The gate valve is
kept open at all times except when the tank is
drained.

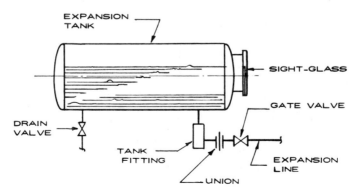

FIGURE 180. Expansion tank hookup.

32.3. FLASH TANK

In heating systems where medium or high pressure steam is used to serve cooking equipment, process equipment, laundries, sterilizers, hot water heat exchangers, etc., the steam condensate is usually collected in a vessel where provision must be made to reduce its temperature. There are two common methods to reduce the high temperature of the steam. One is to use a heat exchanger or condensate cooler, the other is to use a flash tank where the evaporation of the water takes place. When the steam condensate enters the flash tank a portion "flashes" to form steam. The vapor thus formed is then released from the tank to a low-pressure heating main or released to the atmosphere. The remaining condensate then passes through a trap into a return main.

The system is illustrated in Figure 181. The condensate enters the flash tank at the upper left. The steam discharges from the top of the tank and is returned to the supply main (or vented to the atmosphere). The flash water is pumped or moves by vacuum to the return main.

32.4. CODE VESSELS

Many building codes require that for certain applications ASME code vessels must be used. As noted above, this implies a very strict adherence to design, construction, test, and inspection procedures as well as rigorous quality control over materials. Code vessels are stamped with code approval. Papers certifying compliance to the code are provided to the owner both for the building inspector and for purposes of insurance. However, code vessels are more expensive than noncode vessels and are usually not used unless required either by the local building code or by high quality requirements to suit a specific job.

Commercially available tanks, both standard and ASME are shown in Figure 182. Note how the ASME tank end is rounded for greater strength.

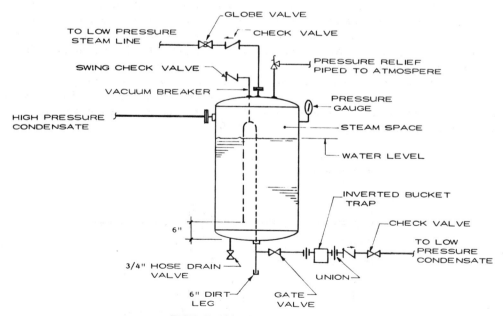

FIGURE 181. Flash tank hookup.

A.S.M.E. CODE TANK

STANDARD TANK

FIGURE 182. Standard and ASME code tanks. Manufactured by H. A. Thrush and Co.)

32.5. ESTIMATED LABOR TO INSTALL TANKS

Description	Capacity (gal.)	Average Weight (lb)	Labor (man-hours)
Expansion tanks	8	30	2.5
(by capacity)	12	55	2.5
	18	65	2.5
	24	75	3
	30	80	3
	40	95	3
	60	130	4
	80	195	4
	100	250	4
	120	350	6
	150	420	6
	175	485	6
	200	550	8
	300	750	8
	400	825	8

Description	Line Size (in.)	Labor (man-hours)
Expansion tanks	$\frac{1}{2}$	5
(by line size), see	$\frac{3}{4}$	5.75
Figure 179.	1	6.25

Description	Capacity (gal.)	Dimensions Diameter × Length		Weight (lb)	Labor (man-hours)
Pressure tanks,	275	36	70	700	8
not ASME,	500	36	123	1000	8
100 psig	500	42	92	1250	8
	1000	42	175	1725	10
	500	48	73	1295	10
	1000	48	125	1795	10
	1500	48	200	2500	12
	2000	48	264	3000	12
	1000	60	92	1750	8
	2000	60	174	2250	12
	3000	60	256	3950	12
	4000	60	338	4550	18
	3000	72	183	3990	12
	4000	72	240	4950	18
	5000	72	296	5950	18
	6000	72	353	6500	18
	7500	72	438	8150	24
	5000	84	223	6050	18
	7500	84	327	8500	30
	10,000	84	433	10,050	30
	5000	96	173	6900	18
	7500	96	256	9250	32
	10,000	96	334	11,500	32
	15,000	96	493	15,600	32

Description	Capacity (gal.)	Dimensions Diameter × Length		Weight (lb)	Labor (man-hours)
Pressure tanks,	100	24	48	300	4
ASME Code,	150	24	72	400	6
125 psig	200	30	60	550	6
	235	30	72	675	10
	310	30	96	850	10
	340	36	72	1050	10
	450	36	96	1250	10
	550	36	120	1550	10
	480	42	72	1300	10
	620	42	96	1600	12
	760	42	120	1800	12
	900	42	144	2200	12
	1000	42	168	2500	12
	825	48	96	2570	12
	1015	48	120	2700	12
	200	48	144	3000	14
	1390	48	168	3550	14
	1580	48	192	3850	14
	1760	48	216	4300	14
Steel storage tanks	90	2	4	320	24
	140	2	6	420	24
	185	2	8	550	24
	280	4	6	1010	24
	375	4	8	1350	32
	475	4	10	1575	32
	560	4	12	1875	32
	725	5	10	2100	32
	880	5	12	2400	32
	1025	5	14	2590	42
	2100	6	10	3500	32
	3100	6	15	6000	42
	4200	6	20	6500	42
	5600	8	15	7500	42
	7500	8	20	9000	42
	9300	8	25	10,000	48
	11,200	8	30	13,500	48
	15,000	8	40	17,500	60
	17,500	10	30	17,500	60
	24,000	10	40	21,000	64
	29,000	10	50	25,000	64
	33,500	12	40	36,500	64
	40,500	12	50	42,500	72
	50,750	12	60	45,000	72

33
VALVES

In the design of piping for hot water, chilled water, refrigeration, and steam condenser water systems important consideration must be given to the selection of valves. Valves are manufactured for specific applications and are designed for long life and low maintenance. Depending upon the application, valves, bodies, and closures may be made of ferrous materials, nonferrous materials, plastics or combinations. Seals and packings include rubber and elastomers, asbestos, teflon, and graphitized fibers.

The choice of valve and seal material depends upon the temperature and pressure of the liquid and whether corrosion requirements are stringent. The type of valve depends upon its function in the system. Is it an on–off device to isolate components for servicing? Does it control flow in some intermediate position? Does it permit flow in one direction but not in the opposite direction, as does a check valve? Is it important for the valve, when fully open to provide straight-through full flow as with a ball valve?

The valves most commonly used in piping systems for HVAC include gate, globe, check, angle, plug, ball, needle, and refrigerant type. Except for the check valve these are shown as hand operated. In practice, they can also be opened and closed by electric motors, solenoids, and pneumatic operators.

33.1. GATE VALVES

Gate valves are used to stop flow. An important feature of the gate valve is that there is less obstruction and turbulence within the valve, therefore, a correspondingly lower pressure drop than most other valves. With the valve wide open, the wedge or disk is lifted entirely out of the fluid stream, thus providing a straight flow area through the valve.

Gate valves are not to be used for flow control. Vibration and chattering of the disk will occur when the valve is used for throttling. This results in damage to the seating surface. The disk in the gate valve is available in several forms: solid wedge, split wedge, flexible wedge, and double disk parallel seat. In larger sizes the solid wedge may not seat as tightly as split or flexible wedge valves.

Figure 183 shows the types of gate valves available in bronze bodies, with pipe threads to connect to the system piping. Figure 184 illustrates iron gate valves which have flanges for bolting to flanged pipe connections. Iron gate valves, though available in 2-in. pipe sizes are available in sizes up to 24 in. Nonrising stem valves are used where clearance between handwheel or stem and another objects does not permit the valve height to "grow" when it is opened.

33.2. GLOBE, ANGLE, AND "Y" VALVES

These three valves are of similar design, use, and construction. They are used for throttling service and give reliable regulation of flow. The method of valve seating reduces wire drawing and seat erosion which is prevalent in gate valves when they are used for throttling.

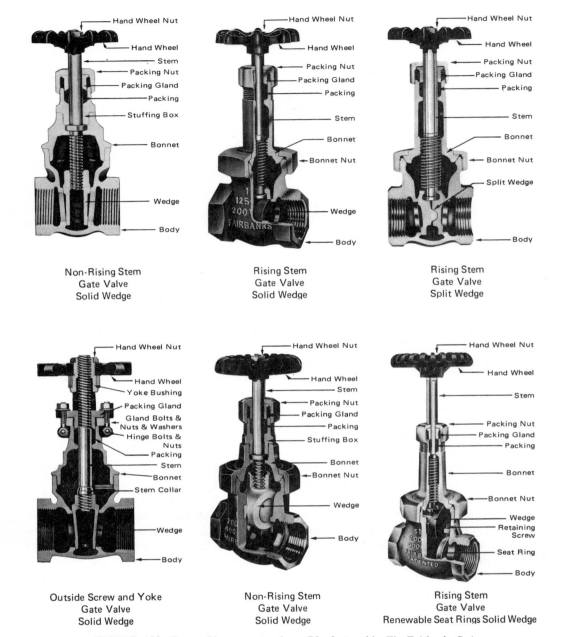

FIGURE 183. Types of bronze gate valves. (Manfactured by The Fairbanks Co.)

The angle or "Y" valve pattern is recommended for full flow service since it has a substantially lower pressure drop than the globe valve. Another advantage of the angle valve is that it can be located to replace an elbow, thus elininating one fitting.

Globe, angle, and "Y" valves can be opened or closed substantially faster than a gate valve because of the shorter lift of the disk. The seating surfaces of the globe, angle, or "Y" valve are less subject to wear and the disks and seats are more easily replaced than those on the gate valve. When valves are to be operated frequently or continuously the globe valve pro-

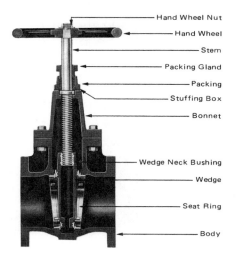

Non-Rising Stem Gate Valve
Solid Wedge

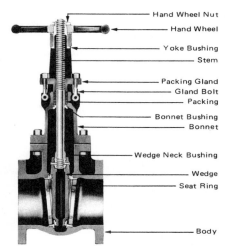

Outside Screw and Yoke Gate Valve
Solid Wedge

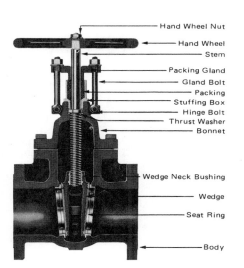

Non-Rising Stem Gate Valve
Solid Wedge

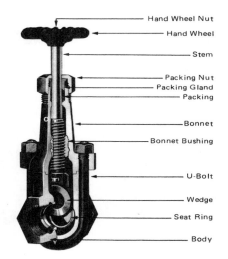

U-Bolt Gate Valve
Solid Wedge

FIGURE 184. Types of iron body gate valves. (Manufactured by The Fairbanks Co.)

vides the more convenient operation. Bronze globe valves are shown in Figure 185.

33.3. CHECK VALVES

There are two basic designs of check valves, the swing check and the lift check. The swing check

valve may be used in a horizontal line or in a vertical line for upward flow. In both cases, they are used to prevent backflow. The flow through the swing check is in a straight line without restriction at the seat. Swing checks are generally used in combination with gate valves. A swing check, seen in Figure 186 top,

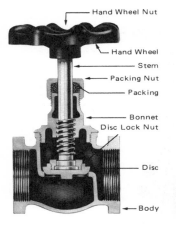

Screw-In
Bonnet Globe Valve
Bronze Disc

Union Bonnet
Globe Valve
Renewable Composition Disc

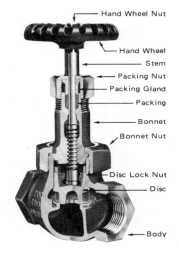

Union Bonnet
Globe Valve
Regrinding Bronze Disc

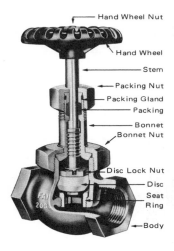

Union Bonnet
Globe Valve
Renewable Seat and Disc

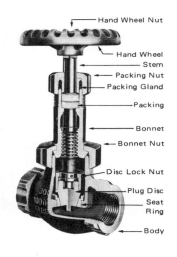

Union Bonnet
Globe Valve
Renewable Cone Plug Disc & Seat

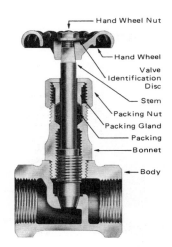

Screw-In Bonnet
Needle Globe Valve

FIGURE 185. Types of bronze globe valves. (Manufactured by The Fairbanks Co.)

rotates on a hinge pin to open and close. The lift check does not swing. It moves away from the seat parallel to the flow on guides.

33.4. BALL AND BUTTERFLY VALVES

Ball and butterfly valves, seen at the bottom of Figure 186, are very easy to operate and turn

on and off rapidly. When open, the ball valve presents a full-open aperture with no internal turbulence or pressure drop.

33.5. PLUG VALVES

Plug valves are used for balancing in a piping system not subject to frequent changes in flow.

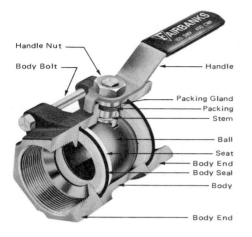

"Sphero" Ball Valve

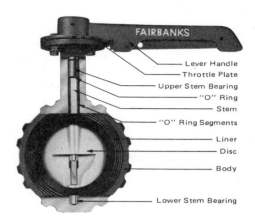

Faircoseal Butterfly Valve
Lever Operated

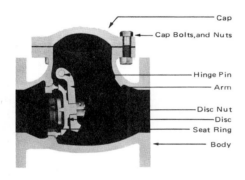

Horizontal Swing Check Valve
Regrinding Bronze Disc

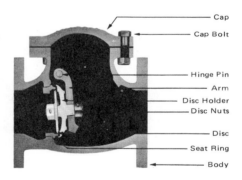

Horizontal Swing Check Valve
Renewable Composition Disc

FIGURE 186. Swing check, ball, and butterfly valves. (Manufactured by The Fairbanks Co.)

They are normally less expensive than globe type valves and the setting cannot be tampered with as easily as a globe valve.

Plug cocks have approximately the same line loss as a gate valve when in the fully open position. When partially closed for balancing, this line loss increases substantially.

ing to the type of diaphragm used, either packed or packless type. The packed valves are available with either a hand-wheel or a wing type seal cap. The wing type seal cap is preferable since it provides the safety of an additional seal. Where frequent operation of the valve is needed the diaphragm packless type is used.

33.6. REFRIGERANT VALVES

Refrigerant valves are back-seating globe valves used with refrigerant gases. They differ accord-

33.7. NEEDLE VALVES

Needle valves, sometimes referred to as expansion valves, are designed to give fine control of flow in small diameter piping. The disk is normally an integral part of the stem and has a sharp point which fits into the reduced area seat opening. Needle valves are essentially an angle valve with a needle type seating arrangement.

Note: For control valves see Chapter 10, Controls.

33.8. ESTIMATED LABOR TO INSTALL VALVES

Description	Size (in.)	Labor (man-hours)
Gate valve, bronze, 125 to 150 lb, threaded, brazed, or soldered installation	$\frac{1}{4}$	0.62
	$\frac{3}{8}$	0.62
	$\frac{1}{2}$	0.62
	$\frac{3}{4}$	0.75
	1	1.10
	$1\frac{1}{4}$	1.25
	$1\frac{1}{2}$	1.50
	2	1.85
	$2\frac{1}{2}$	2.30
	3	3.10
	4	4.20
Gate valve, iron body bronze mounted 250 lb flanged:	2	2.60
	$2\frac{1}{2}$	3.10
	3	3.70
	4	4.70
	5	5.90
	6	7.40
	8	9.50
	10	14.10
	12	16.10
Gate valves, iron body O.S.Y. 125 lb, threaded:	$1\frac{1}{2}$	0.83
	2	1.60
	$2\frac{1}{2}$	1.75
	3	2.90
	4	3.55
flanged:	2	1.80
	$2\frac{1}{2}$	2.00
	3	2.10
	4	3.60
	5	4.75
	6	6.10
	8	7.20
	10	10.20
	12	14.40
	14	19.10
	16	20.10
	18	22.50
	20	26.50
	24	28.00

Description	Size (in.)	Labor (man-hours)
Globe valve, bronze, 125 lb, threaded, brazed, or soldered installation	$\frac{1}{4}$	0.62
	$\frac{3}{8}$	0.62
	$\frac{1}{2}$	0.75
	$\frac{3}{4}$	1.10
	1	1.25
	$1\frac{1}{4}$	1.50
	$1\frac{1}{2}$	1.85
	2	2.30
	$2\frac{1}{2}$	3.10
	3	4.20
Globe valve, iron body, 125 lb, threaded:	2	0.92
	$2\frac{1}{2}$	1.25
	3	2.50
	4	3.10
flanged:	2	1.70
	$2\frac{1}{2}$	1.90
	3	2.05
	4	3.00
	5	3.50
	6	5.10
	8	6.50
	10	8.00
Swing check valve, bronze, 125 lb, threaded, brazed, or soldered installation	$\frac{1}{4}$	0.55
	$\frac{3}{8}$	0.55
	$\frac{1}{2}$	0.55
	$\frac{3}{4}$	0.75
	1	0.90
	$1\frac{1}{4}$	1.10
	$1\frac{1}{2}$	1.20
	2	1.45
	$2\frac{1}{2}$	1.75
	3	2.75
Iron body, 250 lb, threaded:	2	2.75
	$2\frac{1}{2}$	2.80
	3	3.50
	4	3.50
	6	7.10
	8	9.10
	10	10.50
	12	14.10
Plug valves, wrench operated, 150 to 250 lb, lubricated, threaded:	$\frac{3}{4}$	0.90
	1	1.00
	$1\frac{1}{4}$	1.10
	$1\frac{1}{2}$	1.20
	2	1.50
	$2\frac{1}{2}$	2.40
	3	3.00
flanged:	1	0.90
	$1\frac{1}{4}$	1.00
	$1\frac{1}{2}$	1.15
	2	1.75

Gate valve, flanged (continued)	Size (in.)	Labor (man-hours)	Plug valve (continued)	Size (in.)	Labor (man-hours)
	$2\frac{1}{2}$	2.20		4	4.75
	3	2.60		6	4.75
↓	4	4.20		8	6.20
				10	9.20
Worm gear, lubricated flanged	6	12.00		12	10.10
type, 150 to 150 lb	8	14.50	↓	14	12.15
	10	16.00	Butterfly valve 150 to 175 lb	1	0.91
	12	16.00		2	1.25
	14	25.00		$2\frac{1}{2}$	1.75
	16	30.00		3	2.05
	18	32.10		4	3.50
	20	34.00		5	5.75
↓	24	45.00		6	7.50
Ball valve, 150 lb, screw or	$\frac{1}{2}$	0.67		8	12.70
flanged:	$\frac{3}{4}$	0.69		10	15.70
	1	1.33		12	16.00
	$1\frac{1}{2}$	1.97		14	16.80
	2	2.63		16	20.00
	$2\frac{1}{2}$	2.70		18	20.00
↓	3	4.50	↓	20	24.00

34

WATER TREATMENT

Natural water often includes impurities which may be harmful to the components of a hydronic system. Such impurities vary according to geographic location, seasons, precipitation, etc., and are broadly classified as dissolved or suspended organic and inorganic matter, or dissolved gases. The concentration of impurities is expressed in terms of parts of impurities per million parts of water (ppm). This concentration rating is based on weight rather than volume. At times, extremely small concentrations, such as with entrained gasses, may be expressed in terms of parts per billion (ppb).

Once inside a hydronic system, water impurities may form harmful deposits on internal surfaces, cause corrosion, and otherwise damage system components such as pump seals, valve seats, etc. Some form of water treatment usually must be employed to prevent such undesirable occurrences.

In zones of high-heat input, such as boilers, hard scale formations can retard the transfer of heat to the water. This will result in the metal being heated to higher temperatures and may cause overheating and failure of the pressure components. The principal scale-forming impurities are calcium, magnesium, and silica. These impurities may be removed by chemical treatment or ion exchange.

34.1. CHEMICAL TREATMENT

In the lime-soda water softening process, lime, or calcium hydroxide, reacts with the soluble calcium and magnesium bicarbonates. This reaction forms precipitates of calcium and magnesium hydroxides which can be removed as sludge. The soda ash or sodium carbonate, reacts with calcium and magnesium sulfates. This reaction forms insoluble calcium and magnesium carbonate precipitates. Both reactions produce sodium sulfate, a soluble compound which is nonscale forming. When the process is carried out at temperatures above 200°F, a faster reaction rate occurs and some silica may also be removed. These reactions are not 100% effective as some hardness is left in the treated water.

34.2. WATER SOFTENING BY ION EXCHANGE

Certain minerals in water have the ability to exchange sodium cation for calcium and magnesium ions. Softening can then occur if the water is passed through beds of granulated zeolite. The zeolite material returns the calcium and magnesium ions while nonscale forming sodium ions are released into the water. After a time, sodium exhaustion will occur. Before this happens, the softening equipment must be isolated from the incoming water system and regenerated. This regeneration is accomplished by passing a strong brine of sodium chloride waste. Before returning the softening equipment to the system, the excess sodium chloride must be purged from the system. This ion exchange method is often combined with that of chemical treatment to increase the degree of hardness removal.

34.3. DEMINERALIZATION

For nearly complete removal of dissolved solids, a demineralization process is used. In this manner undesirable ions are removed from water solutions by their exchange of hydrogen or hydroxyl ions through the use of synthetic organic resins. The process can be used to remove undesirable cations or anions, and when used in combination, the treated water will be virtually free of mineral solutes. During the process, the released hydroxyl and hydrogen ions combine to form pure water.

The cation exchanger is regenerated by acid to restore hydrogen ions to the resin in exchange for the metallic cations removed from the water. Regeneration of the anion exchanger is accomplished with the use of a base such as caustic soda which restores hydroxyl ions in exchange for the negative chemical radicals, such as sulfates, which were removed from the water during the softening.

34.4. SLUDGE, OIL, SUSPENDED SOLIDS

Sludge or solid particles present in the water will act as an abrasive to prematurely wear out system components, or settle out to restrict flow. Some suspended solids, oil, or sludge may coat internal surfaces and cause overheating because of poor heat transfer.

These impurities may be removed by settling or filtering of large particles, or by coagulation of small particles and oil. In the coagulation process, a floc-forming chemical is used. The floc traps the impurities and is removed by settling or filtering.

34.5. DISSOLVED GASES AND ACIDITY

Water containing acid or dissolved gases is by far the largest problem in hydronic systems. These conditions may cause severe corrosion and eventual failure of a pressure wall. Dissolved air or oxygen is the greatest factor in corrosion of metal surfaces in a hydronic system.

Air may be partially removed by preheating the water in an open vessel, but removal is more effective in deaerators. Air is removed from a closed system by high-pressure air vents which release air but not water. However, all of the air entrained in the water cannot be removed, making water treatment necessary.

34.6. FILMING ADDITIVES

One method of treatment is to introduce a filming additive. These additives coat the internal parts of the system, and some have lubricating qualities. These additives are not the most acceptable since they may form a sludge and prevent proper function of such components as mechanical seals due to the coating formed.

34.7. CHEMICAL ADDITIVES

The most acceptable method of treatment is the addition of chemicals to reduce the acidity. This treatment is carried out according to the pH requirement of system water. The acid or alkali condition of water is measured according to the pH rating. Water which is neither acid nor alkali (neutral) has a pH factor of 7. Any value less than 7 is acid with the concentration of acid increasing as the pH decreases. Any pH value higher than 7 is alkaline with the concentration of alkali increasing as the pH increases. The extremes of pH are from 1 to 14.

If a water test shows the pH less than 7, indicating an acid or entrained gas problem, a base or alkaline chemical is added to neutralize the system. Conversely, if the pH is above 7, the system may be neutralized by adding an acid. In most cases, a base or alkali must be added.

Water softening processes remove impurities from the water. Controlling pH with chemicals, however, adds impurities to the water. For this reason, caution must be exercised in order to keep from overtreating. Additives such as chromates have a slightly abrasive effect, and over-

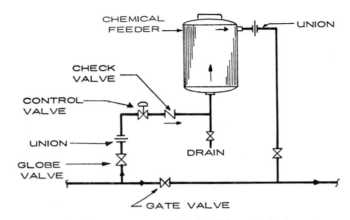

FIGURE 187. Water treatment piping detail for a single pipe system.

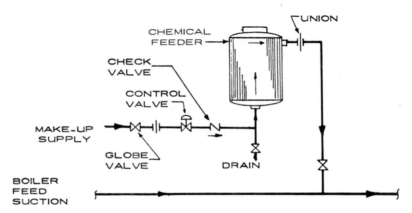

FIGURE 188. Water treatment piping detail for the suction side of a boiler feed pump.

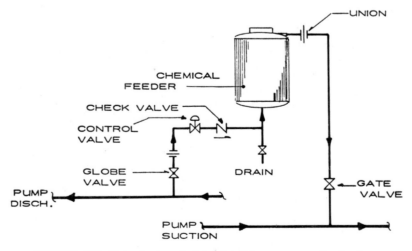

FIGURE 189. Water treatment piping detail for a pump suction hookup.

treatment can cause premature failure of mechanical seals and valve seats. It is, therefore, extremely important to engage a competent water treatment firm to test and treat the water without overdosing the system with additives.

All the above impurities will eventually corrode and clog metal surfaces and pipe systems. Scale and film formations deposited on metal surfaces will reduce the efficiency of the machinery used in an air conditioning apparatus. Most of the impurities found in water used for air conditioning can be neutralized by specially prepared chemicals. "Hard" and "soft" water which is a measure of the solid impurities contained in the water can be controlled by scale inhibitors.

Designers relate the values of 200 and 300 ppm in an approximate equivalent of 6 to 11 grains of hardness per one gallon of water. Soft water is defined as having less than 200 ppm of solids, whereas hard water contains more than 200 ppm.

34.8. WATER TREATMENT PIPING HOOKUPS

Figures 187, 188 and 189 show three types of pipe connections for water treatment pot feeders.

34.9. ESTIMATED LABOR TO INSTALL CHEMICAL WATER TREATMENT EQUIPMENT

Description	Capacity (gal.)	Labor (man-hours)
Chemical water treat-	3	3
ment pot feeders	7	3
	12	3.5
	15	4

Description	Capacity (gal.)	Weight (lb)	Labor (man-hours)
Stainless steel tank,	50	160	8
stand, and mixer	100	330	14
(pump mounted	500	800	16
under tank)			

SECTION III

System Costs and Energy Conservation

1
GUIDELINES FOR ENERGY CONSERVATION

There is presently intense pressure on building designers and engineers, as well as owners and managers of buildings to conserve energy. Every indicator points to more rather than less pressure as time passes. Government regulations, the cost and availability of fuel, and the realization that fossil fuel supplies are finite have made the nation aware that energy conservation must be an integral part of construction planning and of the operation of existing structures. At this writing, a national energy bill has not yet become law, but its passage is inevitable. Even now, the federal government is pressing for conservation in government-sponsored construction. Insulation requirements have been upgraded, energy surveys have been conducted, and energy savings studies have become a part of every architect–engineering contract.

There are four areas where energy conservation can be applied: external design, interior design, HVAC design, and HVAC operation.

1. *External Design.* Besides increasing wall and roof insulation, economies can be effected by reduction in glass area, storm windows, proper sealing and caulking, proper shading, and site orientation. These schemes lend themselves more to new, rather than existing construction.

2. *Interior Design.* Energy can be conserved by lowering temperatures in winter and raising them in summer, reducing lighting levels and reducing the rate of ventilation. In addition, zoning of conditioned space and metering of individual areas can often reduce consumption of energy.

3. *HVAC Design.* We have discussed such systems and components as heat pumps, heat recovery systems, variable air volume systems, and solar energy.

4. *HVAC Operation.* Shutdown of HVAC systems or subsystems during weekends, vacation periods, or nightime setback of comfort conditions can be effective methods of conservation. Computer control of building HVAC conditions or electrical power demand control will save energy and overall cost for larger buildings, or where combined with process control.

This chapter will discuss some HVAC components and systems and their operation for practical ways to enhance energy conservation.

Each HVAC system is of course unique, and its particular characteristics can only be identified by inspection and measurement. However, there will be many characteristics which are common to all systems of a generic group and it is these common characteristics which will be discussed.

The minimum information required to understand the operation of a system and to provide a basis for deciding which modifications are likely to prove beneficial are tabulated as follows:

Air Flow Rates
    total outdoor air

total return air
total supply air
trunk ducts
terminal units
air-cooled condenser

Water Flow Rates
boilers
chillers
cooling towers
heat exchangers
coils and terminal units

Temperatures
outdoor air, DB and WB
return air, DB and WB
mixed air entering coils, DB and WB
supply air leaving coils, DB and WB
hot deck*
cold deck*
air at terminals
conditioned areas DB and WB (typical for
each functional use)
boiler supply and return
chiller supply and return
condenser supply and return
heat exchanger supply and return
coil supply and return

Refrigerant Temperature
hot gas line
suction line

Conservation of energy must be approached in a systematic manner rather than considering individual items. Systems do not operate in isolation but depend upon and interact with others. It is important to recognize this interaction, as modification to one will affect another either beneficially or in a way that can be counterproductive.

Buildings frequently use a combination of systems to meet differing requirements for interior and perimeter zones. It is common practice, for instance, to use a VAV system for interior spaces and multizone system and/or radiation for perimeter spaces. There is an interface between the interior and perimeter zones where mixing and interaction between

*Hot and cold ducts are described in Section I, paragraph 10.2.4.

the two different systems takes place. In cool weather, cold supply air from the interior zone can spill into the perimeter zone resulting in overcooling and shift to a heating mode of operation. Both systems can counteract the other by cooling and overheating. To avoid this situation, increase the supply air temperature for interior zones in cold weather.

1.1. SINGLE-DUCT SYSTEMS

These systems, described in Section I, paragraph 10.2.1, are the simplest and probably the most commonly used today. They can comprise just a single supply system with air intake filters, supply fan, and heating coil, or can become more complex with the addition of a return air duct, return air fan, cooling coil, and various controls to optimize performance. Basically, the system supplies air at a predetermined temperature to one zone or an entire building, the quantity of heating or cooling being controlled either by modulating the supply air temperature or by turning the system on and off. The system is shown in Figure 1.

The energy output of a single-duct system to meet a space load is determined by the volume-temperature differential relationship. For example, to maintain a space temperature of 65°F, the heating load could be met by a system supplying 10,000 cfm at 105°F, or 6000 cfm at 125°F.

To conserve energy in single zone, single-duct system, determine from measurements the system output and the building load. In the heating mode, if the maximum output exceeds the building load, first reduce the air volume as this will show the greatest energy savings. For example, 90% of the initial fan volume requires only 65% of the original fan hp. Next reduce the supply air temperature. This will not conserve as much energy as reduced volume flow, however. In the cooling mode, if the system capacity exceeds the building load, the savings realized by the reduced airflow must be compared with the savings from the increased air temperature since increased performance of the refrigeration equipment may yield savings which exceed the fan power losses.

Install controls on the heating and cooling

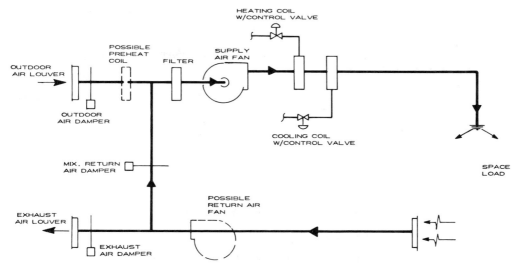

FIGURE 1. Single-duct system (flow diagram).

coils to modulate the supply air temperature. Allow an 8 to 10°F "dead zone" between heating and cooling to prevent simultaneous heating and cooling and rapid cycling. If possible, use an economizer* cycle whenever the total heat of the outdoor air is favorable during occupied periods, and the size of the system makes the installation economical.

Single zone, single-duct systems can be readily converted to VAV systems by adding control boxes on each branch and substituting outlets of the VAV type. Fan volume should preferably be controlled according to demand, either by installing inlet guide vanes or by installing a variable speed motor.

1.2. TERMINAL REHEAT SYSTEMS

Terminal reheat systems as described in Section I, paragraph 10.2.3, were developed to overcome the zoning deficiencies of single duct systems. The terminal reheat system (see Figure 2) allows each different zone to be individually controlled. However, this system wastes energy in the cooling season since all of the supply air must be cooled to a low enough temperature

---

*An economizer cycle uses outdoor air when its temperature is low enough to handle the cooling load, thus, saving energy and wear in the refrigeration unit.

to meet the most critical load zone and be reheated for zones of lesser loads to avoid overcooling. Many reheat systems are controlled at a fixed supply temperature of around 55°F DB/55°F WB.

As with single-duct systems, energy can be conserved by reducing the supply air volume. Another way to conserve energy is to set controls to raise the supply air temperature according to demands of the zone with the greatest cooling load. If one zone has cooling loads grossly in excess of all others, the controlling thermostat should be located in that space.

The greatest quantity of energy can be saved by adding variable air volume boxes to each of the major branch ducts. Each VAV box should be controlled by a local space thermostat. The associated reheat coil should be controlled to prevent reheat until the VAV box has reduced the zone supply air volume to 50% of the original flow.

In buildings where a reheat system supplies zones of different occupancy, add dampers and control valves to enable water to be shut off during unoccupied times. In addition to save energy:

1. Reduce the flow and temperature of the hot water to reheat the coils.
2. Use waste heat from condensate, incinera-

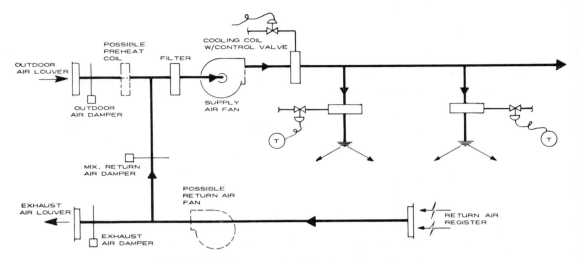

FIGURE 2. Single-zone reheat air system (flow diagram).

tors, diesel or gas engines, or solar energy for reheat. In all cases, raise the cold duct temperature to reduce the refrigeration load.

3. Modify controls to operate terminal reheat systems on a temperature demand cycle only. *Caution*: control of humidity will be eliminated and zone control will be modified.

4. Add or adjust controls to schedule the supply air temperature according to the number of reheat coils in operation. If 90% of the coils are in operation, raise the supply air temperature until the number of reheat coils falls to 10% of the total.

5. Install an interlock between the two control valves to prevent simultaneous heating and cooling.

6. If air conditioning is needed during unoccupied hours or in very lightly occupied areas, install controls to deenergize the reheat coils, raise the cold duct temperature and operate on a demand cycle.

7. For an alteration or expansion, install variable volume rather than terminal reheat or other systems.

8. Deenergize or shut off the terminal reheat coils, raise the chilled water and supply air temperatures of the central system and add recooling coils in ducts in areas where lower temperatures are needed.

1.3. DUAL DUCT LOW-VELOCITY SYSTEMS

Dual duct low-velocity systems supply hot and cold air in individual ducts to various zones of the building, where they are mixed to maintain the desired zone conditions, see Figure 3a. Control of mixing is normally achieved by automatic dampers in the branch ducts, the damper positioned according to the dictates of the space thermostat. Each duct should be analyzed as a single-zone system to determine whether the air volume or its temperature set point can be changed to conserve energy. Reduce the temperature of the hot duct and increase the temperature of the cold duct to that point where the heating and cooling loads of the most critical zone can just be met.

Return air is a mixture from all zones and reflects the average building temperature. In some designs of central station equipment, it is possible to stratify the return air and the outdoor air by installing splitters so that the hottest air favors the hot deck and coldest air favors the cold deck. This will reduce both heating and cooling loads. Dual-duct systems are described in Section I, paragraph 10.2.2.

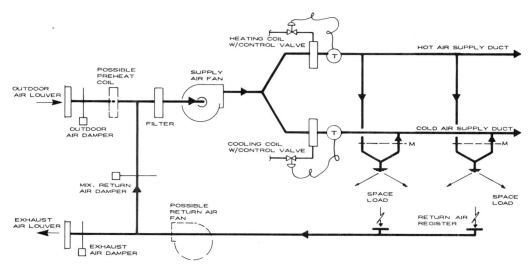

FIGURE 3a. Dual-duct low velocity system (flow diagram). *Note:* T = thermostat, M = mixing valve.

1.4. DUAL DUCT HIGH-VELOCITY SYSTEMS

Dual duct high-velocity systems operate in the same manner as low-velocity systems with the exception that the supply fan runs at a high pressure and that each zone requires a mixing box with sound attenuation, see Figure 3b. Considerable quantities of energy are required to operate the fan at high pressure and careful analyses of pressure drops within the system should be made. The fan pressure should be reduced to the minimum to operate the mixing boxes. To reduce system pressure, replace existing high-pressure mixing boxes with lower pressure types.

Energy conserving techniques for dual-duct systems include the following:

1. For conditions when there are no cooling loads, install controls to close off the cold

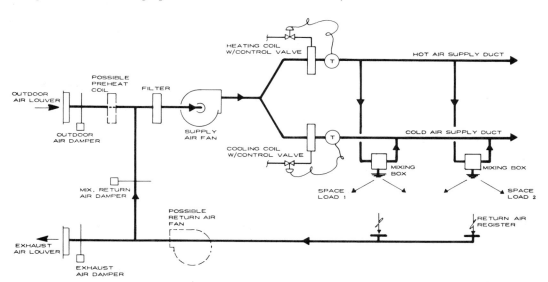

FIGURE 3b. Dual-duct high velocity system (flow diagram). *Note:* T = thermostat, M = mixing valve.

air duct, deenergize chillers and cold water pumps, and operate as a single-duct system, and reschedule the warmer air duct temperature according to heating loads only.

2. Under conditions where there are no heating loads, install controls to close off the warm air ducts, shut off hot water, steam, or electricity to the warm duct, and operate the system with only the cold duct air; and reschedule supply air temperature according to cooling loads.

3. Replace obsolete or defective mixing boxes the eliminate leakage of hot or cold air when the damper is closed.

4. Provide volume control for the supply air fan and reduce capacity preferably by speed reduction when both the hot deck and cold deck air quantities can be reduced to meet peak loads. Reducing the heat loss and heat gain provide an opportunity to reduce the amount of air circulated.

5. When there is more than one air handling unit in a dual-air system, modify duct work if possible, so that each unit supplies a separate zone to provide an opportunity to reduce hot and cold duct temperatures according to shifting loads.

6. Change dual-duct systems to variable volume systems by adding VAV boxes and fan control when energy analysis is favorable and the payback in energy saved is sufficiently attractive.

1.5. MULTIZONE SYSTEMS

Most multizone units currently installed have a single heating coil serving the hot deck and single cooling coil serving cold deck as seen in Figure 4. Each zone supply temperature is adjusted by mixing the required quantities of hot and cold air from these coils. With these units, the hot deck temperature must be sufficiently high to meet the heating demands of the coldest zone, and cold deck air must be sufficiently low to meet the demands of the hottest zone. All intermediate zones are supplied with a mixture of hot and cold air, wasting energy in a manner similar to reheat systems. Multizone systems were described in Section I, paragraph 10.2.4.

New model multizone units are now available which have individual heating and cooling coils for each zone supply duct and the supply air is heated or cooled only to that degree required to meet the zone load. These new units use far less energy than units with common coils. Where renovations are contemplated or the existing multizone unit is at or near the end

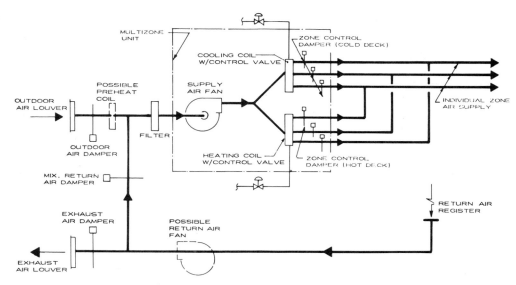

FIGURE 4. Multizone airstream system (flow diagram).

of its useful life, replacement should be considered, using a unit with individual zone coils.

Analyze multizone systems carefully. Treat each zone as a single-zone system and adjust air volumes and temperatures accordingly. Hot and cold deck dampers are often of poor quality and allow considerable leakage even when fully closed. To conserve energy, check these dampers and modify to avoid leakage. Install controls or adjust existing controls to give the minimum hot deck temperatures and maximum cold deck temperatures consistent with the loads of critical zones.

Where a multizone unit serves interior zones that require cooling all of the summer and most of the winter, energy can be conserved by converting to VAV reheat. To achieve this, blank off the hot deck and add low pressure VAV dump boxes with new reheat coils in the branch duct after the VAV box. This system works well in conjunction with an economizer cycle to minimize reheat energy. Careful analysis, however, is required of the existing system and the zone requirements to make the correct selection of equipment. The system can be converted entirely or in part to variable volume by adding terminal units and pressure bypass and/or by adding fan coil units in specific areas requiring constant air volume.

Arrange the controls so that when all hot duct dampers are partially closed, the hot deck temperature will be progressively reduced until one or more of the zone dampers is fully opened and, when all of the cold duct dampers are partially closed, the cold duct temperature will progressively increase until one or more of the zone dampers is fully opened. In addition, install controls to shut off the fan and all heating control valves during unoccupied periods in the cooling season and shut off the cooling valve during unoccupied periods in the heating season.

1.6. VARIABLE VOLUME AIR SYSTEMS

Variable volume air systems are basically a modification of single zone, single-duct systems to allow different conditions in multiple zones to be met by varying quantities of constant temperature air. The air quantity supplied to a zone can be modulated in two different ways by dump boxes which are also called mixing boxes, and by damper control, see Figure 5. Dump boxes installed in the zone supply duct modulate the amount of air through the outlets by bleeding off unused quantities of air and dumping it into the ceiling void and back into the return air system. This type of VAV system maintains constant fan volume and constant flow in all the trunk ducts and can waste energy by mixing the treated dump air with the return air. This system does, however, have practical advantages when used with direct expansion cooling coils since it maintains constant flow through the coils regardless of load. This is necessary in order to obtain correct operation of refrigerating equipment.

The second type of VAV system is where the air volume is controlled at the outlet by closing dampers as the zone load decreases. This type of system varies the total supply volume handled by the fan by increasing the resistance to airflow under light load conditions. To conserve energy, regulate the fan volume according to the demands of the system by a variable speed motor with a SCR controller or by inlet guide vanes.

Provide a control to increase or decrease the fan speed to maintain constant pressure in the supply duct. Reset the supply air temperature in accordance with zone loads. Other energy-conserving guidelines are listed as follows:

1. If not presently installed, provide either variable inlet vanes* or a variable speed drive such as a multispeed motor or SCR control so that the fan volume will be reduced when cortex dampers begin to close until one or more VAV dampers open fully.

2. Provide control to reset supply air temperature at a point where the damper of the variable air volume box serving the zone with the most extreme load is fully opened.

*Variable inlet vanes are volume control devices which can be an integral part of an inlet fan. They can be operated automatically or manually. They are used to sustain long periods of reduced capacity operation and for use with static pressure regulators.

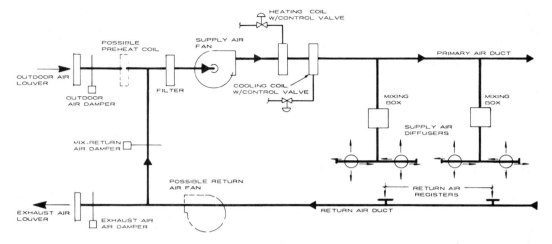

FIGURE 5. Variable volume system (flow diagram).

3. Install an automatic control to reduce the hot water temperature and raise the chilled water temperature in accordance with shifting thermal demands.
4. As with single duct variable volume systems with reheat, set control to delay reheat until airflow is reduced to minimum.
5. Where possible, eliminate the reheat coil and set the box to fully close off the flow.
6. The energy-saving suggestions listed for terminal reheat systems apply similarly to variable volume reheat systems.

1.7. INDUCTION SYSTEMS

Induction systems are commonly used for heating and cooling perimeter zones where large fluctuations of heating and cooling loads occur. Primary air is either heated or cooled and supplied at high pressure to the induction units located within the conditioned space as seen in Figure 6. Primary air is discharged from nozzles that induce room air into the induction unit at approximately four times the volume of the primary air. The induced air is cooled or heated by a secondary water coil. The water coil may be supplied by a two-pipe system, wherein

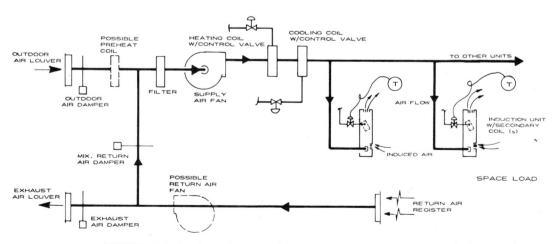

FIGURE 6. Induction unit system (flow diagram). *Note:* T = thermostat.

either chilled or heated water is available, but not simultaneously; or by a three-pipe system, where separate supplies of hot or chilled water are continuously available. After passing through the unit, they are mixed into a common return. In a four-pipe system, a supply and return of hot and chilled water are both continuously available to the secondary water coil.

Energy savings can be achieved in a number of ways, as follows:

1. The primary supply air fan operates at high pressures requiring high horsepower input. By careful analysis, reduce the primary air volume and pressure to the minimum required to operate the induction terminal units.
2. Induction unit nozzles may be worn after many years of cleaning and operation. This may result in increased primary air quantity at lower velocity with resulting lower induced air volumes. Check each induction unit and either repair the nozzles or replace them.
3. Provide controls to:
   a. Set the primary air reheat schedule as low as possible without causing complaints in the occupied spaces.
   b. Lower the set point of secondary heating water temperatures in winter and schedule according to outdoor conditions.
   c. During maximum cooling or heating periods, reduce the secondary water flow rate to the minimum necessary to obtain satisfactory room conditions.
   d. Raise the temperature of the primary cooling air to reduce refrigeration load, but weigh this benefit against increased cooling demand by the secondary coils.
   e. Reduce the water temperature to the heating coils during the heating season to reduce piping loss and improve boiler efficiency.
   f. For night operation during the heating season, shut down primary air fans, raise the hot water temperature and

operate the induction units as gravity convectors.

1.8. FAN COILS—UNIT VENTILATORS

Fan coil and unit ventilator performance can be improved by careful maintenance which in turn will allow energy savings to be made in the associated heating and cooling water distribution systems. Heating and cooling coils should be cleaned and air and water flow reduced to the minimum required to meet space conditions. Consider changing four pipe heating and cooling supply systems to a two-pipe system. This will avoid changeover loss and prevent simultaneous heating and cooling within a given zone.

Other energy-saving techniques include the following:

1. For fan coil systems which have separate coils for heating and air conditioning, install a control to prevent simultaneous heating and cooling.
2. Install a seven-day timer to shut down fans, close valves, and shut off chilled water pumps, compressors and cooling towers or air-cooled condensers during unoccupied periods.
3. Install controls to shut off fans during unoccupied periods during the heating season and to operate the fan coil unit as a gravity convector. The same applies to unit ventilators.
4. Block off outdoor air inlets where no dampers are installed if infiltration meets ventilation requirements.
5. If fan coil units are not located in conditioned areas, insulate the casings to reduce heat loss or gain.
6. With three-pipe systems, set controls for minimum mixing of hot and cold water.
7. Change four-pipe or three-pipe systems to two-pipe systems where possible, to avoid changeover losses.

1.9. WATER-TO-AIR HEAT PUMPS

Air coils should be kept clean to maintain the highest possible seasonal efficiency. Heat pump units are normally controlled by integral thermostats which turn them on and off according to the demands of space load. These units are frequently left in operation 24 hours per day whether the building is occupied or not.

Install a centralized system of on–off controls on a floor-by-floor basis using a remote control contactor on the power circuit to feed the heat pump. Rewire the power supply to the heat pumps in such a manner that one-third, two-thirds, or all of the units in any zone can be turned on and off depending on whether the building is occupied or not and what the expected outdoor conditions are likely to be.

When retrofitting, install one large heat pump rather than multiple smaller units for greater efficiency. All guidelines for condensers, compressors, evaporators, fans, pumps, piping, and duct work listed in other sections are applicable to unitary closed loop water source heat pumps.

1.10. AIR-TO-AIR HEAT PUMPS IN AIR CONDITIONING UNITS

Energy saving tips for air-to-air heat pumps in air conditioning units include:

1. Where possible, direct the warm exhaust air from the building to the inlet of air-to-air heat pumps to raise their coefficient of performance. Waste heat from other processes can also be used as a heat source.
2. Replace air-to-air heat pumps with water-to-air heat pumps where there is a source of heat such as ground water with temperatures above average ambient winter air temperatures.
3. Install a seven-day timer to program operation of compressors in accordance with occupied–unoccupied periods.
4. When replacing incremental air conditioning units equipped with electrical heating coils, install a heat pump model instead.
5. Replacement compressors should have an EER* rating of 9 or better during the

*Energy Efficiency Ratio (EER) is found by dividing Btu/hr output of cooling unit by the watts required. For example, an air conditioner that delivers 7000

cooling cycle. (Applicable also to window air conditioning units).

1.11. EXHAUST SYSTEMS

All guidelines in this chapter detailing methods of reducing fan power requirements by reducing flow rates and resistance apply to exhaust systems as well as supply systems. A number of energy-conserving steps for exhaust systems include the following:

1. Balance exhaust systems so that the exhaust airflow does not exceed the supply airflow. Ideally, exhaust should be 10% less than supply to maintain a positive building pressure.
2. Modulate exhaust fan volume in step with associated VAV supply fan by installing inlet guide vanes or variable speed control.
3. Recirculate toilet room exhaust air through charcoal filters to reduce make-up air requirements.
4. Install controls to operate toilet exhaust fans intermittently, for 10 to 20 min. out of every hour and to automatically shut them off during unoccupied periods.
5. Install a motorized damper at inlet grilles and wire damper to light switches to reduce the air quantity when toilet rooms are unoccupied. Modulate fan volume according to load by monitoring pressure in main exhaust stack.
6. For new installations or when existing fans/or motors are replaced, provide variable speed control or inlet vane control to operate in accordance with static pressure.
7. Shut off supply fans which serve only as make-up air for toilet rooms, and install new door louvers or cut off the bottom of the door to permit air from conditioned areas to migrate into the toilet rooms as

Btu/hr and takes 864 W the EER is calculated as follows:

$$\frac{7000}{864} = 8.1 \text{ EER}$$

This number is a measure of efficiency; the higher the number, the more efficient the unit.

make-up for the exhaust system. Set maximum capacity to provide 1 cfm/sq ft of toilet area.

1.12. THERMAL STORAGE

When making large additions or alterations to existing buildings, the HVAC designer has several options to store heated or chilled water under certain conditions.

For large buildings with interior zones that require cooling all year round, analyze the potential for conserving energy by capturing the heat from the hot condenser line and storing it in water tanks when it is not required for immediate use in the perimeter zones, but can be used during unoccupied times when the cooling machines are not running.

Analyze the cost benefits of providing chilled water storage to reduce peak demands and permit operation of the refrigeration systems at lower condensing temperatures at night or early morning.

2

OWNING AND OPERATING COSTS

A properly engineered system must include economic considerations. Costs are difficult to assess because of the complexities in the effective management of money and the inherent difficulty of predicting operating and maintenance expenses far into the future. Complex tax structures and the time value of money can seriously affect an engineering decision. This does not imply that the choice is either the cheapest or the most expensive system. Instead, decisions must be based on an intelligent analysis of an owner's financial objectives as well as his engineering requirements. Therefore, the engineer is responsible for evaluating the proper use of money as dictated by the specific circumstances. When studying owning and operating costs, the period covering the lifetime of the system should be 20 years.

Although intangible factors may alter the final decision between otherwise equal alternatives, the normal choice should be the alternative with the lowest overall cost. This overall cost may be divided into two main categories; owning costs and operating costs. These may be further subdivided as follows.

Owning Costs:
  Initial cost of the system
  Capital recovery
  Interest and return on the investment
  Property taxes
  Insurance

Operating Costs:
  Fuel and energy

Maintenance allowances
Labor for operation
Water costs
Water treatment cost

A representative form for assembling and tabulating these costs in detail is shown in Table 1.

2.1. OWNING COSTS

In order to calculate the annual cost of ownership, elements must be established: initial cost, capital recovery or service life, and return on investment.

2.1.1. Initial Cost of System

Generally, major decisions affecting annual owning and operating costs for the life of the building must be made prior to the complete development of contract drawings and specifications. Comparisons between alternate solutions of engineering problems peculiar to each project will be made in the early stages of architectural development to achieve the best performance and economies. Estimates based on oversimplification of the alternatives can lead to substantial errors.

A thorough understanding of the installation costs and ancillary requirements must be established. Detailed lists of materials and controls can be prepared, and space and structural requirements, services, installation labor, etc., taken into account to reduce the degree of inaccuracy in preliminary cost estimates. A

TABLE 1. Owning and Operating Cost Data and Summary[a]

OWNING COSTS

I. Initial Cost of System
   A. Equipment (see Table 2 for items included) ___
   B. Control Systems–Complete ___
   C. Wiring and piping costs attributable to system ___
   D. Any increase in building construction cost attributable to system   + ___
   E. Any decrease in building construction cost attributable to system   – ___
   F. Installation Costs ___
   G. Miscellaneous ___
   TOTAL INITIAL COST (IC)

II. Annual Fixed Charges
   A. Amortization period n (number of years during which initial cost is to be recovered) ___
   B. Interest rate, i ___
   C. Capital Recovery Factor (CRF); from Table 4 ___
   D. Equivalent uniform annual cost: (CRF) (IC) ___
   E. Income Taxes ___
   F. Property Taxes ___
   G. Insurance ___
   H. Rent ___
   TOTAL ANNUAL FIXED CHARGES (AFC)

OPERATING COSTS

III. Annual Maintenance Allowances
   A. Replacement or servicing of oil, air, or water filters ___
   B. Contracted Maintenance service ___
   C. Lubricating oil and grease ___
   D. General housekeeping costs ___
   E. Replacement of worn parts (labor and material) ___
   F. Refrigerant ___
   TOTAL ANNUAL MAINTENANCE ALLOWANCE:

IV. Annual Energy and Fuel Costs
   A. Electric Energy Costs
       1. Chiller or compressor ___
       2. Pumps
           a. Chilled water ___
           b. Heating water ___
           c. Condenser or Tower water ___
           d. Well water ___
           d. Boiler auxiliaries (including fuel oil heaters) ___
       3. Fans
           a. Condenser or Tower ___
       b. Inside Air Handling ___
       c. Exhaust ___
       d. Make-up air ___
       d. Boiler auxiliaries and equipment room ventilation ___
       4. Resistance heaters (primary or supplementary) ___
       5. Heat pump ___
       6. Domestic water heating ___
       7. Lighting ___
       8. Cooking and food service equipment ___
       9. Miscellaneous (elevators, escalators, computers, etc.) ___
   B. Gas, Oil, Coal, or Purchased Steam Costs
       1. On-Site generation of the electrical power requirements under *A* this section ___
       2. Heating
           a. Direct heating ___
           b. Ventilation
               1. Preheaters ___
               2. Reheaters ___
           c. Supplementary Heating (i.e., oil preheating) ___
           d. Other ___
       3. Domestic water heating ___
       4. Cooking and food service equipment ___
       5. Air Conditioning
           a. Absorption ___
           b. Chiller or compressor
               1. Gas and Diesel engine driven ___
               2. Gas turbine driven ___
               3. Steam turbine driven ___
       6. Miscellaneous ___
   C. Water
       1. Condenser make-up water ___
       2. Sewer charges ___
       3. Chemicals ___
       4. Miscellaneous ___
   TOTAL ANNUAL FUEL AND ENERGY COSTS:

V. Wages of engineers and operators

SUMMARY

II. Total Annual Fixed Charges ___
III. Total Annual Maintenance Costs ___
IV. Total Annual Fuel and Energy Costs ___
V. Annual Wages for Engineers and Operators ___
   TOTAL ANNUAL OWNING AND OPERATING COSTS: ___

---

[a]Reprinted with permission of ASHRAE

TABLE 2. Initial Costs*

1. Energy and Fuel Service Costs
   a. Fuel service, storage, handling, piping, and distribution costs.
   b. Electrical service entrance and distribution equipment costs.
   c. Total energy plant.
2. Heat Producing Equipment
   a. Boilers and furnaces.
   b. Steam-water converters.
   c. Heat pumps or resistance heaters.
   d. Make-up air heaters.
   e. Heat producing equipment auxiliaries.
3. Refrigeration Equipment
   a. Compressors, chillers, or absorption units.
   b. Cooling towers, condensers, well water supplies.
   c. Refrigeration equipment auxiliaries.
4. Heat Distribution Equipment
   a. Pumps, reducing valves, piping, piping insulation, etc.
   b. Terminal units or devices.
5. Cooling Distribution Equipment
   a. Pumps, piping, piping insulation, condensate drains, etc.
   b. Terminal units, mixing boxes, diffusers, grilles, etc.
6. Air Treatment and Distribution Equipment
   a. Air heaters, humidifiers, dehumidifiers, filters, etc.
   b. Fans, ducts, duct insulation, dampers, etc.
   c. Exhaust and return systems.
7. System and Controls Automation
   a. Terminal or zone controls.
   b. System program control.
   c. Alarms and indicator system.
8. Building Construction and Alteration
   a. Mechanical and electric space.
   b. Chimneys and flues.
   c. Building insulation.
   d. Solar radiation controls.
   e. Acoustical and vibration treatment.
   f. Distribution shafts, machinery foundations, furring.

*Reprinted with permission of ASHRAE.

reasonable estimate of the cost of components may be derived from the cost records of recent installations of comparable design, or from quotations submitted by manufacturers and contractors. Table 2 is representative checklist for initial cost items.

2.1.2. Capital Recovery

Capital recovery is often described by two terms; amortization and depreciation. They are frequently interchanged, but they are different. *Amortization* is the periodic payment of monies to discharge a debt. *Depreciation* is the process of allocating the first cost of a capital asset over the estimated useful life of the asset. (Depreciation is the basis for a deduction against income in calculating income taxes.)

The Internal Revenue Service has published the following discussion of useful life.

The first step in computing depreciation is to determine the estimated useful life of the asset. No average useful life is applicable in all businesses. The useful life of any item depends upon such things as the frequency with which you use it, its age when you acquired it, your policy as to repairs, renewals and replacements, the climate in which it is used, the normal progress of the arts, economic changes, inventions, and other developments within the industry and your trade or business.

You should determine the useful life of the depreciable property on the basis of your particular operating conditions and experience. If your experience is inadequate, you may use the general experience in the industry until your own experience forms in adequate basis for making the determination.

The estimated useful life of the various system components may be obtained from Table 3.

2.1.3. Interest and Return on Investment

Money to be invested in a HVAC system must either be borrowed, obtained from equity investors, or diverted from other uses. In any case, it has value. Its minimum value is the current rate of interest on borrowed money. The cost to the owner is a proper part of the total cost of the installation. That cost (interest on debt, return on equity, or a combination of interest and return) varies widely.

Generally speaking, every investor has set a minimum acceptable rate of return on investments. This is the rate below which the investment would not be made. The prevailing interest rate on borrowed capital is the lowest practical limit to this minimum acceptable rate of return. The prospective building owner should be consulted to establish the proper rate of return on his investment.

TABLE 3. Minimum Depreciation Period*

Item	Years	Item	Years
Air-conditioning systems		Insulation	
Large—over 20 tons	20	Asbestos	15
Medium—5 to 15 tons	15	Cork, cold pipes and tanks	20
Small—under 5 tons	10	Magnesia, hot pipes	15
Air terminals		Wool felt	20
Diffusers	15		
Grilles and registers		Louvres and screens, fresh air	
Ceiling	20	Copper	25
Wall type	20	Steel	15
Window units	15		
Plaques	20	Manometers	15
Air washer (See Dehumidifier)		Motors	
Absorption type liquid chillers[b]	20	Induction, indoor	20
Compressors		Induction, weatherproof for outdoor	20
Air for pneumatic controls	20	Synchronous and exciter set	20
Refrigerating	20	Piping, refrigerant and other	20
Condensers		Pumps	
Double pipe	20	Chilled water	20
Evaporation	15	Condenser water	20
Shell and tube	20	Condensate	20
Coolers, water-tank and coil or shell		Dehumidifiers	22
and tube	20	Evaporative condenser	15
Dehumidifier	10	Sump	25
Drums, purge or surge	20	Well	25
Ducts and other sheet metal work[a]	± life		
	of building	Receivers, refrigerant	25
Engines (Gas, Diesel or Dual Fuel)[b]	20	Regulators, suction or static pressure	5
Fans	15	Silica gel beds	15
Filters, air, oil, self-cleaning	20	Spray pond	15
Dry cleanable	10	Starters, electric	20
Gages	15	Switchboards, electric	25
Heaters		Switchboard, electric panel	20
Boiler, oil burner and tank	20	Thermometers, room type or recording	15
Booster heaters	20	Tower, cooling	15
Electric heaters	15	Transformers	25
Finned tube, steam water cleanable tube	20	Turbines[b]	
Gas	15	Gas	20
Preheaters and reheaters	20	Steam	30
Water heaters, open or closed type	20		
		Valves	
Heating systems		Relief	20
Boilers and furnaces	20	Solenoid	15
Burner equipment		Automatic expansion and bypass	5
Gas	16	Water regulating	20
Oil	10	Wells and well pumps	25
Radiators	25		

[a]From system evaluation study by consultant.
[b]Estimated values, not IRS values.
*Reprinted with permission of ASHRAE.

After determination of the proper rate and amortization period ("n" years) the initial cost can be converted into an equivalent uniform annual cost for "n" years through the use of the capital recovery factor (CRF). The equivalent annual cost of the asset is found by multiplying the initial cost (IC) by the capital recovery factor (CRF). For example, if a HVAC

TABLE 4. Capital Recovery Factors*

Years	\multicolumn{10}{c}{Rate of Return or Interest Rate, %}
	3.5	4.5	6	8	10	12	15	20	25	30
2	0.52640	0.53400	0.54544	0.56077	0.57619	0.59170	0.61512	0.65455	0.69444	0.73478
4	.27225	.27874	.28859	.30192	.31547	.32923	.35027	.38629	.42344	.46163
6	.18767	.19388	.20336	.21632	.22961	.24323	.26424	.30071	.33882	.37839
8	.14548	.15161	.16104	.17401	.18744	.20130	.22285	.26061	.30040	.34192
10	.12024	.12638	.13587	.14903	.16275	.17698	.19925	.23852	.28007	.32346
12	0.10348	0.10967	0.11928	0.13270	0.14676	0.16144	0.18448	0.22526	0.26845	0.31345
14	.09157	.09782	.10758	.12130	.13575	.15087	.17469	.21689	.26150	.30782
16	.08268	.08902	.09895	.11298	.12782	.14339	.16795	.21144	.25724	.30458
18	.07582	.08224	.09236	.10670	.12193	.13794	.16319	.20781	.25459	.30269
20	.07036	.07688	.08718	.10185	.11746	.13388	.15976	.20536	.25292	.30159
25	0.06067	0.06744	0.07823	0.09368	0.11017	0.12750	0.15470	0.20212	0.25095	0.30043
30	.05437	.06139	.07265	.08883	.10608	.12414	.15230	.20085	.25031	.30011
35	.05000	.05727	.06897	.08580	.10369	.12232	.15113	.20034	.25010	.30003
40	.04683	.05434	.06646	.08386	.10226	.12130	.15056	.20014	.25006	.30001

*Reprinted with permission of ASHRAE.

system cost $100,000 and the amortization period was 10 years, the CRF value at an 8% interest rate would be

$100,000 × 0.14903* = $14,903 per year

An abbreviated table of CRF values is given in Table 4. Values not shown can be found in texts on finance or engineering economics.

2.1.4. Property Taxes

The taxes that may be charged to the property will vary according to the practice of the taxing agencies. Tax rates and the basis of taxation can be obtained from such agencies.

2.1.5. Insurance

Insurance rates vary widely and are dependent upon the type and location of the structure, and the nature of the owner's business. The rates are usually set by rating organizations specializing in this work. Exact insurance rates must be determined by consulting the owner's underwriter.

*From Table 4.

2.2. OPERATING COSTS

Operating costs are those costs resulting from actual operation of the system. If operating costs are going to be calculated over a period of years, cost escalation must be incorporated. The cost of labor and materials have increased between 5 and 8% a year for almost a decade.

2.2.1. Fuel and Energy

Energy use for the entire project should be included in an analysis of fuel and energy costs. Only by establishing a base cost for the project can incremental costs for alternate schemes be compared. It often turns out that operating costs of auxiliary systems and accessories are the major factors when comparing operating expenses.

The installed capacity of a system depends in large measure on demand peaks, that is, the simultaneous use of components in the system that require electric power or heat energy. The designer must estimate this demand factor on the basis of occupancy schedules and weather conditions in order to ascertain the heat gain from the equipment, the building heat gain and losses, and the ventilation requirements.

A compilation of sources of electrical and heat energy to be included in an analysis, where applicable, include the following:

Electrical
- indoor and outdoor lighting
- fans
- pumps
- cooking equipment
- elevators
- space heaters
- domestic hot water heaters
- office equipment
- production or process equipment
- refrigeration equipment
- heat dissipation equipment (motors)

Heat Energy (gas, oil, etc.)
- space heating
- domestic hot water heaters
- cooking and process equipment
- absorption refrigeration
- engine drives
- ventilation

The analysis should list each electrical and heat energy source, including the period during which each component is in operation. The magnitude of energy consumption during operation is also essential information to be included. Project occupancy schedule, temperature control programs, heating and air conditioning system, and refrigeration designs will be of major consequence in establishing these loads and the hourly variation in their use.

The most direct and reliable energy cost estimating procedure will require an hourly integration of each load that will take into account individual variations in load because of weather and use schedules. These hourly totals should be accumulated into monthly sums that represent project energy needs irrespective of the energy source. Such data may reveal the comparative cost features of alternate control systems, distribution systems, and other design options. When the monthly energy needs are modified by the appropriate combustion and utilization efficiency factors, these sums determine the monthly energy requirements of the project.

The hourly load calculations must be evaluated to determine the peak demand for utility billing purposes. The monthly demand peak and input energy requirements can then be used with the utility rate schedules to obtain the estimated cost of energy requirements.

2.2.2. Maintenance Allowances

Maintenance allowances are expenses for labor and materials necessary to make repairs and replace parts, as well as the costs for cleaning, painting, inspection, testing etc., to reduce the need for repairs. Generally, routine maintenance labor requirements will be the function of an operating engineer and his staff. If the responsibility of the group extends to facilities beyond the equipment discussed here, it is important that only an equitable share of the group's time be charged to maintenance. Extraordinary repairs are quite often handled by a separate maintenance division and the expense charged back to the system's maintenance costs. In other cases some or all maintenance is handled by outside service firms and the cost considered as maintenance.

These costs vary considerably with the type of system and the proficiency of the servicing organization. The annual maintenance allowance should be based upon the entire period under study rather than the early years of operation when repairs may be expected to be at a minimum.

2.2.3. Labor for Operation

Where operating personnel are required, the expense is often considered maintenance. Since some of their duties may not involve operation of the system, only the applicable portion of labor charges should be applied. Labor charges are the aggregate of the base salary which includes state and federal taxes paid by the employer, plus the cost of such fringe benefits as vacations, insurance, sick leave, pensions, etc.

2.2.4. Water Costs

The quantity of water used in a heating and air conditioning installation for purposes other than heat rejection is a negligible cost item. In a refrigeration plant utilizing water-cooled condensers, heat is rejected either to water that is purchased which is subsequently wasted by discharge to well or surface water, or by means of cooling towers or spray ponds. Often, water conservation equipment must be used regardless of economics because of:

- a lack of adequate water supply
- local regulation intended to conserve existing supply
- taxes or service charges attendant upon wasting to sewer

Because the water evaporated is proportional to the refrigeration accomplished, the seasonal water consumption can be calculated on a basis of operating hours of the refrigeration equipment. The following equation may be used:

$$G = B(TH)$$

where

G = gallons of water evaporated per season

(TH) = ton-hours of refrigeration per season

B = a constant depending upon the refrigeration equipment:

electric motor-driven chillers $B = 3.2$
absorption chillers $B = 6.2$
steam turbine-driven chillers $B = 6.2$
gas and oil-fired turbine-
    driven chillers $B = 3.2$
gas and oil-fired engine-
    driven chillers $B = 3.8$

Knowing the water consumption per season as per year allows calculation of water costs if water is supplied by a utility. If the owner supplies water from his own wells, the operating cost is in power and maintenance.

2.2.5. Water Treatment Costs

Water treatment costs represent a significant operating expense in addition to the cost of the "raw" water. The cost of treatment is directly related to the quantity of make-up and the water chemistry. Local conditions vary considerably which greatly affect the cost for filters, chemicals, and associated equipment that must be maintained and replaced. Because of these variations, local cost estimates are the only reliable source of water treatment costs.

3
ESTIMATING

The estimating procedure is a sequence of steps that an estimator follows to determine the total cost of a HVAC installation. Estimates are made for a number of purposes and depending upon the purpose, will be the extent of detail involved. Ultimately a very careful cost estimate must be made for bidding purposes.

The general approach to a detailed cost estimate begins with the estimator familiarizing himself with the job. He must develop an understanding of site conditions, and a comprehension of the requirements set forth in the bid documents, the drawings, and specifications. This is followed by the take-off, whereby all items covered by the contract are listed on take-off sheets in a systemized way. This system varies between estimators but it is intended as a complete list of all items involving material, labor, and equipment costs. The drawings and specifications are carefully examined for these items and are usually marked in colored pencil to show that a particular item has been entered on the take-off sheet.

The take-off sheet identifies the item by name, size, material, etc., and notes the quantity or length required. It is extremely important to separately list all items of different cost. For example, $\frac{1}{2}$-in. diameter $\times$ 50-ft long copper tubing should be apart from $\frac{1}{2}$-in. diameter $\times$ 50-ft long stainless steel tubing. When the take-off is completed, many estimators summarize the results by bringing all like items together from each subsystem in order to cost them. Some cost factors are fairly straightforward.

Material and equipment costs should be obtained directly from suppliers and manufacturers. Labor costs are based upon productivity. Manhour information is given at the end of each chapter in Section II. Current labor costs are then applied. To these costs are added the cost of owning or renting, special equipment such as scaffolding or a crane. For the bid price to be complete, the estimator must include the contractor's overhead and profit plus something more difficult to predict. Its called contingency and includes such "grey" areas as weather conditions, site conditions, and the degree of completeness of the drawings and specifications.

3.1. TAKE-OFFS

Like any other profession, estimating must be well-organized for best results. The tools of the estimator's trade are as follows:

- accounting ruled paper, for take-off sheets
- adding machine
- rotometer, which is indispensable for measuring duct and pipe runs
- architect's scale
- colored pencils
- updated collection of manufacturer's catalogs

Each estimated job should be in its own folder with a record of both estimated and as-built costs for jobs that were bid and built. Folders for jobs that were estimated but contracted by others contain valuable information and should be retained. Eventually, a body of data will be accumulated that will provide a useful source of information and comparison for newly prepared estimates.

A technique for HVAC take-offs usually divides the total project into five separate stages.

Step One
- sheet metal ductwork by sizes of straight runs and by types and sizes of fittings
- diffusers and registers
- volume dampers, fire dampers, and access doors
- insulation (sq ft)

Step Two
- piping by size, material linear feet, and fittings
- valving sizes and types
- expansion joints and loops
- pipe supports by types and weight of piping

    Note: The first and second steps are done with the use of floor plans and riser drawings up to, but not into the equipment or mechanical rooms.

Step Three
- equipment in equipment and mechanical rooms, such as chillers, pumps, heat exchangers, etc.
- HVAC equipment in isolated areas, such as heaters, fans, collectors, etc.

Step Four
- automatic controls

    Note: Automatic controls can be very vaguely and loosely specified, except for function. The estimator should study the electrical drawings, specifications, and the control diagram and make sure that he understands the quality of components that is expected by the architect or engineer.

Step Five
- transfer take-off totals to price and labor sheets where material and labor costs are applied. A summary sheet, which usually becomes the first sheet of the estimating package, is used to prepare the bid. On this sheet are entered overhead, contingency, and profit. The final figure is the fixed price for the job.

3.2. LABOR COSTS

Labor values used in this manual are the result of personal experience. They have been compared with tables available from a wide range of publications. However, some comparisons made were as much as 200% from agreement with each other and with data in this book. Part of this problem stems from differences in productivity brought about by efficient management, or by differences in efficiency due to the overall size of the job. It appears that another source of ambiguity is the definition of a man-hour as applied to HVAC work.

This book lists the number of man-hours to complete a certain task. What is the nature of this work and what are the elements involved? The following two examples illustrate the point.

Example 1:

A valve in a household heating system needs replacement and a heating contractor is called. His charges will include the valve (material), and his actual installation time and time lost for travel (service charge).

Example 2:

A 100 ton chiller is to be delivered and installed in a mechanical room on the 10th floor of a building. The labor to perform this task can include delivery, rigging, and the actual installation. Delivery and rigging add significantly to the time required by the contractor.

The service and delivery charges and rigging costs mentioned in the examples are not fully covered in the man-hour tables. They are based on the following criteria:

1. All work is to be performed by a knowledgeable mechanic.
2. The material is to be within a reasonable distance of the place of installation.
3. The materials and equipment will not be installed above the second- or third-story level.
4. Neither excessively cold or hot working conditions will impair the workmen's efficiency.

5. Field assembly will take place at productive working hours. This means that the estimate takes into account short work breaks and cleanup time.

It is important to recognize the ground rules for the man-hour estimate so that delivery, rigging and other charges can be listed elsewhere in the estimate. In any event, these values should only be taken as a guide. The reader must develop his own numbers based upon experience and local conditions. That is why individual job files should be maintained.

Assuming that rigging, equipment costs, and service charges are separated from the number of man-hours required to do the job, the estimator must now place a cost against this labor. It was stated above that labor costs are made up of the workman's base wage plus fringe benefits. Added to these are labor insurance costs. The base wage is the employee's salary, including that portion withheld for income tax purposes. Fringe costs include charges for insurance, holidays, pension funds, and other benefits administered by the employer or paid to union trust funds. Labor insurance costs are such expenses to the employer as: Social Security (FICA), Federal Unemployment Insurance Tax (FUIT), State Unemployment Insurance Tax (SUIT), and Workman's Compensation. Put into practical terms, a base salary of $10/hr, as a rough example, could be raised about 25% by the fringe package to $12.50. Labor insurance, at about 15% of the total, would bring the cost per man-hour to $14.38.

3.3. OVERHEAD AND PROFIT

Overhead and profit are two components that must appear in the bid after the estimate has been made. Simply stated, overhead is defined as the cost of doing business. Profit is the return to the businessman for taking a risk.

Overhead can be divided into two categories; field overhead and office overhead. Field overhead includes such expenses as the cost of:

1. field office trailer
2. tools that wear out or are stolen
3. supervision of employees (this may be all part of the salary an owner pays to himself, assuming the productive work is done by others)
4. temporary facilities, such as water, power, and sanitary units (these are usually provided by the general contractor or other trades)
5. preparation of shop drawings or as-built drawings.

Office overhead are costs of running the home office and include the salaries of engineering and clerical employees as well as the rent, power, telephone, heating, legal fees, and other expenses inherent in maintaining a business establishment. The cost of owning and operating a truck should also be included. When

TABLE 5. Budget Estimates for Air Conditioning Loads.

Type	Sq Ft/Ton of Refrigeration
Apartment bldgs.	350–500
Auditoriums–theaters (seats/ton)	18–22
Banks	220–250
Barber shops	200–250
Beauty parlors	200–250
Bowling alleys (tons/alley)	2.10–2.25
Churches (seats/ton)	18–25
Cocktail lounges	150–200
Computer rooms	80–150
Dental offices	200–250
Department stores	300–350
Dormitories	280–350
Dress shops	250–300
Factories	300–350
High-rise office bldgs.	280–320
Hospitals–nursing homes	260–300
Hotels	280–325
Factories	260–325
Libraries	250–280
Medical centers	350–410
Motels	280–320
Post offices	300–320
Residences	500–600
Restaurants	220–250
Schools	250–300
Shopping centers	300–350

bidding the job, usually all of the field overhead can be applied, whereas only part of the office overhead should be used, since the expense of running the office can hopefully be divided among many other jobs. Both field and office overhead are usually calculated as a percentage of the base estimate cost.

There are a number of other costs which can either be applied directly to the estimate or be regarded as an overhead item. In any event they must not be omitted. These include:

1. permits and fees
2. bonds for bidding and performance
3. insurance for public liability, fire, liability, etc.
4. final cleanup
5. interest on loans

Profit is the return to the owner for risking his capital and exercising his skills in a competitive market. Of the essentials that go into an estimate, namely, materials, labor, overhead, and profit, it is only profit that is flexible. If the estimator does his job well the costs of the first three items can vary only slightly since they are fixed by market conditions, labor contracts, and the conditions of a particular job. Profit,

TABLE 6. Budget Estimates of Project and HVAC Costs.

Type	Total Project Cost ($/sq ft)		HVAC Cost ($/sq ft)		HVAC % Cost of Total Project (average)
Apartments	18.00	28.00	1.15	1.75	6.5
Auditoriums	30.00	42.25	4.15	5.70	13.6
Banks	42.00	68.15	4.75	6.90	10.5
Bowling alleys	26.15	33.40			
Churches	24.00	39.15	2.80	4.75	12.0
Clubs (private)	26.10	36.10	4.10	4.50	13.8
Department stores	16.10	24.50	3.05	4.00	17.4
Dormitories	29.50	43.40	3.15	4.20	10.1
Factories	13.50	19.40	2.00	3.50	16.6
Funeral homes	22.00	38.50	2.75	4.10	11.3
Garage, commercial	16.10	26.50	1.15	2.80	9.0
Garage, parking	20.00	26.00	1.00	2.00	6.5
Hangars	10.00	35.00	3.50	8.00	25.6
Hospitals	48.50	62.50	10.50	18.50	27.6
Library	34.00	46.50	4.90	7.80	
Medical centers	36.00	49.50	2.10	6.75	13.2
Motels	22.00	34.00	2.10	4.00	10.9
Nursing homes	36.00	42.50	4.20	6.10	13.1
Office bldgs.	28.00	40.50	3.20	5.80	13.1
Restaurants	29.00	43.50	4.00	6.10	13.9
Retail stores	16.00	24.50	2.00	3.10	12.6
Schools	28.00	42.00	4.10	6.85	15.6
Shopping centers	16.80	24.10	2.10	3.20	13.0
Theaters	26.00	38.00			
Warehouse	12.05	17.10	1.00[a]	2.10[a]	10.7

[a]Heating and ventilating only.

on the other hand, can vary from zero to 50% depending on such factors as the risk involved, competition from other contractors, and the need for work. Marketplace conditions usually limit the profit margin to about 10%.

3.4. BUDGET ESTIMATES

It is essential in the early stages of project development for the owner or his builder, the architect-engineer, and the contractor to understand the cost of building elements so that project costs can be compared with budgeted funds. Since HVAC can account for more than 25% of the total cost for some types of construction, this system merits special attention.

Over the years, records have been kept of total building costs by types of construction and the cost of component systems such as HVAC, electrical, plumbing, foundations, structural, etc. This body of information permits "ball park estimates" for particular building types. In the case of heating, ventilation, and air conditioning, these estimates are available on the basis of HVAC cost per square foot of conditioned space. Another cost index is keyed to the number of tons of refrigeration per square foot of building. This will vary according to the type of building. Given the range and the area of a proposed building it is simple to determine the total tons of refrigeration required, then multiply the total by the current cost to install one ton.

Budget estimates serve a number of useful purposes.

1. It will help determine which type of HVAC system to use early in the design stage.
2. Designers can develop their detail designs within a proper budget framework.
3. Architect-engineer can add or delete portions of the building and know how these will affect costs.
4. It establishes an index or rate of escalation of costs between the early design stage and actual bidding time.

Table 5 is a budget estimate of square feet of area per ton of refrigeration, arranged alphabetically by building type. The range of square feet per ton varies from less than 10%, to 30% according to building type. The larger variation suggests that within certain categories, apartment buildings, for example, there are many individual design layouts that affect HVAC efficiency.

Table 6 also shows budget estimates by building type. Here the total project cost is given on a square foot basis, followed by the cost per square foot of the HVAC and a percentage of the total cost that the HVAC will take. Project and HVAC costs are given in ranges which take into account the differing complexities between jobs, the changes in load requirements because of location, and other factors. Be aware that dollar values will escalate and must be continuously updated. However, the percentage of the whole job taken by HVAC will tend to remain constant.

4

DRAWINGS AND SPECIFICATIONS

Drawings and specifications are the final product of the design team. They identify the scope of the project and designate the materials, sizes, and methods of work required to construct a building. These documents become the basis of a contract between the owner and the general or prime contractor, or between the general contractor and a subcontractor. The contract usually comes about after a competitive bidding process with preference given to the "lowest qualified bidder." It is essential that the HVAC bidder understand the requirements of the drawings and specifications in order to avoid the financial loss that could result from a mistake or omission.

4.1. DRAWINGS

Specifications usually contain a list of drawings for all the trades involved. The estimator will make sure that a complete set of drawings are in his possession at the time of the take-offs, including architectural, structural, electrical, and mechanical drawings. It is important for the HVAC estimator to know the shapes and sizes of structural portions of the building so that duct runs can be established. Architectural drawings will locate partitions and dropped ceilings, name rooms, and identify door and window openings. Electrical drawings will help define the interface between HVAC equipment and the power for their operation and control. Mechanical drawings include HVAC and

plumbing, the latter of importance in locating water supply and drain connections.

The HVAC drawings are assembled in a certain order and marked with a symbol starting with an (H) for Heating, (HV) for Heating and Ventilation, or (AC) for Air Conditioning. Other designations are also used such as (M) for Mechanical and HVAC. The first drawing will usually contain the legend and symbols used in the drawings, abbreviations and general notes. These notes are generally supplemental and not covered in the specifications, or are last moment additions to the design.

The legend and symbol drawing is followed by:

- floor plans
- risers diagrams (air and water)
- details
- sections
- automatic controls
- schedule drawings

Schedule drawings contain tables of equipment and materials and make the take-off procedure for those items more organized.

The estimator should always look for revision notes on the drawings. A revision is an addition to the drawings (or specifications) that is issued after the project has been offered to bidders. The part that has been revised is identified by a heavy irregular outline, as shown in the inset below. The number in the triangle is the number of the addendum and is keyed to the revision list in the title block of the drawing.

The HVAC drawings are to be compared with specifications to insure that wording and design agree and clear definitions of limit of responsibilities are shown among trades. For example, it is not unusual to find that electrical baseboard heating carries the note "furnished by HVAC contractor" where the installation with power connection is not mentioned. This note can be misleading in knowing who does what, and a cross check with the electrical drawings and specifications is in order.

Other cross checking should be made with electrical and plumbing drawings and specifications to determine responsibility for supplying and installing fans, air conditioning units, chillers, pumps, disconnecting switches, etc. If responsibility can not be found then a clarification by the architect-engineer is in order.

In the checking process, mistakes may be uncovered. The electrical drawings should confirm the horsepower load requirements of HVAC equipment. Mechanical drawings should show proper water supply and drainage and other site utilities such as steam and gas. Openings in structural members such as walls and floors should be properly sized and located in such a way as to make the HVAC installation efficient. If mistakes or ambiguities in responsibility affect the cost estimate, they should be brought to the attention of the architect-engineer.

The estimator must be confident that he has the "feel" of the job from the drawings. He should be able to visualize the route and elevation of the ductwork and piping, and identify problem areas that require special attention or where code requirements must be met. This is accomplished by scanning all of the drawings and making notes on matters of concern. Among these concerns will be the areas where coordination with other trades will be required. Even though coordination among trades will take place after the bids have been awarded, the HVAC estimator will have to guide the contrac-

tor to the most practical and economical routes for ducts and piping, before work begins.

One small note of caution. Drawings are made to scale and have dimensions shown. Often the dimension and scale do not agree because of a draftsman's error or because a change was made and only the dimension was changed in order to save time. For this case, the written dimension is held to be correct. This may not be too significant when a few lineal feet of pipe is involved but if a clearance opening in a reinforced concrete wall is too small for the duct because of a scaled dimension, a costly problem would be faced.

Drawing symbols and abbreviations usually found on HVAC drawings are given in paragraph 4.3. Abbreviations used on drawings and in specifications are included in paragraph 4.4.

4.2. SPECIFICATIONS

As previously stated, drawings give the dimensions and location of building elements from which materials take-off lists can be compiled. Specifications describe such things as the specific materials to be used, their quality, construction methods, standards and codes to which materials and workmanship must comply, and the responsibilities of the owner, architect and contractor.

Sometimes changes are made during the preparation of bids which are called addenda. These represent the most up to date and authoritative source of information if a discrepancy is found between the addenda and the specification. Similarly, if a conflict is found between specification and drawing, the specification usually carries. The contractor is advised to bring these differences to the attention of the architect before the bid is submitted to prevent problems from arising.

Specifications are divided into 16 separate divisions based upon the Construction Specifications Institute's CSI Format for Construction Specifications. These are:

Division 1—General Requirements
Division 2—Site Work
Division 3—Concrete
Division 4—Masonry

Division 5—Metals
Division 6—Wood and Plastics
Division 7—Thermal and Moisture
    Protection
Division 8—Doors and Windows
Division 9—Finishes
Division 10—Specialties
Division 11—Equipment
Division 12—Furnishings
Division 13—Special Construction
Division 14—Conveying Systems
Division 15—Mechanical
Division 16—Electrical

Each of these are subdivided into many related subgroups. Of the 16 divisions, the HVAC designer and estimator must be thoroughly conversant with the first, general requirements, and the last two, mechanical and electrical as shown below.

Division 1—General Requirements.

- Description of work
- Schedule of contract documents
- General and special provisions
- Submittal requirements
- Certification of specification
- Compliance—standard form
- Scheduling and progress
- Quality control operations
- Temporary facilities
- Environmental protection during construction
- Execution of work
- Special restrictions
- Contractor's obligations
- Operation and maintenance manuals
- Instructions and training

The general requirements scope the job and set forth the obligations and responsibilities of all parties. It describes the site conditions, meeting schedules, and the type of information that must be submitted by the contractor. There can be items in the general requirements that can cost the contractor money and time. Therefore, this section must receive very careful attention.

Division 15—Mechanical.

- Mechanical and electrical equipment General requirements

- Vibration isolation
- Insulation of mechanical systems
- Water treatment equipment for mechanical systems
- Plumbing
- Sprinkler system, automatic, wet-pipe and standpipe
- Sprinkler system, automatic deluge type
- Sprinkler system, automatic dry-pipe
- Fire protection system
- Fire pumps
- Compressed air equipment
- Heating plant, fire tube, packaged-type boilers
- Fuel oil tanks
- Central refrigeration equipment for air conditioning
- Piping for mechanical systems
- Packaged heating and cooling units
- Air supply systems
- Dust collecting equipment
- Temperature controls for HVAC systems
- Testing and balancing air and water systems
- Carbon monoxide monitoring and alarm system

Division 15, of course is the central division affecting the HVAC contractor. Aside from the obvious equipment that comes under his domain, the HVAC contractor should look for interfaces with the plumbing contractor. Who supplies the water treatment system or pumps? Who digs the hole for the underground fuel oil tank? How about the concrete pad for the boiler?

Division 16—Electrical.

- Electrical systems
- Underfloor duct system
- High voltage electrical feeder system
- Underground electrical distribution
- Lighting fixtures
- Fire detection and alarm system
- Intercommunication and sound system
- Electrical space heating equipment
- Central control system

Electrical specifications are other interfaces with the HVAC. Who shall provide and connect control wiring for HVAC units? Are duct heaters the responsibility of electricians or

HVAC technicians? It is hoped that these questions and others can be answered by studying the specifications in interface division. Specifications in other divisions should also be scanned for related data. Can the freight elevator carry the three-ton heat exchanger? Its capacity and size are found in Division 14. Are there restrictions on the caulking? This is given in Division 7. Division 9 will give information on paints, acoustical treatment and other finishes where touch up is a requirement of the contract. Examples can be found in almost every division.

4.3. SYMBOLS & ABBREVIATIONS MOST WIDELY USED IN HVAC PRACTICES.

4.3.1. Sheet-metal Symbols Double Line Drawing.

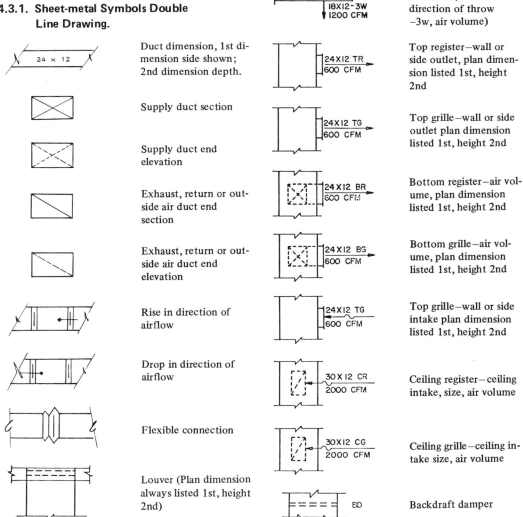

Symbol	Description
24 × 12	Duct dimension, 1st dimension side shown; 2nd dimension depth.
	Supply duct section
	Supply duct end elevation
	Exhaust, return or outside air duct end section
	Exhaust, return or outside air duct end elevation
	Rise in direction of airflow
	Drop in direction of airflow
	Flexible connection
	Louver (Plan dimension always listed 1st, height 2nd)

Symbol	Description
ACCESS DOOR / ACCESS PANEL	Access door or access panel
(← AIR)	Air extractor
10 ∅ NK. 500 CFM	Round ceiling diffuser (neck diameter and air volume)
18X12-3W 1200 CFM	Rectangular ceiling diffuser (neck size 1st dimension in direct ion of airflow, number and direction of throw −3w, air volume)
24X12 TR 600 CFM	Top register—wall or side outlet, plan dimension listed 1st, height 2nd
24X12 TG 600 CFM	Top grille—wall or side outlet plan dimension listed 1st, height 2nd
24X12 BR 600 CFM	Bottom register—air volume, plan dimension listed 1st, height 2nd
24X12 BG 600 CFM	Bottom grille—air volume, plan dimension listed 1st, height 2nd
24X12 TG 600 CFM	Top grille—wall or side intake plan dimension listed 1st, height 2nd
30X12 CR 2000 CFM	Ceiling register—ceiling intake, size, air volume
30X12 CG 2000 CFM	Ceiling grille—ceiling intake size, air volume
BD	Backdraft damper

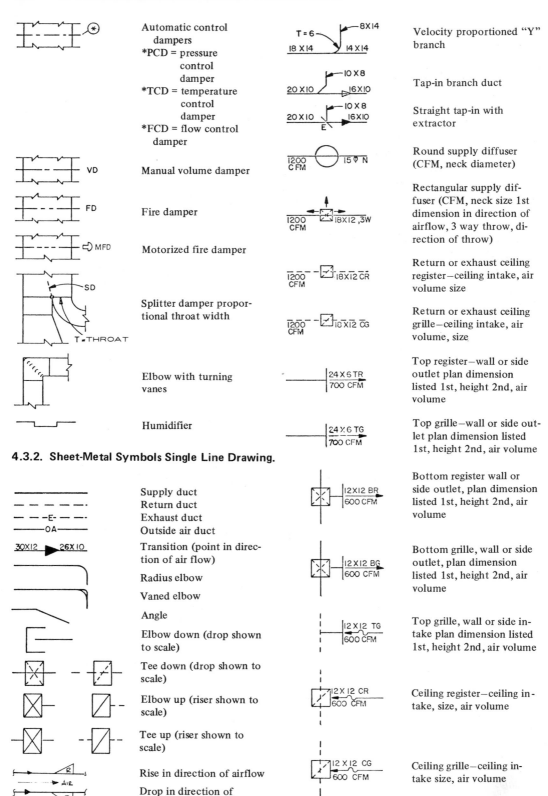

Automatic control dampers
*PCD = pressure control damper
*TCD = temperature control damper
*FCD = flow control damper

VD   Manual volume damper

FD   Fire damper

MFD   Motorized fire damper

SD
T = THROAT   Splitter damper proportional throat width

Elbow with turning vanes

Humidifier

Velocity proportioned "Y" branch

Tap-in branch duct

Straight tap-in with extractor

Round supply diffuser (CFM, neck diameter)

Rectangular supply diffuser (CFM, neck size 1st dimension in direction of airflow, 3 way throw, direction of throw)

Return or exhaust ceiling register—ceiling intake, air volume size

Return or exhaust ceiling grille—ceiling intake, air volume, size

Top register—wall or side outlet plan dimension listed 1st, height 2nd, air volume

Top grille—wall or side outlet plan dimension listed 1st, height 2nd, air volume

4.3.2. Sheet-Metal Symbols Single Line Drawing.

Supply duct
Return duct
Exhaust duct
Outside air duct

Transition (point in direction of air flow)

Radius elbow

Vaned elbow

Angle

Elbow down (drop shown to scale)

Tee down (drop shown to scale)

Elbow up (riser shown to scale)

Tee up (riser shown to scale)

Rise in direction of airflow

Drop in direction of airflow

Bottom register wall or side outlet, plan dimension listed 1st, height 2nd, air volume

Bottom grille, wall or side outlet, plan dimension listed 1st, height 2nd, air volume

Top grille, wall or side intake plan dimension listed 1st, height 2nd, air volume

Ceiling register—ceiling intake, size, air volume

Ceiling grille—ceiling intake size, air volume

Capped duct

==== ==== BD   Backdraft damper

— — — — —   Automatic control damper

— — — — ⌐VD   Manual volume damper

— — — — ▽FD   Fire damper

S   Splitter damper

▽   ▽   Acoustic lined between arrows

⌐AD   Access door

AP   Access panel

—|||—   Flexible connection

→   Air flow line

→   Door louver

Fan

Fan with variable inlet dampers

High pressure blower

Pump

F   Filter

Air washer

Water coil

Steam coil

4.3.3. Equipment Designation.

Roll filter

Louver

Eliminator

Parallel blade

Opposed blade

Backdraft damper

Roof intake or exhaust (power or gravity)

Fan-coil unit

Averaging sensor

4.3.4. Valving Designations.

—▷◁—   GVQC–Gate valve quick closing

—▷◁—   GV–Gate valve

AGV–Angle gate valve

—▷◁—   GLV–Globe valve

—▷/—   CKV–Check valve

—▷◁—   NV–Needle valve

—◇—   PV–Plug valve

—◇—   PV–Three-way plug

AV—Angle valve (elevation)

AV—Angle valve (plan)

BFV—Butterfly valve

DV—Diaphragm valve

BV—Ball valve

SCKV—Stop check valve

SCKV—Stop check valve angle

BDV—blow-down valve
(Y-pattern)

SLV—Sleeve valve

QOV—Quick opening valve

HCV—Hand control valve

FTV—Foot valve with strainer

HV—Angle valve with male hose
end (elevation)

HV—Angle valve with male hose
end (plan)

LCV—Level control valve (float
valve)

PSV—Pressure or temperature
safety valve (elevation)

—Pressure or temperature
safety valve (plan)

PV—Three way plug valve, normal
and alternate flow arrows
shown

PV—Three way plug valve, normal
and alternate flow arrows
shown

Valve with handwheel shown for
space requirements or other
reason

Valve with chain operator

Plug valve gear operated (plan)

Plug valve gear operated
(elevation)

PCV—Pressure control valve, self
contained

Diaphragm operated valve

Diaphragm operated valve with
hand wheel

Diaphragm operated 3-way valve

Solenoid or motor operated valve

Solenoid or motor operated
3-way valve

Solenoid or motor operated
4-way valve

TXV—Thermal expansion valve
self-actuated (refrigerant
type

TCV—Temperature control valve
self-actuated type

Piston operated valve, hydraulic
or pneumatic (vertical piston)

Piston operated valve, hydraulic
or pneumatic (horizontal piston)

Diaphragm operated valve (elev.)

Diaphragm operated valve (plan)

Solenoid, motor, piston or self-
contained valve (plan)

"Y" type strainer

"Y" type strainer with blow-off valve

Basket strainer

Duplex strainer

Filter

Filter air intake

Air eliminator

Blast thermostatic

Boiler return

Bucket

Float

Float & thermostatic

Impulse

Thermostatic

SG—Sight glass

E—Ejector or eductor

DD—Desiccant drier

AAV—Automatic air vent (elevation)

AAV—Automatic air vent (plan)

VB—Vacuum breaker

Floor monument

Interface/match line

Insulated piping (line)

Test connection plugged (line size for smaller than 1".

4.3.5. Piping

HB—Hose bibb
LF—Lawn faucet

BFP—Back flow preventer

Direction of flow

Pitch of line

Screwed cap

Welded cap

Union

Flanged connection

Blind flange connection

Reducer

Eccentric reducer

HC—Hose connection male

HC—Hose connection female

QD—Quick disconnect coupling

QD—Quick disconnect coupling male half

QD—Quick disconnect coupling female half

FH—Flexible hose screwed

FH—Flexible hose flanged

Female conn pluged

EJ—Expansion joint, non-tied axial bellows type

EJ—Expansion joint, tied axial bellows type

EJ—Expansion joint, non-tied offset or universal bellows type

EJ—Expansion joint, tied offset or universal bellows type

EJ—Expansion joint, slip type

EJ—Expansion joint, hinged

BJ—Ball joint (indicate style required)

SJ—Swivel joint plan (indicate style required)

SJ—Swivel joint, elevation (indicate style required)

PA—Pipe anchor

PG—Pipe guide

———IHWR——— Industrial hot water return

———HS——— Hot water supply, heating

————HR———— Hot water return, heating

———CHS——— Chilled water supply

————CHR———— Chilled water return

———P——— Purge

———CWS——— Cooling water supply

————CWR———— Cooling water return

———SW——— Soft water

———DW——— Demineralized or distilled water

———TW——— Treated water (indicate type of treatment)

———PWS——— Process water return (define process.)

———OWD——— Oil and water drain

———FW——— Fire protection water

———CO_2——— Carbon dioxide, fire protection

———HS 100——— Hydraulic supply (number indicate pressure)

———HR 100——— Hydraulic return (number indicates pressure)

———BS 10——— Brine supply (indicate temperature if required such as minus 10°F)

———BR——— Brine return (indicate temperature if required)

———COF——— Clean oil fill

———COS——— Clean oil supply

———COR——— Clean oil return

———CTOS——— Contaminated oil supply

———CTOR——— Contaminated oil return

———DFS——— Diesel fuel oil supply

———DFR——— Diesel fuel oil return

———FOS——— Fuel oil supply (indicate type of fuel if required)

———FOR——— Fuel oil return (indicate type of fuel if required)

———LO——— Engine lube oil supply and return

———LOD——— Engine lube oil drain

———EX——— Exhaust line (define type, diesel, vacuum pump, etc)

———INT——— Air intake (define use: diesel, compressor, etc)

———S100——— Steam (number indicates system pressure)

———C100——— Condensate return (number indicate steam pressure)

———CPD——— Condensate pump discharge

———BFW——— Boiler feed water

———BD——— Blow down

———CFI——— Chemical feed (1) indicates type of chemical required)

———A 100——— Compressed air (number indicates pressure)

———AI 100——— Instrument air (number indicates pressure, 1A indicates dry or filtered air)

———G10——— Natural gas (number indicates pressure)

———VAC——— Vacuum (indicate type or pressure if required)

———N_2 3000——— Nitrogen gas (number indicates pressure)

———H_2 6000——— Helium gas (number indicates pressure)

———O_2 50——— Oxygen gas (number indicates pressure)

———H_2 50——— Hydrogen gas (number indicates pressure)

———LN_2——— Liquid nitrogen (indicate pressure if required)

———LOX———	Liquid oxygen (indicate pressure if required)
———LH$_2$———	Liquid hydrogen (indicate pressure if required)
———LPG———	Liquified petroleum gas
———RD———	Refrigerant discharge gas
———RS———	Refrigerant suction gas
———RL———	Refrigerant liquid
—⫻——⫻——⫻—	Control air line
—✕——✕——✕—	Capillary tubing
— — — — — — —	Electrical lead

SECTION IV

HVAC and Engineering Tables

This portion of the book is a collection of reference tables. Tables 1 through 42, are concerned with HVAC design data and are intended to augment other parts of the book. This is especially true for Section I where cooling and heating load calculations are discussed. These tables will also be useful in Sections II and III to augment equipment data and owning and operating costs. Tables 43 to 60 are more general in nature and will hopefully serve the reader in his work in a number of ways.

Tables 1–42 were reprinted or constructed with data from *ASHRAE Handbooks* and *Product Directory*. The author acknowledges with thanks permission of the American Society of Heating, Refrigerating and Air-Conditioning Engineers to use this copyrighted material.

27. Heat Gain from Commercial Cooking Appliances Located in Air Conditioned Space (Btu/hr)
28. Heat Gain from Miscellaneous Appliances (Btu/hr)
29. Heat Gain from Occupants of Air Conditioned Spaces (Btu/hr)
30. Heat Gain from Electric Motors
31. Solar Heat Gain Factors for Glass Without Inside Shading (Btu/hr)
32. Solar Heat Gain Factors for Glass With Inside Shading (Btu/hr)
33. Shading Coefficients for Glass Without Inside Shading
34. Shading Coefficients for Hollow Glass Block Wall Panels
35. Overhang Shading from April 11 through September 1
36. Shade Factors for Various Types of Outside Shading
37. Shading Coefficients for Domed Skylights
38. Shading Coefficients for Double Glazing with Between-Glass Shading
39. Shading Coefficients for Single Glass with Indoor Shading by Venetian Blinds and Roller Shades
40. Shading Coefficients for Insulating Glass with Indoor Shading by Venetian Blinds and Roller Shades

41. Climatic Conditions for United States and Canada
42. Climatic Conditions for Foreign Countries
43. Water Usage for Water-Cooled Equipment
44. Properties of Mixtures of Air and Saturated Water Vapor
45. Heat Values of Fuels and Electricity
46. Conversion of Engineering Constants
47. Conversion Factors
48. Metric Conversion Tables
49. Decimal Equivalents of an Inch
50. Fractions of an Inch Expressed As Decimals of a Foot
51. Heads of Water in Feet Due to Various Pressures in Pounds Per Square Inch
52. Equivalent Discharge in Cubic Feet Per Second, Gallons Per Minute, Gallons Per 24 Hours
53. Steel Pipe and Tubing Specifications
54. Properties of Steel Pipe
55. Friction of Water in Pipes
56. Properties of Saturated Steam
57. Steam Flow in Pipes
58. Approximate Viscosities and Specific Gravities of Common Liquids
59. Friction of Air in Pipes
60. Wire and Sheet-Metal Gauges

TABLE 1. Efficiency of Fuel Utilization Over the Heating Season (Residential Systems)

Type of Fuel-Burning Unit	Efficiency (%)
Gas	70–80
Oil	70–80

TABLE 3. Correction Factors for Outdoor Design Temperature for Heating Calculations

Outdoor Design Temp. (°F)	–20	–10	0	+10	+20
Correction Factor	0.778	0.875	1.000	1.167	1.400

The multipliers in Table 2, which are high for mild climates and low for cold regions, are not in error as might appear. The unit figures in Table 3 are per square foot of radiator or thousand Btu heat loss per degree day. For equivalent buildings and heating seasons, those in warm climates have lower design heat losses and smaller radiator quantities than those in cold cities. Consequently, the unit figure in quantity of fuel per (sq ft of radiator) (degree day), is larger for warm localities than for colder regions. Since the northern cities have more radiator surface per given building and a higher seasonal degree-day total than cities in the south, the total fuel per season will be larger for the northern city.

TABLE 2. Surface Conductance (f) for Building Structures (Winter–Summer)

Surface	Exposure	Surface Conductance (Btu/hr/sq ft/°F)	
		Winter	Summer
Ceilings	Inside	1.65	1.20
Roofs	Outside	6.00[a]	4.00[b]
Walls	Inside	1.65	1.65
Walls	Outside	6.00[a]	4.00[b]

[a]Average wind velocity 15 mph
[b]Average wind velocity 8 mph

TABLE 4. Unit Fuel Constants (based on 0°F outdoor temp., 70°F indoor temp.)

Fuel and Units	Utilization Efficiency		
	60	70	80
	Unit fuel consumption per degree day per 1000 Btu design heat loss		
Gas in therms[a]	0.00572	0.00490	0.00429
Oil in gallons[b]	0.00405	0.00347	0.00304
Coal in pounds[c]	0.0476	0.0408	0.0357

[a]One therm is equal to 100,000 Btu
[b]Based on a heating value of 141,000 Btu/gal.
[c]Based on a heating value of 12,000 Btu/lb

TABLE 5. Average Monthly and Yearly Degree Days for Cities in the United States and Canada[a,b,c] (Base 65°F)

State	Station	Avg. Winter Temp[d]	July	Aug.	Sept.	Oct.	Nov.	Dec.	Jan.	Feb.	Mar.	Apr.	May	June	Yearly Total
Ala.	Birmingham............A	54.2	0	0	6	93	363	555	592	462	363	108	9	0	2551
	Huntsville.............A	51.3	0	0	12	127	426	663	694	557	434	138	19	0	3070
	Mobile................A	59.9	0	0	0	22	213	357	415	300	211	42	0	0	1560
	Montgomery...........A	55.4	0	0	0	68	330	527	543	417	316	90	0	0	2291
Alaska	Anchorage.............A	23.0	245	291	516	930	1284	1572	1631	1316	1293	879	592	315	10864
	Fairbanks..............A	6.7	171	332	642	1203	1833	2254	2359	1901	1739	1068	555	222	14279
	Juneau.................A	32.1	301	338	483	725	921	1135	1237	1070	1073	810	601	381	9075
	Nome.................A	13.1	481	496	693	1094	1455	1820	1879	1666	1770	1314	930	573	14171
Ariz.	Flagstaff..............A	35.6	46	68	201	558	867	1073	1169	991	911	651	437	180	7152
	Phoenix................A	58.5	0	0	0	22	234	415	474	328	217	75	0	0	1765
	Tucson................A	58.1	0	0	0	25	231	406	471	344	242	75	6	0	1800
	Winslow...............A	43.0	0	0	6	245	711	1008	1054	770	601	291	96	0	4782
	Yuma.................A	64.2	0	0	0	0	108	264	307	190	90	15	0	0	974
Ark.	Fort Smith.............A	50.3	0	0	12	127	450	704	781	596	456	144	22	0	3292
	Little Rock............A	50.5	0	0	9	127	465	716	756	577	434	126	9	0	3219
	Texarkana.............A	54.2	0	0	0	78	345	561	626	468	350	105	0	0	2533
Calif.	Bakersfield.............A	55.4	0	0	0	37	282	502	546	364	267	105	19	0	2122
	Bishop.................A	46.0	0	0	48	260	576	797	874	680	555	306	143	36	4275
	Blue Canyon...........A	42.2	28	37	108	347	594	781	896	795	806	597	412	195	5596
	Burbank...............A	58.6	0	0	6	43	177	301	366	277	239	138	81	18	1646
	Eureka................C	49.9	270	257	258	329	414	499	546	470	505	438	372	285	4643
	Fresno.................A	53.3	0	0	0	84	354	577	605	426	335	162	62	6	2611
	Long Beach............A	57.8	0	0	9	47	171	316	397	311	264	171	93	24	1803
	Los Angeles............A	57.4	28	28	42	78	180	291	372	302	288	219	158	81	2061
	Los Angeles............C	60.3	0	0	6	31	132	229	310	230	202	123	68	18	1349
	Mt. Shasta.............C	41.2	25	34	123	406	696	902	983	784	738	525	347	159	5722
	Oakland...............A	53.5	53	50	45	127	309	481	527	400	353	255	180	90	2870
	Red Bluff..............A	53.8	0	0	0	53	318	555	605	428	341	168	47	0	2515
	Sacramento............A	53.9	0	0	0	56	321	546	583	414	332	178	72	0	2502
	Sacramento............C	54.4	0	0	0	62	312	533	561	392	310	173	76	0	2419
	Sandberg...............C	46.8	0	0	30	202	480	691	778	661	620	426	264	57	4209
	San Diego..............A	59.5	9	0	21	43	135	236	298	235	214	135	90	42	1458
	San Francisco..........A	53.4	81	78	60	143	306	462	508	395	363	279	214	126	3015
	San Francisco..........C	55.1	192	174	102	118	231	388	443	336	319	279	239	180	3001
	Santa Maria............A	54.3	99	93	96	146	270	391	459	370	363	282	233	165	2967
Colo.	Alamosa...............A	29.7	65	99	279	639	1065	1420	1476	1162	1020	696	440	168	8529
	Colorado Springs.......A	37.3	9	25	132	456	825	1032	1128	938	893	582	319	84	6423
	Denver................A	37.6	6	9	117	428	819	1035	1132	938	887	558	288	66	6283
	Denver................C	40.8	0	0	90	366	714	905	1004	851	800	492	254	48	5524
	Grand Junction.........A	39.3	0	0	30	313	786	1113	1209	907	729	387	146	21	5641
	Pueblo................A	40.4	0	0	54	326	750	986	1085	871	772	429	174	15	5462
Conn.	Bridgeport.............A	39.9	0	0	66	307	615	986	1079	966	853	510	208	27	5617
	Hartford...............A	37.3	0	12	117	394	714	1101	1190	1042	908	519	205	33	6235
	New Haven.............A	39.0	0	12	87	347	648	1011	1097	991	871	543	245	45	5897
Del.	Wilmington............A	42.5	0	0	51	270	588	927	980	874	735	387	112	6	4930
D. C.	Washington............A	45.7	0	0	33	217	519	834	871	762	626	288	74	0	4224
Fla.	Apalachicola...........C	61.2	0	0	0	16	153	319	347	260	180	33	0	0	1308
	Daytona Beach.........A	64.5	0	0	0	0	75	211	248	190	140	15	0	0	879
	Fort Myers.............A	68.6	0	0	0	0	24	109	146	101	62	0	0	0	442
	Jacksonville............A	61.9	0	0	0	12	144	310	332	246	174	21	0	0	1239
	Key West..............A	73.1	0	0	0	0	0	28	40	31	9	0	0	0	108
	Lakeland...............C	66.7	0	0	0	0	57	164	195	146	99	0	0	0	661
	Miami.................A	71.1	0	0	0	0	0	65	74	56	19	0	0	0	214

[a] Data for United States cities from a publication of the United States Weather Bureau, *Monthly Normals of Temperature, Precipitation and Heating Degree Days*, 1962, are for the period 1931 to 1960 inclusive. These data also include information from the 1963 Revisions to this publication, where available.

[b] Data for airport stations, A, and city stations, C, are both given where available.

[c] Data for Canadian cities were computed by the Climatology Division, Department of Transport from normal monthly mean temperatures, and the monthly values of heating degree days data were obtained using the National Research Council computer and a method devised by H. C. S. Thom of the United States Weather Bureau. The heating degree days are based on the period from 1931 to 1960.

[d] For period October to April, inclusive.

TABLE 5. (*Continued*)

State	Station	Avg. Winter Temp[d]	July	Aug.	Sept.	Oct.	Nov.	Dec.	Jan.	Feb.	Mar.	Apr.	May	June	Yearly Total
Fla. (Cont'd)	Miami Beach............C	72.5	0	0	0	0	0	40	56	36	9	0	0	0	141
	Orlando................A	65.7	0	0	0	0	72	198	220	165	105	6	0	0	766
	Pensacola..............A	60.4	0	0	0	19	195	353	400	277	183	36	0	0	1463
	Tallahassee.............A	60.1	0	0	0	28	198	360	375	286	202	36	0	0	1485
	Tampa.................A	66.4	0	0	0	0	60	171	202	148	102	0	0	0	683
	West Palm Beach.......A	68.4	0	0	0	0	6	65	87	64	31	0	0	0	253
Ga.	Athens................A	51.8	0	0	12	115	405	632	642	529	431	141	22	0	2929
	Atlanta................A	51.7	0	0	18	124	417	648	636	518	428	147	25	0	2961
	Augusta...............A	54.5	0	0	0	78	333	552	549	445	350	90	0	0	2397
	Columbus..............A	54.8	0	0	0	87	333	543	552	434	338	96	0	0	2383
	Macon.................A	56.2	0	0	0	71	297	502	505	403	295	63	0	0	2136
	Rome..................A	49.9	0	0	24	161	474	701	710	577	468	177	34	0	3326
	Savannah..............A	57.8	0	0	0	47	246	437	437	353	254	45	0	0	1819
	Thomasville............C	60.0	0	0	0	25	198	366	394	305	208	33	0	0	1529
Hawaii	Lihue..................A	72.7	0	0	0	0	0	0	0	0	0	0	0	0	0
	Honolulu...............A	74.2	0	0	0	0	0	0	0	0	0	0	0	0	0
	Hilo...................A	71.9	0	0	0	0	0	0	0	0	0	0	0	0	0
Idaho	Boise..................A	39.7	0	0	132	415	792	1017	1113	854	722	438	245	81	5809
	Lewiston...............A	41.0	0	0	123	403	756	933	1063	815	694	426	239	90	5542
	Pocatello..............A	34.8	0	0	172	493	900	1166	1324	1058	905	555	319	141	7033
Ill.	Cairo..................C	47.9	0	0	36	164	513	791	856	680	539	195	47	0	3821
	Chicago (O'Hare).......A	35.8	0	12	117	381	807	1166	1265	1086	939	534	260	72	6639
	Chicago (Midway).......A	37.5	0	0	81	326	753	1113	1209	1044	890	480	211	48	6155
	Chicago................C	38.9	0	0	66	279	705	1051	1150	1000	868	489	226	48	5882
	Moline.................A	36.4	0	9	99	335	774	1181	1314	1100	918	450	189	39	6408
	Peoria.................A	38.1	0	6	87	326	759	1113	1218	1025	849	426	183	33	6025
	Rockford...............A	34.8	6	9	114	400	837	1221	1333	1137	961	516	236	60	6830
	Springfield.............A	40.6	0	0	72	291	696	1023	1135	935	769	354	136	18	5429
Ind.	Evansville..............A	45.0	0	0	66	220	606	896	955	767	620	237	68	0	4435
	Fort Wayne............A	37.3	0	9	105	378	783	1135	1178	1028	890	471	189	39	6205
	Indianapolis...........A	39.6	0	0	90	316	723	1051	1113	949	809	432	177	39	5699
	South Bend............A	36.6	0	6	111	372	777	1125	1221	1070	933	525	239	60	6439
Iowa	Burlington.............A	37.6	0	0	93	322	768	1135	1259	1042	859	426	177	33	6114
	Des Moines............A	35.5	0	6	96	363	828	1225	1370	1137	915	438	180	30	6588
	Dubuque...............A	32.7	12	31	156	450	906	1287	1420	1204	1026	546	260	78	7376
	Sioux City.............A	34.0	0	9	108	369	867	1240	1435	1198	989	483	214	39	6951
	Waterloo...............A	32.6	12	19	138	428	909	1296	1460	1221	1023	531	229	54	7320
Kans.	Concordia..............A	40.4	0	0	57	276	705	1023	1163	935	781	372	149	18	5479
	Dodge City.............A	42.5	0	0	33	251	666	939	1051	840	719	354	124	9	4986
	Goodland...............A	37.8	0	6	81	381	810	1073	1166	955	884	507	236	42	6141
	Topeka................A	41.7	0	0	57	270	672	980	1122	893	722	330	124	12	5182
	Wichita................A	44.2	0	0	33	229	618	905	1023	804	645	270	87	6	4620
Ky.	Covington..............A	41.4	0	0	75	291	669	983	1035	893	756	390	149	24	5265
	Lexington..............A	43.8	0	0	54	239	609	902	946	818	685	325	105	0	4683
	Louisville..............A	44.0	0	0	54	248	609	890	930	818	682	315	105	9	4660
La.	Alexandria.............A	57.5	0	0	0	56	273	431	471	361	260	69	0	0	1921
	Baton Rouge...........A	59.8	0	0	0	31	216	369	409	294	208	33	0	0	1560
	Lake Charles...........A	60.5	0	0	0	19	210	341	381	274	195	39	0	0	1459
	New Orleans...........A	61.0	0	0	0	19	192	322	363	258	192	39	0	0	1385
	New Orleans...........C	61.8	0	0	0	12	165	291	344	241	177	24	0	0	1254
	Shreveport.............A	56.2	0	0	0	47	297	477	552	426	304	81	0	0	2184
Me.	Caribou................A	24.4	78	115	336	682	1044	1535	1690	1470	1308	858	468	183	9767
	Portland...............A	33.0	12	53	195	508	807	1215	1339	1182	1042	675	372	111	7511
Md.	Baltimore..............A	43.7	0	0	48	264	585	905	936	820	679	327	90	0	4654
	Baltimore..............C	46.2	0	0	27	189	486	806	859	762	629	288	65	0	4111
	Frederich..............A	42.0	0	0	66	307	624	955	995	876	741	384	127	12	5087
Mass.	Boston.................A	40.0	0	9	60	316	603	983	1088	972	846	513	208	36	5634
	Nantucket..............A	40.2	12	22	93	332	573	896	992	941	896	621	384	129	5891
	Pittsfield..............A	32.6	25	59	219	524	831	1231	1339	1196	1063	660	326	105	7578
	Worcester..............A	34.7	6	34	147	450	774	1172	1271	1123	998	612	304	78	6969

TABLE 5. (Continued)

State	Station	Avg. Winter Temp[d]	July	Aug.	Sept.	Oct.	Nov.	Dec.	Jan.	Feb.	Mar.	Apr.	May	June	Yearly Total
Mich.	Alpena..............A	29.7	68	105	273	580	912	1268	1404	1299	1218	777	446	156	8506
	Detroit (City)..........A	37.2	0	0	87	360	738	1088	1181	1058	936	522	220	42	6232
	Detroit (Wayne).........A	37.1	0	0	96	353	738	1088	1194	1061	933	534	239	57	6293
	Detroit (Willow Run)....A	37.2	0	0	90	357	750	1104	1190	1053	921	519	229	45	6258
	Escanaba..............C	29.6	59	87	243	539	924	1293	1445	1296	1203	777	456	159	8481
	Flint.................A	33.1	16	40	159	465	843	1212	1330	1198	1066	639	319	90	7377
	Grand Rapids..........A	34.9	9	28	135	434	804	1147	1259	1134	1011	579	279	75	6894
	Lansing...............A	34.8	6	22	138	431	813	1163	1262	1142	1011	579	273	69	6909
	Marquette.............C	30.2	59	81	240	527	936	1268	1411	1268	1187	771	468	177	8393
	Muskegon..............A	36.0	12	28	120	400	762	1088	1209	1100	995	594	310	78	6696
	Sault Ste. Marie........A	27.7	96	105	279	580	951	1367	1525	1380	1277	810	477	201	9048
Minn.	Duluth................A	23.4	71	109	330	632	1131	1581	1745	1518	1355	840	490	198	10000
	Minneapolis...........A	28.3	22	31	189	505	1014	1454	1631	1380	1166	621	288	81	8382
	Rochester.............A	28.8	25	34	186	474	1005	1438	1593	1366	1150	630	301	93	8295
Miss.	Jackson...............A	55.7	0	0	0	65	315	502	546	414	310	87	0	0	2239
	Meridian..............A	55.4	0	0	0	81	339	518	543	417	310	81	0	0	2289
	Vicksburg.............C	56.9	0	0	0	53	279	462	512	384	282	69	0	0	2041
Mo.	Columbia..............A	42.3	0	0	54	251	651	967	1076	874	716	324	121	12	5046
	Kansas City...........A	43.9	0	0	39	220	612	905	1032	818	682	294	109	0	4711
	St. Joseph............A	40.3	0	6	60	285	708	1039	1172	949	769	348	133	15	5484
	St. Louis.............A	43.1	0	0	60	251	627	936	1026	848	704	312	121	15	4900
	St. Louis.............C	44.8	0	0	36	202	576	884	977	801	651	270	87	0	4484
	Springfield...........A	44.5	0	0	45	223	600	877	973	781	660	291	105	6	4900
Mont.	Billings..............A	34.5	6	15	186	487	897	1135	1296	1100	970	570	285	102	7049
	Glasgow...............A	26.4	31	47	270	608	1104	1466	1711	1439	1187	648	335	150	8996
	Great Falls...........A	32.8	28	53	258	543	921	1169	1349	1154	1063	642	384	186	7750
	Havre.................A	28.1	28	53	306	595	1065	1367	1584	1364	1181	657	338	162	8700
	Havre.................C	29.8	19	37	252	539	1014	1321	1528	1305	1116	612	304	135	8182
	Helena................A	31.1	31	59	294	601	1002	1265	1438	1170	1042	651	381	195	8129
	Kalispell.............A	31.4	50	99	321	654	1020	1240	1401	1134	1029	639	397	207	8191
	Miles City............A	31.2	6	6	174	502	972	1296	1504	1252	1057	579	276	99	7723
	Missoula..............A	31.5	34	74	303	651	1035	1287	1420	1120	970	621	391	219	8125
Neb.	Grand Island..........A	36.0	0	6	108	381	834	1172	1314	1089	908	462	211	45	6530
	Lincoln...............C	38.8	0	6	75	301	726	1066	1237	1016	834	402	171	30	5864
	Norfolk...............A	34.0	9	0	111	397	873	1234	1414	1179	983	498	233	48	6979
	North Platte..........A	35.5	0	6	123	440	885	1166	1271	1039	930	519	248	57	6684
	Omaha.................A	35.6	0	12	105	357	828	1175	1355	1126	939	465	208	42	6612
	Scottsbluff...........A	35.9	0	0	138	459	876	1128	1231	1008	921	552	285	75	6673
	Valentine.............A	32.6	9	12	165	493	942	1237	1395	1176	1045	579	288	84	7425
Nev.	Elko..................A	34.0	9	34	225	561	924	1197	1314	1036	911	621	409	192	7433
	Ely...................A	33.1	28	43	234	592	939	1184	1308	1075	977	672	456	225	7733
	Las Vegas.............A	53.5	0	0	0	78	387	617	688	487	335	111	6	0	2709
	Reno..................A	39.3	43	87	204	490	801	1026	1073	823	729	510	357	189	6332
	Winnemucca............A	36.7	0	34	210	536	876	1091	1172	916	837	573	363	153	6761
N. H.	Concord...............A	33.0	6	50	177	505	822	1240	1358	1184	1032	636	298	75	7383
	Mt. Washington Obsv....	15.2	493	536	720	1057	1341	1742	1820	1663	1652	1260	930	603	13817
N. J.	Atlantic City..........A	43.2	0	0	39	251	549	880	936	848	741	420	133	15	4812
	Newark................A	42.8	0	0	30	248	573	921	983	876	729	381	118	0	4589
	Trenton...............C	42.4	0	0	57	264	576	924	989	885	753	399	121	12	4980
N. M.	Albuquerque...........A	45.0	0	0	12	229	642	868	930	703	595	288	81	0	4348
	Clayton...............A	42.0	0	6	66	310	699	899	986	812	747	429	183	21	5158
	Raton.................A	38.1	9	28	126	431	825	1048	1116	904	834	543	301	63	6228
	Roswell...............A	47.5	0	0	18	202	573	806	840	641	481	201	31	0	3793
	Silver City...........A	48.0	0	0	6	183	525	729	791	605	518	261	87	0	3705
N. Y.	Albany................A	34.6	0	19	138	440	777	1194	1311	1156	992	564	239	45	6875
	Albany................C	37.2	0	9	102	375	699	1104	1218	1072	908	498	186	30	6201
	Binghamton............A	33.9	22	65	201	471	810	1184	1277	1154	1045	645	313	99	7286
	Binghamton............C	36.6	0	28	141	406	732	1107	1190	1081	949	543	229	45	6451
	Buffalo...............A	34.5	19	37	141	440	777	1156	1256	1145	1039	645	329	78	7062
	New York (Cent. Park)..	42.8	0	0	30	233	540	902	986	885	760	408	118	9	4871
	New York (La Guardia)..A	43.1	0	0	27	223	528	887	973	879	750	414	124	6	4811

TABLE 5. (Continued)

State	Station	Avg. Winter Temp[d]	July	Aug.	Sept.	Oct.	Nov.	Dec.	Jan.	Feb.	Mar.	Apr.	May	June	Yearly Total
	New York (Kennedy)....A	41.4	0	0	36	248	564	933	1029	935	815	480	167	12	5219
	Rochester..............A	35.4	9	31	126	415	747	1125	1234	1123	1014	597	279	48	6748
	Schenectady............C	35.4	0	22	123	422	756	1159	1283	1131	970	543	211	30	6650
	Syracuse...............A	35.2	6	28	132	415	744	1153	1271	1140	1004	570	248	45	6756
N. C.	Asheville...............C	46.7	0	0	48	245	555	775	784	683	592	273	87	0	4042
	Cape Hatteras...........	53.3	0	0	0	78	273	521	580	518	440	177	25	0	2612
	Charlotte..............A	50.4	0	0	6	124	438	691	691	582	481	156	22	0	3191
	Greensboro.............A	47.5	0	0	33	192	513	778	784	672	552	234	47	0	3805
	Raleigh................A	49.4	0	0	21	164	450	716	725	616	487	180	34	0	3393
	Wilmington.............A	54.6	0	0	0	74	291	521	546	462	357	96	0	0	2347
	Winston-Salem..........A	48.4	0	0	21	171	483	747	753	652	524	207	37	0	3595
N. D.	Bismarck...............A	26.6	34	28	222	577	1083	1463	1708	1442	1203	645	329	117	8851
	Devils Lake............C	22.4	40	53	273	642	1191	1634	1872	1579	1345	753	381	138	9901
	Fargo.................A	24.8	28	37	219	574	1107	1569	1789	1520	1262	690	332	99	9226
	Williston..............A	25.2	31	43	261	601	1122	1513	1758	1473	1262	681	357	141	9243
Ohio	Akron-Canton..........A	38.1	0	9	96	381	726	1070	1138	1016	871	489	202	39	6037
	Cincinnati.............C	45.1	0	0	39	208	558	862	915	790	642	294	96	6	4410
	Cleveland..............A	37.2	9	25	105	384	738	1088	1159	1047	918	552	260	66	6351
	Columbus..............A	39.7	0	6	84	347	714	1039	1088	949	809	426	171	27	5660
	Columbus..............C	41.5	0	0	57	285	651	977	1032	902	760	396	136	15	5211
	Dayton................A	39.8	0	6	78	310	696	1045	1097	955	809	429	167	30	5622
	Mansfield..............A	36.9	9	22	114	397	768	1110	1169	1042	924	543	245	60	6403
	Sandusky..............C	39.1	0	6	66	313	684	1032	1107	991	868	495	198	36	5796
	Toledo................A	36.4	0	16	117	406	792	1138	1200	1056	924	543	242	60	6494
	Youngstown............A	36.8	6	19	120	412	771	1104	1169	1047	921	540	248	60	6417
Okla.	Oklahoma City.........A	48.3	0	0	15	164	498	766	868	664	527	189	34	0	3725
	Tulsa.................A	47.7	0	0	18	158	522	787	893	683	539	213	47	0	3860
Ore.	Astoria................A	45.6	146	130	210	375	561	679	753	622	636	480	363	231	5186
	Burns.................C	35.9	12	37	210	515	867	1113	1246	988	856	570	366	177	6957
	Eugene................A	45.6	34	34	129	366	585	719	803	627	589	426	279	135	4726
	Meacham...............A	34.2	84	124	288	580	918	1091	1209	1005	983	726	527	339	7874
	Medford...............A	43.2	0	0	78	372	678	871	918	697	642	432	242	78	5008
	Pendleton.............A	42.6	0	0	111	350	711	884	1017	773	617	396	205	63	5127
	Portland..............A	45.6	25	28	114	335	597	735	825	644	586	396	245	105	4635
	Portland..............C	47.4	12	16	75	267	534	679	769	594	536	351	198	78	4109
	Roseburg..............A	46.3	22	16	105	329	567	713	766	608	570	405	267	123	4491
	Salem.................A	45.4	37	31	111	338	594	729	822	647	611	417	273	144.	4754
Pa.	Allentown.............A	38.9	0	0	90	353	693	1045	1116	1002	849	471	167	24	5810
	Erie..................A	36.8	0	25	102	391	714	1063	1169	1081	973	585	288	60	6451
	Harrisburg............A	41.2	0	0	63	298	648	992	1045	907	766	396	124	12	5251
	Philadelphia..........A	41.8	0	0	60	297	620	965	1016	889	747	392	118	40	5144
	Philadelphia..........C	44.5	0	0	30	205	513	856	924	823	691	351	93	0	4486
	Pittsburgh............A	38.4	0	9	105	375	726	1063	1119	1002	874	480	195	39	5987
	Pittsburgh............C	42.2	0	0	60	291	615	930	983	885	763	390	124	12	5053
	Reading...............C	42.4	0	0	54	257	597	939	1001	885	735	372	105	0	4945
	Scranton..............A	37.2	0	19	132	434	762	1104	1156	1028	893	498	195	33	6254
	Williamsport..........A	38.5	0	9	111	375	717	1073	1122	1002	856	468	177	24	5934
R. I.	Block Island..........A	40.1	0	16	78	307	594	902	1020	955	877	612	344	99	5804
	Providence............A	38.8	0	16	96	372	660	1023	1110	988	868	534	236	51	5954
S. C.	Charleston.............A	56.4	0	0	0	59	282	471	487	389	291	54	0	0	2033
	Charleston.............C	57.9	0	0	0	34	210	425	443	367	273	42	0	0	1794
	Columbia..............A	54.0	0	0	0	84	345	577	570	470	357	81	0	0	2484
	Florence..............A	54.5	0	0	0	78	315	552	552	459	347	84	0	0	2387
	Greenville-Spartanburg...A	51.6	0	0	6	121	399	651	660	546	446	132	19	0	2980
S. D.	Huron.................A	28.8	9	12	165	508	1014	1432	1628	1355	1125	600	288	87	8223
	Rapid City............A	33.4	22	12	165	481	897	1172	1333	1145	1051	615	326	126	7345
	Sioux Falls............A	30.6	19	25	168	462	972	1361	1544	1285	1082	573	270	78	7839
Tenn.	Bristol................A	46.2	0	0	51	236	573	828	828	700	598	261	68	0	4143
	Chattanooga...........A	50.3	0	0	18	143	468	698	722	577	453	150	25	0	3254
	Knoxville..............A	49.2	0	0	30	171	489	725	732	613	493	198	43	0	3494
	Memphis...............A	50.5	0	0	18	130	447	698	729	585	456	147	22	0	3232

TABLE 5. (*Continued*)

State or Prov.	Station	Avg. Winter Temp^d	July	Aug.	Sept.	Oct.	Nov.	Dec.	Jan.	Feb.	Mar.	Apr.	May	June	Yearly Total
	Memphis..............C	51.6	0	0	12	102	396	648	710	568	434	129	16	0	3015
	Nashville..............A	48.9	0	0	30	158	495	732	778	644	512	189	40	0	3578
	Oak Ridge.............C	47.7	0	0	39	192	531	772	778	669	552	228	56	0	3817
Tex.	Abilene................A	53.9	0	0	0	99	366	586	642	470	347	114	0	0	2624
	Amarillo...............A	47.0	0	0	18	205	570	797	877	664	546	252	56	0	3985
	Austin.................A	59.1	0	0	0	31	225	388	468	325	223	51	0	0	1711
	Brownsville............A	67.7	0	0	0	0	66	149	205	106	74	0	0	0	600
	Corpus Christi..........A	64.6	0	0	0	0	120	220	291	174	109	0	0	0	914
	Dallas.................A	55.3	0	0	0	62	321	524	601	440	319	90	6	0	2363
	El Paso................A	52.9	0	0	0	84	414	648	685	445	319	105	0	0	2700
	Fort Worth.............A	55.1	0	0	0	65	324	536	614	448	319	99	0	0	2405
	Galveston..............A	62.2	0	0	0	6	147	276	360	263	189	33	0	0	1274
	Galveston..............C	62.0	0	0	0	0	138	270	350	258	189	30	0	0	1235
	Houston................A	61.0	0	0	0	6	183	307	384	288	192	36	0	0	1396
	Houston................C	62.0	0	0	0	0	165	288	363	258	174	30	0	0	1278
	Laredo................A	66.0	0	0	0	0	105	217	267	134	74	0	0	0	797
	Lubbock...............A	48.8	0	0	18	174	513	744	800	613	484	201	31	0	3578
	Midland...............A	53.8	0	0	0	87	381	592	651	468	322	90	0	0	2591
	Port Arthur............A	60.5	0	0	0	22	207	329	384	274	192	39	0	0	1447
	San Angelo............A	56.0	0	0	0	68	318	536	567	412	288	66	0	0	2255
	San Antonio...........A	60.1	0	0	0	31	204	363	428	286	195	39	0	0	1546
	Victoria...............A	62.7	0	0	0	6	150	270	344	230	152	21	0	0	1173
	Waco.................A	57.2	0	0	0	43	270	456	536	389	270	66	0	0	2030
	Wichita Falls...........A	53.0	0	0	0	99	381	632	698	518	378	120	6	0	2832
Utah	Milford................A	36.5	0	0	99	443	867	1141	1252	988	822	519	279	87	6497
	Salt Lake City..........A	38.4	0	0	81	419	849	1082	1172	910	763	459	233	84	6052
	Wendover.............A	39.1	0	0	48	372	822	1091	1178	902	729	408	177	51	5778
Vt.	Burlington.............A	29.4	28	65	207	539	891	1349	1513	1333	1187	714	353	90	8269
Va.	Cape Henry............C	50.0	0	0	0	112	360	645	694	633	536	246	53	0	3279
	Lynchburg.............A	46.0	0	0	51	223	540	822	849	731	605	267	78	0	4166
	Norfolk................A	49.2	0	0	0	136	408	698	738	655	533	216	37	0	3421
	Richmond..............A	47.3	0	0	36	214	495	784	815	703	546	219	53	0	3865
	Roanoke...............A	46.1	0	0	51	229	549	825	834	722	614	261	65	0	4150
Wash.	Olympia...............A	44.2	68	71	198	422	636	753	834	675	645	450	307	177	5236
	Seattle-Tacoma.........A	44.2	56	62	162	391	633	750	828	678	657	474	295	159	5145
	Seattle................C	46.9	50	47	129	329	543	657	738	599	577	396	242	117	4424
	Spokane...............A	36.5	9	25	168	493	879	1082	1231	980	834	531	288	135	6655
	Walla Walla............C	43.8	0	0	87	310	681	843	986	745	589	342	177	45	4805
	Yakima................A	39.1	0	12	144	450	828	1039	1163	868	713	435	220	69	5941
W. Va.	Charleston.............A	44.8	0	0	63	254	591	865	880	770	648	300	96	9	4476
	Elkins.................A	40.1	9	25	135	400	729	992	1008	896	791	444	198	48	5675
	Huntington............A	45.0	0	0	63	257	585	856	880	764	636	294	99	12	4446
	Parkersburg...........C	43.5	0	0	60	264	606	905	942	826	691	339	115	6	4754
Wisc.	Green Bay.............A	30.3	28	50	174	484	924	1333	1494	1313	1141	654	335	99	8029
	La Crosse.............A	31.5	12	19	153	437	924	1339	1504	1277	1070	540	245	69	7589
	Madison...............A	30.9	25	40	174	474	930	1330	1473	1274	1113	618	310	102	7863
	Milwaukee............A	32.6	43	47	174	471	876	1252	1376	1193	1054	642	372	135	7635
Wyo.	Casper................A	33.4	6	16	192	524	942	1169	1290	1084	1020	657	381	129	7410
	Cheyenne..............A	34.2	28	37	219	543	909	1085	1212	1042	1026	702	428	150	7381
	Lander................A	31.4	6	19	204	555	1020	1299	1417	1145	1017	654	381	153	7870
	Sheridan..............A	32.5	25	31	219	539	948	1200	1355	1154	1051	642	366	150	7680
Alta.	Banff.................C	—	220	295	498	797	1185	1485	1624	1364	1237	855	589	402	10551
	Calgary...............A	—	109	186	402	719	1110	1389	1575	1379	1268	798	477	291	9703
	Edmonton.............A	—	74	180	411	738	1215	1603	1810	1520	1330	765	400	222	10268
	Lethbridge............A	—	56	112	318	611	1011	1277	1497	1291	1159	696	403	213	8644
B. C.	Kamloops.............A	—	22	40	189	546	894	1138	1314	1057	818	462	217	102	6799
	Prince George*.........A	—	236	251	444	747	1110	1420	1612	1319	1122	747	468	279	9755
	Prince Rupert..........C	—	273	248	339	539	708	868	936	808	812	648	493	357	7029
	Vancouver*............A	—	81	87	219	456	657	787	862	723	676	501	310	156	5515
	Victoria*.. ....A	—	136	140	225	462	663	775	840	718	691	504	341	204	5699
	Victoria...... C	—	172	184	243	426	607	723	805	668	660	487	354	250	5579

State or Prov.	Station	Avg. Winter Temp[d]	July	Aug.	Sept.	Oct.	Nov.	Dec.	Jan.	Feb.	Mar.	Apr.	May	June	Yearly Total
Man.	Brandon*...............A	—	47	90	357	747	1290	1792	2034	1737	1476	837	431	198	11036
	Churchill................A	—	360	375	681	1082	1620	2248	2558	2277	2130	1569	1153	675	16728
	The Pas.................C	—	59	127	429	831	1440	1981	2232	1853	1624	969	508	228	12281
	Winnipeg................A	—	38	71	322	683	1251	1757	2008	1719	1465	813	405	147	10679
N. B.	Fredericton*.............A	—	78	68	234	592	915	1392	1541	1379	1172	753	406	141	8671
	Moncton.................C	—	62	105	276	611	891	1342	1482	1336	1194	789	468	171	8727
	St. John.................C	—	109	102	246	527	807	1194	1370	1229	1097	756	490	249	8219
Nfld.	Argentia.................A	—	260	167	294	564	750	1001	1159	1085	1091	879	707	483	8440
	Corner Brook...........C	—	102	133	324	642	873	1194	1358	1283	1212	885	639	333	8978
	Gander..................A	—	121	152	330	670	909	1231	1370	1266	1243	939	657	366	9254
	Goose*..................A	—	130	205	444	843	1227	1745	1947	1689	1494	1074	741	348	11887
	St. John's*..............A	—	186	180	342	651	831	1113	1262	1170	1187	927	710	432	8991
N. W. T.	Aklavik..................C	—	273	459	807	1414	2064	2530	2632	2336	2282	1674	1063	483	18017
	Fort Norman............C	—	164	341	666	1234	1959	2474	2592	2209	2058	1386	732	294	16109
	Resolution Island.......C	—	843	831	900	1113	1311	1724	2021	1850	1817	1488	1181	942	16021
N. S.	Halifax..................C	—	58	51	180	457	710	1074	1213	1122	1030	742	487	237	7361
	Sydney..................A	—	62	71	219	518	765	1113	1262	1206	1150	840	567	276	8049
	Yarmouth...............A	—	102	115	225	471	696	1029	1156	1065	1004	726	493	258	7340
Ont.	Cochrane................C	—	96	180	405	760	1233	1776	1978	1701	1528	963	570	222	11412
	Fort William............A	—	90	133	366	694	1140	1597	1792	1557	1380	876	543	237	10405
	Kapuskasing............C	—	74	171	405	756	1245	1807	2037	1735	1562	978	580	222	11572
	Kitchener...............C	—	16	59	177	505	855	1234	1342	1226	1101	663	322	66	7566
	London..................A	—	12	43	159	477	837	1206	1305	1198	1066	648	332	66	7349
	North Bay..............C	—	37	90	267	608	990	1507	1680	1463	1277	780	400	120	9219
	Ottawa..................C	—	25	81	222	567	936	1469	1624	1441	1231	708	341	90	8735
	Toronto.................C	—	7	18	151	439	760	1111	1233	1119	1013	616	298	62	6827
P. E. I.	Charlottetown..........C	—	40	53	198	518	804	1215	1380	1274	1169	813	496	204	8164
	Summerside.............C	—	47	84	216	546	840	1246	1438	1291	1206	841	518	216	8488
Que.	Arvida...................C	—	102	136	327	682	1074	1659	1879	1619	1407	891	521	231	10528
	Montreal*...............A	—	9	43	165	521	882	1392	1566	1381	1175	684	316	69	8203
	Montreal................C	—	16	28	165	496	864	1355	1510	1328	1138	657	288	54	7899
	Quebec*.................A	—	56	84	273	636	996	1516	1665	1477	1296	819	428	126	9372
	Quebec..................C	—	40	68	243	592	972	1473	1612	1418	1228	780	400	111	8937
Sasks.	Prince Albert...........A	—	81	136	414	797	1368	1872	2108	1763	1559	867	446	219	11630
	Regina..................A	—	78	93	360	741	1284	1711	1965	1687	1473	804	409	201	10806
	Saskatoon...............C	—	56	87	372	750	1302	1758	2006	1689	1463	798	403	186	10870
Y. T.	Dawson.................C	—	164	326	645	1197	1875	2415	2561	2150	1838	1068	570	258	15067
	Mayo Landing..........C	—	208	366	648	1135	1794	2325	2427	1992	1665	1020	580	294	14454

* The data for these normals were from the full ten-year period 1951–1960, adjusted to the standard normal period 1931–1960.

TABLE 6. Resistance (R) of Common Building and Insulating Materials

Material	Description	Density (lb/cu ft)	Thermal Resistance, R^a per inch thickness (1/k)	Thermal Resistance, R^a for thickness listed (1/C)
Building Board				
Boards	Asbestos-cement	120	0.25	
	Asbestos-cement $\frac{1}{8}''$	120		0.03
	Asbestos-cement $\frac{1}{4}''$	120		0.06
	Gypsum or plaster $\frac{3}{8}''$	50		0.32
	Gypsum or plaster $\frac{1}{2}''$	50		0.45
	Plywood	34	1.25	
	Plywood $\frac{1}{4}''$	34		0.31
	Plywood $\frac{3}{8}''$	34		0.47
	Plywood $\frac{1}{2}''$	34		0.62
	Plywood or wood panel $\frac{3}{4}''$			0.93

TABLE 6. (*Continued*)

Material	Description	Density (lb/cu ft)	Thermal Resistance, R^a per inch thickness (1/k)	for thickness listed (1/C)
Insulating board				
	Sheathing, reg. density $\frac{1}{2}''$	18		1.32
	Sheathing, intermediate density $\frac{1}{2}''$	22		1.22
	Nail-base sheathing $\frac{1}{2}''$	25		1.14
	Shingle backer $\frac{3}{8}''$	18		0.94
	Shingle backer $\frac{5}{16}''$	18		0.78
	Sound deadening board $\frac{1}{2}''$	15		1.35
	Tile and lay-in panel, plain or acoustic	18	2.50	
	$\frac{1}{2}''$	18		1.25
	$\frac{3}{4}''$	18		1.89
	Laminated paper	30	2.00	
	Homogeneous board from repulped paper	30	2.00	
Hardboard	Medium density siding $\frac{7}{16}''$	40		0.67
	Other medium density	50	1.37	
	High density, service temp. service, underlay	55	1.22	
	High density, std. tempered	63	1.00	
Board and slabs	Cellular glass	9	2.50	
	Glass fiber, organic bonded	4–9	4.00	
	Expanded rubber (rigid)	4.5	4.55	
	Expanded polystyrene, extruded plain	1.8	4.00	
	Expanded polystyrene, extruded (R-12 exp.)	2.2	5.00	
	Expanded polystyrene, extruded (R-12 exp.) (thickness 1'' and greater)	3.5	5.26	
	Expanded polystyrene, molded beads	1.0	3.57	
	Expanded polyurethane (R-11 exp.)	1.5	6.25	
	(1'' or greater thickness)	2.5		
	Mineral fiber with resin binder	15	3.45	
	Mineral fiberboard, wet felted core or roof insulation	16–17	2.94	
	acoustical tile	18	2.86	
	acoustical tile	21	2.70	
	Mineral fiberboard, wet molded, acoustical tile	23	2.38	
	Wood or cane fiberboard acoustical tile $\frac{1}{2}''$			1.25
	acoustical tile $\frac{3}{4}''$			1.89
	Interior finish (plank, tile)	15	2.86	
	Insulating roof deck approx. 1-$\frac{1}{2}''$			4.17
	approx. 2''			5.56
	approx. 3''			8.33

TABLE 6. (*Continued*)

Material	Description	Density (lb/cu ft)	Thermal Resistance, R^a per inch thickness (1/k)	for thickness listed (1/C)
Board and slabs	Wood shredded (cemented in preformed slabs)	22	1.67	
Particleboard	Low density	37	1.85	
	Medium density	50	1.06	
	High density	62.5	0.85	
	Underlayment $\frac{5}{8}''$	40		0.82
	Wood subfloor $\frac{3}{4}''$			0.94
Building Paper	Vapor permeable felt			0.06
	Vapor seal 2 layers of mopped 15 lb felt			0.12
	Vapor seal, plastic film			Negl.
Finish Flooring Materials	Carpet and fibrous pad			2.08
	Carpet and rubber pad			1.23
	Cork tile $\frac{1}{8}''$			0.28
	Terrazzo $1''$			0.08
	Tile-asphalt, linoleum vinyl, rubber			0.05
	Wood, hardwood finish $\frac{3}{4}''$			0.68
Insulating Materials				
Blanket and batt	Mineral fiber, fibrous form processed from rock, slag, or glass			
	approx. $2-2\frac{3}{4}''$			7
	approx. $3-3\frac{1}{2}''$			11
	approx. $5\frac{1}{4}-6\frac{1}{2}$			19
Loose fill	Cellulose insulation (milled paper or wood pulp)	2.5–3	3.70	
	Sandust or shavings	0.8–1.5	2.22	
	Wood fiber, softwoods	2.0–3.5	3.33	
	Perlite, expanded	5.0–8.0	2.70	
	Mineral fiber (rock, slag or glass)			
	approx. $3''$			9
	approx. $4\frac{1}{2}''$			13
	approx. $6\frac{1}{4}''$			19
	approx. $7\frac{1}{4}''$			24
	Silica aerogel	7.6	5.88	
	Vermiculite (expanded)	7.0–8.2	2.13	
		4.0–6.0	2.27	
Roof insulation	Preformed, for use above deck			
	approx. $\frac{1}{2}''$			1.39
	approx. $1''$			2.78
	approx. $1\frac{1}{2}''$			4.17
	approx. $2''$			5.56
	approx. $2\frac{1}{2}''$			6.67
	approx. $3''$			8.33
	Cellular glass	9	2.50	

TABLE 6. (*Continued*)

| | | | Thermal Resistance, R^a | |
Material	Description	Density (lb/cu ft)	per inch thickness (1/k)	for thickness listed (1/C)
Masonry Materials				
Concretes	Cement mortar	116	0.20	
	Gypsum-fiber concrete			
	$87\frac{1}{2}$% gypsum,			
	$12\frac{1}{2}$% wood chips	51	0.60	
	Light weight aggregates including			
	expanded slags, cinders	120	0.19	
	pumice; vermiculite; also	100	0.28	
	cellular concretes	80	0.40	
		60	0.59	
		40	0.86	
		30	1.11	
		20	1.43	
	Sand and gravel or stone aggregate			
	(oven dried)	140	0.11	
	Sand and gravel or stone aggregate			
	(not dried)	140	0.08	
	Stucco	116	0.20	
Masonry Units	Brick, common	120	0.20	
	Brick, face	130	0.11	
	Clay tile, hollow:			
	1 cell deep, 3″			0.80
	2 cell deep, 4″			1.11
	2 cell deep, 6″			1.52
	2 cells deep, 8″			1.85
	2 cells deep, 10″			2.22
	2 cells deep, 12″			2.50
	Concrete blocks, 3 oval cores;			
	sand and gravel aggregate			
	4″			0.71
	8″			1.11
	10″			1.20
	12″			1.28
	Cinder aggregate			
	3″			0.86
	4″			1.11
	8″			1.72
	12″			1.89
	Lightweight			
	aggregate (expanded shale, 3″			1.27
	clay slate or slag; pumice) 4″			1.50
	8″			2.00
	12″			2.27
	Gypsum partition tile:			
	3 × 12 × 30″ solid			1.26
	3 × 12 × 30″ 4-cell			1.35
	4 × 12 × 30″ 3-cell			1.67
Plastering Materials	Cement plaster,	116	0.20	
	Sand aggregate $\frac{3}{8}$″			0.08
	Sand aggregate $\frac{3}{4}$			0.15

TABLE 6. (*Continued*)

Material	Description	Density (lb/cu ft)	Thermal Resistance, R^a per inch thickness (1/k)	for thickness listed (1/C)
	Gypsum plaster:			
	lightweight aggregate $\frac{1}{2}''$	45		0.32
	lightweight aggregate $\frac{5}{8}''$	45		0.39
	lightweight aggregate on metal lath $\frac{3}{4}''$			0.47
	Perlite aggregate	45	0.67	
	Sand aggregate	105	0.18	
	Sand aggregate $\frac{1}{2}''$	105		0.09
	Sand aggregate $\frac{5}{8}''$	105		0.11
	Sand aggregate on metal lath $\frac{3}{4}''$			0.1
	Vermiculite aggregate	45	0.59	
Roofing	Asbestos-cement shingles	120		0.21
	Asphalt roofing	70		0.15
	Asphalt shingles	70		0.44
	Built-up roofing $\frac{3}{8}''$	70		0.33
	Slate $\frac{1}{2}''$			0.05
	Wood shingles, plain plastic film faced			0.94
Siding Materials (on Flat Surface)	Shingles			
	Asbestos-cement	120		0.21
	Wood 16", $7\frac{1}{2}''$ exposure			0.87
	Wood double 16", 12" exposure			1.19
	Wood plus insul. backer board $\frac{5}{16}''$			1.40
	Siding			
	Asbestos-cement $\frac{1}{4}''$, lapped			1.21
	Asphalt roll siding			0.15
Siding Materials (on Flat Surface)	Asphalt insulating siding ($\frac{1}{2}''$ bd.)			1.46
	Wood, drop $1'' \times 8''$			0.79
	Wood, bevel $\frac{1}{2}'' \times 8''$, lapped			0.81
	Wood, bevel $\frac{3}{4}'' \times 10''$, lapped			1.05
	Wood, plywood, $\frac{3}{8}''$, lapped			0.59
	Aluminum or steel, over sheathing hollow-backed			0.61
	Insulating-board, backed nominal $\frac{3}{8}''$			1.82
	Insulating-board, backed nominal $\frac{3}{8}''$ foil backed			2.96
	Architectural Glass			0.10
Wood	Maple, oak, and similar hardwoods	45	0.91	
	Fir, pine, and similar softwoods	32	1.25	
	Fir, pine, and similar softwoods			
	$\frac{3}{4}''$	32		0.94
	$1\frac{1}{2}''$	32		1.89
	$2\frac{1}{2}''$	32		3.12
	$3\frac{1}{2}''$	32		4.35

[a]All resistance factors are based on a 75°F mean temperature.

TABLE 7. Coefficients of Transmission $(U)^a$ of Wall and Roof Assemblies.

"U" VALUES

Example 1

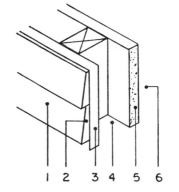

```
Construction              Resistance    (R)
1.Outside surface (15 mph wind)......... 0.17
2.Siding,wood,1/2" x 8" lapped (avg) ... 0.81
3.Sheathing, 1/2" asphalt impregnated.... 1.32
4.Air space ........................... 0.97
5.Gypsum wallboard (1/2").............. 0.45
6.Inside surface (still air)............... 0.68
                                        _____
   Total resistance..................... 4.40
   U = 1/R = 1/4.40 =                    0.23
   Adjustment for framing ( 2x 4 in.@ 16 in.
   o.c.) = U x 0.96 - 0.23 x 0.96 = 0.22
```

Example 2

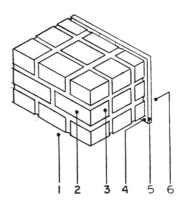

```
Construction              Resistance    (R)
1.Outside surface (15 mph wind)......... 0.17
2.Face brick (4") ...................... 0.44
3.Common brick (4").................... 0.80
4.Air space............................ 0.97
5.Gypsum wallboard ( 1/2").............. 0.45
6.Inside surface (still air)............... 0.68
                                        _____
   Total resistance................... 3.51
   U = 1/R = 1/3.51 = ................ 0.29
   Adjustment for furring (1 x 2" @ 16" o.c.)
   = U x 1.00 = 0.29 x 1.00 = 0.29
```

Example 3

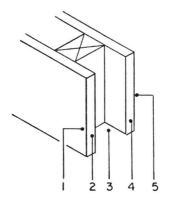

```
Construction              Resistance    (R)
1.Surface (still air)...................... 0.68
2.Gypsum bd. (1/2")./.................. 0.45
3.Air space............................ 0.97
4.Gypsum wall board (1/2") ............ 0.45
5.Surface (still air)...................... 0.68
                                        _____
   Total resistance ................... 3.23
   U = 1/R = 1/3.23.................. 0.31
   Adjustment for framing (2 x 4 " @ 16"o.c.)
   = U x 0.95 = 0.31 x 0.95 = 0.29
```

TABLE 7. (*Continued*)

Example 4

Construction	Resistance	(R)
1. Outside surface (15 mph wind)	....	0.17
2. Face brick (4")	....	0.44
3. Cement mortar (1/2")	....	0.10
4. Concrete block (cinder agg.)(8")	......	1.72
5. Air space (reflective)	....	2.80
6. Gypsum wallboard, Foil back (1/2")	...	0.45
7. Inside surface (still air)	....	0.68

Total resistance 6.36
U = 1/R = 1/6.36 = ... 0.16
Adjustment for furring (1 x 2" @ 16" o.c.)
= U x 1.04 = 0.16 x 1.04 = 0.17

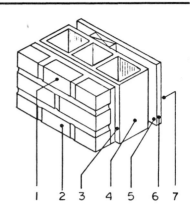

1 2 3 4 5 6 7

Example 5

Construction	Resistance	(R)
1. Outside surface (15 mph wind)	....	0.17
2. Common brick (4")	....	0.80
3. Air space	....	0.97
4. Concrete block (stone agg.)(4")	....	0.71
5. Air space	....	0.97
6. Gypsum wallboard (1/2")	....	0.45
7. Inside surface (still air)	....	0.68

Total resistance   4.75
U = 1/R = 1/4.75 =   0.21
Adjustment for surring (1 x 2" @ 16" o.c.)
= U x 1.00 = 0.21 x 1.00 = 0.21

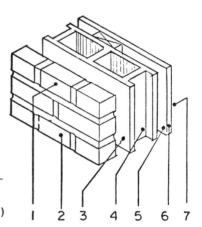

1 2 3 4 5 6 7

Example 6

Construction	Resistance	(R)
1. Inside surface (still air)	....	0.68
2. Plas. (lt.wt.agg.) 5/8"	....	0.39
3. Cement block (cinder agg.) (4")	....	1.11
4. Plas. (lt.wt.agg.)5/8"	....	0.39
5. Inside surface (still air)	....	0.68

Total resistance 3.25
U = 1/R = 1/3.25 .... 0.31

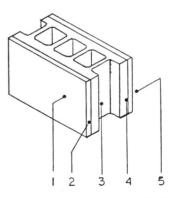

1 2 3 4 5

TABLE 7. (*Continued*)

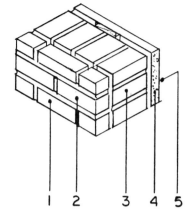

Example 7

Construction	Resistance	(R)
1. Outside Air Film		0.25
2. 4" Face Brick		0.44
3. 8" Common Brick		1.60
4. 5/8" Plaster		0.11
5. Inside Air Film		0.68
	Total resistance	3.08
	U = 1/R = 1/3.08	0.33

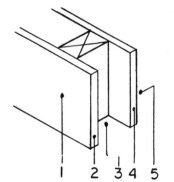

Example 8

Construction	Resistance	(R)
1. Inside Air Film		0.68
2. Metal Lath & Plaster		0.10
3. Air Space		0.91
4. Metal Lath & Plaster		0.10
5. Inside Air Film		0.68
	Total resistance	2.47
	U = 1/R = 1/2.47	0.40

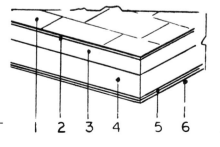

Example 9

Construction	Resistance	(R)
1. Outside Air Film		0.25
2. Built up Roof		0.33
3. 2" Rigid Insulation		5.26
4. 2" Gypsum Conc.		1.20
5. 1/4" Asbestos cement Brd.		.06
6. Inside Air Film		0.92
	Total resistance	8.02
	U = 1/R = 1/8.02	0.12

TABLE 7. *(Continued)*

Example 10

```
Construction (heat flow up)      Resistance (R)
1. Top surface (still air)................ 0.61
2. Linoleum or tile (avg.R).............. 0.05
3. Felt.................................. 0.06
4. Plywood (5/8")....................... 0.78
5. Wood subfloor (3/4")................. 0.94
6. Air space............................ 0.85
7. Metal lath and 3/4"plas. (lt.wt.agg.)... 0.47
8. Bottom surface (still air)............ 0.61
                                         _____
            Total resistance............. 4.37
            U = 1/R = 1/4.37............. 0.23
            Adjustment for framing (2 × 8" @
            16"o.c.)
            = U × 0.94 = 0.23 × 0.94 = 0.22
```

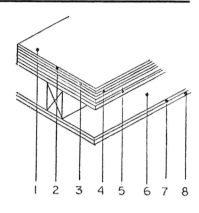

```
1  2  3  4  5  6  7  8
```

Example 11

```
Construction (heat flow up)      Resistance (R)
1. Outside surface (15 mph wind)......... 0.17
2. Built-up roofing 3/8"................. 0.33
3. Roof insulation (none)................  –
4. Concrete slab (lt.wt.agg.) (2")....... 2.22
5. Corrugated metal...................... 0
6. Air space............................ 0.85
7. Metal lath and 3/4" plas.(lt.wt.agg.)... 0.47
8. Inside surface (still air)............ 0.61
                                         _____
            Total resistance ........... 4.65
            U = 1/R = 1/4.65 .......... 0.22
```

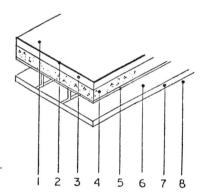

```
1  2  3  4  5  6  7  8
```

Example 12

```
Construction          Resistance      (R)
(heat flow up)
1. Outside surface (15 mph wind)......... 0.17
2. Built-up roofing (3/8")............... 0.33
3. Roof insulation (C = 0.72) ........... 1.39
4. Plywood deck (5/8").................. 0.78
5. Air space ........................... 0.85
6. Gypsum wallboard (1/2").............. 0.45
7. Acoustical tile (1/2")-glued.......... 1.25
8. Inside surface (still air)............ 0.61
                                         _____
            Total resistance ........... 5.83
            U = 1/R = 1/5.83........... 0.17
```

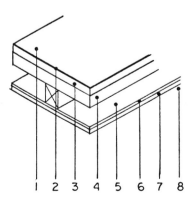

```
1  2  3  4  5  6  7  8
```

TABLE 7. (*Continued*)

Example 13

Construction	Resistance	(R)
1. Outside surface (15 mph wind)		0.17
2. Built-up roofing (3/8")		0.33
3. Roof insulation (C = 0.24)		4.17
4. Metal deck		0.00
5. Air space		0.99
6. Metal lath and		
7. 3/4" plas. (sand agg.)		0.13
8. Inside surface (still air)		0.92
	Total resistance	6.71
	U = 1/R = 1/6.71	0.15

Example 14

Construction	Resistance	(R)
(heat flow up)		
1. Outside surface (15 mph wind)		0.17
2. Asphalt shingle roofing		0.44
3. Building paper		0.06
4. Plywood deck (5/8")		0.78
5. Air space (3.5", reflective surface)		2.06
6. Gypsum wallboard (1/2")		0.45
7. Inside surface (still air)		0.62
	Total resistance	4.58
	U = 1/R = 1/4.58	0.22

[a] U is expressed in Btu per (hr) (sq ft) (ΔF between air on two sides of wall)

TABLE 7a. Coefficient of Transmission (U) of Wall Assemblies by Groups [Btu/(hr) (sq ft) (°F)]

Group*	Components	Wt. lb/sq.ft.	U Value
A	1"stucco+4"l.w.concrete block+air space	28.6	0.267
	1"stucco+air space+2" insulation	16.3	0.106
B	1"stucco+4'common brick	55.9	0.393
	1"stucco+4"h.w. concrete	62.5	0.481
C	4"face brick+4"l.w.concrete block+1"insulation	62.5	0.158
	1"stucco+4"h.w.concrete+2"insulation	62.9	0.114
D	1"stucco+8"l.w.concrete block+1"insulation	41.4	0.141
	1"stucco+2"insulation+4"h.w.concrete block	36.6	0.111
E	4"face brick+4"l.w. concrete block	62.2	0.333
	1"stucco+8"h.w.concrete block	56.6	0.349

TABLE 7a. (*Continued*)

Group*	Components	Wt. lb/sq. ft.	U Value
F	4"face brick+4"common brick	80.5	0.360
	4"face brick+2"insulation+4"l.w.concrete block	62.5	0.103
G	1"stucco+8"clay tile+1"insulation	62.8	0.141
	1"stucco+2"insulation+4"common brick	56.2	0.108
H	4"face brick+8"clay tile+1"insulation	96.4	0.137
	4"face brick+8"common brick	129.6	0.280
	1"stucco+12"h.w.concrete	155.9	0.365
	4"face brick+2"insulation+4"common brick	89.8	0.106
	4"face brick+2"insulation+4"h.w. concrete	96.5	0.111
	4"face brick+2"insulation+8"h.w. concrete bl.	90.6	0.102
I	1"stucco+8" clay tile+air space	62.6	0.209
	4"face brick+air space+4"h.w. concrete bl.	69.9	0.282
J	face brick+8" common brick+1"insulation	129.8	0.145
	4"face brick+2"insulation+8"clay tile	96.5	0.094
	1"stucco+2"insulation+8"common brick	96.3	0.100
K	4"face brick+air space+8"clay tile	96.2	0.200
	4"face brick+2"insulation+8"common brick	129.9	0.098
	4"face brick+2"insulation+8"h.w.concrete	143.3	0.107
L	4"face brick+8"clay tile+air space	96.2	0.200
	4"face brick+air space+4"common brick	89.5	0.265
	4"face brick+air space+4" h.w.concrete	96.2	0.301
	4"face brick+air space+8"h.w. concrete bl.	90.2	0.246
	1"stucco+2"insulation+12" h.w.concrete	156.3	0.106
M	4"face brick+air space+8"common brick	129.6	0.218
	4"face brick+air space+12"h.w.concrete	189.5	0.251
	4"face brick+2"insulation+12"h.w.concrete	189.9	0.104

In addition to the structure components listed above all walls had an:
a) Outside Air Resistance (fi) R = 0.333
b) Inside Air Resistance (fo) R = 0.685

*Group designation A through M is used in Table 8 to determine total equivalent temperature differential.

TABLE 8. Total Equivalent Temperature Differential for Wall Assemblies[a]

	SUN TIME												
	A.M.						P.M.						
	8		10		12		2		4		6		8	
	EXTERIOR COLOR OF WALL — D=Dark, L=Light												
Direction Wall Faces	D	L	D	L	D	L	D	L	D	L	D	L	D	L
GROUP A[b]													
NE	14	8	23	13	27	17	26	17	25	18	23	17	21	16
E	16	9	30	17	37	22	33	21	30	20	27	20	24	19
SE	13	8	27	15	35	21	35	21	31	21	28	19	24	17
S	7	5	14	9	23	15	30	19	31	21	28	19	23	17
SW	9	6	15	9	22	14	29	19	37	23	39	26	32	22
W	9	6	15	9	22	14	27	18	36	24	43	28	38	25
NW	7	5	13	8	19	13	21	16	27	19	34	23	31	21
N	8	5	11	7	16	11	18	13	20	15	20	15	19	15

TABLE 8. (*Continued*)

Direction Wall Faces	SUN TIME												
	A.M.						P.M.						
	8		10		12		2		4		6		8	
	EXTERIOR COLOR OF WALL – D=Dark, L=Light												
	D	L	D	L	D	L	D	L	D	L	D	L	D	L
GROUP B													
NE	6	4	16	10	27	15	30	18	31	20	30	21	29	21
E	8	4	23	12	37	20	42	24	41	25	37	25	33	23
SE	6	3	16	8	31	16	40	23	41	25	39	25	34	24
S	3	2	5	3	13	8	24	15	35	22	39	26	36	26
SW	3	2	5	3	9	6	17	11	21	20	46	29	50	32
W	3	2	5	3	9	6	14	10	27	18	43	28	54	34
NW	3	2	4	3	10	6	14	10	21	15	33	23	42	28
N	3	2	7	5	11	7	15	10	19	14	23	18	26	20
GROUP C													
NE	5	3	13	7	21	12	26	15	28	17	29	19	29	20
E	6	4	15	8	28	15	30	20	39	22	38	24	35	23
SE	4	3	11	7	23	13	32	18	36	22	38	24	36	23
S	4	3	6	4	10	6	17	11	26	16	33	22	35	24
SW	5	3	6	4	9	6	13	8	22	14	34	21	43	27
W	5	4	7	4	10	6	12	8	19	13	31	20	40	28
NW	4	3	6	4	9	6	11	8	16	12	24	17	34	23
N	4	3	6	4	9	6	12	8	16	10	20	14	23	18
GROUP D													
NE	4	3	13	7	21	12	27	15	30	18	30	21	29	21
E	5	3	15	8	29	16	38	21	40	23	39	24	35	23
SE	4	3	11	6	22	12	33	19	39	23	39	25	36	24
S	3	2	4	3	9	6	17	11	28	17	34	22	36	24
SW	4	3	6	3	8	5	13	8	23	15	36	23	46	29
W	4	3	6	4	8	6	12	8	20	13	33	21	47	29
NW	4	3	5	3	8	5	11	7	16	11	25	17	36	24
N	3	2	6	4	9	6	12	8	16	12	20	15	23	18
GROUP E													
NE	5	3	15	8	25	13	29	16	30	19	30	20	29	21
E	6	3	18	10	33	18	40	22	41	24	38	24	34	23
SE	4	3	13	7	27	15	36	21	40	24	40	25	36	24
S	2	2	4	3	11	7	21	13	31	20	37	24	36	24
SW	3	2	5	3	9	5	15	9	27	17	40	25	48	32
W	4	3	5	3	9	5	13	9	23	17	38	25	50	31
NW	3	2	5	4	8	6	12	8	19	13	28	20	38	26
N	3	2	6	4	10	6	13	9	18	13	22	16	25	19

TABLE 8. (*Continued*)

Direction Wall Faces	8 D	8 L	10 D	10 L	12 D	12 L	2 D	2 L	4 D	4 L	6 D	6 L	8 D	8 L
	A.M.						P.M.							
GROUP F														
NE	5	4	10	6	17	10	22	13	25	16	28	18	28	19
E	5	4	12	7	22	13	30	16	35	20	37	22	37	23
SE	5	4	9	6	18	10	26	14	32	18	36	21	36	23
S	5	4	6	4	9	6	14	8	21	13	28	17	32	20
SW	6	5	8	5	10	7	12	7	18	13	27	17	36	23
W	7	5	9	6	11	7	11	7	16	10	25	16	36	23
NW	6	4	7	5	10	6	10	7	14	9	20	13	28	19
N	4	4	6	5	9	6	11	7	13	9	17	12	20	15
GROUP G														
NE	6	5	8	10	16	12	21	14	24	15	26	16	27	18
E	7	5	12	8	22	13	28	16	32	19	34	20	33	22
SE	7	5	10	7	18	11	24	14	30	17	33	20	33	21
S	6	5	9	6	12	8	14	9	19	13	25	16	29	19
SW	8	6	10	7	14	9	14	9	17	11	24	16	32	21
W	9	6	12	8	15	10	15	9	16	11	23	15	32	20
NW	7	5	9	6	12	8	13	9	15	10	18	13	25	17
N	5	4	8	6	10	7	11	8	13	9	16	11	19	14
GROUP H														
NE	8	6	12	8	17	12	19	13	21	14	23	15	24	16
E	9	7	14	9	20	13	24	15	27	16	29	18	31	19
SE	9	7	13	9	18	12	21	13	25	15	27	17	29	18
S	8	6	11	8	14	10	15	10	17	11	21	13	23	16
SW	11	7	15	9	18	12	17	11	17	12	21	14	26	16
W	12	8	16	10	19	13	18	12	18	12	20	14	25	16
NW	10	7	13	9	16	11	15	10	15	11	17	12	20	14
N	7	5	9	7	12	9	12	9	13	10	14	11	16	12
GROUP I														
NE	8	6	13	9	19	13	21	14	22	15	24	16	24	16
E	10	7	16	11	23	15	27	16	29	18	30	19	29	19
SE	10	7	14	10	21	14	24	15	26	17	28	18	29	19
S	8	6	12	9	16	11	17	12	19	13	22	15	24	16
SW	10	7	15	10	19	13	19	13	21	14	24	16	28	18
W	11	7	16	10	20	13	20	13	21	14	24	16	28	18
NW	9	6	13	9	16	11	17	11	18	12	20	13	23	15
N	7	5	9	7	13	9	13	10	14	11	15	12	17	13

EXTERIOR COLOR OF WALL — D=Dark, L=Light

TABLE 8. (Continued)

Direction Wall Faces	8 D	8 L	10 D	10 L	12 D	12 L	2 D	2 L	4 D	4 L	6 D	6 L	8 D	8 L
	A.M.						P.M.							
GROUP J														
NE	9	7	13	9	18	12	19	13	20	13	21	14	23	15
E	11	8	16	11	21	14	23	15	25	16	27	17	28	18
SE	11	8	15	11	20	14	21	14	23	15	25	16	27	17
S	10	7	13	10	17	12	16	11	17	12	19	13	21	14
SW	13	8	17	12	21	14	20	13	19	13	20	14	22	15
W	14	9	19	12	23	15	21	14	20	13	20	14	23	15
NW	11	8	15	10	19	13	18	12	17	11	17	12	18	14
N	8	6	11	9	14	11	13	10	13	10	14	10	15	11
GROUP K														
NE	10	7	14	10	19	13	20	13	20	14	21	14	22	15
E	12	8	17	12	22	13	24	16	25	16	26	17	27	17
SE	12	8	17	11	22	15	22	15	23	15	24	16	26	17
S	10	7	15	10	19	13	18	12	18	13	19	13	20	14
SW	13	8	18	12	23	15	23	14	21	14	22	15	23	16
W	14	9	19	12	24	16	23	15	22	15	23	15	24	16
NW	11	6	15	8	20	11	19	10	18	10	19	11	20	11
N	8	6	11	8	14	11	14	10	14	10	14	11	15	11
GROUP L														
NE	9	7	14	10	19	13	20	13	21	14	22	15	23	15
E	11	8	17	11	23	15	24	15	26	18	27	18	28	18
SE	11	7	16	9	22	12	23	12	24	13	26	14	27	15
S	10	7	13	9	17	12	18	12	19	13	20	14	22	15
SW	12	8	17	11	22	14	21	14	21	14	22	15	24	16
W	13	8	18	12	23	15	22	14	22	14	23	15	25	16
NW	10	7	15	10	19	11	18	12	18	12	19	13	20	14
N	7	6	11	8	14	10	13	10	14	10	15	11	16	12
GROUP M														
NE	10	7	15	11	20	14	20	13	20	14	20	14	21	14
E	13	10	19	12	24	16	24	16	25	16	24	16	25	17
SE	13	8	18	12	23	15	23	15	23	15	24	16	23	16
S	11	7	15	11	20	12	19	13	18	13	19	13	20	14
SW	13	9	19	12	25	16	24	16	23	15	22	15	23	15
W	14	9	20	13	26	17	25	16	25	16	23	15	24	15
NW	11	8	16	11	21	14	20	14	20	13	19	14	19	13
N	8	6	11	9	15	11	14	11	14	11	14	11	15	11

[a]For calculating heat gain.
[b]See Table 7a for group designations.

TABLE 8a. Total Equivalent Temperature Differential for Roofs*

Description of Roof Construction	Wt, lb Per Sq. Ft.	SUN TIME						
		A.M.			P.M.			
		8	10	12	2	4	6	8
		Dark	Dark	Dark	Dark	Dark	Dark	Dark
Light Construction Roofs — Exposed to Sun								
1" Insulation + Steel Siding	7.4	15	43	73	89	84	61	29
2" Insulation + Steel Siding	7.8	12	40	70	88	86	65	32
1" Insulation + 1" Wood	8.4	6	29	58	80	86	73	45
2" Insulation + 2" Wood	8.5	4	25	53	77	86	75	49
1" Insulation + 2.5" Wood	12.7	1	14	34	56	71	73	60
2" Insulation + 2.5" Wood	13.1	1	11	30	51	67	72	62
Medium Construction Roofs — Exposed to Sun								
1" Insulation + 4" Wood	17.3	3	9	22	38	53	62	59
2" Insulation + 4" Wood	17.8	3	9	21	35	49	59	58
1" Insulation + 2" h.w. Concrete	28.3	2	16	39	61	75	73	58
2" Insulation + 2" h.w. Concrete	28.8	1	14	34	57	72	73	60
4" l.w. Concrete	17.8	1	16	41	67	81	79	59
6" l.w. Concrete	24.5	0	5	20	41	61	72	68
8" l.w. Concrete	31.2	3	4	12	23	39	53	60
Heavy Construction Roofs — Exposed to Sun								
1" Insulation + 4" h.w. Concrete	51.6	4	11	24	40	53	60	57
2" Insulation + 4" h.w. Concrete	52.1	4	10	22	37	49	58	57
1" Insulation + 6" h.w. Concrete	75.0	7	12	22	32	42	49	51
2" Insulation + 6" h.w. Concrete	75.4	8	13	21	30	40	47	50

*For calculating heat gain.

TABLE 9. Estimated Heat Loss from Building by Infiltration

Room or Building Type	No. of walls w/windows	Temp. difference (°F)			
		25	50	75	100
Offices, apartments, hotels, multistory in general	None	0.23	0.45	0.68	0.90
	1	0.34	0.68	1.02	1.36
	2	0.68	1.35	2.02	2.70
	3 or 4	0.90	1.80	2.70	3.60
Entrance halls or vestibules	Any	1.35	2.70	4.05	5.40
Industrial buildings	Any	0.90–1.35	1.80–2.70	2.70–4.05	3.60–5.40
Houses, all types all rooms except vestibules	Any	0.45–0.68	0.90–1.35	1.35–2.02	1.80–2.70
Public or institutional buildings	Any	0.68–1.35	1.35–2.70	2.03–4.05	2.70–5.40

Note: The tabulated factors when multiplied by room or building volume (cu ft) will result in estimated heat loss (Btu/hr) due to infiltration and does not include the heat needed to warm ventilating air.

TABLE 10. Coefficients of Transmission (*U*) for Doors [Btu/(hr) (sq ft) (°F)]

Thickness[a]	Winter			Summer
	Solid Wood, No Storm Door	Storm Door[b] Wood	Metal	No Storm Door
1″	0.64	0.30	0.39	0.61
$1\frac{1}{4}$″	0.55	0.28	0.34	0.53
$1\frac{1}{2}$″	0.49	0.27	0.33	0.47
2″	0.43	0.24	0.29	0.42
	Steel Door			
$1\frac{3}{4}$″				
Mineral fiber core (2 lb/cu ft)	0.59			0.58
Solid urethane foam core	0.40			0.39
Solid polystyrene core	0.47			0.46

[a]Nominal thickness.
[b]Values for wood storm doors are for approximately 50% glass; for metal storm doors values apply for any percent of glass.

TABLE 11. Coefficients of Transmission (*U*) for Windows, Glass Block, and Plastic Sheet [(Btu/(hr) (sq ft) (°F))]

DESCRIPTION	Exterior Glass			Interior Glass
	Winter	Summer		
		No Indoor Shading	Indoor Shading	
Flat Glass				
single glass	1.13	1.06	0.81	0.73
insulating glass-double				
3/16 in. air space	0.69	0.64	0.54	0.51
1/4 in. air space	0.65	0.61	0.52	0.49
1/2 in. air space	0.58	0.56	0.48	0.46
insulating glass-triple				
1/4 in. air spaces	0.47	0.45		0.38
1/2 in. air spaces	0.36	0.35		0.30
storm windows				
1 to 4 in. air space	0.56	0.54	0.47	0.44
Glass Block				
6 x 6 x 4 in. thick	0.60	0.57		0.46
8 x 8 x 4 in. thick	0.56	0.54		0.44
—with cavity divider	0.48	0.46		0.38
12 x 12 x 4 in. thick	0.52	0.50		0.41
—with cavity divider	0.44	0.42		0.36
12 x 12 x 2 in. thick	0.60	0.57		0.46
Single Plastic Sheet	1.09	1.00		0.70

TABLE 12. Coefficients of Transmission (U) for Horizontal Skylights [(Btu/(hr) (sq ft) ($^\circ$F))]

DESCRIPTION	Exterior Glass		Interior Glass
	Winter	Summer	
Flat Glass			
single glass	1.22	0.83	0.96
insulating glass-double			
3/16 in. air space	0.75	0.49	0.62
1/4 in. air space	0.70	0.46	0.59
1/2 in. air space	0.66	0.44	0.56
Glass Block			
11 x 11 x 3 in. thick			
—with cavity divider	0.53	0.35	0.44
12 x 12 x 4 in. thick			
—with cavity divider	0.51	0.34	0.42
Plastic Bubbles			
single walled	1.15	0.80	—
double walled	0.70	0.46	—

TABLE 13. Correction Factor for Glass Windows[a]

Window Description	Single Glass	Double or Triple Glass	Storm Windows
All Glass	1.00	1.00	1.00
Wood Sash—80% Glass	0.90	0.95	0.90
Wood Sash—60% Glass	0.80	0.85	0.80
Metal Sash—80% Glass	1.00	1.20	1.20

[a]Multiply U-Factor found in Tables 11 and 12 by these values.

TABLE 14. Heat Loss—Below Grade (Btu/sq ft)

OUTDOOR TEMP.	BASEMENT FLOOR LOSS ① BTU/SQ. FT.	BELOW GRADE WALL LOSS ② BTU/SQ. FT
20°	1.0	2.0
0°	2.0	4.0
−20°	3.0	6.0

① Based on basement temperature of 70°F and a "U" Value of 0.10.
② Assumed twice basement floor loss.

TABLE 15. Heat Loss—Unheated Slab on Grade
(Btu/hr)

OUTDOOR DESIGN TEMPERATURE, °F.	TOTAL WIDTH OF INSULATION, IN.	HEAT LOSS PER FOOT OF EXPOSED SLAB EDGE BTUH		
		R = 5.0	R = 3.33	R = 2.50
−30 and colder	24	46	69	92
−25 to −29	24	44	66	88
−20 to −24	24	41	61	82
−15 to −19	24	39	59	78
−10 to −14	24	37	55	74
−5 to −9	24	35	52	70
0 to −4	24	32	48	64
+5 to +1	24	30	45	60
+10 to +6	18	25	38	50
+15 to +11	12	25	38	50
+20 to +16	Edge only	25	38	50

R = Thermal resistance of insulation.

TABLE 16. Heat Loss—Heated Slab on Grade
(Btu/hr)

OUTDOOR DESIGN TEMPERATURE, °F.	TOTAL WIDTH OR DEPTH OF INSULATION, IN.	HEAT LOSS PER FOOT OF EXPOSED SLAB EDGE, BTUH			
		R = 5.0	R = 3.75	R = 2.50	NONE
−30 and colder	24	34	51	67	90
−25 to −29	24	32	48	64	85
−20 to −24	24	30	45	60	80
−15 to −19	24	28	43	57	75
−10 to −14	24	27	40	54	70
− 5 to − 9	24	25	38	51	65
0 to − 4	24	24	36	48	60
+ 5 to + 1	24	22	33	44	55
+10 to + 6	18	21	31	42	50
+15 to +11	12	21	31	42	45
+20 to +16	Edge only	21	31	42	45

TABLE 17. Recommended Infiltration Values
through Outside Doors
(cfm per door opening per hour)

Wind Velocity (mph)	Swinging Door 3 × 7 ft (cfm)	Revolving Door (cfm)
15 (Avg. winter design)	10–15	5–7
7.5 (Avg. summer design)	5–7.5	5–7

TABLE 18. Recommended Infiltration of Building Spaces
(in air changes per hour)

Kind of Room or Building	Summer		Winter	
	Regular	Weather Stripping or Storm Sash	Regular	Weather Stripping or Storm Sash
Rooms with no windows or outside doors	0.3		0.50	
Entrance halls	1.25–1.75	0.60–0.90	2.00–3.00	1.00–1.50
Reception halls	1.2	0.60	2.00	1.00
Bath rooms	1.2	0.60	2.00	1.00
Infiltration through walls with windows:				
space with 1 side exposed	0.6	0.30	1.00	0.50
space with 2 sides exposed	0.9	0.45	1.50	0.75
space with 3 or 4 sides exposed	1.2	0.60	2.00	1.00

TABLE 19. Infiltration Through Windows
(cfh/sq ft)[a]

Type of Window	Description of Window	Wind Velocity (mph)	
		$7\frac{1}{2}$	15
Double hung	Total for average window, nonweather stripped, $\frac{1}{16}''$. Crack and $\frac{3}{64}''$ clearance	14	27
Wood sash	ditto, weather stripped	8	24
(Unlocked)	Total for poorly fitted window nonweather stripped, $\frac{3}{32}''$ crack and $\frac{3}{32}''$ clearance	48	111
	ditto, weather stripped	13	34
Double hung	Nonweather stripped, locked	33	70
Metal sash	Nonweather stripped, unlocked	34	74
	Weather stripped, unlocked	13	32
Rolled	Industrial pivoted, $\frac{1}{16}''$ crack	80	176
Section	Architectural projected, $\frac{3}{64}''$ crack	36	88
Steel	Residential casement, $\frac{1}{32}''$ crack	23	52
Sash	Heavy casement section, projected $\frac{1}{32}''$ crack	16	38

[a]Cubic feet of air per hour per lineal foot of crack.

TABLE 20. Heat Transfer Factors for Ducts
Heat Transfer—Loss or Gain
(mbh per 100 ft of duct)[a]

Airflow (cfm)	Duct Insulation (in.)	Temperature Difference (°F)					
		20	40	60	80	100	120
	None	5.70	11.02	13.68	20.52	25.08	30.40
250	1	1.35	2.61	3.24	4.86	5.94	7.20
	2	0.75	1.45	1.80	2.70	3.30	4.00
	None	6.08	11.78	16.34	22.80	28.12	34.96
500	1	1.44	2.79	3.87	5.40	6.60	8.28
	2	0.80	1.55	2.15	3.00	3.70	4.60
	None	7.22	14.44	19.00	28.12	34.20	42.56
1000	1	1.71	3.42	4.50	6.66	8.10	10.08
	2	0.95	1.90	2.50	3.70	4.50	5.60
	None	8.36	16.72	23.56	32.68	40.28	50.16
1500	1	1.98	3.96	5.58	7.74	9.54	11.88
	2	1.10	2.20	3.10	4.30	5.30	6.60
	None	10.64	19.00	27.36	37.24	46.36	56.24
2000	1	2.52	4.50	6.48	8.82	10.98	13.32
	2	1.40	2.50	3.60	4.90	6.10	7.40
	None	10.64	22.80	33.44	45.60	56.24	74.48
3000	1	2.70	5.40	7.92	10.80	13.32	17.64
	2	1.50	3.00	4.40	6.00	7.40	9.80
	None	13.68	27.36	39.52	53.20	65.74	79.80
4000	1	3.24	6.48	9.36	12.60	15.57	18.90
	2	1.80	3.60	5.20	7.00	8.65	10.50
	None	15.20	30.40	44.84	60.04	74.10	89.68
5000	1	3.60	7.20	10.62	14.22	17.55	21.24
	2	2.00	4.00	5.90	7.90	9.75	11.80
	None	16.72	33.06	49.40	65.74	81.32	98.04
6000	1	3.96	7.83	11.70	15.57	19.26	23.22
	2	2.20	4.35	6.50	8.65	10.70	12.90
	None	17.48	35.72	52.82	70.68	88.16	105.64
7000	1	4.14	8.46	12.51	16.74	20.88	25.02
	2	2.30	4.70	6.95	9.30	11.60	13.90
	None	18.62	37.24	55.48	75.24	93.48	112.48
8000	1	4.41	8.82	13.14	17.82	22.14	26.64
	2	2.45	4.90	7.30	9.90	12.30	14.80
	None	18.62	39.52	57.76	78.28	98.04	117.04
9000	1	4.50	9.45	13.70	18.54	23.22	27.72
	2	2.45	5.20	7.60	10.30	12.90	15.40
	None	19.00	40.28	59.28	79.04	101.84	121.60
10,000	1	4.55	9.60	14.04	18.80	24.12	28.80
	2	2.50	5.30	7.80	10.40	13.40	16.00

[a]Based on: 1) Air Velocity 1300–1600 fpm, 2) Duct Aspect Ratio 2:1.

TABLE 21. Approximate kW Input for Equipment per 12,000 Btu/hr Cooling Capacity

Item of Power Load	Type of Heat Rejection Water-Cooled Condenser[a]			
Refrigeration compressor	1.00	1.00	1.00	1.15[b]
Blower for air conditioner	0.10	0.10	0.10	0.10
Blower for rejection air			0.15	0.15
Pump, cooling water		0.15	0.15	
Total kW per 12,000 Btu/hr	1.10	1.25	1.40	1.40

[a]Other heat rejection means include wells, spray ponds, evaporative condensers, water coils in the ground, finned water coils in air etc.
[b]In split systems with air-cooled condensing units this figure is between 1.35 and 1.55.

TABLE 22. Recommended Winter Inside Design Conditions

Application	DB (°F)	WB (°F)	RH (%)
General: Homes, apartments, offices, schools, theatres, hotel and hospital bedrooms, bathrooms, restaurants. Healthy adults and children, normally clothed, seated at rest	76 74	* 57.4	* 35
Elderly people and invalids normally clothed, seated at rest	80 77	* 59.7	* 35
Occupations requiring light work	72 70	* 54.3	* 35
Occupations requiring heavy work	68 66	* 51.5	* 35
Garages	65	*	*
Gymnasiums	65	*	*
Hospital Operating Rooms	73	63.5	60
Kitchens	70	*	*
Laundries	70	*	*
Locker Rooms	70	*	*
Stores: Customers wearing street clothes	70 68	* 52.9	* 35
Swimming Pools	80	69.7	60
Toilets, Public	72	*	*

*Heating only without humidification. 15% RH assumed at design condition.

TABLE 23. Average Optimum Effective Temperatures

SEASON	SEX	AGE	EFFECTIVE TEMPERATURE (DEG F)		
			North	Middle	South
Winter	Men and women	all	58	69	70
	Men	under 40	67	68	69
		over 40	68	69	70
	Women	under 40	68	69	70
		over 40	69	70	71
Summer	Men and women	all	70	71	72
	Men	under 40	69	70	71
		over 40	70	71	72
	Women	under 40	70	71	72
		over 40	71	72	73

TABLE 24. Radiant and Convective Heat Gain

Types	Radiant (%)	Convective (%)
Solar, without inside shading	100	
Solar, with inside shading	58	42
Transmission through walls and roofs	60	40

TABLE 25. Outdoor Air Requirements for Ventilation

Application	Smoking	Cfm per Person[b]		Cfm per Sq Ft of Floor[b]
		Recommended	Minimum[c]	Minimum[c]
Apartment				
Average	Some	20	10	
DeLuxe	Some	20	10	
Banking space	Occasional	10	7½	
Barber shops	Considerable	15	10	
Beauty parlors	Occasional	10	7½	
Brokers' board rooms	Very heavy	50	20	
Cocktail bars		40	25	
Corridors (supply or exhaust)				0.25
Department stores	None	7½	5	0.05
Directors' rooms	Extreme	50	30	

TABLE 25. (*Continued*)

Application	Smoking	Cfm per Person		Cfm per Sq Ft of Floor
		Recom-mended	Mini-mum	Mini-mum
Drug stores[e]	Considerable	10	$7\frac{1}{2}$	
Factories[d,f]	None	10	$7\frac{1}{2}$	0.10
Five and Ten Cent stores	None	$7\frac{1}{2}$	5	
Funeral parlors	None	10	$7\frac{1}{2}$	
Garages[d]				1.0
Hospitals				
Operating rooms[f,g]	None			2.0
Private rooms	None	30	25	0.33
Wards	None	20	10	
Hotel rooms	Heavy	30	25	0.33
Kitchens				
Restaurant				4.0
Residence				2.0
Laboratories[e]	Some	20	15	
Meeting rooms	Very heavy	50	30	1.25
Offices				
General	Some	15	10	0.25
Private	None	25	15	0.25
Private	Considerable	30	25	0.25
Restaurants				
Cafeteria[e]	Considerable	12	10	
Dining room[e]	Considerable	15	12	
Schoolrooms[d]	None			
Shop, retail	None	10	$7\frac{1}{2}$	
Theater[d]	None	$7\frac{1}{2}$	5	
Theater	Some	15	10	
Toilets[d] (exhaust)				2.0

[a] Taken from present-day practice.
[b] This is contaminant-free air.
[c] *When minimum is used, take the larger of the two.*
[d] See local codes which may govern.
[e] May be governed by exhaust.
[f] May be governed by special sources of contamination or local codes.
[g] All outside air recommended to overcome explosion hazard of anesthetics,

TABLE 26. Estimated Annual Hours of Operation for Properly Sized Equipment in Typical Cities During Normal Cooling Season[a]

City	Hours	City	Hours
Atlanta, Ga.	750	Jacksonville, Fla.	1600
Boston, Mass.	200	Minneapolis, Minn.	350
Chicago, Ill.	400	New Orleans, La.	1500
Cleveland, Ohio	450	New York, N.Y.	350
Dallas, Texas	1400	St. Louis, Mo.	1000
Fresno, Calif.	900	Washington, D.C.	800

[a]Based on average indoor temperature of 80°F.

TABLE 27. Heat Gain from Commercial Cooking Appliances Located in Air Conditioned Space[a] (Btu/hr)

Appliance	Capacity	Overall Dim., Inches Width×Depth×Height	Miscellaneous Data (Dimensions in Inches)	Manufacturer's Input Rating — Boiler hp or Watts	Manufacturer's Input Rating — Btu/Hr	Probable Max. Hourly Input Btu/Hr	Recommended Rate of Heat Gain, Btu/Hr — Without Hood Sensible	Without Hood Latent	Without Hood Total	With Hood[b] All Sensible
Electric, Floor Mounted Type										
Broiler, no oven			23 wide×25 deep grid	12,000	40,900	20,500	Exhaust hood required			6,500
With oven			23×27×12 oven	18,000	61,400	30,700				9,800
Deep fat fryer	28 lb fat	20×38×36	14 wide×15 deep kettle	12,000	40,900	20,500				6,500
	60 lb fat	24×36×36	20 wide×20 deep kettle	18,000	61,400	30,700				9,800
Oven, baking, per sq ft of hearth			Compartment 8-in. high	500	1,700	850				270
Oven, roasting, per sq ft of hearth			Compartment 12-in. high	900	3,070	1,500				490
Range, heavy duty Top section		36×36×36		15,000	51,100	25,600				8,200
Oven				6,000	20,400	10,200				3,200
Range, medium duty Top section		30×32×36		8,000	27,300	13,600				4,300
Oven				3,600	12,300	6,200				1,900
Range, light duty Top section		30×29×36		6,600	22,500	11,200				3,600
Oven				3,000	10,200	5,100				1,600
Steam Heated										
Coffee urn	3 gal.			0.2	6,600	3,300	2,180	1,120	3,300	1,000
	5 gal.			0.3	10,000	5,000	3,300	1,700	5,000	1,600
	8 gal. twin			0.4	13,200	6,600	4,350	2,250	6,600	2,100
Steam table per sq ft of top			With insets	0.05	1,650	825	500	325	825	260
Bain marie per sq ft of top			Open tank	0.10	3,300	1,650	825	825	1,650	520
Oyster steamer			Jacketed type	0.5	16,500	8,250	5,000	3,250	8,250	2,600
Steam kettles per gal. capacity				0.06	2,000	1,000	600	400	1,000	320
Compartment steamer per compartment		24×25×12 compartment	Floor mounted	1.2	40,000	20,000	12,000	8,000	20,000	6,400
Compartment steamer per compartment	3 pans 12×20×2½		Single counter unit	0.5	16,500	8,250	5,000	3,250	8,250	2,600
Plate warmer per cu ft				0.05	1,650	825	550	275	825	260

[a] Heat gain from cooking appliances located in the conditioned area (but not included in the table) should be estimated as follows:
1. Obtain *probable maximum hourly input* by multiplying the manufacturer's hourly input rating by the usage factor of 0.50.
2. If appliances are installed without an exhaust hood, the estimated latent heat gain is 34 percent of the *probable maximum hourly input* and the sensible heat gain is 66 percent.
3. If appliances are installed under an effective exhaust hood, the estimate heat gain is all sensible heat.

[b] For poorly designed or undersized exhaust systems the heat gains in this column should be doubled and half of the increase assumed as latent heat.

TABLE 28. Heat Gain from Miscellaneous Appliances
(Btu/hr)

Appliance	Manufacturers' Rating		Recommended Rate of Heat Gain (Btu/hr)		
	Watts	Btu/hr	Sensible	Latent	Total
Electrical appliances					
Hair dryer					
Blower type	1580	5400	2300	400	2700
Helmet type	705	2400	1870	330	2200
Permanent wave machine					
60 heaters @ 25 W, 36″ normal use	1500	5000	850	150	1000
Neon sign, per linear ft of tube					
$\frac{1}{2}$″ diameter			30		30
$\frac{3}{8}$″ diameter			60		60
Sterilizer	1100	3750	650	1200	1850
Gas-burning appliances					
Lab burners					
Bunsen ($\frac{7}{16}$″ barrel)		3000	1680	420	2100
Fishtail ($1\frac{1}{2}$″ wide)	5000	5000	2800	700	3500
Meeker (1″ diameter)	6000	6000	3360	840	4200
Gas light, per burner (mantle type)		2000	1800	200	2000
Cigar lighter (continuous flame)		2500	900	100	1000

TABLE 29. Heat Gain from Occupants of Conditioned Spaces[a]
(Btu/hr)

Degree of Activity	Typical Application	Total Heat Adults, Male, Btu/Hr	Total Heat Adjusted,[b] Btu/Hr	Sensible Heat, Btu/Hr	Latent Heat, Btu/Hr
Seated at rest	Theater—Matinee	390	330	225	105
	Theater—Evening	390	350	245	105
Seated, very light work	Offices, hotels, apartments	450	400	245	155
Moderately active office work	Offices, hotels, apartments	475	450	250	200
Standing, light work; or walking slowly	Department store, retail store, dime store	550	450	250	200
Walking; seated Standing; walking slowly	Drug store, Bank	550	500	250	250
Sedentary work	Restaurant[c]	490	550	275	275
Light bench work	Factory	800	750	275	475
Moderate dancing	Dance hall	900	850	305	545
Walking 3 mph; moderately heavy work	Factory	1000	1000	375	625
Bowling[d] Heavy work	Bowling alley Factory	1500	1450	580	870

[a] *Note:* Tabulated values are based on 75 F room dry-bulb temperature. For 80 F room dry-bulb, the total heat remains the same, but the sensible heat values should be decreased by approximately 20 percent, and the latent heat values increased accordingly.

[b] *Adjusted total heat gain* is based on normal percentage of men, women, and children for the application listed, with the postulate that the gain from an adult female is 85 percent of that for an adult male, and that the gain from a child is 75 percent of that for an adult male.

[c] Adjusted total heat value for *sedentary work, restaurant,* includes 60 Btu per hour for food per individual (30 Btu sensible and 30 Btu latent).

[d] For *bowling* figure one person per alley actually bowling, and all others as sitting (400 Btu per hour) or standing (550 Btu per hour).

TABLE 30. Heat Gain from Electric Motors[a] (Btu/hr)

Nameplate[b] or Brake Horsepower	Full Load Motor Efficiency Percent	Location of Equipment with Respect to Conditioned Space or Air Stream[c]		
		Motor In-Driven Machines in $HP \times 2545 / \% \text{ Eff}$	Motor Out-Driven Machine in $HP \times 2545$	Motor In-Driven Machine out $HP \times 2545\,(1 - \% \text{ Eff}) / \% = \text{Eff}$
		Btu per Hour		
1/20	40	320	130	190
1/12	49	430	210	220
1/8	55	580	320	260
1/6	60	710	430	280
1/4	64	1,000	640	360
1/3	66	1,290	850	440
1/2	70	1,820	1,280	540
3/4	72	2,680	1,930	750
1	79	3,220	2,540	680
1 1/2	80	4,770	3,820	950
2	80	6,380	5,100	1,280
3	81	9,450	7,650	1,800
5	82	15,600	12,800	2,800
7 1/2	85	22,500	19,100	3,400
10	85	30,000	25,500	4,500
15	86	44,500	38,200	6,300
20	87	58,500	51,000	7,500
25	88	72,400	63,600	8,800
30	89	85,800	76,400	9,400
40	89	115,000	102,000	13,000
50	89	143,000	127,000	16,000
60	89	172,000	153,000	19,000
75	90	212,000	191,000	21,000
100	90	294,000	255,000	29,000
125	90	354,000	318,000	36,000
150	91	420,000	382,000	38,000
200	91	560,000	510,000	50,000
250	91	700,000	636,000	64,000

[a] For intermittent operation, an appropriate usage factor should be used, preferably measured.

[b] If motors are overloaded and amount of overloading is unknown, multiply the above heat gain factors by the following maximum service factors:

Maximum Service Factors

Horsepower	1/20–1/8	1/6–1/3	1/2–3/4	1	1½–2	3–250
AC Open Type	1.4	1.35	1.25	1.25	1.20	1.15
DC Open Type	—	—	—	1.15	1.15	1.15

No overload is allowable with enclosed motors.

[c] For a fan or pump in air conditioned space, exhausting air, and pumping fluid to outside of space, use values in last column.

TABLE 31. Solar Heat Gain Factors for Glass Without Inside Shading (Btu/hr)

August 21

Latitude	Solar Time	N	NE	E	SE	S	SW	W	NW	HOR
24° N	8:00 AM	10	65	88	59	9	8	8	8	38
	10:00 AM	23	106	159	120	28	21	21	21	124
	12:00 PM	33	88	140	125	52	37	33	33	220
	2:00 PM	37	46	70	75	63	75	70	46	257
	4:00 PM	33	33	33	37	52	125	140	88	220
	6:00 PM	23	21	21	21	28	120	159	106	124
32° N	8:00 AM	10	66	94	66	10	8	8	8	39
	10:00 AM	23	98	164	136	38	21	21	21	122
	12:00 PM	32	75	139	146	77	41	32	32	212
	2:00 PM	35	40	69	91	97	91	69	40	247
	4:00 PM	32	32	32	41	77	146	139	75	212
	6:00 PM	23	21	21	21	38	136	164	98	122
40° N	8:00 AM	10	65	98	73	13	8	8	8	39
	10:00 AM	22	90	167	149	51	21	20	20	116
	12:00 PM	30	64	136	165	105	48	30	30	199
	2:00 PM	33	35	67	107	132	107	67	35	230
	4:00 PM	30	30	31	48	105	165	136	64	199
	6:00 PM	22	20	20	21	51	149	167	90	116
48° N	8:00 AM	10	64	100	77	16	8	8	8	38
	10:00 AM	21	82	167	160	65	20	20	20	108
	12:00 PM	28	54	132	180	130	55	29	28	180
	2:00 PM	31	32	64	121	162	121	64	32	208
	4:00 PM	28	28	29	55	130	180	132	54	180
	6:00 PM	21	20	20	20	65	160	167	82	108

Solar Heat Gain Through Glass = Area × Solar Factor × Shading Factor
(no inside shading)   (Table 8)   (Tables 33, 34, 37)

This chart has been extracted by permission from **ASHRAE Handbook 1972** using Table 4 with the following recommended criteria:

IN—Examination of ASHRAE Table 4 shows that for south windows at 40° north latitude, the solar heat gain factors at 4:00 p.m. sun time are highest in November and lowest in June. Some judgment is therefore required to select the heat gain through windows which, combined with the other loads, will result in the maximum total heat gain. The equivalent temperature differential method for calculating roof loads is shown to apply for the months of April through August without a reduction in heat flow. Therefore, August is selected for the calculation of heat gains through windows.

From Table 4, the solar heat gain factor for south-facing glass at 4:00 p.m. is 40 Btu/hr. The average solar heat gain factor for the five hour period up to and including the design hour (4:00 p.m.) is:

$$\frac{40 + 79 + 116 + 140 + 149}{5} = 105 \text{ Btu/hr}$$

≡or north facing glass, the solar heat gain factor is 23 Btu/hr at 4:00 p.m., and the average for the same five hour period is:

$$\frac{23 + 28 + 32 + 34 + 35}{5} = 30 \text{ Btu/hr}$$

TABLE 32. Solar Heat Gain Factors for Glass with Inside Shading (Btu/hr)

August 21

LATITUDE	SOLAR TIME	N	NE	E	SE	S	SW	W	NW	HOR
24° N	8:00 AM	12	78	106	74	12	10	10	10	55
	10:00 AM	28	94	157	130	39	27	27	27	172
	12:00 PM	35	67	98	97	60	46	36	36	244
	2:00 PM	23	41	55	58	60	104	105	59	249
	4:00 PM	30	29	29	31	41	138	174	116	182
	6:00 PM	15	13	13	13	17	78	107	74	74
32° N	8:00 AM	12	74	110	81	14	11	11	11	55
	10:00 AM	27	80	159	149	57	26	26	26	167
	12:00 PM	34	59	97	118	91	57	35	34	234
	2:00 PM	34	37	54	67	92	123	104	46	239
	4:00 PM	29	28	28	34	57	156	173	103	176
	6:00 PM	17	13	13	13	23	93	120	78	74
40° N	8:00 AM	12	70	111	88	17	10	10	10	53
	10:00 AM	26	69	159	165	78	26	25	25	157
	12:00 PM	32	52	95	140	124	72	34	32	219
	2:00 PM	33	34	52	76	125	141	102	37	223
	4:00 PM	27	27	27	37	78	171	170	90	167
	6:00 PM	18	14	14	14	32	107	131	80	72
48° N	8:00 AM	11	65	111	92	22	10	10	10	50
	10:00 AM	25	61	158	178	98	25	24	24	144
	12:00 PM	30	45	91	158	151	86	32	30	198
	2:00 PM	31	31	50	84	155	155	98	32	202
	4:00 PM	25	25	26	41	98	183	165	78	152
	6:00 PM	18	14	14	14	41	118	139	82	69

Solar Heat Gain Through Glass = Area × Solar Factor × Shading Factor

(inside shading)   (Table 32)   (Tables 38, 39, 40)

Notes:

1. Use this table when it is known that some type of inside shading will be used.
2. All data presented in Tables 31 and 32 are based on August 21. Local experience or conditions, particularly in the lower latitudes may suggest the use of a more appropriate month. For variance from this date refer to *ASHRAE Handbook of Fundamentals 1972.*

TABLE 33. Shading Coefficients for Glass Without Inside Shading

Single Glass			Insulating Glass[b]		
Type of Glass	Nominal Thickness[a]	Shading Coefficient	Type of Glass	Nominal Thickness[c]	Shading Coefficient
Regular sheet	$\frac{3}{32}, \frac{1}{8}$	1.00	Regular sheet out,		
Regular plate/float	$\frac{1}{4}$	0.95	regular sheet in	$\frac{3}{32}, \frac{1}{8}$	0.90
	$\frac{3}{8}$	0.91	Regular plate out,		
	$\frac{1}{2}$	0.88	regular plate in	$\frac{1}{4}$	0.83
Heat-absorbing	$\frac{3}{16}$	0.72	Heat-absorbing plate out,		
	$\frac{1}{4}$	0.70	regular plate in	$\frac{1}{4}$	0.56
	$\frac{3}{8}$	0.56	Grey plate out,		
	$\frac{1}{2}$	0.50	regular plate in	$\frac{1}{4}$	0.56
Grey sheet	$\frac{1}{8}$	0.78	Grey plate with sun		
	$\frac{3}{16}$	0.90	control film out,		
	$\frac{7}{32}$	0.66	regular plate in	$\frac{1}{4}$	0.42
	$\frac{7}{32}$	0.88	Regular plate out,		
	$\frac{1}{4}$	0.86	regular plate with sun		
Grey plate	$\frac{13}{64}$	0.72	control film in	$\frac{1}{4}$	0.29
	$\frac{1}{4}$	0.70			
	$\frac{3}{8}$	0.56			
	$\frac{1}{2}$	0.50			

[a]refer to manufacturers' literature for values.
[b]refers to factory-fabricated units with $\frac{3}{16}$, $\frac{1}{4}$, or $\frac{1}{2}$-in. air space or to prime windows plus storm windows.
[c]thickness of each pane of glass, not thickness of assembled unit.

TABLE 34. Shading Coefficients for Hollow Glass Block Wall Panels

Type of Glass Block*	Description of Glass Block	Shading Coefficient	
		**Panels in the Sun	***Panels in Shade (N,NW,W,SW)
Type I	Glass Colorless or Aqua Smooth Face A,D: Smooth B,C: Smooth or wide ribs, or flutes horizontal or vertical, or shallow configuration E: None	0.65	0.40
Type IA	Same as Type I except A: Ceramic Enamel on exterior face	0.27	0.20
Type II	Same as Type I except E: Glass fiber screen	0.44	0.34
Type III	Glass Colorless or Aqua A,D: Narrow vertical ribs or flutes B,C: Horizontal light-diffusing prisms or horizontal light-directing prisms E: Glass fiber screen	0.33	0.27
Type IIIA	Same as Type III except E: Glass fiber screen with green ceramic spray coating, or glass fiber screen and gray glass, or glass fiber screen with light-selecting prisms	0.25	0.18

* All values are for 7 3/4 × 7 3/4 × 3 7/8 in. block, set in light-colored mortar. For 11 3/4 × 11 3/4 × 3 7/8 in. block, increase coefficients by 15% and for 5 3/4 × 5 3/4 × 3 7/8 in. block reduce coefficients by 15%.

** Shading coefficients are for peak load condition, but provide a close approximation for other conditions.

*** For NE, E, and SE panels in the shade add 50% to the values listed for panels in the shade.

TABLE 35. Overhang Shading from April 11 Through September 1

LATITUDE	SUN ① TIME AM ↓→	SHADING DOWN FROM OVERHANG FOR EACH FOOT OF PROJECTION				SUN ② TIME PM
		E	SE	S	SW	
24° North	6 AM	†	†	**	**	6 PM
	7	†	†	**	**	5
	8	0.6	0.6	6.2	**	4
	9	1.0	0.9	4.3	**	3
	10	2.0	1.3	3.7	**	2
	11	3.5	2.7	3.7	**	1
	12 N	**	4.5	3.5	4.5	12 N
32° North	6 AM	†	†	**	**	6 PM
	7	†	†	6.0	**	5
	8	0.6	0.7	3.3	**	4
	9	1.0	1.0	2.8	**	3
	10	1.4	1.3	2.3	**	2
	11	3.5	1.9	2.3	8.3	1
	12 N	**	3.3	2.3	3.3	12 N
40° North	6 AM	†	†	**	**	6 PM
	7	†	†	5.6	**	5
	8	0.5	0.6	2.3	**	4
	9	0.9	0.9	1.8	**	3
	10	1.5	1.0	1.7	**	2
	11	3.2	1.3	1.6	4.5	1
	12 N	**	2.3	1.4	2.3	12 N
48° North	6 AM	†	†	**	**	6 PM
	7	†	†	2.8	**	5
	8	0.5	0.5	1.5	**	4
	9	0.7	0.7	1.3	**	3
	10	1.3	0.9	1.2	**	2
	11	3.1	1.2	1.2	3.1	1
	12 N	**	1.4	1.2	1.6	↑ 12 N
		W	SW	S	SE	← PM

① AM — Read from top down   ② PM — Read from bottom up   † — No shade   ** — Completely shaded

$$\left(\begin{array}{c}\text{Factor for Hour of Day and}\\\text{Direction Window Faces (3A)}\end{array}\right) \times \left(\begin{array}{c}\text{Width of}\\\text{Overhang}\end{array}\right) - \left(\begin{array}{c}\text{Distance from Top of}\\\text{Windows to Overhang}\end{array}\right) = \text{Shade Height on Window}$$

Note: Use North Solar Factor for shaded portion of glass (Table 31 or 32).

TABLE 36. Shade Factors for Various Types of Outside Shading

TYPE OF SHADING	FINISH ON SIDE EXPOSED TO SUN	OUTSIDE SHADING FACTOR	
Canvas awning sides open	Dark or Medium	0.25	
Canvas awning top and sides tight against building	Dark or Medium	0.35	
Outside venetian blind, slats set at 45 deg. ①	White, cream	0.15	
Outside venetian blind, slats set at 45 deg. ① ② extended as awning fully covering window	White, cream	0.15	
Outside venetian blind, slats set at 45 deg. extended as awning covering 2/3 of window	White, cream	0.43	
Overhang (Use Table 35 to find shaded portion)		Use north solar factor for shaded portion.	
		DARK ③	GREEN TINT ④
Outside shading screen, solar altitude 10 deg.		0.52	0.46
Outside shading screen, solar altitude 20 deg.		0.40	0.35
Outside shading screen, solar altitude 30 deg.		0.25	0.24
Outside shading screen, solar altitude, above 40 deg.		0.15	0.22

① Venetian blinds are fully drawn and cover window. It is assumed that the occupant will adjust slats to prevent direct rays from passing between slats. If slats are fully closed (slats set at 90 deg) use same factors as used for roller shade fully drawn.

② Commercial shade with wide slats. The sun may shine on window through sides of shade. Estimate the exposed portion of glass as unshaded.

③ Commercial shade, bronze. Metal slats 0.05 inches wide 17 per inch and set at 17 deg. angle with horizontal. At solar altitudes below 40 deg. some direct solar rays are allowed to pass between slats, and this amount becomes progressively greater at low solar altitudes.

④ Commercial aluminum shade. Slats 0.057 inches wide, 17.5 per inch, set at 17 deg. angle with horizontal. At solar altitudes below 40 deg. some direct solar rays are allowed to pass between slats and this amount becomes progressively greater at low solar altitudes.

Sq. ft × Solar Factor × Inside Shade Factor × Outside Shade Factor = Btuh Gain

TABLE 37. Shading Coefficients for Domed Skylights

Dome	Light Diffuser (Translucent)	Curb Height, in.	Width to Height Ratio	Shading Coefficient	U-Value
Clear $\tau = 0.86$	yes $\tau = 0.58$	0	∞	0.61	0.46
		9	5	0.58	0.43
		18	2.5	0.50	0.40
Clear $\tau = 0.86$	None	0	∞	0.99	0.80
		9	5	0.88	0.75
		18	2.5	0.80	0.70
Translucent $\tau = 0.52$	None	0	∞	0.57	0.80
		9	5	0.51	0.75
		18	2.5	0.46	0.70
Translucent $\tau = 0.27$	None	0	∞	0.34	0.80
		9	5	0.30	0.75
		18	2.5	0.28	0.70

NOTE: t = transmittance of glass (.86 for clear glass, .52 for ¼ inch grey sheet, .27 for ½ inch bronze plate.)

TABLE 38. Shading Coefficients for Double Glazing With Between-Glass Shading

TYPE OF GLASS	NOMINAL THICKNESS EACH PANE	DESCRIPTION OF AIR SPACE	TYPE OF SHADING		
			Venetian Blinds		Louvered Sun Screen
			Light	Medium	
Regular Sheet Out Regular Sheet In	3/32, 1/8	Shade in contact with glass or shade separated from glass by air space. Shade in contact with glass-voids filled with plastic.	0.33	0.36	0.43
Regular Plate Out Regular Plate In	1/4		—	—	0.49
Heat-Abs. Plate Out Regular Plate In	1/4	Shade in contact with glass or shade separated from glass by air space. Shade in contact with glass-voids filled with plastic.	0.28	0.30	0.37
Grey Plate Out Regular Plate In	1/4		—	—	0.41

TABLE 39. Shading Coefficients for Single Glass With Indoor Shading by Venetian Blinds and Roller Shades

TYPE OF GLASS	NOMINAL THICKNESS	TYPE OF SHADING				
		Venetian Blinds		Roller Shades		
				Opaque		Translucent
		Medium	Light	Dark	White	Light
Regular Sheet Regular Plate Regular Pattern Heat-Absorbing Pattern Grey Sheet	3/32 to 1/4 1/4 to 1/2 1/8 to 9/32 1/8 3/12, 7/32	0.64	0.55	0.59	0.25	0.39
Heat-Absorbing Sheet Heat-Absorbing Plate Heat-Absorbing Pattern Grey Sheet Grey Plate	7/32 1/4 3/16, 1/4 1/8, 7/32 13/64, 1/4	0.57	0.53	0.45	0.30	0.36
Heat-Absorbing Sheet, Plate or Pattern Heat-Absorbing Plate Grey Plate	— 3/8 3/8	0.54	0.52	0.40	0.28	0.32
Heat-Absorbing Sheet, Plate or Pattern Grey Plate	— 1/2	0.51	0.50	0.36	0.28	0.31

TABLE 40. Shading Coefficients for Insulating Glass With Indoor Shading by Venetian Blinds and Roller Shades

TYPE OF GLASS	NOMINAL THICKNESS EACH LIGHT*	TYPE OF SHADING				
		Venetian Blinds		Roller Shade		
				Opaque		Translucent
		Medium	Light	Dark	White	Light
Regular Sheet Out Regular Sheet In	3/32, 1/8	0.57	0.51	0.60	0.25	0.37
Regular Plate Out Regular Plate In	1/4	0.57	0.51	0.60	0.25	0.37
Heat-Absorbing Plate Out Regular Plate In	1/4	0.39	0.36	0.40	0.22	0.30
Grey Plate Out Regular Plate In	1/4	0.39	0.36	0.40	0.22	0.30

*Refers to factory-fabricated units with 3/16, 1/4, or 1/2 in. air space, or to prime windows plus storm windows.

For vertical blinds with opaque white or beige louvers, tightly closed, shading coefficient is approximately the same as for opaque white roller shades.

TABLE 41. Climatic Conditions for United States and Canada*, a

Col. 1 State and Station[b]	Col. 2 Latitude[c] ° '		Col. 3 Elev,[d] Ft	Winter				Summer							
				Col. 4			Col. 5 Coincident Wind Velocity[e]	Col. 6 Design Dry-Bulb			Col. 7 Outdoor Daily Range[f]	Col. 8 Design Wet-Bulb		
				Median of Annual Extremes	99%	97½%		1%	2½%	5%		1%	2½%	5%
ALABAMA														
Alexander City	33	0	660	12	16	20	L	96	94	93	21	79	78	77
Anniston AP	33	4	599	12	17	19	L	96	94	93	21	79	78	77
Auburn	32	4	730	17	21	25	L	98	96	95	21	80	79	78
Birmingham AP	33	3	610	14	19	22	L	97	94	93	21	79	78	77
Decatur	34	4	580	10	15	19	L	97	95	94	22	79	78	77
Dothan AP	31	2	321	19	23	27	L	97	95	94	20	81	80	79
Florence AP	34	5	528	8	13	17	L	97	95	94	22	79	78	77
Gadsden	34	0	570	11	16	20	L	96	94	93	22	78	77	76
Huntsville AP	34	4	619	−6	13	17	L	97	95	94	23	78	77	76
Mobile AP	30	4	211	21	26	29	M	95	93	91	18	80	79	79
Mobile CO	30	4	119	24	28	32	M	96	94	93	16	80	79	79
Montgomery AP	32	2	195	18	22	26	L	98	95	93	21	80	79	78
Selma-Craig AFB	32	2	207	18	23	27	L	98	96	94	21	81	80	79
Talladega	33	3	565	11	15	19	L	97	95	94	21	79	78	77
Tuscaloosa AP	33	1	170r	14	19	23	L	98	96	95	22	81	80	79
ALASKA														
Anchorage AP	61	1	90	−29	−25	−20	VL	73	70	67	15	63	61	59
Barrow	71	2	22	−49	−45	−42	M	58	54	50	12	54	51	48
Fairbanks AP	64	5	436	−59	−53	−50	VL	82	78	75	24	64	63	61
Juneau AP	58	2	17	−11	−7	−4	L	75	71	68	15	66	64	62
Kodiak	57	3	21	4	8	12	M	71	66	63	10	62	60	58
Nome AP	64	3	13	−37	−32	−28	L	66	62	59	10	58	56	54
ARIZONA†														
Douglas AP	31	3	4098	13	18	22	VL	100	98	96	31	70	69	68
Flagstaff AP	35	1	6973	−10	0	5	VL	84	82	80	31	61	60	59
Fort Huachuca AP	31	3	4664	18	25	28	VL	95	93	91	27	69	68	67
Kingman AP	35	2	3446	18	25	29	VL	103	100	97	30	70	69	69
Nogales	31	2	3800	15	20	24	VL	100	98	96	31	72	71	70
Phoenix AP	33	3	1117	25	31	34	VL	108	106	104	27	77	76	75
Prescott AP	34	4	5014	7	15	19	VL	96	94	91	30	67	66	65
Tucson AP	33	1	2584	23	29	32	VL	105	102	100	26	74	73	72
Winslow AP	35	0	4880	2	9	13	VL	97	95	92	32	66	65	64
Yuma AP	32	4	199	32	37	40	VL	111	109	107	27	79	78	77
ARKANSAS														
Blytheville AFB	36	0	264	6	12	17	L	98	96	93	21	80	79	78
Camden	33	4	116	13	19	23	L	99	97	96	21	81	80	79
El Dorado AP	33	1	252	13	19	23	L	98	96	95	21	81	80	79
Fayetteville AP	36	0	1253	3	9	13	M	97	95	93	23	77	76	75
Fort Smith AP	35	2	449	9	15	19	M	101	99	96	24	79	78	77
Hot Springs Nat. Pk.	34	3	710	12	18	22	M	99	97	96	22	79	78	77
Jonesboro	35	5	345	8	14	18	M	98	96	95	21	80	79	78
Little Rock AP	34	4	257	13	19	23	M	99	96	94	22	80	79	78
Pine Bluff AP	34	1	204	14	20	24	L	99	96	95	22	81	80	79
Texarkana AP	33	3	361	16	22	26	M	99	97	96	21	80	79	78
CALIFORNIA†														
Bakersfield AP	35	2	495	26	31	33	VL	103	101	99	32	72	71	70
Barstow AP	34	5	2142	18	24	28	VL	104	102	99	37	73	72	71
Blythe AP	33	4	390	26	31	35	VL	111	109	106	28	78	77	76
Burbank AP	34	1	699	30	36	38	VL	97	94	91	25	72	70	69
Chico	39	5	205	23	29	33	VL	102	100	97	36	71	70	69
Concord	38	0	195	27	32	36	VL	96	92	88	32	69	67	66

* Data for U. S. stations extracted from *Evaluated Weather Data for Cooling Equipment Design. Addendum No. 1, Winter and Summer Data*, with the permission of the publisher, Fluor Products Company, Inc , Box 1267, Santa Rosa, California.
  a Data compiled from official weather stations, where hourly weather observations are made by trained observers, and from other sources. Table 1 prepared by ASHRAE Technical Committee 2.2, Weather Data and Design Conditions. Percentage of *winter* design data show the percent of 3-month period, December through February. Canadian data are based on January only. Percentage of *summer* design data show the percent of 4-month period, June through September. Canadian data are based on July only. Also see References 1 to 7.
  b When airport temperature observations were used to develop design data, "AP" follows station name, and "AFB" follows Air Force Bases. Data for stations followed by "CO" came from office locations within an urban area and generally reflect an influence of the surrounding area. Stations without designation can be considered semirural and may be directly compared with most airport data.
  c Latitude is given to the nearest 10 minutes, for use in calculating solar loads. For example, the latitude for Anniston, Alabama is given as 33 4, or 33°40'.
  d Elevations are elevations for each station as of 1964. Temperature readings are generally made at an elevation of 5 ft above ground, except for locations marked r, indicating roof exposure of thermometer.
  e Coincident wind velocities derived from approximately coldest 600 hours out of 20,000 hours of December through February data per station. Also see References 5 and 6. The four classifications are:
  VL = Very Light, 70 percent or more of cold extreme hours ≤7 mph.   M = Moderate, 50 to 74 percent cold extreme hours >7 mph.
  L = Light, 50 to 69 percent cold extreme hours ≤7 mph.   H = High, 75 percent or more cold extreme hours >7 mph, and 50 percent are >12 mph.
  f The difference between the average maximum and average minimum temperatures during the warmest month.
  † More detailed data on Arizona, California, and Nevada may be found in *Recommended Design Temperatures, Northern California*, published by the Golden Gate Chapter; and *Recommended Design Temperatures, Southern California, Arizona, Nevada*, published by the Southern California Chapter.

TABLE 41. (*Continued*)

Col. 1 State and Station[b]	Col. 2 Latitude[c] ° ′		Col. 3 Elev,[d] Ft	Col. 4 Median of Annual Extremes	Col. 4 99%	Col. 4 97½%	Col. 5 Coincident Wind Velocity[e]	Col. 6 Design Dry-Bulb 1%	Col. 6 2½%	Col. 6 5%	Col. 7 Outdoor Daily Range[f]	Col. 8 Design Wet-Bulb 1%	Col. 8 2½%	Col. 8 5%
CALIFORNIA† (continued)														
Covina	34	0	575	32	38	41	VL	100	97	94	31	73	72	71
Crescent City AP	41	5	50	28	33	36	L	72	69	65	18	61	60	59
Downey	34	0	116	30	35	38	VL	93	90	87	22	72	71	70
El Cajon	32	4	525	26	31	34	VL	98	95	92	30	74	73	72
El Centro AP	32	5	−30	26	31	35	VL	111	109	106	34	81	80	79
Escondido	33	0	660	28	33	36	VL	95	92	89	30	73	72	71
Eureka/Arcata AP	41	0	217	27	32	35	L	67	65	63	11	60	59	58
Fairfield-Travis AFB	38	2	72	26	32	34	VL	98	94	90	34	71	69	67
Fresno AP	36	5	326	25	28	31	VL	101	99	97	34	73	72	71
Hamilton AFB	38	0	3	28	33	35	VL	89	85	81	28	71	68	66
Laguna Beach	33	3	35	32	37	39	VL	83	80	77	18	69	68	67
Livermore	37	4	545	23	28	30	VL	99	97	94	24	70	69	68
Lompoc, Vandenburg AFB	34	4	552	32	36	38	VL	82	79	76	20	65	63	61
Long Beach AP	33	5	34	31	36	38	VL	87	84	81	22	72	70	69
Los Angeles AP	34	0	99	36	41	43	VL	86	83	80	15	69	68	67
Los Angeles CO	34	0	312	38	42	44	VL	94	90	87	20	72	70	69
Merced-Castle AFB	37	2	178	24	30	32	VL	102	99	96	36	73	72	70
Modesto	37	4	91	26	32	36	VL	101	98	96	36	72	71	70
Monterey	36	4	38	29	34	37	VL	82	79	76	20	64	63	61
Napa	38	2	16	26	31	34	VL	94	92	89	30	69	68	67
Needles AP	34	5	913	27	33	37	VL	112	110	107	27	76	75	74
Oakland AP	37	4	3	30	35	37	VL	85	81	77	19	65	63	62
Oceanside	33	1	30	33	38	40	VL	84	81	78	13	69	68	67
Ontario	34	0	995	26	32	34	VL	100	97	94	36	72	71	70
Oxnard AFB	34	1	43	32	35	37	VL	84	80	78	19	70	69	67
Palmdale AP	34	4	2517	18	24	27	VL	103	101	98	35	70	68	67
Palm Springs	33	5	411	27	32	36	VL	110	108	105	35	79	78	77
Pasadena	34	1	864	31	36	39	VL	96	93	90	29	72	70	69
Petaluma	38	1	27	24	29	32	VL	94	90	87	31	70	68	67
Pomona CO	34	0	871	26	31	34	VL	99	96	93	36	73	72	71
Redding AP	40	3	495	25	31	35	VL	103	101	98	32	70	69	67
Redlands	34	0	1318	28	34	37	VL	99	96	93	33	72	71	70
Richmond	38	0	55	28	35	38	VL	85	81	77	17	66	64	63
Riverside-March AFB	33	5	1511	26	32	34	VL	99	96	94	37	72	71	69
Sacramento AP	38	3	17	24	30	32	VL	100	97	94	36	72	70	69
Salinas AP	36	4	74	27	32	35	VL	87	85	82	24	67	65	64
San Bernardino, Norton AFB	34	1	1125	26	31	33	VL	101	98	96	38	75	73	71
San Diego AP	32	4	19	38	42	44	VL	86	83	80	12	71	70	68
San Fernando	34	1	977	29	34	37	VL	100	97	94	38	73	72	71
San Francisco AP	37	4	8	32	35	37	L	83	79	75	20	65	63	62
San Francisco CO	37	5	52	38	42	44	VL	80	77	73	14	64	62	61
San Jose AP	37	2	70r	30	34	36	VL	90	88	85	26	69	67	65
San Luis Obispo	35	2	315	30	35	37	VL	89	85	82	26	65	64	63
Santa Ana AP	33	4	115r	28	33	36	VL	92	89	86	28	72	71	70
Santa Barbara CO	34	3	100	30	34	36	VL	87	84	81	24	67	66	65
Santa Cruz	37	0	125	28	32	34	VL	87	84	80	28	66	65	63
Santa Maria AP	34	5	238	28	32	34	VL	85	82	79	23	65	64	63
Santa Monica CO	34	0	57	38	43	45	VL	80	77	74	16	69	68	67
Santa Paula	34	2	263	28	33	36	VL	91	89	86	36	72	71	70
Santa Rosa	38	3	167	24	29	32	VL	95	93	90	34	70	68	67
Stockton AP	37	5	28	25	30	34	VL	101	98	96	37	72	70	69
Ukiah	39	1	620	22	27	30	VL	98	96	93	40	70	69	67
Visalia	36	2	354	26	32	36	VL	102	100	97	38	73	72	70
Yreka	41	4	2625	7	13	17	VL	96	94	91	38	68	66	65
Yuba City	39	1	70	24	30	34	VL	102	100	97	36	71	70	69
COLORADO														
Alamosa AP	37	3	7536	−26	−17	−13	VL	84	82	79	35	62	61	60
Boulder	40	0	5385	− 5	4	8	L	92	90	87	27	64	63	62
Colorado Springs AP	38	5	6173	− 9	− 1	4	L	90	88	86	30	63	62	61
Denver AP	39	5	5283	− 9	− 2	3	L	92	90	89	28	65	64	63
Durango	37	1	6550	−10	0	4	VL	88	86	83	30	64	63	62

TABLE 41. (*Continued*)

Col: 1 State and Station[b]	Col. 2 Latitude[e] ° ′		Col. 3 Elev,[d] Ft	Winter				Summer							
				Col. 4			Col. 5 Coincident Wind Velocity[e]	Col. 6 Design Dry-Bulb			Col. 7 Outdoor Daily Range[f]	Col. 8 Design Wet-Bulb		
				Median of Annual Extremes	99%	97½%		1%	2½%	5%		1%	2½%	5%
COLORADO (*continued*)														
Fort Collins	40	4	5001	−18	−9	−5	L	91	89	86	28	63	62	61
Grand Junction AP	39	1	4849	−2	8	11	VL	96	94	92	29	64	63	62
Greeley	40	3	4648	−18	−9	−5	L	94	92	89	29	65	64	63
La Junta AP	38	0	4188	−14	−6	−2	M	97	95	93	31	72	71	69
Leadville	39	2	10177	−18	−9	−4	VL	76	73	70	30	56	55	54
Pueblo AP	38	2	4639	−14	−5	−1	L	96	94	92	31	68	67	66
Sterling	40	4	3939	−15	−6	−2	M	95	93	90	30	67	66	65
Trinidad AP	37	2	5746	−9	1	5	L	93	91	89	32	66	65	64
CONNECTICUT														
Bridgeport AP	41	1	7	−1	4	8	M	90	88	85	18	77	76	75
Hartford, Brainard Field	41	5	15	−4	1	5	M	90	88	85	22	77	76	74
New Haven AP	41	2	6	0	5	9	H	88	86	83	17	77	76	75
New London	41	2	60	0	4	8	H	89	86	83	16	77	75	74
Norwalk	41	1	37	−5	0	4	M	91	89	86	19	77	76	75
Norwich	41	3	20	−7	−2	2	M	88	86	83	18	77	76	75
Waterbury	41	3	605	−5	0	4	M	90	88	85	21	77	76	75
Windsor Locks, Bradley Field	42	0	169	−7	−2	2	M	90	88	85	22	76	75	73
DELAWARE														
Dover AFB	39	0	38	8	13	15	M	93	90	88	18	79	78	77
Wilmington AP	39	4	78	6	12	15	M	93	90	87	20	79	77	76
DISTRICT OF COLUMBIA														
Andrews AFB	38	5	279	9	13	16	M	94	91	88	18	79	77	76
Washington National AP	38	5	14	12	16	19	M	94	92	90	18	78	77	76
FLORIDA														
Belle Glade	26	4	16	31	35	39	M	93	91	90	16	80	79	79
Cape Kennedy AP	28	1	16	33	37	40	L	90	89	88	15	81	80	79
Daytona Beach AP	29	1	31	28	32	36	L	94	92	91	15	81	80	79
Fort Lauderdale	26	0	13	37	41	45	M	91	90	89	15	81	80	79
Fort Myers AP	26	4	13	34	38	42	M	94	92	91	18	80	80	79
Fort Pierce	27	3	10	33	37	41	M	93	91	90	15	81	80	79
Gainesville AP	29	4	155	24	28	32	L	96	94	93	18	80	79	79
Jacksonville AP	30	3	24	26	29	32	L	96	94	92	19	80	79	79
Key West AP	24	3	6	50	55	58	M	90	89	88	9	80	79	79
Lakeland CO	28	0	214	31	35	39	M	95	93	91	17	80	79	78
Miami AP	25	5	7	39	44	47	M	92	90	89	15	80	79	79
Miami Beach CO	25	5	9	40	45	48	M	91	89	88	10	80	79	79
Ocala	29	1	86	25	29	33	L	96	94	93	18	80	79	79
Orlando AP	28	3	106r	29	33	37	L	96	94	93	17	80	79	78
Panama City, Tyndall AFB	30	0	22	28	32	35	M	92	91	90	14	81	80	80
Pensacola CO	30	3	13	25	29	32	M	92	90	89	14	82	81	80
St. Augustine	29	5	15	27	31	35	L	94	92	90	16	81	80	79
St. Petersburg	28	0	35	35	39	42	M	93	91	90	16	81	80	79
Sanford	28	5	14	29	33	37	L	95	93	92	17	80	79	79
Sarasota	27	2	30	31	35	39	M	93	91	90	17	80	80	79
Tallahassee AP	30	2	58	21	25	29	L	96	94	93	19	80	79	79
Tampa AP	28	0	19	32	36	39	M	92	91	90	17	81	80	79
West Palm Beach AP	26	4	15	36	40	44	M	92	91	89	16	81	80	80
GEORGIA														
Albany, Turner AFB	31	3	224	21	26	30	L	98	96	94	20	80	79	78
Americus	32	0	476	18	22	25	L	98	96	93	20	80	79	78
Athens	34	0	700	12	17	21	L	96	94	91	21	78	77	76
Atlanta AP	33	4	1005	14	18	23	H′	95	92	90	19	78	77	76
Augusta AP	33	2	143	17	20	23	L	98	95	93	19	80	79	78
Brunswick	31	1	14	24	27	31	L	97	95	92	18	81	80	79
Columbus, Lawson AFB	32	3	242	19	23	26	L	98	96	94	21	80	79	78
Dalton	34	5	720	10	15	19	L	97	95	92	22	78	77	76
Dublin	32	3	215	17	21	25	L	98	96	93	20	80	79	78
Gainesville	34	2	1254	11	16	20	L	94	92	89	21	78	77	76
Griffin	33	1	980	13	17	22	L	95	93	90	21	79	78	77
La Grange	33	0	715	12	16	20	L	96	94	92	21	79	78	77
Macon AP	32	4	356	18	23	27	L	98	96	94	22	80	79	78
Marietta, Dobbins AFB	34	0	1016	12	17	21	L	95	93	91	21	78	77	76

TABLE 41. (*Continued*)

Col. 1 State and Station[b]	Col. 2 Latitude[c] ° '		Col. 3 Elev,[d] Ft	Winter				Summer						
				Col. 4			Col. 5 Coincident Wind Velocity[e]	Col. 6 Design Dry-Bulb			Col. 7 Outdoor Daily Range[f]	Col. 8 Design Wet-Bulb		
				Median of Annual Extremes	99%	97½%		1%	2½%	5%		1%	2½%	5%
GEORGIA (continued)														
Moultrie	31	1	340	22	26	30	L	97	95	93	20	80	79	78
Rome AP	34	2	637	11	16	20	L	97	95	93	23	78	77	76
Savannah-Travis AP	32	1	52	21	24	27	L	96	94	92	20	81	80	79
Valdosta-Moody AFB	31	0	239	24	28	31	L	96	94	92	20	80	79	78
Waycross	31	2	140	20	24	28	L	97	95	93	20	80	79	78
HAWAII														
Hilo AP	19	4	31	56	59	61	L	85	83	82	15	74	73	72
Honolulu AP	21	2	7	58	60	62	L	87	85	84	12	75	74	73
Kaneohe	21	2	198	58	60	61	L	85	83	82	12	74	73	73
Wahiawa	21	3	215	57	59	61	L	86	84	83	14	75	74	73
IDAHO														
Boise AP	43	3	2842	0	4	10	L	96	93	91	31	68	66	65
Burley	42	3	4180	− 5	4	8	VL	95	93	89	35	68	66	64
Coeur d'Alene AP	47	5	2973	− 4	2	7	VL	94	91	88	31	66	65	63
Idaho Falls AP	43	3	4730r	−17	−12	− 6	VL	91	88	85	38	65	64	62
Lewiston AP	46	2	1413	1	6	12	VL	98	96	93	32	67	66	65
Moscow	46	4	2660	−11	− 3	1	VL	91	89	86	32	64	63	61
Mountain Home AFB	43	0	2992	− 3	2	9	L	99	96	93	36	68	66	64
Pocatello AP	43	0	4444	−12	− 8	− 2	VL	94	91	88	35	65	63	62
Twin Falls AP	42	3	4148	− 5	4	8	L	96	94	91	34	68	64	63
ILLINOIS														
Aurora	41	5	744	−13	− 7	− 3	M	93	91	88	20	78	77	75
Belleville, Scott AFB	38	3	447	0	6	10	M	97	95	92	21	79	78	77
Bloomington	40	3	775	− 7	− 1	3	M	94	92	89	21	79	78	77
Carbondale	37	5	380	1	7	11	M	98	96	94	21	80	79	78
Champaign/Urbana	40	0	743	− 6	0	4	M	96	94	91	21	79	78	77
Chicago, Midway AP	41	5	610	− 7	− 4	1	M	95	92	89	20	78	76	75
Chicago, O'Hare AP	42	0	658	− 9	− 4	0	M	93	90	87	20	77	75	74
Chicago, CO	41	5	594	− 5	− 3	1	M	94	91	88	15	78	76	75
Danville	40	1	558	− 6	− 1	4	M	96	94	91	21	79	78	76
Decatur	39	5	670	− 6	0	4	M	96	93	91	21	79	78	77
Dixon	41	5	696	−13	− 7	− 3	M	93	91	89	23	78	77	76
Elgin	42	0	820	−14	− 8	− 4	M	92	90	87	21	78	76	75
Freeport	42	2	780	−16	−10	− 6	M	92	90	87	24	78	77	75
Galesburg	41	0	771	−10	− 4	0	M	95	92	89	22	79	78	76
Greenville	39	0	563	− 3	3	7	M	96	94	92	21	79	78	77
Joliet AP	41	3	588	−11	− 5	− 1	M	94	92	89	20	78	77	75
Kankakee	41	1	625	−10	− 4	1	M	94	92	89	21	78	77	76
La Salle/Peru	41	2	520	− 9	− 3	1	M	94	93	90	22	78	77	76
Macomb	40	3	702	− 5	− 3	1	M	95	93	90	22	79	78	77
Moline AP	41	3	582	−12	− 7	− 3	M	94	91	88	23	79	77	76
Mt. Vernon	38	2	500	0	6	10	M	97	95	92	21	79	78	77
Peoria AP	40	4	652	− 8	− 2	2	M	94	92	89	22	78	77	76
Quincy AP	40	0	762	− 8	− 2	2	M	97	95	92	22	80	79	77
Rantoul, Chanute AFB	40	2	740	− 7	− 1	3	M	94	92	89	21	78	77	76
Rockford	42	1	724	−13	− 7	− 3	M	92	90	87	24	77	76	75
Springfield AP	39	5	587	− 7	− 1	4	M	95	92	90	21	79	78	77
Waukegan	42	2	680	−11	− 5	− 1	M	92	90	87	21	77	76	75
INDIANA														
Anderson	40	0	847	− 5	0	5	M	93	91	88	22	78	77	76
Bedford	38	5	670	− 3	3	7	M	95	93	90	22	79	78	77
Bloomington	39	1	820	− 3	3	7	M	95	92	90	22	79	78	76
Columbus, Bakalar AFB	39	2	661	− 3	3	7	M	95	92	90	22	79	78	76
Crawfordsville	40	0	752	− 8	− 2	2	M	95	93	90	22	79	77	76
Evansville AP	38	0	381	1	6	10	M	96	94	91	22	79	78	77
Fort Wayne AP	41	0	791	− 5	0	5	M	93	91	88	24	77	76	75
Goshen AP	41	3	823	−10	− 4	0	M	92	90	87	23	77	76	74
Hobart	41	3	600	−10	− 4	0	M	93	91	88	21	78	76	75
Huntington	40	4	802	− 8	− 2	2	M	94	92	89	23	78	76	75
Indianapolis AP	39	4	793	− 5	0	4	M	93	91	88	22	78	77	76
Jeffersonville	38	2	455	3	9	13	M	96	94	91	23	79	78	77
Kokomo	40	3	790	− 6	0	4	M	94	92	89	22	78	76	75
Lafayette	40	2	600	− 7	− 1	3	M	94	92	89	22	78	77	76

TABLE 41. (Continued)

Col. 1 State and Station[b]	Col. 2 Latitude[e] ° '		Col. 3 Elev.[d] Ft	Winter				Summer						
				Col. 4			Col. 5 Coinci- dent Wind Ve- locity[e]	Col. 6 Design Dry-Bulb			Col. 7 Out- door Daily Range[f]	Col. 8 Design Wet-Bulb		
				Median of Annual Ex- tremes	99%	97½%		1%	2½%	5%		1%	2½%	5%
INDIANA (continued)														
La Porte...................	41	3	810	−10	− 4	0	M	93	91	88	22	77	76	74
Marion....................	40	3	791	− 8	− 2	2	M	93	91	88	23	78	76	75
Muncie....................	40	1	955	− 8	− 2	2	M	93	91	88	22	78	77	75
Peru, Bunker Hill AFB........	40	4	804	− 9	− 3	1	M	91	89	86	22	77	76	74
Richmond AP...............	39	5	1138	− 7	− 1	3	M	93	91	88	22	78	77	75
Shelbyville................	39	3	765	− 4	2	6	M	94	92	89	22	78	77	76
South Bend AP.............	41	4	773	− 6	− 2	3	M	92	89	87	22	77	76	74
Terre Haute AP.............	39	3	601	− 3	3	7	M	95	93	91	22	79	78	77
Valparaiso................	41	2	801	−12	− 6	− 2	M	92	90	87	22	78	76	75
Vincennes.................	38	4	420	− 1	5	9	M	96	94	91	22	79	78	77
IOWA														
Ames.....................	42	0	1004	−17	−11	− 7	M	94	92	89	23	79	78	76
Burlington AP..............	40	5	694	−10	− 4	0	M	95	92	89	22	80	78	77
Cedar Rapids AP...........	41	5	863	−14	− 8	− 4	M	92	90	87	23	78	76	75
Clinton...................	41	5	595	−13	− 7	− 3	M	92	90	87	23	78	77	76
Council Bluffs.............	41	2	1210	−14	− 7	− 3	M	97	94	91	22	79	78	76
Des Moines AP.............	41	3	948r	−13	− 7	− 3	M	95	92	89	23	79	77	76
Dubuque..................	42	2	1065	−17	−11	− 7	M	92	90	87	22	78	76	75
Fort Dodge................	42	3	1111	−18	−12	− 8	M	94	92	89	23	78	77	75
Iowa City.................	41	4	645	−14	− 8	− 4	M	94	91	88	22	79	77	76
Keokuk...................	40	2	526	− 9	− 3	1	M	95	93	90	22	79	78	77
Marshalltown..............	42	0	898	−16	−10	− 6	M	93	91	88	23	79	77	76
Mason City AP.............	43	1	1194	−20	−13	− 9	M	91	88	85	24	77	75	74
Newton...................	41	4	946	−15	− 9	− 5	M	95	93	90	23	79	77	76
Ottumwa AP...............	41	1	842	−12	− 6	− 2	M	95	93	90	22	79	78	76
Sioux City AP..............	42	2	1095	−17	−10	− 6	M	96	93	90	24	79	77	76
Waterloo..................	42	3	868	−18	−12	− 8	M	91	89	86	23	78	76	75
KANSAS														
Atchison..................	39	3	945	− 9	− 2	2	M	97	95	92	23	79	78	77
Chanute AP...............	37	4	977	− 3	3	7	H	99	97	95	23	79	78	77
Dodge City AP.............	37	5	2594	− 5	3	7	M	99	97	95	25	74	73	72
El Dorado.................	37	5	1282	− 3	4	8	H	101	99	96	24	78	77	76
Emporia..................	38	2	1209	− 4	3	7	H	99	97	94	25	78	77	76
Garden City AP............	38	0	2882	−10	− 1	3	M	100	98	96	28	74	73	72
Goodland AP..............	39	2	3645	−10	− 2	4	M	99	96	93	31	71	70	69
Great Bend................	38	2	1940	− 5	2	6	M	101	99	96	28	77	76	75
Hutchinson AP.............	38	0	1524	− 5	2	6	H	101	99	96	28	77	76	75
Liberal...................	37	0	2838	− 4	4	8	M	102	100	99	28	74	73	71
Manhattan, Fort Riley........	39	0	1076	− 7	− 1	4	H	101	98	95	24	79	78	77
Parsons..................	37	2	908	− 2	5	9	H	99	97	94	23	79	78	77
Russell AP................	38	5	1864	− 7	0	4	M	102	100	97	29	78	76	75
Salina....................	38	5	1271	− 4	3	7	H	101	99	96	26	78	76	75
Topeka AP................	39	0	877	− 4	3	6	M	99	96	94	24	79	78	77
Wichita AP................	37	4	1321	− 1	5	9	H	102	99	96	23	77	76	75
KENTUCKY														
Ashland..................	38	3	551	1	6	10	L	94	92	89	22	77	76	75
Bowling Green AP..........	37	0	535	1	7	11	L	97	95	93	21	79	78	77
Corbin AP................	37	0	1175	0	5	9	L	93	91	89	23	79	77	76
Covington AP.............	39	0	869	− 3	3	8	L	93	90	88	22	77	76	75
Hopkinsville, Campbell AFB....	36	4	540	4	10	14	L	97	95	92	21	79	78	77
Lexington AP..............	38	0	979	0	6	10	'M	94	92	90	22	78	77	76
Louisville AP..............	38	1	474	1	8	12	L	96	93	91	23	79	78	77
Madisonville..............	37	2	439	1	7	11	L	96	94	92	22	79	78	77
Owensboro................	37	5	420	0	6	10	L	96	94	92	23	79	78	77
Paducah AP...............	37	0	398	4	10	14	L	97	95	94	20	80	79	78
LOUISIANA														
Alexandria AP.............	31	2	92	20	25	29	L	97	95	94	20	80	80	79
Baton Rouge AP...........	30	3	64	22	25	30	L	96	94	92	19	81	80	79
Bogalusa.................	30	5	103	20	24	28	L	96	94	93	19	80	79	78
Houma...................	29	3	13	25	29	33	L	94	92	91	15	81	80	79
Lafayette AP..............	30	1	38	23	28	32	L	95	93	92	18	81	81	80
Lake Charles AP...........	30	1	14	25	29	33	M	95	93	91	17	80	79	79
Minden...................	32	4	250	17	22	26	L	98	96	95	20	81	80	79

TABLE 41. (Continued)

Col. 1 State and Station[b]	Col. 2 Latitude[c] ° '		Col. 3 Elev.[d] Ft	Winter			Col. 5 Coincident Wind Velocity[e]	Summer						
				Col. 4				Col. 6 Design Dry-Bulb			Col. 7 Outdoor Daily Range[f]	Col. 8 Design Wet-Bulb		
				Median of Annual Extremes	99%	97½%		1%	2½%	5%		1%	2½%	5%
LOUISIANA (continued)														
Monroe AP	32	3	78	18	23	27	L	98	96	95	20	81	81	80
Natchitoches	31	5	120	17	22	26	L	99	97	96	20	81	80	79
New Orleans AP	30	0	3	29	32	35	M	93	91	90	16	81	80	79
Shreveport AP	32	3	252	18	22	26	M	99	96	94	20	81	80	79
MAINE														
Augusta AP	44	2	350	−13	− 7	− 3	M	88	86	83	22	74	73	71
Bangor, Dow AFB	44	5	162	−14	− 8	− 4	M	88	85	81	22	75	73	71
Caribou AP	46	5	624	−24	−18	−14	L	85	81	78	21	72	70	68
Lewiston	44	0	182	−14	− 8	− 4	M	88	86	83	22	74	73	71
Millinocket AP	45	4	405	−22	−16	−12	L	87	85	82	22	74	72	70
Portland AP	43	4	61	−14	− 5	0	L	88	85	81	22	75	73	71
Waterville	44	3	89	−15	− 9	− 5	M	88	86	82	22	74	73	71
MARYLAND														
Baltimore AP	39	1	146	8	12	15	M	94	91	89	21	79	78	77
Baltimore CO	39	2	14	12	16	20	M	94	92	89	17	79	78	77
Cumberland	39	4	945	0	5	9	L	94	92	89	22	76	75	74
Frederick AP	39	2	294	2	7	11	M	94	92	89	22	78	77	76
Hagerstown	39	4	660	1	6	10	L	94	92	89	22	77	76	75
Salisbury	38	2	52	10	14	18	M	92	90	87	18	79	78	77
MASSACHUSETTS														
Boston AP	42	2	15	− 1	6	10	H	91	88	85	16	76	74	73
Clinton	42	2	398	− 8	− 2	2	M	87	85	82	17	75	74	72
Fall River	41	4	190	− 1	5	9	H	88	86	83	18	75	74	73
Framingham	42	2	170	− 7	− 1	3	M	91	89	86	17	76	74	73
Gloucester	42	3	10	− 4	2	6	H	86	84	81	15	74	73	72
Greenfield	42	3	205	−12	− 6	− 2	M	89	87	84	23	75	74	73
Lawrence	42	4	57	− 9	− 3	1	M	90	88	85	22	76	74	72
Lowell	42	3	90	− 7	− 1	3	M	91	89	86	21	76	74	72
New Bedford	41	4	70	3	9	13	H	86	84	81	19	75	73	72
Pittsfield AP	42	3	1170	−11	− 5	− 1	M	86	84	81	23	74	72	71
Springfield, Westover AFB	42	1	247	− 8	− 3	2	M	91	88	85	19	76	74	73
Taunton	41	5	20	− 9	− 4	0	H	88	86	83	18	76	75	74
Worcester AP	42	2	986	− 8	− 3	1	M	89	87	84	18	75	73	71
MICHIGAN														
Adrian	41	5	754	− 6	0	4	M	93	91	88	23	76	75	74
Alpena AP	45	0	689	−11	− 5	− 1	M	87	85	82	27	74	73	71
Battle Creek AP	42	2	939	− 6	1	5	M	92	89	86	23	76	74	73
Benton Harbor AP	42	1	649	− 7	− 1	3	M	90	88	85	20	76	74	73
Detroit Met. CAP	42	2	633	0	4	8	M	92	88	85	20	76	75	74
Escanaba	45	4	594	−13	− 7	− 3	M	82	80	77	17	73	71	69
Flint AP	43	0	766	− 7	− 1	3	M	89	87	84	25	76	75	74
Grand Rapids AP	42	5	681	− 3	2	6	M	91	89	86	24	76	74	73
Holland	42	5	612	− 4	2	6	M	90	88	85	22	76	74	73
Jackson AP	42	2	1003	− 6	0	4	M	92	89	86	23	76	75	74
Kalamazoo	42	1	930	− 5	1	5	M	92	89	86	23	76	75	74
Lansing AP	42	5	852	− 4	2	6	M	89	87	84	24	76	75	73
Marquette CO	46	3	677	−14	− 8	− 4	L	88	86	83	18	73	71	69
Mt. Pleasant	43	4	796	− 9	− 3	1	M	89	87	84	24	75	74	73
Muskegon AP	43	1	627	− 2	4	8	M	87	85	82	21	75	74	73
Pontiac	42	4	974	− 6	0	4	M	90	88	85	21	76	75	73
Port Huron	43	0	586	− 6	− 1	3	M	90	88	85	21	76	74	73
Saginaw AP	43	3	662	− 7	− 1	3	M	88	86	83	23	76	75	73
Sault Ste. Marie AP	46	3	721	−18	−12	− 8	L	83	81	78	23	73	71	69
Traverse City AP	44	4	618	− 6	0	4	M	89	86	83	22	75	73	72
Ypsilanti	42	1	777	− 3	− 1	5	M	92	89	86	22	76	74	73
MINNESOTA														
Albert Lea	43	4	1235	−20	−14	−10	M	91	89	86	24	77	76	74
Alexandria AP	45	5	1421	−26	−19	−15	L	90	88	85	24	76	74	72
Bemidji AP	47	3	1392	−38	−32	−28	L	87	84	81	24	73	72	71
Brainerd	46	2	1214	−31	−24	−20	L	88	85	82	24	74	73	72
Duluth AP	46	5	1426	−25	−19	−15	M	85	82	79	22	73	71	69
Faribault	44	2	1190	−23	−16	−12	L	90	88	85	24	77	75	74
Fergus Falls	46	1	1210	−28	−21	−17	L	92	89	86	24	75	74	72
International Falls AP	48	3	1179	−35	−29	−24	L	86	82	79	26	72	69	68

TABLE 41. (Continued)

Col. 1 State and Station[b]	Col. 2 Latitude[c] ° '		Col. 3 Elev,[d] Ft	Col. 4 Median of Annual Extremes	99%	97½%	Col. 5 Coincident Wind Velocity[e]	Col. 6 Design Dry-Bulb 1%	2½%	5%	Col. 7 Outdoor Daily Range[f]	Col. 8 Design Wet-Bulb 1%	2½%	5%
MINNESOTA (continued)														
Mankato	44	1	785	−23	−16	−12	L	91	89	86	24	77	75	74
Minneapolis/St. Paul AP	44	5	822	−19	−14	−10	L	92	89	86	22	77	75	74
Rochester AP	44	0	1297	−23	−17	−13	M	90	88	85	24	77	75	74
St. Cloud AP	45	4	1034	−26	−20	−16	L	90	88	85	24	77	75	73
Virginia	47	3	1435	−32	−25	−21	L	86	83	80	23	73	71	69
Willmar	45	1	1133	−25	−18	−14	L	91	88	85	24	77	75	73
Winona	44	1	652	−19	−12	− 8	M	91	89	86	24	77	76	74
MISSISSIPPI														
Biloxi, Keesler AFB	30	2	25	26	30	32	M	93	92	90	16	82	81	80
Clarksdale	34	1	178	14	20	24	L	98	96	95	21	81	80	79
Columbus AFB	33	4	224	13	18	22	L	97	95	93	22	79	79	79
Greenville AFB	33	3	139	16	21	24	L	98	96	94	21	81	80	79
Greenwood	33	3	128	14	19	23	L	98	96	94	21	81	80	79
Hattiesburg	31	2	200	18	22	26	L	97	95	94	21	80	79	78
Jackson AP	32	2	330	17	21	24	L	98	96	94	21	79	78	78
Laurel	31	4	264	18	22	26	L	97	95	94	21	80	79	78
McComb AP	31	2	458	18	22	26	L	96	94	93	18	80	79	78
Meridian AP	32	2	294	15	20	24	L	97	95	94	22	80	79	78
Natchez	31	4	168	18	22	26	L	96	94	93	21	80	80	79
Tupelo	34	2	289	13	18	22	L	98	96	95	22	80	79	78
Vicksburg CO	32	2	234	18	23	26	L	97	95	94	21	80	80	79
MISSOURI														
Cape Girardeau	37	1	330	2	8	12	M	98	96	94	21	80	79	78
Columbia AP	39	0	778	− 4	2	6	M	97	95	92	22	79	78	77
Farmington AP	37	5	928	− 2	4	8	M	97	95	93	22	79	78	77
Hannibal	39	4	489	− 7	− 1	4	M	96	94	91	22	79	78	77
Jefferson City	38	4	640	− 4	2	6	M	97	95	93	23	79	78	77
Joplin AP	37	1	982	1	7	11	M	97	95	93	24	79	78	77
Kansas City AP	39	1	742	− 2	4	8	M	100	97	94	20	79	77	76
Kirksville AP	40	1	966	−13	− 7	− 3	M	96	94	91	24	79	78	77
Mexico	39	1	775	− 7	− 1	3	M	96	94	91	22	79	78	77
Moberly	39	3	850	− 8	− 2	2	M	96	94	91	23	79	78	77
Poplar Bluff	36	5	322	3	9	13	M	98	96	94	22	80	79	78
Rolla	38	0	1202	− 3	3	7	M	97	95	93	22	79	78	77
St. Joseph AP	39	5	809	− 8	− 1	3	M	97	95	92	23	79	78	77
St. Louis AP	38	5	535	− 2	4	8	M	98	95	92	21	79	78	77
St. Louis CO	38	4	465	1	7	11	M	96	94	92	18	79	78	77
Sedalia, Whiteman AFB	38	4	838	− 2	4	9	M	97	94	92	22	79	77	76
Sikeston	36	5	318	4	10	14	M	98	96	94	21	80	79	78
Springfield AP	37	1	1265	0	5	10	M	97	94	91	23	78	77	76
MONTANA														
Billings AP	45	5	3567	−19	−10	− 6	L	94	91	88	31	68	66	65
Bozeman	45	5	4856	−25	−15	−11	L	88	85	82	32	61	60	59
Butte AP	46	0	5526r	−34	−24	−16	VL	86	83	80	35	60	59	57
Cut Bank AP	48	4	3838r	−32	−23	−17	L	89	86	82	35	65	63	61
Glasgow AP	48	1	2277	−33	−25	−20	L	96	93	89	29	69	67	65
Glendive	47	1	2076	−28	−20	−16	L	96	93	90	29	71	69	68
Great Falls AP	47	3	3664r	−29	−20	−16	L	91	88	85	28	64	63	61
Havre	48	3	2488	−32	−22	−15	M	91	87	84	33	66	64	63
Helena AP	46	4	3893	−27	−17	−13	L	90	87	84	32	65	63	61
Kalispell AP	48	2	2965	−17	− 7	− 3	VL	88	84	81	34	65	63	62
Lewiston AP	47	0	4132	−27	−18	−14	L	89	86	83	30	65	63	62
Livingston AP	45	4	4653	−26	−17	−13	L	91	88	85	32	63	62	61
Miles City AP	46	3	2629	−27	−19	−15	L	97	94	91	30	71	69	68
Missoula AP	46	5	3200	−16	− 7	− 3	VL	92	89	86	36	65	63	61
NEBRASKA														
Beatrice	40	2	1235	−10	− 3	1	M	99	97	94	24	78	77	76
Chadron AP	42	5	3300	−21	−13	− 9	M	97	95	92	30	72	70	69
Columbus	41	3	1442	−14	− 7	− 3	M	98	96	93	25	78	76	75
Fremont	41	3	1203	−14	− 7	− 3	M	99	97	94	22	78	77	76
Grand Island AP	41	0	1841	−14	− 6	− 2	M	98	95	92	28	76	75	74
Hastings	40	4	1932	−11	− 3	1	M	98	96	94	27	77	75	74
Kearney	40	4	2146	−14	− 6	− 2	M	97	95	92	28	76	75	74
Lincoln CO	40	5	1150	−10	− 4	0	M	100	96	93	24	78	77	76

TABLE 41. (*Continued*)

Col. 1 State and Station[b]	Col. 2 Latitude[c] ° '		Col. 3 Elev,[d] Ft	Winter				Summer						
				Col. 4			Col. 5 Coincident Wind Velocity[e]	Col. 6 Design Dry-Bulb			Col. 7 Outdoor Daily Range[f]	Col. 8 Design Wet-Bulb		
				Median of Annual Extremes	99%	97½%		1%	2½%	5%		1%	2½%	5%
NEBRASKA (continued)														
McCook	40	1	2565	−12	−4	0	M	99	97	94	28	74	72	71
Norfolk	42	0	1532	−18	−11	−7	M	97	95	92	30	78	76	75
North Platte AP	41	1	2779	−13	−6	−2	M	97	94	90	28	74	73	72
Omaha AP	41	2	978	−12	−5	−1	M	97	94	91	22	79	78	76
Scottsbluff AP	41	5	3950	−16	−8	−4	M	96	94	91	31	70	69	67
Sidney AP	41	1	4292	−15	−7	−2	M	95	92	89	31	70	69	67
NEVADA†														
Carson City	39	1	4675	−4	3	7	VL	93	91	88	42	62	61	60
Elko AP	40	5	5075	−21	−13	−7	VL	94	92	90	42	64	62	61
Ely AP	39	1	6257	−15	−6	−2	VL	90	88	86	39	60	59	58
Las Vegas AP	36	1	2162	18	23	26	VL	108	106	104	30	72	71	70
Lovelock AP	40	0	3900	0	7	11	VL	98	96	93	42	65	64	62
Reno AP	39	3	4404	−2	2	7	VL	95	92	90	45	64	62	61
Reno CO	39	3	4490	8	12	17	VL	94	92	89	45	64	62	61
Tonopah AP	38	0	5426	2	9	13	VL	95	92	90	40	64	63	62
Winnemucca AP	40	5	4299	−8	1	5	VL	97	95	93	42	64	62	61
NEW HAMPSHIRE														
Berlin	44	3	1110	−25	−19	−15	L	87	85	82	22	73	71	70
Claremont	43	2	420	−19	−13	−9	L	89	87	84	24	74	73	72
Concord AP	43	1	339	−17	−11	−7	M	91	88	85	26	75	73	72
Keene	43	0	490	−17	−12	−8	M	90	88	85	24	75	73	72
Laconia	43	3	505	−22	−16	−12	M	89	87	84	25	74	73	72
Manchester, Grenier AFB	43	0	253	−11	−5	1	M	92	89	86	24	76	74	73
Portsmouth, Pease AFB	43	1	127	−8	−2	3	M	88	86	83	22	75	73	72
NEW JERSEY														
Atlantic City CO	39	3	11	10	14	18	H	91	88	85	18	78	77	76
Long Branch	40	2	20	4	9	13	H	93	91	88	18	77	76	75
Newark AP	40	4	11	6	11	15	M	94	91	88	20	77	76	75
New Brunswick	40	3	86	3	8	12	M	91	89	86	19	77	76	75
Paterson	40	5	100	3	8	12	M	93	91	88	21	77	76	75
Phillipsburg	40	4	180	1	6	10	L	93	91	88	21	77	76	75
Trenton CO	40	1	144	7	12	16	M	92	90	87	19	78	77	76
Vineland	39	3	95	7	12	16	M	93	90	87	19	78	77	76
NEW MEXICO														
Alamagordo, Holloman AFB	32	5	4070	12	18	22	L	100	98	96	30	70	69	68
Albuquerque AP	35	0	5310	6	14	17	L	96	94	92	27	66	65	64
Artesia	32	5	3375	9	16	19	L	101	99	97	30	71	70	69
Carlsbad AP	32	2	3234	11	17	21	L	101	99	97	28	72	71	70
Clovis AP	34	3	4279	2	14	17	L	99	97	95	28	70	69	68
Farmington AP	36	5	5495	−3	6	9	VL	95	93	91	30	66	65	64
Gallup	35	3	6465	−13	−5	−1	VL	92	90	87	32	64	63	62
Grants	35	1	6520	−15	−7	−3	VL	91	89	86	32	64	63	62
Hobbs AP	32	4	3664	9	15	19	L	101	99	96	29	72	71	70
Las Cruces	32	2	3900	13	19	23	L	102	100	97	30	70	69	68
Los Alamos	35	5	7410	−4	5	9	L	88	86	83	32	64	63	62
Raton AP	36	5	6379	−11	−2	2	L	92	90	88	34	66	65	64
Roswell, Walker AFB	33	2	3643	5	16	19	L	101	99	97	33	71	70	69
Santa Fe CO	35	4	7045	−2	7	11	L	90	88	85	28	65	63	62
Silver City AP	32	4	5373	8	14	18	VL	95	93	91	30	68	67	66
Socorro AP	34	0	4617	6	13	17	L	99	97	94	30	67	66	65
Tucumcari AP	35	1	4053	1	9	13	L	99	97	95	28	71	70	69
NEW YORK														
Albany AP	42	5	277	−14	−5	0	L	91	88	85	23	76	74	73
Albany CO	42	5	19	−5	1	5	L	91	89	86	20	76	74	73
Auburn	43	0	715	−10	−2	2	M	89	87	84	22	75	73	72
Batavia	43	0	900	−7	−1	3	M	89	87	84	22	75	74	72
Binghamton CO	42	1	858	−8	−2	2	L	91	89	86	20	74	72	71
Buffalo AP	43	0	705r	−3	3	6	M	88	86	83	21	75	73	72
Cortland	42	4	1129	−11	−5	−1	L	90	88	85	23	75	73	72
Dunkirk	42	3	590	−2	4	8	M	88	86	83	18	75	74	72
Elmira AP	42	1	860	−5	1	5	L	92	90	87	24	75	73	72
Geneva	42	5	590	−8	−2	2	M	91	89	86	22	75	73	72
Glens Falls	43	2	321	−17	−11	−7	L	88	86	83	23	74	72	71
Gloversville	43	1	770	−12	−6	−2	L	89	87	84	23	75	73	71
Hornell	42	2	1325	−15	−9	−5	L	87	85	82	24	74	72	71

TABLE 41. (Continued)

Col. 1 State and Station[b]	Col. 2 Latitude[c] ° '		Col. 3 Elev.[d] Ft	Winter				Summer							
				Col. 4			Col. 5 Coincident Wind Velocity[e]	Col. 6 Design Dry-Bulb			Col. 7 Outdoor Daily Range[f]	Col. 8 Design Wet-Bulb		
				Median of Annual Extremes	99%	97½%		1%	2½%	5%		1%	2½%	5%
NEW YORK (continued)														
Ithaca	42	3	950	−10	− 4	0	L	91	88	85	24	75	73	72
Jamestown	42	1	1390	− 5	1	5	M	88	86	83	20	75	73	72
Kingston	42	0	279	− 8	− 2	2	L	92	90	87	22	76	74	73
Lockport	43	1	520	− 4	2	6	M	87	85	82	21	75	74	72
Massena AP	45	0	202r	−22	−16	−12	M	86	84	81	20	75	74	72
Newburgh-Stewart AFB	41	3	460	− 4	2	6	M	92	89	86	21	78	76	74
NYC-Central Park	40	5	132	6	11	15	H	94	91	88	17	77	76	75
NYC-Kennedy AP	40	4	16	12	17	21	H	91	87	84	16	77	76	75
NYC-LaGuardia AP	40	5	19	7	12	16	H	93	90	87	16	77	76	75
Niagara Falls AP	43	1	596	− 2	4	7	M	88	86	83	20	75	74	73
Olean	42	1	1420	−13	− 8	− 3	L	87	85	82	23	74	72	71
Oneonta	42	3	1150	−13	− 7	− 3	L	89	87	84	24	74	72	71
Oswego CO	43	3	300	− 4	2	6	M	86	84	81	20	75	74	72
Plattsburg AFB	44	4	165	−16	−10	− 6	L	86	84	81	22	74	73	71
Poughkeepsie	41	4	103	− 6	− 1	3	L	93	90	87	21	77	75	74
Rochester AP	43	1	543	− 5	2	5	M	91	88	85	22	75	74	72
Rome-Griffiss AFB	43	1	515	−13	− 7	− 3	L	90	87	84	22	76	74	73
Schenectady	42	5	217	−11	− 5	− 1	L	90	88	85	22	75	73	72
Suffolk County AFB	40	5	57	4	9	13	H	87	84	81	16	76	75	74
Syracuse AP	43	1	424	−10	− 2	2	M	90	87	85	20	76	74	73
Utica	43	1	714	−12	− 6	− 2	L	89	87	84	22	75	73	72
Watertown	44	0	497	−20	−14	−10	M	86	84	81	20	75	74	72
NORTH CAROLINA														
Asheville AP	35	3	2170r	8	13	17	L	91	88	86	21	75	74	73
Charlotte AP	35	1	735	13	18	22	L	96	94	92	20	78	77	76
Durham	36	0	406	11	15	19	L	94	92	89	20	78	77	76
Elizabeth City AP	36	2	10	14	18	22	M	93	91	89	18	80	79	78
Fayetteville, Pope AFB	35	1	95	13	17	20	L	97	94	92	20	80	79	78
Goldsboro, Seymour-Johnson AFB	35	2	88	14	18	21	M	95	92	90	18	80	79	78
Greensboro AP	36	1	897	9	14	17	L	94	91	89	21	77	76	75
Greenville	35	4	25	14	18	22	M	95	93	90	19	81	80	79
Henderson	36	2	510	8	12	16	L	94	92	89	20	79	78	77
Hickory	35	4	1165	9	14	18	L	93	91	88	21	77	76	75
Jacksonville	34	5	24	17	21	25	M	94	92	89	18	81	80	79
Lumberton	34	4	132	14	18	22	L	95	93	90	20	81	80	79
New Bern AP	35	1	17	14	18	22	L	94	92	89	18	81	80	79
Raleigh/Durham AP	35	5	433	13	16	20	L	95	92	90	20	79	78	77
Rocky Mount	36	0	81	12	16	20	L	95	93	90	19	80	79	78
Wilmington AP	34	2	30	19	23	27	L	93	91	89	18	82	81	80
Winston-Salem AP	36	1	967	9	14	17	L	94	91	89	20	77	76	75
NORTH DAKOTA														
Bismarck AP	46	5	1647	−31	−24	−19	VL	95	91	88	27	74	72	70
Devil's Lake	48	1	1471	−30	−23	−19	M	93	89	86	25	73	71	69
Dickinson AP	46	5	2595	−31	−23	−19	L	96	93	90	25	72	70	68
Fargo AP	46	5	900	−28	−22	−17	L	92	88	85	25	76	74	72
Grand Forks AP	48	0	832	−30	−26	−23	L	91	87	84	25	74	72	70
Jamestown AP	47	0	1492	−29	−22	−18	L	95	91	88	26	75	73	71
Minot AP	48	2	1713	−31	−24	−20	M	91	88	84	25	72	70	68
Williston	48	1	1877	−28	−21	−17	M	94	90	87	25	71	69	67
OHIO														
Akron/Canton AP	41	0	1210	− 5	1	6	M	89	87	84	21	75	73	72
Ashtabula	42	0	690	− 3	3	7	M	89	87	84	18	76	75	74
Athens	39	2	700	− 3	3	7	M	93	91	88	22	77	76	75
Bowling Green	41	3	675	− 7	− 1	3	M	93	91	88	23	77	75	74
Cambridge	40	0	800	− 6	0	4	M	91	89	86	23	77	76	75
Chillicothe	39	2	638	− 1	5	9	M	93	91	88	22	77	76	75
Cincinnati CO	39	1	761	2	8	12	L	94	92	90	21	78	77	76
Cleveland AP	41	2	777r	− 2	2	7	M	91	89	86	22	76	75	74
Columbus AP	40	0	812	− 1	2	7	M	92	88	86	24	77	76	75
Dayton AP	39	5	997	− 2	0	6	M	92	90	87	20	77	75	74
Defiance	41	2	700	− 7	− 1	1	M	93	91	88	24	77	76	74
Findlay AP	41	0	797	− 6	0	4	M	92	90	88	24	77	76	75

TABLE 41. (Continued)

Col. 1 State and Station[b]	Col. 2 Latitude[c] °	′	Col. 3 Elev.[d] Ft	Winter Col. 4 Median of Annual Extremes	99%	97½%	Col. 5 Coincident Wind Velocity[e]	Summer Col. 6 Design Dry-Bulb 1%	2½%	5%	Col. 7 Outdoor Daily Range[f]	Col. 8 Design Wet-Bulb 1%	2½%	5%
OHIO (continued)														
Fremont	41	2	600	− 7	− 1	3	M	92	90	87	24	76	75	74
Hamilton	39	2	650	− 2	4	8	M	94	92	90	22	78	77	76
Lancaster	39	4	920	− 5	1	5	M	93	91	88	23	77	76	75
Lima	40	4	860	− 6	0	4	M	93	91	88	24	77	76	75
Mansfield AP	40	5	1297	− 7	1	3	M	91	89	86	22	76	75	74
Marion	40	4	920	− 5	1	6	M	93	91	88	23	77	76	75
Middletown	39	3	635	− 3	3	7	M	93	91	88	22	77	76	75
Newark	40	1	825	− 7	− 1	3	M	92	90	87	23	77	76	75
Norwalk	41	1	720	− 7	− 1	3	M	92	90	87	22	76	75	74
Portsmouth	38	5	530	0	5	9	L	94	92	89	22	77	76	75
Sandusky CO	41	3	606	− 2	4	8	M	92	90	87	21	76	75	74
Springfield	40	0	1020	− 3	3	7	M	93	90	88	21	77	76	75
Steubenville	40	2	992	− 2	4	9	M	91	89	86	22	76	75	74
Toledo AP	41	4	676 r	− 5	1	5	M	92	90	87	25	77	75	74
Warren	41	2	900	− 6	0	4	M	90	88	85	23	75	74	73
Wooster	40	5	1030	− 7	− 1	3	M	90	88	85	22	76	75	74
Youngstown AP	41	2	1178	− 5	1	6	M	89	86	84	23	75	74	73
Zanesville AP	40	0	881	− 7	− 1	3	M	92	89	87	23	77	76	75
OKLAHOMA														
Ada	34	5	1015	6	12	16	H	102	100	98	23	79	78	77
Altus AFB	34	4	1390	7	14	18	H	103	101	99	25	77	76	75
Ardmore	34	2	880	9	15	19	H	103	101	99	23	79	78	77
Bartlesville	36	5	715	− 1	5	9	H	101	99	97	23	79	78	77
Chickasha	35	0	1085	5	12	16	H	103	101	99	24	77	76	75
Enid-Vance AFB	36	2	1287	3	10	14	H	103	100	98	24	78	77	76
Lawton AP	34	3	1108	6	13	16	H	103	101	98	24	78	77	76
McAlester	34	5	760	7	13	17	H	102	100	98	23	79	78	77
Muskogee AP	35	4	610	6	12	16	M	102	99	96	23	79	78	77
Norman	35	1	1109	5	11	15	H	101	99	97	24	78	77	76
Oklahoma City AP	35	2	1280	4	11	15	H	100	97	95	23	78	77	76
Ponca City	36	4	996	1	8	12	H	102	100	97	24	78	77	76
Seminole	35	2	865	6	12	16	H	102	100	98	23	78	77	76
Stillwater	36	1	884	2	9	13	H	101	99	97	24	78	77	76
Tulsa AP	36	1	650	4	12	16	H	102	99	96	22	79	78	77
Woodward	36	3	1900	− 3	4	8	H	103	101	98	26	76	74	73
OREGON														
Albany	44	4	224	17	23	27	VL	91	88	84	31	69	67	65
Astoria AP	46	1	8	22	27	30	M	79	76	72	16	61	60	59
Baker AP	44	5	3368	−10	− 3	1	VL	94	92	89	30	66	65	63
Bend	44	0	3599	− 7	0	4	VL	89	87	84	33	64	62	61
Corvallis	44	3	221	17	23	27	VL	91	88	84	31	69	67	65
Eugene AP	44	1	364	16	22	26	VL	91	88	84	31	69	67	65
Grants Pass	42	3	925	16	22	26	VL	94	92	89	33	68	66	65
Klamath Falls AP	42	1	4091	− 5	1	5	VL	89	87	84	36	63	62	61
Medford AP	42	2	1298	15	21	23	VL	98	94	91	35	70	68	66
Pendleton AP	45	4	1492	− 2	3	10	VL	97	94	91	29	66	65	63
Portland AP	45	4	21	17	21	24	L	89	85	81	23	69	67	66
Portland CO	45	3	57	21	26	29	L	91	88	84	21	69	68	67
Roseburg AP	43	1	505	19	25	29	VL	93	91	88	30	69	67	65
Salem AP	45	0	195	15	21	25	VL	92	88	84	31	69	67	66
The Dalles	45	4	102	7	13	17	VL	93	91	88	28	70	68	67
PENNSYLVANIA														
Allentown AP	40	4	376	− 2	3	5	M	92	90	87	22	77	75	74
Altoona CO	40	2	1468	− 4	1	5	L	89	87	84	23	74	73	72
Butler	40	4	1100	− 8	− 2	2	L	91	89	86	22	75	74	73
Chambersburg	40	0	640	0	5	9	L	94	92	89	23	76	75	74
Erie AP	42	1	732	1	7	11	M	88	85	82	18	76	74	73
Harrisburg AP	40	1	335	4	9	13	L	92	89	86	21	76	75	74
Johnstown	40	2	1214	− 4	1	5	L	91	87	85	23	74	73	72
Lancaster	40	1	255	− 3	2	6	L	92	90	87	22	77	76	75
Meadville	41	4	1065	− 6	0	4	M	88	86	83	21	75	73	72
New Castle	41	0	825	− 7	− 1	4	M	91	89	86	23	75	74	73

TABLE 41. (Continued)

Col. 1 State and Station[b]	Col. 2 Latitude[c] ° '		Col. 3 Elev.[d] Ft	Winter Col. 4 Median of Annual Extremes	99%	97½%	Col. 5 Coincident Wind Velocity[e]	Summer Col. 6 Design Dry-Bulb 1%	2½%	5%	Col. 7 Outdoor Daily Range[f]	Col. 8 Design Wet-Bulb 1%	2½%	5%
PENNSYLVANIA (continued)														
Philadelphia AP	39	5	7	7	11	15	M	93	90	87	21	78	77	76
Pittsburgh AP	40	3	1137	− 1	5	9	M	90	87	85	22	75	74	73
Pittsburgh CO	40	3	749 r	1	7	11	M	90	88	85	19	75	74	73
Reading CO	40	2	226	1	6	9	M	92	90	87	19	77	76	75
Scranton/Wilkes-Barre	41	2	940	− 3	2	6	L	89	87	84	19	75	74	73
State College	40	5	1175	− 3	2	6	L	89	87	84	23	74	73	72
Sunbury	40	5	480	− 2	3	7	L	91	89	86	22	76	75	74
Uniontown	39	5	1040	− 1	4	8	L	90	88	85	22	75	74	73
Warren	41	5	1280	− 8	− 3	1	L	89	87	84	24	75	73	72
West Chester	40	0	440	4	9	13	M	92	90	87	20	77	76	75
Williamsport AP	41	1	527	− 5	1	5	L	91	89	86	23	76	75	74
York	40	0	390	− 1	4	8	L	93	91	88	22	77	76	75
RHODE ISLAND														
Newport	41	3	20	1	5	11	H	86	84	81	16	75	74	73
Providence AP	41	4	55	0	6	10	M	89	86	83	19	76	75	74
SOUTH CAROLINA														
Anderson	34	3	764	13	18	22	L	96	94	91	21	77	76	75
Charleston AFB	32	5	41	19	23	27	L	94	92	90	18	81	80	79
Charleston CO	32	5	9	23	26	30	L	95	93	90	13	81	80	79
Columbia AP	34	0	217	16	20	23	L	98	96	94	22	79	79	78
Florence AP	34	1	146	16	21	25	L	96	94	92	21	80	79	78
Georgetown	33	2	14	19	23	26	L	93	91	88	18	81	80	79
Greenville AP	34	5	957	14	19	23	L	95	93	91	21	77	76	75
Greenwood	34	1	671	15	19	23	L	97	95	92	21	78	77	76
Orangeburg	33	3	244	17	21	25	L	97	95	92	20	80	79	78
Rock Hill	35	0	470	13	17	21	L	97	95	92	20	78	77	76
Spartanburg AP	35	0	816	13	18	22	L	95	93	90	20	77	76	75
Sumter-Shaw AFB	34	0	291	18	23	26	L	96	94	92	21	80	79	78
SOUTH DAKOTA														
Aberdeen AP	45	3	1296	−29	−22	−18	L	95	92	89	27	77	75	74
Brookings	44	2	1642	−26	−19	−15	M	93	90	87	25	77	75	74
Huron AP	44	3	1282	−24	−16	−12	L	97	93	90	28	77	75	74
Mitchell	43	5	1346	−22	−15	−11	M	96	94	91	28	77	76	74
Pierre AP	44	2	1718 r	−21	−13	− 9	M	98	96	93	29	76	74	73
Rapid City AP	44	0	3165	−17	− 9	− 6	M	96	94	91	28	72	71	69
Sioux Falls AP	43	4	1420	−21	−14	−10	M	95	92	89	24	77	75	74
Watertown AP	45	0	1746	−27	−20	−16	L	93	90	87	26	76	74	73
Yankton	43	0	1280	−18	−11	− 7	M	96	94	91	25	78	76	75
TENNESSEE														
Athens	33	3	940	10	14	18	L	96	94	91	22	77	76	75
Bristol-Tri City AP	36	3	1519	− 1	11	16	L	92	90	88	22	76	75	74
Chattanooga AP	35	0	670	11	15	19	L	97	94	92	22	78	78	77
Clarksville	36	4	470	6	12	16	L	98	96	94	21	79	78	77
Columbia	35	4	690	8	13	17	L	97	95	93	21	79	78	77
Dyersburg	36	0	334	7	13	17	L	98	96	94	21	80	79	78
Greenville	35	5	1320	5	10	14	L	93	91	88	22	76	75	74
Jackson AP	35	4	413	8	14	17	L	97	95	94	21	80	79	78
Knoxville AP	35	5	980	9	13	17	L	95	92	90	21	77	76	75
Memphis AP	35	0	263	11	17	21	L	98	96	94	21	80	79	78
Murfreesboro	35	5	608	7	13	17	L	97	94	92	22	79	78	77
Nashville AP	36	1	577	6	12	16	L	97	95	92	21	79	78	77
Tullahoma	35	2	1075	7	13	17	L	96	94	92	22	79	78	77
TEXAS														
Abilene AP	32	3	1759	12	17	21	M	101	99	97	22	76	75	74
Alice AP	27	4	180	26	30	34	M	101	99	97	20	81	80	79
Amarillo AP	35	1	3607	2	8	12	M	98	96	93	26	72	71	70
Austin AP	30	2	597	19	25	29	M	101	98	96	22	79	78	77
Bay City	29	0	52	25	29	33	M	95	93	91	16	81	80	79
Beaumont	30	0	18	25	29	33	M	96	94	93	19	81	80	79
Beeville	28	2	225	24	28	32	M	99	97	96	18	81	80	79
Big Spring AP	32	2	2537	12	18	22	M	100	98	96	26	75	73	72
Brownsville AP	25	5	16	32	36	40	M	94	92	91	18	80	80	79
Brownwood	31	5	1435	15	20	25	M	102	100	98	22	76	75	74
Bryan AP	30	4	275	22	27	31	M	100	98	96	20	79	78	78

TABLE 41. (Continued)

Col. 1 State and Station[b]	Col. 2 Latitude[c] ° '		Col. 3 Elev,[d] Ft	Winter Col. 4 Median of Annual Extremes	99%	97½%	Col. 5 Coincident Wind Velocity[e]	Summer Col. 6 Design Dry-Bulb 1%	2½%	5%	Col. 7 Outdoor Daily Range[f]	Col. 8 Design Wet-Bulb 1%	2½%	5%
TEXAS (continued)														
Corpus Christi AP	27	5	43	28	32	36	M	95	93	91	19	81	80	80
Corsicana	32	0	425	16	21	25	M	102	100	98	21	79	78	77
Dallas AP	32	5	481	14	19	24	H	101	99	97	20	79	78	78
Del Rio, Laughlin AFB	29	2	1072	24	28	31	M	101	99	98	24	79	77	76
Denton	33	1	655	12	18	22	H	102	100	98	22	79	78	77
Eagle Pass	28	5	743	23	27	31	L	106	104	102	24	80	79	78
El Paso AP	31	5	3918	16	21	25	L	100	98	96	27	70	69	68
Fort Worth AP	32	5	544 r	14	20	24	H	102	100	98	22	79	78	77
Galveston AP	29	2	5	28	32	36	M	91	89	88	10	82	81	81
Greenville	33	0	575	13	19	24	H	101	99	97	21	79	78	78
Harlingen	26	1	37	30	34	38	M	96	95	94	19	80	80	79
Houston AP	29	4	50	23	28	32	M	96	94	92	18	80	80	79
Houston CO	29	5	158 r	24	29	33	M	96	94	92	18	80	80	79
Huntsville	30	4	494	22	27	31	M	99	97	96	20	80	79	78
Killeen-Gray AFB	31	0	1021	17	22	26	M	100	99	97	22	78	77	76
Lamesa	32	5	2965	7	14	18	M	100	98	96	26	74	73	72
Laredo AFB	27	3	503	29	32	36	L	103	101	100	23	79	78	78
Longview	32	2	345	16	21	25	M	100	98	96	20	81	80	79
Lubbock AP	33	4	3243	4	11	15	M	99	97	94	26	73	72	71
Lufkin AP	31	1	286	19	24	28	M	98	96	95	20	81	80	79
McAllen	26	1	122	30	34	38	M	102	100	98	21	80	79	78
Midland AP	32	0	2815 r	13	19	23	M	100	98	96	26	74	73	72
Mineral Wells AP	32	5	934	12	18	22	H	102	100	98	22	78	77	76
Palestine CO	31	5	580	16	21	25	M	99	97	96	20	80	79	78
Pampa	35	3	3230	0	7	11	M	100	98	95	26	73	72	71
Pecos	31	2	2580	10	15	19	L	102	100	97	27	72	71	70
Plainview	34	1	3400	3	10	14	M	100	98	95	26	73	72	71
Port Arthur AP	30	0	16	25	29	33	M	94	92	91	19	81	80	80
San Angelo, Goodfellow AFB	31	2	1878	15	20	25	M	101	99	97	24	76	75	74
San Antonio AP	29	3	792	22	25	30	L	99	97	96	19	77	77	76
Sherman-Perrin AFB	33	4	763	12	18	23	H	101	99	97	22	79	78	77
Snyder	32	4	2325	9	15	19	M	102	100	97	26	75	74	73
Temple	31	1	675	18	23	27	M	101	99	97	22	79	78	77
Tyler AP	32	2	527	15	20	24	M	99	97	96	21	80	79	78
Vernon	34	1	1225	7	14	18	H	103	101	99	24	77	76	75
Victoria AP	28	5	104	24	28	32	M	98	96	95	18	80	79	79
Waco AP	31	4	500	16	21	26	M	101	99	98	22	79	78	78
Wichita Falls AP	34	0	994	9	15	19	H	103	100	98	24	77	76	75
UTAH														
Cedar City AP	37	4	5613	−10	− 1	6	VL	94	91	89	32	65	64	62
Logan	41	4	4775	− 7	3	7	VL	93	91	89	33	66	65	63
Moab	38	5	3965	2	12	16	VL	100	98	95	30	66	65	64
Ogden CO	41	1	4400	− 3	7	11	VL	94	92	89	33	66	65	64
Price	39	4	5580	− 7	3	7	L	93	91	88	33	65	64	63
Provo	40	1	4470	− 6	2	6	L	96	93	91	32	67	66	65
Richfield	38	5	5300	−10	− 1	3	L	94	92	89	34	66	65	64
St. George CO	37	1	2899	13	22	26	VL	104	102	99	33	71	70	69
Salt Lake City AP	40	5	4220	− 2	5	9	L	97	94	92	32	67	66	65
Vernal AP	40	3	5280	−20	−10	− 6	VL	90	88	84	32	64	63	62
VERMONT														
Barre	44	1	1120	−23	−17	−13	L	86	84	81	23	73	72	70
Burlington AP	44	3	331	−18	−12	− 7	M	88	85	83	23	74	73	71
Rutland	43	3	620	−18	−12	− 8	L	87	85	82	23	74	73	71
VIRGINIA														
Charlottsville	38	1	870	7	11	15	L	93	90	88	23	79	77	76
Danville AP	36	3	590	9	13	17	L	95	92	90	21	78	77	76
Fredericksburg	38	2	50	6	10	14	M	94	92	89	21	79	78	76
Harrisonburg	38	3	1340	0	5	9	L	92	90	87	23	78	77	76
Lynchburg AP	37	2	947	10	15	19	L	94	92	89	21	77	76	75
Norfolk AP	36	5	26	18	20	23	M	94	91	89	18	79	78	78
Petersburg	37	1	194	10	15	18	L	96	94	91	20	80	79	78

TABLE 41. (Continued)

Col. 1 State and Station[b]	Col. 2 Latitude[c] °	'	Col. 2 Elev.[d] Ft	Col. 4 Median of Annual Extremes	99%	97½%	Col. 5 Coincident Wind Velocity[e]	Col. 6 Design Dry-Bulb 1%	2½%	5%	Col. 6 Outdoor Daily Range[f]	Col. 8 Design Wet-Bulb 1%	2½%	5%
VIRGINIA (continued)														
Richmond AP	37	3	162	10	14	18	L	96	93	91	21	79	78	77
Roanoke AP	37	2	1174r	9	15	18	L	94	91	89	23	76	75	74
Staunton	38	2	1480	3	8	12	L	92	90	87	23	78	77	75
Winchester	39	1	750	1	6	10	L	94	92	89	21	78	76	75
WASHINGTON														
Aberdeen	47	0	12	19	24	27	M	83	80	77	16	62	61	60
Bellingham AP	48	5	150	8	14	18	L	76	74	71	19	67	65	63
Bremerton	47	3	162	17	24	29	L	85	81	77	20	68	66	65
Ellensburg AP	47	0	1729	− 5	2	6	VL	91	89	86	34	67	65	63
Everett-Paine AFB	47	5	598	13	19	24	L	82	78	74	20	67	65	63
Kennewick	46	0	392	4	11	15	VL	98	96	93	30	69	68	66
Longview	46	1	12	14	20	24	L	88	86	83	30	68	66	65
Moses Lake, Larson AFB	47	1	1183	−14	− 7	− 1	VL	96	93	90	32	68	66	65
Olympia AP	47	0	190	15	21	25	L	85	83	80	32	67	65	63
Port Angeles	48	1	99	20	26	29	M	75	73	70	18	60	58	57
Seattle-Boeing Fld	47	3	14	17	23	27	L	82	80	77	24	67	65	64
Seattle CO	47	4	14	22	28	32	L	81	79	76	19	67	65	64
Seattle-Tacoma AP	47	3	386	14	20	24	L	85	81	77	22	66	64	63
Spokane AP	47	4	2357	− 5	− 2	4	VL	93	90	87	28	66	64	63
Tacoma-McChord AFB	47	1	350	14	20	24	L	85	81	78	22	68	66	64
Walla Walla AP	46	1	1185	5	12	16	VL	98	96	93	27	69	68	66
Wenatchee	47	2	634	− 2	5	9	VL	95	92	89	32	68	66	64
Yakima AP	46	3	1061	− 1	6	10	VL	94	92	89	36	69	67	65
WEST VIRGINIA														
Beckley	37	5	2330	− 4	0	6	L	91	88	86	22	74	73	72
Bluefield AP	37	2	2850	1	6	10	L	88	86	83	22	74	73	72
Charleston AP	38	2	939	1	9	14	L	92	90	88	20	76	75	74
Clarksburg	39	2	977	− 2	3	7	L	92	90	87	21	76	75	74
Elkins AP	38	5	1970	− 4	1	5	L	87	84	82	22	74	73	72
Huntington CO	38	2	565r	4	10	14	L	95	93	91	22	77	76	75
Martinsburg AP	39	2	537	1	6	10	L	96	94	91	21	78	77	76
Morgantown AP	39	4	1245	− 2	3	7	L	90	88	85	22	76	74	73
Parkersburg CO	39	2	615r	2	8	12	L	93	91	88	21	77	76	75
Wheeling	40	1	659	0	5	9	L	91	89	86	21	76	75	74
WISCONSIN														
Appleton	44	2	742	−16	−10	− 6	M	89	87	84	23	75	74	72
Ashland	46	3	650	−27	−21	−17	L	85	83	80	23	73	71	69
Beloit	42	3	780	−13	− 7	− 3	M	92	90	87	24	77	76	75
Eau Claire AP	44	5	888	−21	−15	−11	L	90	88	85	23	76	74	72
Fond du Lac	43	5	760	−17	−11	− 7	M	89	87	84	23	76	74	73
Green Bay AP	44	3	683	−16	−12	− 7	M	88	85	82	23	75	73	72
La Crosse AP	43	5	652	−18	−12	− 8	M	90	88	85	22	78	76	75
Madison AP	43	1	858	−13	− 9	− 5	M	92	88	85	22	77	75	73
Manitowoc	44	1	660	−11	− 5	− 1	M	88	86	83	21	75	74	72
Marinette	45	0	605	−14	− 8	− 4	M	88	86	83	20	74	72	70
Milwaukee AP	43	0	672	−11	− 6	− 2	M	·90	87	84	21	77	75	73
Racine	42	4	640	−10	− 4	0	M	90	88	85	21	77	75	73
Sheboygan	43	4	648	−10	− 4	0	M	89	87	84	20	76	74	72
Stevens Point	44	3	1079	−22	−16	−12	M	89	87	84	23	75	73	71
Waukesha	43	0	860	−12	− 6	− 2	M	91	89	86	22	77	75	74
Wausau AP	44	6	1196	−24	−18	−14	M	89	86	83	23	74	72	70
WYOMING														
Casper AP	42	5	5319	−20	−11	− 5	L	92	90	87	31	63	62	60
Cheyenne AP	41	1	6126	−15	− 6	− 2	M	89	86	83	30	63	62	61
Cody AP	44	3	5090	−23	−13	− 9	L	90	87	84	32	61	60	59
Evanston	41	2	6860	−22	−12	− 8	VL	84	82	79	32	58	57	56
Lander AP	42	5	5563	−26	−16	−12	VL	92	90	87	32	63	62	60
Laramie AP	41	2	7266	−17	− 6	− 2	M	82	80	77	28	61	59	58
Newcastle	43	5	4480	−18	− 9	− 5	M	92	89	86	30	68	67	66
Rawlins	41	5	6736	−24	−15	−11	L	86	84	81	40	62	61	60
Rock Springs AP	41	4	6741	−16	− 6	− 1	VL	86	84	82	32	58	57	56
Sheridan AP	44	5	3942	−21	−12	− 7	L	95	92	89	32	67	65	64
Torrington	42	0	4098	−20	−11	− 7	M	94	92	89	30	68	67	66

TABLE 41. (Continued)

CANADA

Col. 1 Province and Station[b]	Col. 2 Latitude[c] °	′	Col. 3 Elev.[d] Ft	Winter Col. 4 Average Annual Minimum	99%	97½%	Col. 5 Coincident Wind Velocity[e]	Summer Col. 6 Design Dry-Bulb 1%	2½%	5%	Col. 7 Outdoor Daily Range[f]	Col. 8 Design Wet-Bulb 1%	2½%	5%
ALBERTA														
Calgary AP	51	1	3540	−30	−29	−25	M	87	85	82	26	66	64	63
Edmonton AP	53	3	2219	−30	−29	−26	VL	86	83	80	23	69	67	65
Grande Prairie AP	55	1	2190	−44	−43	−37	VL	84	81	78	23	66	64	63
Jasper CO	52	5	3480	−38	−32	−28	VL	87	84	81	28	66	64	63
Lethbridge AP	49	4	3018	−31	−31	−24	M	91	88	85	28	68	66	64
McMurray AP	56	4	1216	−44	−42	−39	VL	87	84	81	28	69	67	65
Medicine Hat AP	50	0	2365	−33	−30	−26	M	96	93	90	28	72	69	67
Red Deer AP	52	1	2965	−38	−33	−28	VL	88	86	83	25	67	65	64
BRITISH COLUMBIA														
Dawson Creek	55	5	2200	−47	−40	−35	L	84	81	78	25	66	64	63
Fort Nelson AP	58	5	1230	−43	−44	−41	VL	87	84	81	23	66	64	63
Kamloops CO	50	4	1150	−15	−16	−10	VL	97	94	91	31	71	69	68
Nanaimo CO	49	1	100	16	17	20	VL	81	78	75	20	66	64	62
New Westminster CO	49	1	50	12	15	19	VL	86	84	82	20	68	66	65
Penticton AP	49	3	1121	0	−1	3	L	94	91	88	31	71	69	68
Prince George AP	53	5	2218	−38	−37	−31	VL	85	82	79	26	68	65	63
Prince Rupert CO	54	2	170	9	11	15	L	73	71	69	13	62	60	59
Trail	49	1	1400	−3	−2	3	VL	94	91	88	30	70	68	67
Vancouver AP	49	1	16	13	15	19	L	80	78	76	17	68	66	65
Victoria CO	48	3	228	20	20	23	M	80	76	72	16	64	62	60
MANITOBA														
Brandon CO	49	5	1200	−36	−29	−26	M	90	87	84	26	75	73	71
Churchill AP	58	5	115	−43	−40	−38	H	79	75	72	18	68	66	63
Dauphin AP	51	1	999	−35	−29	−26	M	89	86	83	24	74	72	70
Flin Flon CO	54	5	1098	−38	−40	−36	L	85	81	78	19	71	69	67
Portage la Prairie AP	49	5	867	−28	−25	−22	M	90	87	84	22	75	74	72
The Pas AP	54	0	894	−41	−35	−32	M	85	81	78	20	73	71	69
Winnipeg AP	49	5	786	−31	−28	−25	M	90	87	84	23	75	74	72
NEW BRUNSWICK														
Campbellton CO	48	0	25	−20	−18	−14	L	87	84	81	20	74	71	69
Chatham AP	47	0	112	−17	−15	−10	M	90	87	84	22	74	71	69
Edmundston CO	47	2	500	−29	−20	−16	M	84	81	78	21	75	72	70
Fredericton AP	45	5	74	−19	−16	−10	L	89	86	83	23	73	70	68
Moncton AP	46	1	248	−16	−12	−7	H	88	85	82	21	74	71	69
Saint John AP	45	2	352	−15	−12	−7	M	81	79	77	18	71	68	66
NEWFOUNDLAND														
Corner Brook CO	49	0	40	−9	−10	−5	H	84	81	79	18	69	68	66
Gander AP	49	0	482	−5	−5	−1	H	85	82	79	20	69	68	66
Goose Bay AP	53	2	144	−28	−27	−25	M	86	81	77	18	69	67	65
St. John's AP	47	4	463	1	2	6	H	79	77	75	17	69	68	66
Stephenville	48	3	44	−4	−6	−1	H	79	76	74	13	69	68	66
NORTHWEST TERRITORIES														
Fort Smith AP	60	0	665	−51	−49	−46	VL	85	83	80	25	67	65	64
Frobisher Bay AP	63	5	68	−45	−45	−42	H	63	59	56	14			
Inuvik	68	2	75	−54	−50	−48	VL	80	77	75	23	63	61	60
Resolute AP	74	4	209	−52	−49	−47	M	54	51	49	10			
Yellowknife AP	62	3	682	−51	−49	−47	VL	78	76	74	17	65	63	62
NOVA SCOTIA														
Amherst	45	5	63	−15	−10	−5	H	85	82	79	21	72	70	68
Halifax AP	44	4	136	−4	0	4	H	83	80	77	16	69	68	67
Kentville CO	45	0	50	−8	−4	0	M	86	83	80	23	72	70	69
New Glasgow	45	4	317	−16	−10	−5	H	84	81	79	21	72	70	68
Sydney AP	46	1	197	−3	0	5	H	84	82	80	20	72	70	68
Truro CO	45	2	77	−17	−12	−7	M	84	81	79	22	72	70	69
Yarmouth AP	43	5	136	2	5	9	H	76	73	71	15	69	68	67
ONTARIO														
Belleville CO	44	1	250	−15	−11	−7	M	89	86	84	21	77	75	73
Chatham CO	42	2	600	−1	3	6	M	92	90	88	20	77	75	74
Cornwall	45	0	210	−22	−14	−9	M	89	86	84	23	77	75	74
Fort William AP	48	2	644	−31	−27	−23	L	86	83	80	23	72	70	68
Hamilton	43	2	303	−2	0	3	M	91	88	86	21	77	75	73
Kapuskasing AP	49	3	752	−37	−31	−28	M	87	84	81	23	73	71	69
Kenora AP	49	5	1345	−33	−31	−28	M	86	83	80	20	75	73	71
Kingston CO	44	2	300	−16	−10	−7	M	85	82	80	20	77	75	73

TABLE 41. (Continued)

Col. 1 Province and Station[b]	Col. 2 Latitude[c] °	'	Col. 3 Elev.[d] Ft	Col. 4 Average Annual Minimum	99%	97½%	Col. 5 Coincident Wind Velocity[e]	Col. 6 Design Dry-Bulb 1%	2½%	5%	Col. 7 Outdoor Daily Range[f]	Col. 8 Design Wet-Bulb 1%	2½%	5%
ONTARIO (continued)														
Kitchener	43	3	1125	−11	− 3	1	M	88	85	83	24	76	75	74
London AP	43	0	912	− 9	− 1	3	M	90	88	86	22	76	75	74
North Bay AP	46	2	1210	−27	−21	−17	M	87	84	82	18	71	70	69
Oshawa	43	5	370	−11	− 5	− 2	M	90	87	85	21	77	75	73
Ottawa AP	45	2	339	−21	−17	−13	M	90	87	84	21	75	74	73
Owen Sound	44	3	597	− 9	− 5	− 1	M	87	84	82	21	74	72	71
Peterborough CO	44	2	648	−20	−13	− 9	M	90	87	85	22	76	74	73
St. Catharines CO	43	1	325	1	2	5	M	91	88	86	20	77	75	73
Sarnia	43	0	625	− 6	2	6	M	92	90	88	19	76	74	73
Sault Ste. Marie CO	46	3	675	−21	−20	−15	M	88	85	83	22	72	70	68
Sudbury	46	3	850	−25	−20	−15	VL	89	86	84	25	72	70	69
Timmins CO	48	3	1100	−37	−33	−28	M	90	87	84	24	73	71	69
Toronto AP	43	4	578	−10	− 3	1	M	90	87	85	22	77	75	73
Windsor AP	42	2	637	− 1	4	7	M	92	90	88	20	77	75	74
PRINCE EDWARD ISLAND														
Charlottetown AP	46	2	186	−11	− 6	− 3	H	84	81	79	16	72	70	68
Summerside AP	46	3	78	−10	− 8	− 3	H	84	81	79	16	72	70	68
QUEBEC														
Bagotville	48	2	536	−35	−26	−22	VL	88	84	81	20	72	71	69
Chicoutimi CO	48	3	150	−31	−24	−20	VL	87	83	80	20	72	71	69
Drummondville CO	45	5	270	−26	−18	−13	M	88	85	82	22	76	74	72
Granby	45	2	550	−23	−17	−12	L	87	84	82	21	76	74	72
Hull	45	3	200	−21	−17	−13	M	90	87	84	21	75	74	73
Mégantic AP	45	4	1362	−27	−20	−16	M	84	81	78	19	75	73	71
Montréal AP	45	3	98	−20	−16	−10	M	88	86	84	18	76	74	73
Québec AP	46	5	245	−25	−19	−13	M	86	82	79	21	75	73	71
Rimouski	48	3	117	−18	−16	−12	H	78	74	71	18	71	69	68
St. Jean	45	2	129	−21	−15	−10	M	87	85	83	20	76	74	73
St. Jérôme	45	5	310	−30	−18	−13	L	87	84	82	23	76	74	73
Sept Îles AP	50	1	190	−29	−27	−22	L	80	78	75	17	66	64	63
Shawinigan	46	3	306	−27	−20	−15	L	88	85	83	21	76	74	72
Sherbrooke CO	45	2	595	−25	−18	−13	L	87	84	81	20	75	73	71
Thetford Mines	46	0	1020	−25	−19	−14	M	86	83	80	22	75	73	71
Trois Rivières CO	46	2	200	−30	−18	−13	M	88	85	82	23	76	74	72
Val d'Or AP	48	0	1108	−37	−31	−27	L	88	85	82	22	72	71	69
Valleyfield	45	2	150	−20	−14	− 9	M	87	85	83	21	76	74	73
SASKATCHEWAN														
Estevan AP	49	0	1884	−32	−30	−25	M	93	89	86	25	75	73	71
Moose Jaw AP	50	2	1857	−33	−32	−27	M	93	89	86	27	73	71	69
North Battleford AP	52	5	1796	−33	−33	−29	L	90	86	83	25	71	69	67
Prince Albert AP	53	1	1414	−45	−41	−35	VL	88	84	81	25	72	70	68
Regina AP	50	3	1884	−38	−34	−29	M	92	88	85	27	73	71	69
Saskatoon AP	52	1	1645	−37	−34	−30	M	90	86	83	25	71	69	67
Swift Current AP	50	2	2677	−31	−29	−25	M	93	89	86	24	72	70	68
Yorkton AP	51	2	1653	−38	−33	−28	M	89	85	82	23	74	72	70
YUKON TERRITORY														
Whitehorse AP	60	4	2289	−45	−45	−42	VL	78	75	72	22	62	60	59

* Data for U. S. stations extracted from *Evaluated Weather Data for Cooling Equipment Design, Addendum No. 1, Winter and Summer Data*, with the permission of the publisher, Fluor Products Company, Inc., Box 1267, Santa Rosa, California.

ᵃ Data compiled from official weather stations, where hourly weather observations are made by trained observers, and from other sources. Table 1 prepared by ASHRAE Technical Committee 2.2, Weather Data and Design Conditions. Percentage of *winter* design data show the percent of 3-month period, December through February. Canadian data are based on January only. Percentage of *summer* design data show the percent of 4-month period, June through September. Canadian data are based on July only. Also see References 1 to 7.

ᵇ When airport temperature observations were used to develop design data, "AP" follows station name, and "AFB" follows Air Force Bases. Data for stations followed by "CO" came from office locations within an urban area and generally reflect an influence of the surrounding area. Stations without designation can be considered semirural and may be directly compared with most airport data.

ᶜ Latitude is given to the nearest 10 minutes, for use in calculating solar loads. For example, the latitude for Anniston, Alabama is given as 33 4, or 33°40′.

ᵈ Elevations are ground elevations for each station as of 1964. Temperature readings are generally made at an elevation of 5 ft above ground, except for locations marked r, indicating roof exposure of thermometer.

ᵉ Coincident wind velocities derived from approximately coldest 600 hours out of 20,000 hours of December through February data per station. Also see References 5 and 6. The four classifications are:
VL = Very Light, 70 percent or more of cold extreme hours ≤7 mph.   M = Moderate, 50 to 74 percent cold extreme hours >7 mph.
L = Light, 50 to 69 percent cold extreme hours ≤7 mph.   H = High, 75 percent or more cold extreme hours >7 mph., and 50 percent are >12 mph.
ᶠ The difference between the average maximum and average minimum temperatures during the warmest month.

† More detailed data on Arizona, California, and Nevada may be found in *Recommended Design Temperatures, Northern, California*, published by the Golden Gate Chapter; and *Recommended Design Temperatures, Southern California, Arizona, Nevada*, published by the Southern California Chapter.

TABLE 42. Climatic Conditions for Foreign Countries

Col. 1 Country and Station	Col. 2 Latitude and Longitude ° ′	Col. 3 Eleva-tion, Ft	Winter — Col. 4 Mean of Annual Ex-tremes	99%	97½%	Summer — Col. 5 Design Dry-Bulb 1%	2½%	5%	Col. 6 Out-door Daily Range F deg	Col. 7 Design Wet-Bulb 1%	2½%	5%
ADEN												
Aden.....................	12 50N/45 02E	10	63	68	70	102	100	98	11	83	82	82
AFGHANISTAN												
Kabul....................	34 35N/69 12E	5955	2	6	9	98	96	93	32	66	65	64
ALGERIA												
Algiers...................	36 46N/3 03E	194	38	43	45	95	92	89	14	77	76	75
ARGENTINA												
Buenos Aires..............	34 35S/58 29W	89	27	32	34	91	89	86	22	77	76	75
Córdoba..................	31 22S/64 15W	1388	21	28	32	100	96	93	27	76	75	74
Tucuman.................	26 50S/65 10W	1401	24	32	36	102	99	96	23	76	75	74
AUSTRALIA												
Adelaide.................	34 56S/138 35E	140	36	38	40	98	94	91	25	72	70	68
Alice Springs.............	23 48S/133 53E	1795	28	34	37	104	102	100	27	75	74	72
Brisbane.................	27 28S/153 02E	137	39	44	47	91	88	86	18	77	76	75
Darwin..................	12 28S/130 51E	88	60	64	66	94	93	91	16	82	81	81
Melbourne...............	37 49S/144 58E	114	31	35	38	95	91	86	21	71	69	68
Perth...................	31 57S/115 51E	210	38	40	42	100	96	93	22	76	74	73
Sydney..................	33 52S/151 12E	138	38	40	42	89	84	80	13	74	73	72
AUSTRIA												
Vienna..................	48 15N/16 22E	644	− 2	6	11	88	86	83	16	71	69	67
AZORES												
Lajes (Terceira)...........	38 45N/27 05W	170	42	46	49	80	78	77	11	73	72	71
BAHAMAS												
Nassau..................	25 05N/77 21W	11	55	61	63	90	89	88	13	80	80	79
BELGIUM												
Brussels.................	50 48N/4 21E	328	13	15	19	83	79	77	19	70	68	67
BERMUDA												
Kindley AFB.............	33 22N/64 41W	129	47	53	55	87	86	85	12	79	78	78
BOLIVIA												
La Paz..................	16 30S/68 09W	12001	28	31	33	71	69	68	24	58	57	56
BRAZIL												
Belem...................	1 27S/48 29W	42	67	70	71	90	89	87	19	80	79	78
Belo Horizonte............	19 56S/43 57W	3002	42	47	50	86	84	83	18	76	75	75
Brasilia.................	15 52S/47 55W	3442	46	49	51	89	88	86	17	76	75	75
Curitiba.................	25 25S/49 17W	3114	28	34	37	86	84	82	21	75	74	74
Fortaleza................	3 46S/38 33W	89	66	69	70	91	90	89	17	79	78	78
Porto Alegre.............	30 02S/51 13W	33	32	37	40	95	92	89	20	76	76	75
Recife...................	8 04S/34 53W	97	67	69	70	88	87	86	10	78	77	77
Rio de Janeiro............	22 55S/43 12W	201	56	58	60	94	92	90	11	80	79	78
Salvador.................	13 00S/38 30W	154	65	67	68	88	87	86	12	79	79	78
São Paulo................	23 33S/46 38W	2608	36	42	46	86	84	82	18	75	74	74
BRITISH HONDURAS												
Belize..................	17 31N/88 11W	17	55	60	62	90	90	89	13	82	82	81
BULGARIA												
Sofia...................	42 42N/23 20E	1805	− 2	3	8	89	86	84	26	71	70	69
BURMA												
Mandalay................	21 59N/96 06E	252	50	54	56	104	102	101	30	81	80	80
Rangoon................	16 47N/96 09E	18	59	62	63	100	98	95	25	83	82	82
CAMBODIA												
Phnom Penh.............	11 33N/104 51E	36	62	66	68	98	96	94	19	83	82	82
CEYLON												
Colombo.................	6 54N/79 52E	24	65	69	70	90	89	88	15	81	80	80
CHILE												
Punta Arenas.............	53 10S/70 54W	26	22	25	27	68	66	64	14	56	55	54
Santiago................	33 27S/70 42W	1706	27	32	35	90	89	88	32	71	70	69
Valparaiso...............	33 01S/71 38W	135	39	43	46	81	79	77	16	67	66	65
CHINA												
Chungking...............	29 33N/106 33E	755	34	37	39	99	97	95	18	81	80	79
Shanghai................	31 12N/121 26E	23	16	23	26	94	92	90	16	81	81	80
COLOMBIA												
Baranquilla..............	10 59N/74 48W	44	66	70	72	95	94	93	17	83	82	82
Bogotá..................	4 36N/74 05W	8406	42	45	46	72	70	69	19	60	59	58
Cali....................	3 25N/76 30W	3189	53	57	58	84	82	79	15	70	69	68
Medellin................	6 13N/75 36W	4650	48	53	55	87	85	84	25	73	72	72
CONGO												
Brazzaville..............	4 15S/15 15E	1043	54	60	62	93	92	91	21	81	81	80
Kinasha (Leopoldville)......	4 20S/15 18E	1066	54	60	62	92	91	90	19	81	80	80
Stanleyville.............	0 26N/15 14E	1370	65	67	68	92	91	90	19	81	80	80

TABLE 42. (*Continued*)

Col. 1 Country and Station	Col. 2 Latitude and Longitude ° ′	Col. 3 Elevation, Ft	Winter Col. 4 Mean of Annual Extremes	99%	97½%	Summer Col. 5 Design Dry-Bulb 1%	2½%	5%	Col. 6 Outdoor Daily Range F deg	Col. 7 Design Wet-Bulb 1%	2½%	5%
CUBA												
Guantanamo Bay	19 54N/75 09W	21	60	64	66	94	93	92	16	82	81	80
Havana	23 08N/82 21W	80	54	59	62	92	91	89	14	81	81	80
CZECHOSLOVAKIA												
Prague	50 05N/14 25E	662	3	4	9	88	85	83	16	66	65	64
DENMARK												
Copenhagen	55 41N/12 33E	43	11	16	19	79	76	74	17	68	66	64
DOMINICAN REPUBLIC												
Santo Domingo	18 29N/69 54W	57	61	63	65	92	90	88	16	81	80	80
ECUADOR												
Guayaquil	2 10S/79 53W	20	61	64	65	92	91	89	20	80	80	79
Quito	0 13S/78 32W	9446	30	36	39	73	72	71	32	63	62	62
EL SALVADOR												
San Salvador	13 42N/89 13W	2238	51	54	56	98	96	95	32	77	76	75
ETHIOPIA												
Addis Ababa	9 02N/38 45E	7753	35	39	41	84	82	81	28	66	65	64
Asmara	15 17N/38 55E	7628	36	40	42	83	81	80	27	65	64	63
FINLAND												
Helsinki	60 10N/24 57E	30	−11	− 7	− 1	77	74	72	14	66	65	63
FRANCE												
Lyon	45 42N/4 47E	938	− 1	10	14	91	89	86	23	71	70	69
Marseilles	43 18N/5 23E	246	23	25	28	90	87	84	22	72	71	69
Nantes	47 15N/1 34W	121	17	22	26	86	83	80	21	70	69	67
Nice	43 42N/7 16E	39	31	34	37	87	85	83	15	73	72	72
Paris	48 49N/2 29E	164	16	22	25	89	86	83	21	70	68	67
Strasbourg	48 35N/7 46E	465	9	11	16	86	83	80	20	70	69	67
FRENCH GUIANA												
Cayenne	4 56N/52 27W	20	69	71	72	92	91	90	17	83	83	82
GERMANY												
Berlin	52 27N/13 18E	187	6	7	12	84	81	78	19	68	67	66
Hamburg	53 33N/9 58E	66	10	12	16	80	76	73	13	68	66	65
Hannover	52 24N/9 40E	561	7	16	20	82	78	75	17	68	67	65
Mannheim	49 34N/8 28E	359	2	8	11	87	85	82	18	71	69	68
Munich	48 09N/11 34E	1729	− 1	5	9	86	83	80	18	68	66	64
GHANA												
Accra	5 33N/0 12W	88	65	68	69	91	90	89	13	80	79	79
GIBRALTAR												
Gibraltar	36 09N/5 22W	11	38	42	45	92	89	86	14	76	75	74
GREECE												
Athens	37 58N/23 43E	351	29	33	36	96	93	91	18	72	71	71
Thessaloniki	40 37N/22 57E	78	23	28	32	95	93	91	20	77	76	75
GREENLAND												
Narssarssuaq	61 11N/45 25W	85	−23	−12	− 8	66	63	61	20	56	54	52
GUATEMALA												
Guatemala City	14 37N/90 31W	4855	45	48	51	83	82	81	24	69	68	67
GUYANA												
Georgetown	6 50N/58 12W	6	70	72	73	89	88	87	11	80	79	79
HAITI												
Port au Prince	18 33N/72 20W	121	63	65	67	97	95	93	20	82	81	80
HONDURAS												
Tegucigalpa	14 06N/87 13W	3094	44	47	50	89	87	85	28	73	72	71
HONG KONG												
Hong Kong	22 18N/114 10E	109	43	48	50	92	91	90	10	81	80	80
HUNGARY												
Budapest	47 31N/19 02E	394	8	10	14	90	86	84	21	72	71	70
ICELAND												
Reykjavik	64 08N/21 56E	59	8	14	17	59	58	56	16	54	53	53
INDIA												
Ahmenabad	23 02N/72 35E	163	49	53	56	109	107	105	28	80	79	78
Bangalore	12 57N/77 37E	3021	53	56	58	96	94	93	26	75	74	74
Bombay	18 54N/72 49E	37	62	65	67	96	94	92	13	82	81	81
Calcutta	22 32N/88 20E	21	49	52	54	98	97	96	22	83	82	82
Madras	13 04N/80 15E	51	61	64	66	104	102	101	19	84	83	83
Nagpur	21 09N/79 07E	1017	45	51	54	110	108	107	30	79	79	78
New Delhi	28 35N/77 12E	703	35	39	41	110	107	105	26	83	82	82
INDONESIA												
Djakarta	6 11S/106 50E	26	69	71	72	90	89	88	14	80	79	78
Kupang	10 10S/123 34E	148	63	66	68	94	93	92	20	81	80	80
Makassar	5 08S/119 28E	61	64	66	68	90	89	88	17	80	80	79
Medan	3 35N/98 41E	77	66	69	71	92	91	90	17	81	80	79
Palembang	3 00S/104 46E	20	67	70	71	92	91	90	17	80	79	79
Surabaya	7 13S/112 43E	10	64	66	68	91	90	89	18	80	79	79

TABLE 42. (*Continued*)

Col. 1 Country and Station	Col. 2 Latitude and Longitude ° '	Col. 3 Eleva- tion, Ft	Winter Col. 4 Mean of Annual Extremes	99%	97½%	Summer Col. 5 Design Dry-Bulb 1%	2½%	5%	Col. 6 Outdoor Daily Range F deg	Col. 7 Design Wet-Bulb 1%	2½%	5%
IRAN												
Abadan	30 21N/48 16E	7	32	39	41	116	113	110	32	82	81	81
Meshed	36 17N/59 36E	3104	3	10	14	99	96	93	29	68	67	66
Tehran	35 41N/51 25E	4002	15	20	24	102	100	98	27	75	74	73
IRAQ												
Baghdad	33 20N/44 24E	111	27	32	35	113	111	108	34	73	72	72
Mosul	36 19N/43 09E	730	23	29	32	114	112	110	40	73	72	72
IRELAND												
Dublin	53 22N/6 21W	155	19	24	27	74	72	70	16	65	64	62
Shannon	52 41N/8 55W	8	19	25	28	76	73	71	14	65	64	63
ISRAEL												
Jerusalem	31 47N/35 13E	2485	31	36	38	95	94	92	24	70	69	69
Tel Aviv	32 06N/34 47E	36	33	39	41	96	93	91	16	74	73	72
ITALY												
Milan	45 27N/09 17E	341	12	18	22	89	87	84	20	76	75	74
Naples	40 53N/14 18E	220	28	34	36	91	88	86	19	74	73	72
Rome	41 48N/12 36E	377	25	30	33	94	92	89	24	74	73	72
IVORY COAST												
Abidjan	5 19N/4 01W	65	64	67	69	91	90	88	15	83	82	81
JAPAN												
Fukuoka	33 35N/130 27E	22	26	29	31	92	90	89	20	82	80	79
Sapporo	43 04N/141 21E	56	− 7	1	5	86	83	80	20	76	74	72
Tokyo	35 41N/139 46E	19	21	26	28	91	89	87	14	81	80	79
JORDAN												
Amman	31 57N/35 57E	2548	29	33	36	97	94	92	25	70	69	68
KENYA												
Nairobi	1 16S/36 48E	5971	45	48	50	81	80	78	24	66	65	65
KOREA												
Pyongyang	39 02N/125 41E	186	−10	− 2	3	89	87	85	21	77	76	76
Seoul	37 34N/126 58E	285	− 1	7	9	91	89	87	16	81	79	78
LEBANON												
Beirut	33 54N/35 28E	111	40	42	45	93	91	90	15	78	77	76
LIBERIA												
Monrovia	6 18N/10 48W	75	64	68	69	90	89	88	19	82	82	81
LIBYA												
Bengasi	32 06N/20 04E	82	41	46	48	97	94	91	13	77	76	75
MADAGASCAR												
Tananarive	18 55S/47 33E	4531	39	43	46	86	84	83	23	73	72	71
MALAYSIA												
Kuala Lumpur	3 07N/101 42E	127	67	70	71	94	93	92	20	82	82	81
Penang	5 25N/100 19E	17	69	72	73	93	93	92	18	82	81	80
Singapore	1 18N/103 50E	33	69	71	72	92	91	90	14	82	81	80
MARTINIQUE												
Fort de France	14 37N/61 05W	13	62	64	66	90	89	88	14	81	81	80
MEXICO												
Guadalajara	20 41N/103 20W	5105	35	39	42	93	91	89	29	68	67	66
Mérida	20 58N/89 38W	72	56	59	61	97	95	94	21	80	79	77
Mexico City	19 24N/99 12W	7575	33	37	39	83	81	79	25	61	60	59
Monterrey	25 40N/100 18W	1732	31	38	41	98	95	93	20	79	78	77
Vera Cruz	19 12N/96 08W	184	55	60	62	91	89	88	12	83	83	82
MOROCCO												
Casablanca	33 35N/7 39W	164	36	40	42	94	90	86	50	73	72	70
NEPAL												
Katmandu	27 42N/85 12E	4388	30	33	35	89	87	86	25	78	77	76
NETHERLANDS												
Amsterdam	52 23N/4 55E	5	17	20	23	79	76	73	10	65	64	63
NEW GUINEA												
Manokwari	0 52S/134 05E	62	70	71	72	89	88	87	12	82	81	81
Point Moresby	9 29S/147 09E	126	62	67	69	92	91	90	14	80	80	79
NEW ZEALAND												
Auckland	36 51S/174 46E	140	37	40	42	78	77	76	14	67	66	65
Christ Church	43 32S/172 37E	32	25	28	31	82	79	76	17	68	67	66
Wellington	41 17S/174 46E	394	32	35	37	76	74	72	14	66	65	64
NICARAGUA												
Managua	12 10N/86 15W	135	62	65	67	94	93	92	21	81	80	79
NIGERIA												
Lagos	6 27N/3 24E	10	67	70	71	92	91	90	12	82	82	81
NORWAY												
Bergen	60 24N/5 19E	141	14	17	20	75	74	73	21	67	66	65
Oslo	59 56N/10 44E	308	− 2	0	4	79	77	74	17	67	66	64

TABLE 42. (Continued)

Col. 1 Country and Station	Col. 2 Latitude and Longitude ° ′	Col. 3 Elevation, Ft	Winter Col. 4 Mean of Annual Extremes	Col. 4 99%	Col. 4 97½%	Summer Col. 5 Design Dry-Bulb 1%	Col. 5 2½%	Col. 5 5%	Col. 6 Outdoor Daily Range F deg	Col. 7 Design Wet-Bulb 1%	Col. 7 2½%	Col. 7 5%
PAKISTAN												
Chittagong...............	22 21N/91 50E	87	48	52	54	93	91	89	20	82	81	81
Karachi..................	24 48N/66 59E	13	45	49	51	100	98	95	14	82	82	81
Lahore...................	31 35N/74 20E	702	32	35	37	109	107	105	27	83	82	81
Peshwar..................	34 01N/71 35E	1164	31	35	37	109	106	103	29	81	80	79
PANAMA AND CANAL ZONE												
Panama City..............	8 58N/79 33W	21·	69	72	73	93	92	91	18	81	81	80
PARAGUAY												
Asunción.................	25 17S/57 30W	456	35	43	46	100	98	96	24	81	81	80
PERU												
Lima....................	12 05S/77 03W	394	51	53	55	86	85	84	17	76	75	74
PHILIPPINES												
Manila..................	14 35N/120 59E	47	69	73	74	94	92	91	20	82	81	81
POLAND												
Kraków..................	50 04N/19 57E	723	− 2	2	6	84	81	78	19	68	67	66
Warsaw..................	52 13N/21 02E	394	− 3	3	8	84	81	78	19	71	70	68
PORTUGAL												
Lisbon..................	38 43N/9 08W	313	32	37	39	89	86	83	16	69	68	67
PUERTO RICO												
San Juan.................	18 29N/66 07W	82	65	67	68	89	88	87	11	81	80	79
RUMANIA												
Bucharest................	44 25N/26 06E	269	− 2	3	8	93	91	89	26	72	71	70
SAUDI ARABIA												
Dhahran.................	26 17N/50 09E	80	39	45	48	111	110	108	32	86	85	84
Jedda...................	21 28N/39 10E	20	52	57	60	106	103	100	22	85	84	83
Riyadh..................	24 39N/46 42E	1938	29	37	40	110	108	106	32	78	77	76
SENEGAL												
Dakar...................	14 42N/17 29W	131	58	61	62	95	93	91	13	81	80	80
SOMALIA												
Mogadiscio...............	2 02N/49 19E	39	67	69	70	91	90	89	12	82	82	81
SOUTH AFRICA												
Capetown................	33 56S/18 29E	55	36	40	42	93	90	86	20	72	71	70
Johannesburg.............	26 11S/78 03E	5463	26	31	34	85	83	81	24	70	69	69
Pretoria.................	25 45S/28 14E	4491	27	32	35	90	87	85	23	70	69	68
SOVIET UNION												
Alma Ata................	43 14N/76 53E	2543	−18	−10	− 6	88	86	83	21	69	68	67
Archangel................	64 33N/40 32E	22	−29	−23	−18	75	71	68	13	60	58	57
Kaliningrad..............	54 43N/20 30E	23	− 3	+ 1	6	83	80	77	17	67	66	65
Krasnoyarsk..............	56 01N/92 57E	498	−41	−32	−27	84	80	76	12	64	62	60
Kiev....................	50 27N/30 30E	600	−12	− 5	+ 1	87	84	81	22	69	68	67
Kharkov.................	50 00N/36 14E	472	−19	−10	− 3	87	84	82	23	69	68	67
Kuibyshev...............	53 11N/50 06E	190	−23	−19	−13	89	85	81	20	69	67	66
Leningrad................	59 56N/30 16E	16	−14	− 9	− 5	78	75	72	15	65	64	63
Minsk...................	53 54N/27 33E	738	−19	−11	− 4	80	77	74	16	67	66	65
Moscow..................	55 46N/37 40E	505	−19	−11	− 6	84	81	78	21	69	67	65
Odessa..................	46 29N/30 44E	214	− 1	4	8	87	84	82	14	70	69	68
Petropavlovsk.............	52 53N/158 42E	286	− 9	− 3	0	70	68	65	13	58	57	56
Rostov on Don............	47 13N/39 43E	159	− 9	− 2	4	90	87	84	20	70	69	68
Sverdlovsk...............	56 49N/60 38E	894	−34	−25	−20	80	76	72	16	63	62	60
Tashkent.................	41 20N/69 18E	1569	− 4	3	8	95	93	90	29	71	70	69
Tbilisi..................	41 43N/44 48E	1325	12	18	22	87	85	83	18	68	67	66
Vladivostok..............	43 07N/131 55E	94	−15	−10	− 7	80	77	74	11	70	69	68
Volgograd...............	48 42N/44 31E	136	−21	−13	− 7	93	89	86	19	71	70	69
SPAIN												
Barcelona................	41 24N/2 09E	312	31	33	36	88	86	84	13	75	74	73
Madrid..................	40 25N/3 41W	2188	22	25	28	93	91	89	25	71	69	67
Valencia.................	39 28N/0 23W	79	31	33	37	92	90	88	14	75	74	73
SUDAN												
Khartoum................	15 37N/32 33E	1279	47	53	56	109	107	104	30	77	76	75
SURINAM												
Paramaribo..............	5 49N/55 09W	12	66	68	70	93	92	90	18	82	82	81
SWEDEN												
Stockholm................	59 21N/18 04E	146	3	5	8	78	74	72	15	64	62	60
SWITZERLAND												
Zurich..................	47 23N/8 33E	1617	4	9	14	84	81	78	21	68	67	66
SYRIA												
Damascus...............	33 30N/36 20E	2362	25	29	32	102	100	98	35	72	71	70
TAIWAN												
Tainan..................	22 57N/120 12E	70	40	46	49	92	91	90	14	84	83	82
Taipei..................	25 02N/121 31E	30	41	44	47	94	92	90	16	83	82	81

TABLE 43. Water Usage for Water-Cooled Equipment

Summer Water Temperature (°F) (normal extremes)	Gallons per Hour per Ton	Summer Water Temperature (°F) (normal extremes)	Gallons per Hour per Ton
55	34	75	58
60	39	80	70
65	44	85	88
70	50		

TABLE 45. Heat Values of Fuels and Electricity

Product	Btu	Unit
Anthracite	25,400,000	ton
Bituminous	26,200,000	ton
Coke	24,800,000	ton
Natural gas (Dry)	103,500	Cft^3
Butane	102,000	gal.
Propane	91,500	gal.
Crude oil	5,800,000	barrel
Diesel fuel	138,238	gal.
Distillate fuel oil	138,690	gal.
Gasoline	125,071	gal.
Jet fuel	135,000	gal.
Kerosene	135,000	gal.
Electricity	3,412	kWh

TABLE 44. Properties of Mixtures of Air and Saturated Water Vapor
Table Based on Barometric Pressure of 29.92 Inches.

TEMP. F	HUMIDITY RATIO - WEIGHT OF SATURATED VAPOR PER POUND OF DRY AIR POUNDS	GRAINS	ENTHALPY OF 1 LB. OF DRY AIR ABOVE 0 F IN BTU	ENTHALPY OF (SATURATED) VAPOR, BTU	ENTHALPY OF MIXTURE OF 1 LB OF DRY AIR WITH VAPOR TO SATURATE IT IN BTU
0	0.000787	5.51	0.0	0.835	0.835
2	.000874	6.12	0.480	0.929	1.408
4	.000969	6.78	0.961	1.030	1.991
6	.001074	7.52	1.441	1.142	2.583
8	.001189	8.32	1.922	1.266	3.188
10	.001315	9.21	2.402	1.401	3.803
12	.001454	10.18	2.882	1.550	4.432
14	.001606	11.24	3.363	1.713	5.076
16	.001772	12.40	3.843	1.892	5.735
18	.001953	13.67	4.324	2.088	6.412
20	.002152	15.06	4.804	2.302	7.106
22	.002369	16.58	5.284	2.536	7.820
24	.002606	18.24	5.765	2.792	8.557
26	.002865	20.06	6.245	3.072	9.317
28	.003147	22.03	6.726	3.377	10.103
30	.003454	24.18	7.206	3.709	10.915
32	.003788	26.52	7.686	4.072	11.758
33	.003944	27.61	7.927	4.242	12.169
34	.004107	28.75	8.167	4.418	12.585
35	.004275	29.93	8.407	4.601	13.008
36	.004450	31.15	8.647	4.791	13.438
37	.004631	32.42	8.887	4.987	13.874
38	.004818	33.73	9.128	5.191	14.319
39	.005012	35.08	9.368	5.403	14.771
40	.005213	36.49	9.608	5.662	15.230
41	.005421	37.95	9.848	5.849	15.697
42	.005638	39.47	10.088	6.084	16.172
43	.005860	41.02	10.329	6.328	16.657
44	.006091	42.64	10.569	6.580	17.149
45	.00633	44.31	10.809	6.841	17.650
46	.00658	46.06	11.049	7.112	18.161
47	.00684	47.88	11.289	7.391	18.680
48	.00710	49.70	11.530	7.681	19.211
49	.00737	51.59	11.770	7.981	19.751
50	.00766	53.62	12.010	8.291	20.301
51	.00795	55.65	12.250	8.612	20.862
52	.00826	57.82	12.491	8.945	21.436
53	.00857	59.99	12.731	9.289	22.020
54	.00889	62.23	12.971	9.644	22.615
55	.00923	64.61	13.211	10.01	23.22
56	.00958	67.06	13.452	10.39	23.84
57	.00993	69.51	13.692	10.79	24.48
58	.01030	72.10	13.932	11.19	25.12
59	.01069	74.83	14.172	11.61	25.78
60	.01108	77.56	14.413	12.05	26.46
61	.01149	80.43	14.653	12.50	27.15
62	.01191	83.37	14.893	12.96	27.85
63	.01235	86.45	15.134	13.44	28.57
64	.01280	89.60	15.374	13.94	29.31
65	.01326	92.82	15.614	14.45	30.06
66	.01374	96.16	15.855	14.98	30.83
67	.01424	99.68	16.095	15.53	31.62
68	.01475	103.3	16.335	16.09	32.42
69	.01528	107.0	16.576	16.67	33.25
70	.01582	110.7	16.816	17.27	34.09
71	.01639	114.7	17.056	17.89	34.95
72	.01697	118.8	17.297	18.53	35.83
73	.01757	123.0	17.537	19.20	36.74
74	.01819	127.3	17.778	19.88	37.66
75	.01882	131.7	18.018	20.59	38.61
76	.01948	136.4	18.259	21.31	39.57
77	.02016	141.1	18.499	22.07	40.57
78	.02086	146.0	18.740	22.84	41.58
79	.02158	151.1	18.980	23.64	42.62
80	.02233	156.3	19.221	24.47	43.69
81	.02310	161.7	19.461	25.32	44.78
82	.02389	167.2	19.702	26.20	45.90
83	.02471	173.0	19.942	27.10	47.04
84	.02555	178.9	20.183	28.04	48.22
85	.02642	184.9	20.423	29.01	49.43
86	.02731	191.2	20.663	30.00	50.66
87	.02824	197.7	20.904	31.03	51.93
88	.02919	204.3	21.144	32.09	53.23
89	.03017	211.2	21.385	33.18	54.56
90	.03118	218.3	21.625	34.31	55.93
91	.03223	225.6	21.865	35.47	57.33
92	.03330	233.1	22.106	36.67	58.78
93	.03441	240.9	22.346	37.90	60.25
94	.03556	248.9	22.587	39.18	61.77
95	.03673	257.1	22.827	40.49	63.32
96	.03795	265.7	23.068	41.85	64.92
97	.03920	274.4	23.308	43.24	66.55
98	.04049	283.4	23.548	44.68	68.23
99	.04182	292.7	23.789	46.17	69.96
100	.04319	302.3	24.029	47.70	71.73
101	.04460	312.2	24.270	49.28	73.55
102	.04606	322.4	24.510	50.91	75.42
103	.04756	332.9	24.751	52.59	77.34
104	.04911	343.8	24.991	54.32	79.31
105	.0507	355.	25.232	56.11	81.34
106	.0523	366.	25.472	57.95	83.42
107	.0540	378.	25.713	59.85	85.56
108	.0558	391.	25.953	61.80	87.76
109	.0576	403.	26.194	63.82	90.03
110	.0594	416.	26.434	65.91	92.34
111	.0614	430.	26.675	68.05	94.72
112	.0633	443.	26.915	70.27	97.18
113	.0654	458.	27.156	72.55	99.71
114	.0675	473.	27.397	74.91	102.31
115	.0696	487.	27.637	77.34	104.98
116	.0719	503.	27.878	79.85	107.73
117	.0742	519.	28.119	82.43	110.55
118	.0765	536.	28.359	85.10	113.46
119	.0790	553.	28.600	87.86	116.46
120	.0815	570.	28.841	90.70	119.54
125	.0954	668.	30.044	106.4	136.44
130	.1116	781.	31.248	124.7	155.9
135	.1308	916.	32.452	146.4	178.9
140	.1534	1074.	33.655	172.0	205.7
145	.1803	1262.	34.059	202.5	237.4
150	.2125	1488.	36.063	239.2	275.3
155	.2514	1760.	37.267	283.5	320.8
160	.2990	2093.	38.472	337.8	376.3
165	.3581	2507.	39.677	405.3	445.0
170	.4327	3028.9	40.882	490.6	531.5
175	.5292	3704.4	42.087	601.1	643.2
180	.6578	4604.6	43.292	748.5	791.8
185	.8363	5854.1	44.498	953.2	997.7
190	1.099	7693.	45.704	1255.0	1301.0
200	2.295	16065.	48.119	2629.0	2677.0

TABLE 46. Conversions of Engineering Constants

Conversions of Engineering Constants

Units of Power

1 kw. = 1.3415 hp. = 738 *ft.-lb./sec. = 44,268 ft.-lb./min. = 2,656,100 ft.-lb./hr. = .948 Btu./sec. = 56.9 Btu./min. = 3413 Btu./hr.
1 hp. = .7455 kw. = 550 ft.-lb./sec. = 33,000 ft.-lb./min. = 1,980,000 ft.-lb./hr. = .707 Btu./sec. = 42.4 Btu./min. = 2544 Btu./hr.
1 ft.-lb./sec. = .001355 kw. = .001818 hp. = 60 ft.-lb./min. = 3600 ft.-lb./hr. = .001284 Btu./sec. = 0.771 Btu./min. = 4.62 Btu./hr.
1 ft.-lb./min. = .00002259 kw. = .0000303 hp. = .01667 ft.-lb./sec. = 60 ft.-lb./hr. = .00002141 Btu./sec. = .001284 Btu./min. = .0771 Btu./hr.
1 ft.-lb. = .000000376 kw. = .000000505 hp. = .000278 ft.-lb./sec. = .01667 ft.-lb./min. = .000000357 Btu./sec. = .00002141 Btu./min. = .001284 Btu./hr.
1 Btu./sec. = 1.055 kw. = 1.416 hp. = 778 ft.-lb./sec. = 46,700 ft.-lb./min. = 2,802,000 ft.-lb./hr. = 60 Btu./min. = 3600 Btu./hr.
1 Btu./min. = .01759 kw. = .02359 hp. = 12.98 ft.-lb./sec. = 778 ft.-lb./min. = 46,700 ft.-lb./hr. = .01667 Btu./sec. = 60 Btu./hr.
1 Btu./hr. = .0002931 kw. = .0003932 hp. = .2163 ft.-lb./sec. = 12.98 ft.-lb./min. = 778 ft.-lb./hr. = .0002778 Btu./sec. = .01667 Btu./min.

*Ft.-lb. above, and in.-lb. table below, mean foot-pound and inch-pound, the work done in moving against one lb. resistance a distance on one ft. and one in., respectively.

Units of Work, Energy, Heat

1 Btu. = 9340 in.-lb. = 778.3 ft.-lb. = .0002930 *kwhr. = .0003931 hp.-hr.
1 in.-lb. = .0001070 Btu. = .0833 ft.-lb. = .00000003137 kwhr. = .0000000421 hp.-hr.
1 ft.-lb. = .001284 Btu. = 12 in.-lb. = .000000376 kwhr. = .000000505 hp.-hr.
1 kwhr. = 3413 Btu. = 31,873,000 in.-lb. = 2,656,100 ft.-lb. = 1.342 hp.-hr.
1 hp.-hr. = 2544 Btu. = 23,760,000 in.-lb. = 1,980,000 ft.-lb. = 0.7455 kwhr.

*1 kilowatthour = 3413 Btu. and 1 Btu. = 778.3 ft.-lb.

Units of Length

1 inch = .0833 foot = .0277 yard = .0000158 miles
1 foot = 12 inches = .333 yard = .000189 miles
1 yard = 36 inches = 3 feet = .000568 miles
1 mile = 63,360 inches = 5280 ft. = 1760 yards

Units of Weight

1 gr. = .00229* oz. = .0001429 lb. = .0000000714 tons
1 oz. = 437.5 gr. = .0625 lb. = .00003215 tons
1 lb. = 7000 gr. = 16 oz. = .000500 tons
1 ton = 14,000,000 gr. = 32,000 oz. = 2000 lb.

*Avoirdupois oz. and lb. short ton of 2000 lb.

Units of Volume

1 cu. in. = .00433 gal. = .00579 cu. ft. = .0000214 cu. yd.
1 gal. = 234 cu. in. = .1337 cu. ft. = .00495 cu. yd. = .00000307* acre-ft.
1 cu. ft. = 1728 cu. in. = 7.48 gal. = .0370 cu. yd. = .0000230 acre-ft.
1 cu. yd. = 46,656 cu. in. = 202.0 gal. = 27 cu. ft. = .000620 acre-ft.
1 acre-ft. = 325,800 gal. = 43,560 cu. ft. = 1613 cu. yd.

*Acre-ft. of water is the volume in 1 ft. of depth covering 1 acre.

Units of Time

1 sec. = .01667 min. = .0002778 hr. = .00001157 days = .0000003805* mo. = .0000000317 yr.
1 min. = 60 sec. = .01667 hr. = .000694 days = .0000228 mo. = .000001903 yr.
1 hr. = 3600 sec. = 60 min. = .0417 days = .001370 mo. = .0001142 yr.
1 day = 86,400 sec. = 1440 min. = 24 hr. = .0329 mo. = .002740 yr.
1 mo. = 2,628,000 sec. = 43,800 min. = 730 hr. = 30.4 days = .0833 yr.
1 yr. = 31,536,000 sec. = 525,600 min. = 8760 hr. = 365 days = 12 mo.

*Month used is exactly 1/12 year.

Weight-Time Rates

1 lb./sec. = 60 lb./min. = 3600 lb./hr. = 86,400 lb./day = 2,628,000 lb./*mo. = 31,536,000 lb./yr.
1 lb./min. = .01667 lb./sec. = 60 lb./hr. = 1440 lb./day = 43,000 lb./mo. = 525,600 lb./yr.
1 lb./hr. = .0002778 lb./sec. = .01667 lb./min. = 24 lb./day = 730 lb./mo. = 8760 lb./yr.
1 lb./day = .00001157 lb./sec. = .000694 lb./min. = .0417 lb./hr. = 30.4 lb./mo. = 365 lb./yr.
1 lb./mo. = .000000381 lb./sec. = .0000228 lb./min. = .001370 lb./hr. = .0329 lb./day = 12 lb./yr.
1 lb./yr. = .0000000317 lb./sec. = .000001903 lb./min. = .0001142 lb./hr. = .002740 lb./day = .0833 lb./mo.

*Month used is exactly 1/12 year = 30.4 days.

Units of Area

1 *cir. mil. = 0.000000785 sq. in.
1 sq. in. = 1,273,200 cir. mils. = .00694 sq. ft. = .000772 sq. yd.
1 sq. ft. = 144 sq. in. = .1111 sq. yd. = .00002296 acres
1 sq. yd. = 1296 sq. in. = 9 sq. ft. = .0002066 acres
1 acre = 43,500 sq. ft. = 4840 sq. yd.

*A cir. (circular) mil is the area of a circle of 1/1000 in. diam. Thus, a round rod of 1-in. diam. has an area of 1,000,000 cir. mils.

Units of Velocity

1 fps. = 60 fpm. = 3600 fph. = .01136 mpm. = .682 mph.
1 fpm. = .01667 fps. = 60 fph. = .0001894 mpm. = .01136 mph.
1 fph. = .0002778 fps. = .01667 fpm. = .00000316 mpm. = .0001894 mph.
1 mpm. = 88 fps. = 5280 fpm. = 316,800 fph. = 60 mph.
1 mph. = 1.467 fps. = 88 fpm. = 5280 fph. = .01667 mpm.

Units of Density

1 lb./cu. in. = 1728 lb./cu. ft. = 0.864 *tons/cu. ft. = 23.3 tons/cu. yd. = 231 lb./gal.
1 lb./cu. ft. = .000579 lb./cu. in. = .000500 tons/cu. ft. = .0135 tons/cu. yd. = .1337 lb./gal.
1 ton/cu. ft. = 1.157 lb./cu. in. = 2000 lb./cu. ft. = 27 tons/cu. yd. = 267 lb./gal.
1 ton/cu. yd. = .0429 lb./cu. in. = 74.1 lb./cu. ft. = .0370 tons/cu. ft. = 9.90 lb./gal.
1 lb./gal. = .00433 lb./cu. in. = 7.48 lb./cu. ft. = .00374 tons/cu. ft. = .1010 tons/cu. yd.

*Tons are short = 2000 lb.

Units of Pressure

1 in. water* = .0833 ft. water = .0735 in. Hg. = .577 oz./sq. in. = 83.1 oz./sq. ft. = .0361 lb./sq. in. = 5.20 lb./sq. ft.
1 ft. water = 12 in. water = .882 in. Hg. = 6.93 oz./sq. in. = 998 oz./sq. ft. = .433 lb./sq. in. = 62.4 lb./sq. ft.
1 in. Hg. = 13.61 in. water = 1.134 ft. water = 7.86 oz./sq. in. = 1131 oz./sq. ft. = .491 lb./sq. in. = 70.7 lb./sq. ft.
1 oz./sq. in. = 1.732 in. water = .1443 ft. water = .1276 in. Hg. = 144 oz./sq. ft. = .0625 lb./sq. in. = 9 lb./sq. ft.
1 oz./sq. ft. = .01203 in. water = .001002 ft. water = .000884 in. Hg. = .00694 oz./sq. in. = .0004434 lb./sq. in. = .0625 lb./sq. ft.
1 lb./sq. in. = 27.71 in. water = 2.31 ft. water = 2.04 in. Hg. = 16 oz./sq. in. = 2304 oz./sq. ft. = 144 lb./sq. ft.
1 lb./sq. ft. = .1924 in. water = .01604 ft. water = .01414 in. Hg. = .1111 oz./sq. in. = 16 oz./sq. ft. = .00694 lb./sq. in.

*In. water means inches head of water at 60°F; in. Hg. is inches head mercury at 32°F.

Volume-Flow Rates

1 cu. ft./sec. = 60 cu. ft./min. = 3600 cu. ft./hr. = 7.48 gal./sec. = 448.8 gal./min. = 26,930 gal./hr.
1 cu. ft./min. = .01667 cu. ft./sec. = 60 cu. ft./hr. = .1247 gal./sec. = 7.48 gal./min. = 448.8 gal./hr.
1 cu. ft./hr. = .0002778 cu. ft./sec. = .01667 cu. ft./min. = .002078 gal./sec. = .1247 gal./min. = 7.48 gal./hr.
1 gal./sec. = .1337 cu. ft./sec. = 8.02 cu. ft./min. = 481 cu. ft./hr. = 60 gal./min. = 3600 gal./hr.
1 gal./min. = .002228 cu. ft./sec. = .1337 cu. ft./min. = 8.02 cu. ft./hr. = .01667 gal./sec. = 60 gal./hr.
1 gal./hr. = .0000371 cu. ft./sec. = .002228 cu. ft./min. = .1337 cu. ft./hr. = .0002778 gal./sec. = .01667 gal./min.

TABLE 47. Conversion Factors

Multiply	by	To Obtain	Multiply	by	To Obtain
Absolute viscosity (poise)	1	Gram/second centimeter	Calories (Kg)/Kg	1.8	BTU/pound
Absolute viscosity (centipoise)	0.01	Poise	Calories (Kg)/minute	51.43	Foot pounds/second
				0.09351	Horse power
Acceleration due to gravity (g)	32.174	Feet/second²		0.06972	Kilowatts
	980.6	Centimeters/second²	Centimeters	0.3937	Inches
Acres	0.4047	Hectares		0.032808	Feet
	10	Square Chains		0.01	Meters
	43,560	Square Feet		10	Millimeters
	4047	Square Meters	Centimeters of Hg at 32°F	0.01316	Atmospheres
	0.001562	Square Miles		0.4461	Feet of water at 62°F
	4840	Square Yards		136	Kgs/Square meter
	160	Square Rods		27.85	Pounds/Square foot
Atmospheres	76.0	Cms of Hg at 32°F		0.1934	Pounds/Square inch
	29.921	Inches of Hg at 32°F	Centimeters/second	1.969	Feet/minute
	33.94	Feet of Water at 62°F		0.03281	Feet/second
	10,333	Kgs/Square meter		0.036	Kilometers/hour
	14.6963	Pounds/Square inch		0.6	Meters/minute
	1.058	Tons/Square foot		0.02237	Miles/hour
	1013.15	Millibars		0.0003728	Miles/minute
	235.1408	Ounces/Square inch	Centimeters/second²	0.03281	Feet/second²
Bags of cement	94	Pounds of cement	Centipoise	0.000672	Pounds/sec foot
Barrels of oil	42	Gallons of oil (US)		2.42	Pounds/hour foot
Barrels of cement	376	Pounds of cement		0.01	Poise
Barrels (not legal) or	31	Gallons (US)	Cheval-vapeur	1	Metric horse power
	31.5	Gallons (US)		75	Kilogram meters/second
Board feet	144 × 1 in.*	Cubic inches		0.98632	Horse power
Boiler horse power	33,479	BTU/hour	Circular inches	10⁶	Circular mils
	9.803	Kilowatts		0.7854	Square inches
	34.5	Pounds of water evaporated/hour at 212°F		785,400	Square mils
BTU	252.016	Calories (gm)	Circular mils	0.7854	Square mils
	0.252	Calories (Kg)		10⁻⁶	Circular inches
	777.54	Foot pounds		7.854 × 10⁻⁷	Square inches
	0.0003927	Horse power hours	Cubic centimeters	3.531 × 10⁻⁵	Cubic feet
	1054.2	Joules		0.06102	Cubic inches
	107.5	Kilogram meters		10⁻⁶	Cubic meters
	0.0002928	Kilowatt hours		1.308 × 10⁻⁶	Cubic yards
BTU/Cu foot	8.89	Calories (Kg)/Cu meter at 32°F		0.0002642	Gallons (US)
				0.001	Liters
BTU/Hr/ft²/°F/ft	0.00413	Cal (gm)/Sec/cm²/°C/cm		0.002113	Pints (liq. US)
	1.49	Cal (Kg)/Hr/M²/°C/Meter		0.001057	Quarts (liq. US)
BTU/minute	12.96	Foot pounds/second		0.0391	Ounces (fluid)
	0.02356	Horse power	Cubic feet	28,320	Cubic centimeters
	0.01757	Kilowatts		1728	Cubic inches
BTU/minute	17.57	Watts		0.02832	Cubic meters
BTU/pound	0.556	Calories (Kg)/Kilogram		0.03704	Cubic yards
Calories (gm)	0.003968	BTU		7.48052	Gallons (US)
	0.001	Calories (Kg)		28.32	Liters
	3.088	Foot pounds		59.84	Pints (liq. US)
	1.558 × 10⁻⁶	Horse power hours		29.92	Quarts (liq. US)
	4.185	Joules		2.296 × 10⁻⁵	Acre feet
	0.4265	Kilogram meters		0.803564	Bushels
	1.1628 × 10⁻⁶	Kilowatt hours	Cubic feet of water	62.4266	Pounds at 39.2°F
	0.0011628	Watt hours		62.3554	Pounds at 62°F
Cal (gm)/sec/cm²/°C/cm	242.13	BTU/Hr/ft²/°F/ft	Cubic feet/minute	472	Cubic centimeters/sec
Calories (Kg)	3.968	BTU		0.1247	Gallons (US)/second
	1000	Calories (gm)		0.472	Liters/second
	3088	Foot pounds		62.36	Pounds water/min at 62°F
	0.001558	Horse power hours		7.4805	Gallons (US)/minute
	4185	Joules		10,772	Gallons/24 hours
	426.5	Kilogram meters		0.033058	Acre feet/24 hours
	0.0011628	Kilowatt hours	Cubic feet/second	646,317	Gallons (US) 24 hours
	1.1628	Watt hours		448.831	Gallons/minute
Calories (Kg)/Cu meter	0.1124	BTU/Cu foot at 0°C		1.98347	Acre feet/24 hours
Cal (Kg)/Hr/M²/°C/M	0.671	BTU/Hr/ft²/°F/foot	Cubic inches	16.387	Cubic centimeters
				0.0005787	Cubic feet
				1.639 × 10⁻⁵	Cubic meters
				2.143 × 10⁻⁵	Cubic yards
				0.004329	Gallons (US)
				0.01639	Liters
				0.03463	Pints (liq. US)
				0.01732	Quarts (liq. US)
			Cubic meters	10⁶	Cubic centimeters
				35.31	Cubic feet
				61,023	Cubic inches
				1.308	Cubic yards
				264.2	Gallons (US)
				1000	Liters

* For thickness less than 1 in. use actual thickness in decimals of an inch.

TABLE 47. (Continued)

Multiply	by	To Obtain	Multiply	by	To Obtain
	2113	Pints (liq. US)		1.3558	Joules
	1057	Quarts (liq. US)		0.13826	Kilogram meters
Cubic yards	764,600	Cubic centimeters		3.766 × 10⁻⁷	Kilowatt hours
	27	Cubic feet		0.0003766	Watt hours
	46,656	Cubic inches	Foot pounds/minute	0.001286	BTU/minute
	0.7646	Cubic meters		0.01667	Foot pounds/second
	202	Gallons (US)		3.03 × 10⁻⁵	Horse power
	764.6	Liters		0.0003241	Calories (Kg)/minute
	1616	Pints (liq. US)		2.26 × 10⁻⁵	Kilowatts
	807.9	Quarts (liq. US)	Foot pounds/second	0.07717	BTU/minute
Cubic yards/minute	0.45	Cubic feet/second		0.001818	Horse power
	3.367	Gallons (US)/second		0.01945	Calories (Kg)/minute
	12.74	Liters/second		0.001356	Kilowatts
Degrees (angle)	60	Minutes	Gallons (Imperial)	277.42	Cubic inches
	0.01745	Radians		4.543	Liters
	3600	Seconds		1.20095	Gallons (US)
Degrees F [less 32]	0.5556	Degrees C	Gallons (US)	3785	Cubic centimeters
Degrees F	1 [plus 460]	Degrees F above absolute 0		0.13368	Cubic feet
				231	Cubic inches
Degrees C	1.8 [plus 32]	Degrees F		0.003785	Cubic meters
	1 [plus 273]	Degrees C above absolute 0		0.004951	Cubic yards
				3.785	Liters
Degrees/second	0.01745	Radians/second		8	Pints (liq. US)
	0.1667	Revolutions/minute		4	Quarts (liq. US)
	0.002778	Revolutions/second		0.83267	Gallons (Imperial)
Diameter (circle) (approx)	3.14159265359	Circumference		3.069 × 10⁻⁶	Acre feet
(approx)	3.1416		Gallons (US) of water at 62° F	8.3357	Pounds of water
(approx)	3.14				
(approx)	²²⁄₇		Gallons (US) of water/minute	6.0086	Tons of water/24 hours
Diameter (circle)	0.88623	Side of equal square	Gallons (US)/minute	0.002228	Cubic feet/second
	0.7071	Side of inscribed square		0.13368	Cubic feet/minute
Diameter³ (sphere)	0.5236	Volume (sphere)		8.0208	Cubic feet/hour
Diam (major) × diam (minor)	0.7854	Area of ellipse		0.06309	Liters/second
				3.78533	Liters/minute
Diameter² (circle)	0.7854	Area (circle)		0.0044192	Acre feet/24 hours
Diameter² (sphere)	3.1416	Surface (sphere)	Grains	1	Grains (avoirdupois)
Diam (inches) × RPM	0.262	Belt speed ft/minute		1	Grains (apothecary)
Drams (avoirdupois)	27.34375	Grains		1	Grains (troy)
	0.0625	Ounces (avoir.)		0.0648	Grams
	1.771845	Grams		0.0020833	Ounces (troy)
				0.0022857	Ounces (avoir.)
Fathoms	6	Feet	Grains/gallon (US)	17.118	Parts/million
Feet	30.48	Centimeters		142.86	Pounds/million gallons (US)
	12	Inches	Grams	980.7	Dynes
	0.3048	Meters		15.43	Grains
	⅓	Yards		0.001	Kilograms
	0.06061	Rods		1000	Milligrams
Feet of water at 62° F	0.029465	Atmospheres		0.03527	Ounces (avoir.)
	0.88162	Inches of Hg at 32° F		0.03215	Ounces (troy)
	62.3554	Pounds/square foot		0.002205	Pounds
	0.43302	Pounds, square inch	Grams/centimeter	0.0056	Pounds/inch
	304.44	Kilogram, sq meter	Grams/cubic centimeter	62.43	Pounds/cubic foot
Feet/minute	0.5080	Centimeters/second		0.03613	Pounds/cubic inch
	0.01667	Feet/second		4.37	Grains/100 cubic ft
	0.01829	Kilometers/hour	Grams/liter	58.417	Grains/gallon (US)
Feet/minute	0.3048	Meters/minute		8.345	Pounds/100 gallons (US)
	0.01136	Miles/hour		0.062427	Pounds/cubic foot
Feet/second	30.48	Centimeters/second		1000	Parts/million
	1.097	Kilometers/hour	Gravity (g)	32.174	Feet/second²
	0.5921	Knots		980.6	Centimeters/second²
	18.29	Meters/minute	Horse power	42.44	BTU/minute
	0.6818	Miles/hour		33,000	Foot pounds/minute
	0.01136	Miles/minute		550	Foot pounds/second
Feet/second²	30.48	Centimeters/second²		1.014	Metric horse power (Cheval vapeur)
	0.3048	Meters/second²		10.7	Calories (Kg)/min
Flat of a hexagon	1.155	Distance across corners		0.7457	Kilowatts
Flat of a square	1.414	Distance across corners		745.7	Watts
Foot pounds	0.0012861	BTU	Horse power (boiler)	33,479	BTU/hour
	0.32412	Calories (gm)		9.803	Kilowatts
	0.0003241	Calories (Kg)		34.5	Pounds of water evaporated/hour at 212° F
	5.05 × 10⁻⁷	Horse power hours	Horse power hours	2546.5	BTU
				641,700	Calories (gm)
				641.7	Calories (Kg)

TABLE 47. (Continued)

Multiply	by	To Obtain	Multiply	by	To Obtain
	1,980,000	Foot pounds	Kilowatt hours	3413	BTU
	2,684,500	Joules	Kilowatt hours	860,500	Calories (gm)
	273,740	Kilogram meters		860.5	Calories (Kg)
	0.7455	Kilowatt hours		2,655,200	Foot pounds
	745.5	Watt hours		1.341	Horse power hours
Inches	2.54	Centimeters		3,600,000	Joules
	0.08333	Feet		367,100	Kilogram meters
	1000	Mils		1000	Watt hours
	12	Lines	Knots	1	Nautical miles/hour
	72	Points		1.1516	Miles/hour
Inches of Hg at 32° F	0.03342	Atmospheres		1.8532	Kilometers/hour
	345.3	Kilograms/square meter	Liters	1000	Cubic centimeters
	70.73	Pounds/square foot		0.03531	Cubic feet
	0.49117	Pounds/square inch		61.02	Cubic inches
	1.1343	Feet of water at 62° F		0.001	Cubic meters
Inches of Hg at 32° F	13.6114	Inches of water at 62° F		0.001308	Cubic yards
	7.85872	Ounces/square inch		0.2642	Gallons (US)
Inches of water at 62° F	0.002455	Atmospheres		0.22	Gallons (Imp)
	25.37	Kilograms/square meter		2.113	Pints (liq. US)
	0.5771	Ounces/square inch		1.057	Quarts (liq. US)
	5.1963	Pounds/square foot		8.107 × 10⁻⁷	Acre Feet
	0.03609	Pounds/square inch		2.2018	Pounds of water at 62° F
	0.07347	Inches of Hg at 32° F	Liters/minute	0.0005886	Cubic feet/second
Joules	0.00094869	BTU		0.004403	Gallons (US)/second
	0.239	Calories (gm)		0.26418	Gallons (US)/minute
	0.000239	Calories (Kg)	Meters	100	Centimeters
	0.73756	Foot pounds		3.281	Feet
	3.72 × 10⁻⁷	Horse power hours		39.37	Inches
	0.10197	Kilogram meters		1.094	Yards
	2.778 × 10⁻⁷	Kilowatt hours		0.001	Kilometers
	0.0002778	Watt hours		1000	Millimeters
	1	Watt second	Meters/minute	1.667	Centimeters/second
Kilograms	980,665	Dynes		3.281	Feet/minute
	2.205	Pounds		0.05468	Feet/second
	0.001102	Tons (short)		0.06	Kilometers/hour
	1000	Grams		0.03728	Miles/hour
	35.274	Ounces (avoir.)	Meters/second	196.8	Feet/minute
	32.1507	Ounces (troy)		3.281	Feet/second
Kilogram meters	0.009302	BTU		3.6	Kilometers/hour
	2.344	Calories (gm)		0.06	Kilometers/minute
	0.002344	Calories (Kg)		2.237	Miles/hour
	7.233	Foot pounds		0.03728	Miles/minute
	3.653 × 10⁻⁶	Horse power hours	Microns	10⁻⁶	Meters
	9.806	Joules		0.001	Millimeters
	2.724 × 10⁻⁶	Kilowatt hours		0.03937	Mils
	0.002724	Watt hours	Mils	0.001	Inches
Kilograms/cubic meter	0.06243	Pounds/cubic foot		0.0254	Millimeters
Kilograms/meter	0.6720	Pounds/foot		25.4	Microns
Kilograms/sq centimeter	14.223	Pounds/sq inch	Miles	160,934	Centimeters
	1	Metric atmosphere		5280	Feet
Kilogram/sq meter	9.678 × 10⁻⁵	Atmospheres		63,360	Inches
	0.003285	Feet of water at 62° F		1.609	Kilometers
	0.002896	Inches of Hg at 32° F		1760	Yards
	0.2048	Pounds/square foot		80	Chains
	0.001422	Pounds/square inch		320	Rods
	0.007356	Centimeters of Hg at 32° F		0.8684	Nautical miles
Kiloliters	1000	Liters	Miles/hour	44.70	Centimeters/second
Kilometers	100,000	Centimeters		88	Feet/minute
	1000	Meters		1.467	Feet/second
	3281	Feet		1.609	Kilometers/hour
	0.6214	Miles		0.8684	Knots
	1094	Yards		26.82	Meters/minute
Kilometers/hour	27.78	Centimeters/second	Miles/minute	2682	Centimeters/second
	54.68	Feet/minute		88	Feet/second
	0.9113	Feet/second		1.609	Kilometers/minute
	16.67	Meters/minute		60	Miles/hour
	0.6214	Miles/hour	Millibars	0.000987	Atmosphere
	0.5396	Knots	Milliers	1000	Kilograms
Kilometers/hr/sec	27.78	Centimeters/sec/sec	Milligrams	0.001	Grams
	0.9113	Feet/sec/sec		0.01543	Grains
	0.2778	Meters/sec/sec	Milligrams/liter	1	Parts/million
Kilowatts	56.92	BTU/minute	Milliliters	0.031	Liters
	44,250	Foot pounds/minute	Million gals/24 hours	1.54723	Cubic feet/second
	737.6	Foot pounds/second	Millimeters	0.1	Centimeters
	1.341	Horse power		0.03937	Inches
	14.34	Calories (Kg)/min			
	1000	Watts			

TABLE 47. (Continued)

Multiply	by	To Obtain	Multiply	by	To Obtain
Minutes (angle)	0.0002909	Radians		0.006944	Pounds/square inch
Nautical miles	6080.2	Feet		0.014139	Inches of Hg at 32° F
	1.1516	Miles		0.0004725	Atmospheres
Ounces (avoirdupois)	16	Drams (avoir.)	Pounds/square inch	0.068044	Atmospheres
	437.5	Grains		2.30934	Feet of water at 62° F
	0.0625	Pounds (avoir.)		2.0360	Inches of Hg at 32° F
	28.349527	Grams		703.067	Kilograms/square meter
	0.9115	Ounces (troy)		27.912	Inches of water at 62° F
Ounces (fluid)	1.805	Cubic inches	Quarts (dry)	67.20	Cubic inches
	0.02957	Liters	Quarts (liq. US)	2	Pints (liq. US)
	29.57	Cubic centimeters		0.9463	Liters
	0.25	Gills		32	Ounces (fluid)
Ounces (troy)	480	Grains		57.75	Cubic inches
	20	Pennyweights (troy)		946.3	Cubic centimeters
	0.08333	Pounds (troy)	Quires	25	Sheets
	31.103481	Grams	Radians	57.30	Degrees
	1.09714	Ounces (avoir.)		3438	Minutes
Ounces/square inch	0.0625	Pounds/square inch		206,625	Seconds
	1.732	Inches of water at 62° F		0.637	Quadrants
	4.39	Centimeters of water at 62° F	Radians/second	57.30	Degrees/second
	0.12725	Inches of Hg at 32° F		0.1592	Revolutions/second
	0.004253	Atmospheres		9.549	Revolutions/minute
Parts/million	0.0584	Grains/gallon (US)	Radians/second²	573.0	Revolutions/minute²
	0.07016	Grains/gallon (Imp)		0.1592	Revolutions/second²
	8.345	Pounds/million gal (US)	Reams	500	Sheets
Pennyweights (troy)	24	Grains	Revolutions	360	Degrees
	1.55517	Grams		4	Quadrants
	0.05	Ounces (troy)		6.283	Radians
	0.0041667	Pounds (troy)	Revolutions/minute	6	Degrees/second
Pints (liq. US)	4	Gills		0.1047	Radians/second
	16	Ounces (fluid)		0.01667	Revolutions/second
	0.5	Quarts (liq. US)	Revolutions/minute²	0.001745	Radians/second²
	28.875	Cubic inches		0.0002778	Revolutions/second²
	473.1	Cubic centimeters	Revolutions/second	360	Degrees/second
Poise	0.0672	Pounds/sec foot		6.283	Radians/second
	242	Pounds/hour foot		60	Revolutions/minute
	100	Centipoise	Revolutions/second²	6.283	Radians/second²
Pounds (avoirdupois)	16	Ounces (avoir.)		3600	Revolutions/minute²
	256	Drams (avoir.)	Rods	16.5	Feet
	7000	Grains		5.5	Yards
	0.0005	Tons (short)	Seconds (angle)	4.848 × 10⁻⁶	Radians
	453.5924	Grams	Sections	1	Square miles
	1.21528	Pounds (troy)	Side of a square	1.4142	Diameter of inscribed circle
	14.5833	Ounces (troy)		1.1284	Diameter of circle with equal area
Pounds (troy)	5760	Grains	Square centimeters	0.001076	Square feet
	240	Pennyweights (troy)		0.1550	Square inches
	12	Ounces (troy)		0.0001	Square meters
	373.24177	Grams		100	Square millimeters
	0.822857	Pounds (avoir.)	Square feet	2.296 × 10⁻⁵	Acres
	13.1657	Ounces (avoir.)		929.0	Square centimeters
	0.00036735	Tons (long)		144	Square inches
	0.00041143	Tons (short)		0.0929	Square meters
	0.00037324	Tons (metric)		3.587 × 10⁻⁸	Square miles
Pounds of water at 62° F	0.01604	Cubic feet		0.1111	Square yards
	27.72	Cubic inches	Square inches	6.452	Square centimeters
	0.120	Gallons (US)		0.006944	Square feet
Pounds of water/min at 62° F	0.0002673	Cubic feet/second		645.2	Square millimeters
				1.27324	Circular inches
Pounds/cubic foot	0.01602	Grams/cubic centimeter		1,273,239	Circular mils
	16.02	Kilograms/cubic meter		1,000,000	Square mils
	0.0005787	Pounds/cubic inch	Square kilometers	247.1	Acres
Pounds/cubic inch	27.68	Grams/cubic centimeter		10,760,000	Square feet
	27,680	Kilograms/cubic meter		1,000,000	Square meters
	1728	Pounds/cubic foot		0.3861	Square miles
Pounds/foot	1.488	Kilograms/meter		1,196,000	Square yards
Pounds/inch	178.6	Grams/centimeter	Square meters	0.0002471	Acres
Pounds/hour foot	0.4132	Centipoise		10.764	Square feet
	0.004132	Poise grams/sec cm		1.196	Square yards
Pounds/sec foot	14.881	Poise grams/sec cm		1	Centares
	1488.1	Centipoise	Square miles	640	Acres
Pounds/square foot	0.016037	Feet of water at 62° F			
	4.882	Kilograms/square meter			

TABLE 47. *(Continued)*

Multiply	by	To Obtain	Multiply	by	To Obtain
Square miles	27,878,400	Square feet	Tons of water/24 hours		
	2.590	Square kilometers	at 62° F	83.33	Pounds of water/hour
	259	Hectares		0.16510	Gallons (US)/minute
	3,097,600	Square yards		1.3263	Cubic feet/hour
	102,400	Square rods			
	1	Sections	Watts	0.05692	BTU/minute
Square millimeters	0.01	Square centimeters		44.26	Foot pounds/minute
	0.00155	Square inches		0.7376	Foot pounds/second
	1550	Square mils		0.001341	Horse power
	1973	Circular mils		0.01434	Calories (Kg)/minute
				0.001	Kilowatts
Square mils	1.27324	Circular mils		1	Joule/second
	0.0006452	Square millimeters			
	10^{-6}	Square inches	Watt hours	3.413	BTU
				860.5	Calories (gm)
Square yards	0.0002066	Acres		0.8605	Calories (Kg)
	9	Square feet		2655	Foot pounds
	0.8361	Square meters		0.001341	Horse power hours
	3.228×10^{-7}	Square miles		3600	Joules
				367.1	Kilogram meters
Tons (long)	1016	Kilograms		0.001	Kilowatt hours
	2240	Pounds			
	1.12	Tons (short)	Watts/square inch	8.2	BTU/square foot/ minute
Tons (metric)	1000	Kilograms		6373	Foot pounds/sq ft/ minute
	2205	Pounds			
	1.1023	Tons (short)		0.1931	Horse power/square foot
Tons (short)	2000	Pounds	Yards	91.44	Centimeters
	32,000	Ounces		3	Feet
	907.185	Kilograms		36	Inches
	0.90718	Tons (metric)		0.9144	Meters
	0.89286	Tons (long)		0.1818	Rods
Tons of refrigeration	12,000	BTU/hour			
	288,000	BTU/24 hours	Year (365 days)	8760	Hours

TABLE 48. Metric Conversion Table

inches	millimeters
0	0.0000
1/128	0.1984
1/64	0.3969
3/128	0.5953
1/32	0.7937
5/128	0.9921
3/64	1.1906
7/128	1.3890

Convert 3.7643 meters to feet, inches and fractions

$$\begin{aligned}
3.7643 \text{ meters} \\
\underline{3.6556} &= 12 \text{ ft.} \\
108.70 \text{ mm} \\
\underline{107.95} &= 4\tfrac{1}{4} \text{ in.} \\
.75 &= \tfrac{1}{32}" \\
3.7643 \text{ meters} &= 12'-4\tfrac{9}{32}"
\end{aligned}$$

Convert 15'-6 7/16" to meters

$$\begin{aligned}
15' &= 4.5720 \text{ meters} \\
6\tfrac{7}{16}" &= .163513 \text{ meters} \\
15'-6\tfrac{7}{16}" &= 4.735513 \text{ meters}
\end{aligned}$$

inches and fractions — millimeters

inches	millimeters	inches	millimeters	inches	millimeters	inches	millimeters	inches	millimeters	inches	millimeters
1/16	1.5875	2 1/16	52.3876	4 1/16	103.188	6 1/16	153.988	8 1/16	204.788	10 1/16	255.588
1/8	3.1750	2 1/8	53.9751	4 1/8	104.775	6 1/8	155.575	8 1/8	206.375	10 1/8	257.176
3/16	4.7625	2 3/16	55.5626	4 3/16	106.363	6 3/16	157.163	8 3/16	207.963	10 3/16	258.763
1/4	6.3500	2 1/4	57.1501	4 1/4	107.950	6 1/4	158.750	8 1/4	209.550	10 1/4	260.351
5/16	7.9375	2 5/16	58.7376	4 5/16	109.538	6 5/16	160.338	8 5/16	211.138	10 5/16	261.938
3/8	9.5250	2 3/8	60.3251	4 3/8	111.125	6 3/8	161.925	8 3/8	212.725	10 3/8	263.526
7/16	11.1125	2 7/16	61.9126	4 7/16	112.713	6 7/16	163.513	8 7/16	214.313	10 7/16	265.113
1/2	12.7000	2 1/2	63.5001	4 1/2	114.300	6 1/2	165.100	8 1/2	215.900	10 1/2	266.701
9/16	14.2875	2 9/16	65.0876	4 9/16	115.888	6 9/16	166.688	8 9/16	217.488	10 9/16	268.288
5/8	15.8750	2 5/8	66.6751	4 5/8	117.475	6 5/8	168.275	8 5/8	219.075	10 5/8	269.876
11/16	17.4625	2 11/16	68.2626	4 11/16	119.063	6 11/16	169.863	8 11/16	220.663	10 11/16	271.463
3/4	19.0500	2 3/4	69.8501	4 3/4	120.650	6 3/4	171.450	8 3/4	222.250	10 3/4	273.051
13/16	20.6375	2 13/16	71.4376	4 13/16	122.238	6 13/16	173.038	8 13/16	223.838	10 13/16	274.638
7/8	22.2250	2 7/8	73.0251	4 7/8	123.825	6 7/8	174.625	8 7/8	225.425	10 7/8	276.226
15/16	23.8125	2 15/16	74.6126	4 15/16	125.413	6 15/16	176.213	8 15/16	227.013	10 15/16	277.813
1	25.4001	3	76.2002	5	127.000	7	177.800	9	228.600	11	279.401
1 1/16	26.9876	3 1/16	77.7877	5 1/16	128.588	7 1/16	179.388	9 1/16	230.188	11 1/16	280.988
1 1/8	28.5751	3 1/8	79.3752	5 1/8	130.175	7 1/8	180.975	9 1/8	231.775	11 1/8	282.576
1 3/16	30.1626	3 3/16	80.9627	5 3/16	131.763	7 3/16	182.563	9 3/16	233.363	11 3/16	284.163
1 1/4	31.7501	3 1/4	82.5502	5 1/4	133.350	7 1/4	184.150	9 1/4	234.950	11 1/4	285.751
1 5/16	33.3376	3 5/16	84.1377	5 5/16	134.938	7 5/16	185.738	9 5/16	236.538	11 5/16	287.338
1 3/8	34.9251	3 3/8	85.7252	5 3/8	136.525	7 3/8	187.325	9 3/8	238.125	11 3/8	288.926
1 7/16	36.5126	3 7/16	87.3127	5 7/16	138.113	7 7/16	188.913	9 7/16	239.713	11 7/16	290.513
1 1/2	38.1001	3 1/2	88.9002	5 1/2	139.700	7 1/2	190.500	9 1/2	241.300	11 1/2	292.101
1 9/16	39.6876	3 9/16	90.4877	5 9/16	141.288	7 9/16	192.088	9 9/16	242.888	11 9/16	293.688
1 5/8	41.2751	3 5/8	92.0752	5 5/8	142.875	7 5/8	193.675	9 5/8	244.475	11 5/8	295.276
1 11/16	42.8626	3 11/16	93.6627	5 11/16	144.463	7 11/16	195.263	9 11/16	246.063	11 11/16	296.863
1 3/4	44.4501	3 3/4	95.2502	5 3/4	146.051	7 3/4	196.850	9 3/4	247.650	11 3/4	298.451
1 13/16	46.0376	3 13/16	96.8377	5 13/16	147.638	7 13/16	198.438	9 13/16	249.238	11 13/16	300.038
1 7/8	47.6251	3 7/8	98.4252	5 7/8	149.225	7 7/8	200.025	9 7/8	250.825	11 7/8	301.626
1 15/16	49.2126	3 15/16	100.013	5 15/16	150.813	7 15/16	201.613	9 15/16	252.413	11 15/16	303.213
2	50.8001	4	101.600	6	152.400	8	203.200	10	254.001	12	304.801

feet into meters

feet	meters	feet	meters	feet	meters	feet	meters	feet	meters	feet	meters	feet	meters
1	0.3048	16	4.8768	31	9.4488	46	14.021	61	18.593	76	23.165	91	27.736
2	0.6096	17	5.1816	32	9.7536	47	14.326	62	18.898	77	23.470	92	28.041
3	0.9144	18	5.4864	33	10.058	48	14.630	63	19.202	78	23.774	93	28.346
4	1.2192	19	5.7912	34	10.363	49	14.935	64	19.507	79	24.079	94	28.650
5	1.5240	20	6.0960	35	10.668	50	15.240	65	19.812	80	24.384	95	28.955
6	1.8288	21	6.4008	36	10.973	51	15.545	66	20.117	81	24.689	96	29.260
7	2.1336	22	6.7056	37	11.278	52	15.850	67	20.422	82	24.994	97	29.565
8	2.4384	23	7.0104	38	11.582	53	16.154	68	20.726	83	25.297	98	29.870
9	2.7432	24	7.3152	39	11.887	54	16.459	69	21.031	84	25.602	99	30.174
10	3.0480	25	7.6200	40	12.192	55	16.764	70	21.336	85	25.907	100	30.480
11	3.3528	26	7.9248	41	12.497	56	17.069	71	21.641	86	26.212		
12	3.6556	27	8.2296	42	12.802	57	17.374	72	21.946	87	26.517		
13	3.9624	28	8.5344	43	13.106	58	17.678	73	22.250	88	26.822		
14	4.2672	29	8.8392	44	13.411	59	17.983	74	22.555	89	27.126		
15	4.5720	30	9.1440	45	13.716	60	18.288	75	22.860	90	27.432		

TABLE 49. Decimal Equivalents of An Inch

Fractions	Decimals of an Inch	Fractions	Decimals of an Inch	Fractions	Decimals of an Inch	Fractions	Decimals of an Inch
1/64	0.015625	17/64	0.265625	33/64	0.515625	49/64	0.765625
1/32	0.03125	9/32	0.28125	17/32	0.53125	25/32	0.78125
3/64	0.046875	19/64	0.296875	35/64	0.546875	51/64	0.796875
1/16	0.0625	5/16	0.3125	9/16	0.5625	13/16	0.8125
5/64	0.078125	21/64	0.328125	37/64	0.578125	53/64	0.828125
3/32	0.09375	11/32	0.34375	19/32	0.59375	27/32	0.84375
7/64	0.109375	23/64	0.359375	39/64	0.609375	55/64	0.859375
1/8	0.125	3/8	0.375	5/8	0.625	7/8	0.875
9/64	0.140625	25/64	0.390625	41/64	0.640625	57/64	0.890625
5/32	0.15625	13/32	0.40625	21/32	0.65625	29/32	0.90625
11/64	0.171875	27/64	0.421875	43/64	0.671875	59/64	0.921875
3/16	0.1875	7/16	0.4375	11/16	0.6875	15/16	0.9375
13/64	0.203125	29/64	0.453125	45/64	0.703125	61/64	0.953125
7/32	0.21875	15/32	0.46875	23/32	0.71875	31/32	0.96875
15/64	0.234375	31/64	0.484375	47/64	0.734375	63/64	0.984375
1/4	0.25	1/2	0.5	3/4	0.75	1	1.000

TABLE 50. Fractions of An Inch Expressed As Decimals of A Foot

Lineal Inches	Lineal Foot	Lineal Inches	Lineal Foot	Lineal Inches	Lineal Foot	Lineal Inches	Lineal Foot	Lineal Inches	Lineal Foot	Lineal Inches	Lineal Foot
1/64	0.00130208	11/16	0.057292	1 7/8	0.15625	3 3/4	0.3125	6 3/4	0.5625	9 3/4	0.8125
1/32	0.00260417	3/4	0.0625	2	0.16667	4	0.33333	7	0.58333	10	0.83333
1/16	0.00520833	13/16	0.067708	2 1/8	0.17708	4 1/4	0.35417	7 1/4	0.60416	10 1/4	0.85417
1/8	0.01041667	7/8	0.072917	2 1/4	0.1875	4 1/2	0.375	7 1/2	0.625	10 1/2	0.875
3/16	0.015625	15/16	0.078125	2 3/8	0.19792	4 3/4	0.39583	7 3/4	0.64583	10 3/4	0.89583
1/4	0.020833	1	0.083333	2 1/2	0.20833	5	0.41667	8	0.66667	11	0.91667
5/16	0.026042	1 1/8	0.09375	2 5/8	0.21875	5 1/4	0.4375	8 1/4	0.6875	11 1/4	0.9375
3/8	0.03125	1 1/4	0.104167	2 3/4	0.22917	5 1/2	0.45834	8 1/2	0.70833	11 1/2	0.95833
7/16	0.036458	1 3/8	0.114583	2 7/8	0.23958	5 3/4	0.47917	8 3/4	0.72917	11 3/4	0.97917
1/2	0.041667	1 1/2	0.125	3	0.25	6	0.5	9	0.75	12	1.000
9/16	0.046875	1 5/8	0.135417	3 1/4	0.27083	6 1/4	0.52083	9 1/4	0.77083		
5/8	0.052083	1 3/4	0.145833	3 1/2	0.29167	6 1/2	0.54167	9 1/2	0.79167		

TABLE 51. Heads of Water in Feet Due To Various Pressures in Pounds Per Square Inch

Pressure in lbs. per sq. Inch	Head in Feet	Pressure in lbs. per sq. Inch	Head in Feet	Pressure in lbs. per sq. Inch	Head in Feet	Pressure in lbs. per sq. Inch	Head in Feet
1	2.307	40	92.280	79	182.253	185	426.795
2	4.614	41	94.587	80	184.560	190	438.330
3	6.921	42	96.894	81	186.867	195	449.865
4	9.228	43	99.201	82	189.174	200	461.400
5	11.535	44	101.508	83	191.481	210	484.470
6	13.842	45	103.815	84	193.788	220	507.540
7	16.149	46	106.122	85	196.095	230	530.610
8	18.456	47	108.429	86	198.402	240	553.680
9	20.763	48	110.736	87	200.709	250	576.750
10	23.070	49	113.043	88	203.116	260	599.820
11	25.377	50	115.350	89	205.423	270	622.890
12	27.684	51	117.657	90	207.730	280	645.960
13	29.991	52	119.964	91	210.037	290	669.030
14	32.298	53	122.271	92	212.344	300	692.100
15	34.605	54	124.578	93	214.651	320	738.240
16	36.912	55	126.885	94	216.958	340	784.380
17	39.219	56	129.192	95	219.265	360	830.520
18	41.526	57	131.499	96	221.572	380	876.660
19	43.833	58	133.806	97	223.879	400	922.800
20	46.140	59	136.113	98	226.186	420	968.940
21	48.447	60	138.420	99	228.493	440	1,015.080
22	50.754	61	140.727	100	230.700	460	1,061.220
23	53.061	62	143.034	105	242.235	480	1,107.360
24	55.368	63	145.341	110	253.770	500	1,153.500
25	57.675	64	147.648	115	265.305	550	1,268.850
26	59.982	65	149.955	120	276.840	600	1,384.200
27	62.289	66	152.262	125	288.375	650	1,499.550
28	64.596	67	154.569	130	299.910	700	1,614.900
29	66.903	68	156.876	135	311.445	750	1,730.250
30	69.210	69	159.183	140	322.980	800	1,845.600
31	71.517	70	161.490	145	334.515	850	1,960.950
32	73.824	71	163.797	150	346.050	900	2,076.300
33	76.131	72	166.104	155	357.585	950	2,191.650
34	78.438	73	168.411	160	369.120	1,000	2,307.000
35	80.745	74	170.718	165	380.655	2,000	4,614.000
36	83.052	75	173.025	170	392.190	3,000	6,921.000
37	85.359	76	175.332	175	403.725	4,000	9,228.000
38	87.666	77	177.639	180	415.260	5,000	11,535.000
39	89.973	78	179.946				

TABLE 52. Equivalent Discharge in Cubic Feet Per Second, Gallons Per Minute, Gallons Per 24 Hours

Cu. Ft. per Sec.	Gal. per Min.	Gal. per 24 Hrs.	Cu. Ft. per Sec.	Gal. per Min.	Gal. per 24 Hrs.	Cu. Ft. per Sec.	Gal. per Min.	Gal. per 24 Hrs.
	DISCHARGE			DISCHARGE			DISCHARGE	
0.1	44.88	64,627	4.3	1,930.2	2,778,969	8.5	3,816.0	5,493,311
0.2	89.76	129,254	4.4	1,975.1	2,843,596	8.6	3,860.9	5,557,938
0.3	134.64	193,882	4.5	2,020.0	2,908,223	8.7	3,905.8	5,622,565
0.4	179.52	258,509	4.6	2,064.9	2,972,850	8.8	3,950.7	5,687,192
0.5	224.40	323,136	4.7	2,109.8	3,037,477	8.9	3,995.6	5,751,819
0.6	269.28	387,763	4.8	2,154.7	3,102,104	9.0	4,040.5	5,816,446
0.7	314.16	452,390	4.9	2,199.6	3,166,731	9.1	4,085.4	5,881,073
0.8	359.04	517,017	5.0	2,244.5	3,231,358	9.2	4,130.3	5,945,700
0.9	403.92	581,645	5.1	2,289.4	3,295,985	9.3	4,175.2	6,010,327
1.0	448.80	646,272	5.2	2,334.3	3,360,612	9.4	4,220.1	6,074,954
1.1	493.68	710,899	5.3	2,379.2	3,425,239	9.5	4,265.0	6,139,581
1.2	538.56	775,526	5.4	2,424.1	3,489,866	9.6	4,309.9	6,204,208
1.3	583.44	840,153	5.5	2,469.0	3,554,493	9.7	4,354.8	6,268,835
1.4	628.32	904,780	5.6	2,513.9	3,619,120	9.8	4,399.7	6,333,462
1.5	673.20	969,407	5.7	2,558.8	3,683,747	9.9	4,444.6	6,398,089
1.6	718.08	1,034,034	5.8	2,603.7	3,748,374	10.0	4,488.0	6,462,720
1.7	762.96	1,098,662	5.9	2,648.6	3,813,001	12.0	5,386.0	7,755,260
1.8	807.84	1,163,290	6.0	2,693.5	3,877,628	14.0	6,283.0	9,047,810
1.9	852.72	1,227,917	6.1	2,738.4	3,942,255	16.0	7,181.0	10,340,340
2.0	897.60	1,292,544	6.2	2,783.3	4,006,882	18.0	8,078.0	11,632,900
2.1	942.48	1,357,171	6.3	2,828.2	4,071,509	20.0	8,976.0	12,925,440
2.2	987.36	1,421,798	6.4	2,873.1	4,136,136	25.0	11,220.0	16,156,800
2.3	1,032.2	1,486,425	6.5	2,918.0	4,200,764	30.0	13,464.0	19,388,140
2.4	1,077.1	1,551,052	6.6	2,962.9	4,265,392	35.0	15,708.0	22,619,520
2.5	1,122.0	1,615,679	6.7	3,007.8	4,330,020	40.0	17,952.0	25,850,890
2.6	1,166.9	1,680,306	6.8	3,052.7	4,394,648	45.0	20,196.0	29,082,230
2.7	1,211.8	1,744,933	6.9	3,097.6	4,459,276	50.0	22,440.0	32,313,580
2.8	1,256.7	1,809,560	7.0	3,142.5	4,523,904	60.0	26,928.0	38,776,280
2.9	1,301.6	1,874,187	7.1	3,187.4	4,588,532	70.0	31,416.0	45,239,040
3.0	1,346.5	1,938,814	7.2	3,232.3	4,653,160	80.0	35,904.0	51,701,760
3.1	1,391.4	2,003,441	7.3	3,277.2	4,717,787	90.0	40,392.0	58,164,460
3.2	1,436.3	2,068,068	7.4	3,322.1	4,782,414	100.0	44,880.0	64,627,200
3.3	1,481.2	2,132,696	7.5	3,367.0	4,847,041	110.0	49,368.0	71,089,900
3.4	1,526.1	2,197,324	7.6	3,411.9	4,911,668	120.0	53,856.0	77,552,600
3.5	1,571.0	2,261,952	7.7	3,456.8	4,976,295	130.0	58,344.0	84,015,300
3.6	1,615.9	2,326,580	7.8	3,501.7	5,040,922	140.0	62,832.0	90,478,000
3.7	1,660.8	2,391,207	7.9	3,546.6	5,105,549	150.0	67,320.0	96,940,700
3.8	1,705.7	2,455,834	8.0	3,591.5	5,170,176	160.0	71,808.0	103,403,400
3.9	1,750.6	2,520,461	8.1	3,636.4	5,234,803	170.0	76,296.0	109,866,200
4.0	1,795.5	2,585,088	8.2	3,681.3	5,299,430	180.0	80,784.0	116,329,000
4.1	1,840.4	2,649,715	8.3	3,726.2	5,364,057	190.0	85,272.0	122,791,700
4.2	1,885.3	2,714,342	8.4	3,771.1	5,428,684	200.0	89,760.0	129,254,400

Based on: 1 cu. ft. of water = 7.48 gallons.

1 cu. ft. per sec. = (7.48)(60) = 448.80 gal. per min.

1 cu. ft. per sec. = (7.48)(60)(60)(24) = 646,272 gal. per 24 hours.

TABLE 53. Steel Pipe and Tubing Specifications

ASTM–A-53

This specification is most generally called for by engineers on process piping and refinery installations (except on stills and high pressure boiler installations—see Specification A-106).

The specification covers both black and galvanized, welded or seamless steel pipe. This pipe is average wall and is intended for coiling, bending and flanging; and is suitable for welding. (Note: Buttweld pipe is not intended for flanging or Van Stoning).

The steel for both welded and seamless pipe shall be made by either open hearth, electric furnace or a special grade Bessemer process.

Pipe produced to this specification is given the following tests: tensile test, bending test, flattening test and hydrostatic test.

ASTM–A-83

This specification covers seamless carbon steel pipe $1\frac{1}{2}$-inch size and under. For higher temperature and pressure service; commonly used by refineries, shipbuilding, and power piping contractors where a better grade of carbon steel pipe is needed.

Pipe ordered under this specification is nominal (average) wall and shall be suitable for bending, flanging, and similar forming operations.

Sizes $\frac{1}{8}$-inch through and including $1\frac{1}{2}$-inch are made in cold drawn tubing (I.P.S.). Sizes $1\frac{1}{4}$ and $1\frac{1}{2}$-inch are also made in hot rolled tubing (I.P.S.).

This tubing is normally furnished in random lengths, plain end square cut unless otherwise specified.

Tubing produced from this specification is given the following tests: flattening test, flange test, crush test, and hydrostatic test.

ASTM–A-106

This specification covers seamless carbon steel pipe for high temperature service. Pipe ordered under these specifications is nominal (average) wall and shall be suitable for bending, flanging, and similar forming operations.

The material for this specification shall be killed steel made by either open hearth or electric furnace, or both.

Pipe produced from this specification is given the following tests: tensile test, bending test, flattening test–1 end, and hydrostatic test.

Supplementary requirements of an optional nature are provided for seamless pipe intended for use in central stations having steam service pressures of 400 psi. and over, and high temperatures; or other applications where a superior grade of pipe is required.

ASTM–A-120

This specification covers black and hot-dipped galvanized welded and seamless steel pipe.

Pipe ordered under this specification is nominal (average) wall and is intended for ordinary uses in steam, water, gas, and air lines, but is not intended for close coiling or bending, or high-temperature service.

No mechanical tests are specified on pipe made to this specification, except hydrostatic tests which shall be made at the mills, as this specification is intended to cover pipe purchased mainly from jobber's stock.

The steel for both welded and seamless pipe under this specification may be produced by either the open hearth, electric furnace or acid Bessemer process.

ASTM–API 5-L

Covers the specifications of the American Petroleum Institute.

All of the above specifications can be supplied in either Grade A or B. Tensile strength: Grade A, 48,000 psi.; grade B, 60,000 psi.

Chemistry of Various Specification of Pipe According to Grades

		Max. Carbon	Manganese	Silicon	Phosphorus
ASTM–A-53:	Lap or Buttweld	—	—	—	.07 max.
	Grade A or B Seamless or Electric Weld	—	—	—	.040 max.
*ASTM–A-83:	Grade A	.08–.16	.30–.60	—	—
	Grade B	.03	.03 max.	—	—
*ASTM–A-106:	Grade A	.23	.30–.90	.12 min.	—
	Grade B	.27	.35–1.00	.12 min.	—
ASTM–A-120:	No Chemistry	—	—	—	—

*$\frac{1}{8}$ through $1\frac{1}{2}$-inch seamless tubing.

TABLE 53. (Continued)

Permissible Temperature Limits

ASTM–A-53 Lap or Buttweld   –20 to 750°F.
ASTM–A-53 Seamless or
             Electric Weld   –20 to 1000°F.
ASTM–A-106   –20 to 1000°F.
ASTM–A-120   –20 to 450°F.

Extra Charge

ASTM–A-53: Welded, $1\frac{1}{2}$ points extra.
         Seamless, no extra.
ASTM–A-83: See schedule covering seamless boiler
         tube. Price predicated on size and
         quantity purchased.
ASTM–A-106: Seamless 10 per cent extra.
ASTM–A-120: No extra.
ASTM–API-5L: No extra.

Specifications Applying to Various Grades of Pipe

ASTM–A-120 meets Federal Specification WWP-406.
ASTM–A-53 meets the following specifications:
  Federal Spec. WWP-404 (June 3, 1944 Am. #1
    June 9, 1945)
  Navy Specification 44P-10K (Jan. 2, 1945)
  Coast Guard Specification, Marine Engineering
    Regulation, Par. 51.37 (July, 1948)
  American Bureau Shipping Regulation; Section 40,
    Grade 1 (1948)
  Association of American Railroads AARM111
    (1946)
  General Electric Co. B-30E5S9 (Nov. 28, 1947)
  Westinghouse Electric 7483A
  N. Y.-N. H. & H. R. R. 156-B (July 15, 1936)
  Norfolk & Western R. R. 67-G (Oct. 1, 1938)
  P. R. R. 19-E (March 21, 1929)
  Southern R. R. 115-A (June 23, 1938)
  Virginian R. R. N-133-G (July 29, 1939)

TABLE 54. Properties of Steel Pipe

The following formulas are used in the computation of the values shown in the table:

† weight of pipe per foot (pounds)	$= 10.6802t(D-t)$
weight of water per foot (pounds)	$= 0.3405d^2$
square feet outside surface per foot	$= 0.2618D$
square feet inside surface per foot	$= 0.2618d$
inside area (square inches)	$= 0.785d^2$
area of metal (square inches)	$= 0.785(D^2-d^2)$
moment of inertia (inches⁴)	$= 0.0491(D^4-d^4)$
	$= A_m R_g{}^2$
section modulus (inches³)	$= \dfrac{0.0982(D^4-d^4)}{D}$
radius of gyration (inches)	$= 0.25\sqrt{D^2+d^2}$

$A_m =$ area of metal (square inches)
$d \quad=$ inside diameter (inches)
$D \quad=$ outside diameter (inches)
$R_g =$ radius of gyration (inches)
$t \quad=$ pipe wall thickness (inches)

† The ferritic steels may be about 5% less, and the austenitic stainless steels about 2% greater than the values shown in this table which are based on weights for carbon steel.

* schedule numbers

Standard weight pipe and schedule 40 are the same in all sizes through 10-inch; from 12-inch through 24-inch, standard weight pipe has a wall thickness of ⅜-inch.

Extra strong weight pipe and schedule 80 are the same in all sizes through 8-inch; from 8-inch through 24-inch, extra strong weight pipe has a wall thickness of ½-inch.

Double extra strong weight pipe has no corresponding schedule number.

a: ASA B36.10 steel pipe schedule numbers

b: ASA B36.10 steel pipe nominal wall thickness designation

c: ASA B36.19 stainless steel pipe schedule numbers

nominal pipe size outside diameter, in.	schedule number* a	b	c	wall thick-ness, in.	inside diam-eter, in.	inside area, sq. in.	metal area, sq. in.	sq ft outside surface, per ft	sq ft inside surface, per ft	weight per ft, lb†	weight of water per ft, lb	moment of inertia, in.⁴	section modu-lus, in.³	radius gyra-tion, in.
⅛ 0.405	......	......	10S	0.049	0.307	0.0740	0.0548	0.106	0.0804	0.186	0.0321	0.00088	0.00437	0.1271
	40	Std	40S	0.068	0.269	0.0568	0.0720	0.106	0.0705	0.245	0.0246	0.00106	0.00525	0.1215
	80	XS	80S	0.095	0.215	0.0364	0.0925	0.106	0.0563	0.315	0.0157	0.00122	0.00600	0.1146
¼ 0.540	......	......	10S	0.065	0.410	0.1320	0.0970	0.141	0.1073	0.330	0.0572	0.00279	0.01032	0.1694
	40	Std	40S	0.088	0.364	0.1041	0.1250	0.141	0.0955	0.425	0.0451	0.00331	0.01230	0.1628
	80	XS	80S	0.119	0.302	0.0716	0.1574	0.141	0.0794	0.535	0.0310	0.00378	0.01395	0.1547
⅜ 0.675	......	......	SS	0.065	0.710	0.396	0.1582	0.220	0.1859	0.538	0.1716	0.01197	0.0285	0.2750
	......	......	10S	0.065	0.545	0.2333	0.1246	0.177	0.1427	0.423	0.1011	0.00586	0.01737	0.2169
	40	Std	40S	0.091	0.493	0.1910	0.1670	0.177	0.1295	0.568	0.0827	0.00730	0.02160	0.2090
	80	XS	80S	0.126	0.423	0.1405	0.2173	0.177	0.1106	0.739	0.0609	0.00862	0.02554	0.1991

TABLE 54. (*Continued*)

nominal pipe size outside diameter, in.	schedule number* a	b	c	wall thickness, in.	inside diameter, in.	inside area, sq. in.	metal area, sq. in.	sq ft outside surface, per ft	sq ft inside surface, per ft	weight per ft, lb†	weight of water per ft, lb	moment of inertia, in.⁴	section modulus, in.³	radius gyration, in.
½ 0.840	......	......	5S	0.065	0.710	0.3959	0.1583	0.220	0.1859	0.538	0.171	0.0120	0.0285	0.2750
	......	......	10S	0.083	0.674	0.357	0.1974	0.220	0.1765	0.671	0.1547	0.01131	0.0341	0.2692
	40	Std	40S	0.109	0.622	0.304	0.2503	0.220	0.1628	0.851	0.1316	0.01710	0.0407	0.2613
	80	XS	80S	0.147	0.546	0.2340	0.320	0.220	0.1433	1.088	0.1013	0.02010	0.0478	0.2505
	160			0.187	0.466	0.1706	0.383	0.220	0.1220	1.304	0.0740	0.02213	0.0527	0.2402
	......	XXS		0.294	0.252	0.0499	0.504	0.220	0.0660	1.714	0.0216	0.02425	0.0577	0.2192
¾ 1.050	......	......	5S	0.065	0.920	0.665	0.2011	0.275	0.2409	0.684	0.2882	0.02451	0.0467	0.349
	......	......	10S	0.083	0.884	0.614	0.2521	0.275	0.2314	0.857	0.2661	0.02970	0.0566	0.343
	40	Std	40S	0.113	0.824	0.533	0.333	0.275	0.2157	1.131	0.2301	0.0370	0.0706	0.334
	80	XS	80S	0.154	0.742	0.432	0.435	0.275	0.1943	1.474	0.1875	0.0448	0.0853	0.321
	160			0.218	0.614	0.2961	0.570	0.275	0.1607	1.937	0.1284	0.0527	0.1004	0.304
	......	XXS	......	0.308	0.434	0.1479	0.718	0.275	0.1137	2.441	0.0641	0.0579	0.1104	0.2840
1 1.315	......	......	5S	0.065	1.185	1.103	0.2553	0.344	0.310	0.868	0.478	0.0500	0.0760	0.443
	......	......	10S	0.109	1.097	0.945	0.413	0.344	0.2872	1.404	0.409	0.0757	0.1151	0.428
	40	Std	40S	0.133	1.049	0.864	0.494	0.344	0.2746	1.679	0.374	0.0874	0.1329	0.421
	80	XS	80S	0.179	0.957	0.719	0.639	0.344	0.2520	2.172	0.311	0.1056	0.1606	0.407
	160			0.250	0.815	0.522	0.836	0.344	0.2134	2.844	0.2261	0.1252	0.1903	0.387
	......	XXS		0.356	0.599	0.2818	1.076	0.344	0.1570	3.659	0.1221	0.1405	0.2137	0.361
1¼ 1.660	......	......	5S	0.065	1.530	1.839	0.326	0.434	0.401	1.107	0.797	0.1038	0.1250	0.564
	...	......	10S	0.109	1.442	1.633	0.531	0.434	0.378	1.805	0.707	0.1605	0.1934	0.550
	40	Std	40S	0.140	1.380	1.496	0.669	0.434	0.361	2.273	0.648	0.1948	0.2346	0.540
	80	XS	80S	0.191	1.278	1.283	0.881	0.434	0.335	2.997	0.555	0.2418	0.2913	0.524
	160	.		0.250	1.160	1.057	1.107	0.434	0.304	3.765	0.458	0.2839	0.342	0.506
	......	XXS	......	0.382	0.896	0.631	1.534	0.434	0.2346	5.214	0.2732	0.341	0.411	0.472
1½ 1.900	..	......	5S	0.065	1.770	2.461	0.375	0.497	0.463	1.274	1.067	0.1580	0.1663	0.649
	......	......	10S	0.109	1.682	2.222	0.613	0.497	0.440	2.085	0.962	0.2469	0.2599	0.634
	40	Std	40S	0.145	1.610	2.036	0.799	0.497	0.421	2.718	0.882	0.310	0.326	0.623
	80	XS	80S	0.200	1.500	1.767	1.068	0.497	0.393	3.631	0.765	0.391	0.412	0.605
1½ 1.900	160			0.281	1.338	1.406	1.429	0.497	0.350	4.859	0.608	0.483	0.508	0.581
	......	XXS	......	0.400	1.100	0.950	1.885	0.497	0.288	6.408	0.412	0.568	0.598	0.549
	......	......		0.525	0.850	0.567	2.267	0.497	0.223	7.710	0.246	0.6140	0.6470	0.5200
	......	......		0.650	0.600	0.283	2.551	0.497	0.157	8.678	0.123	0.6340	0.6670	0.4980
2 2.375	......	......	5S	0.065	2.245	3.96	0.472	0.622	0.588	1.604	1.716	0.315	0.2652	0.817
	......	......	10S	0.109	2.157	3.65	0.776	0.622	0.565	2.638	1.582	0.499	0.420	0.802
	40	Std	40S	0.154	2.067	3.36	1.075	0.622	0.541	3.653	1.455	0.666	0.561	0.787
	80	XS	80S	0.218	1.939	2.953	1.477	0.622	0.508	5.022	1.280	0.868	0.731	0.766
	160			0.343	1.689	2.240	2.190	0.622	0.442	7.444	0.971	1.163	0.979	0.729
	......	XXS		0.436	1.503	1.774	2.656	0.622	0.393	9.029	0.769	1.312	1.104	0.703
	......	......	......	0.562	1.251	1.229	3.199	0.622	0.328	10.882	0.533	1.442	1.2140	0.6710
	......	......	......	0.687	1.001	0.787	3.641	0.622	0.262	12.385	0.341	1.5130	1.2740	0.6440
2½ 2.875	......	......	5S	0.083	2.709	5.76	0.728	0.753	0.709	2.475	2.499	0.710	0.494	0.988
	......	......	10S	0.120	2.635	5.45	1.039	0.753	0.690	3.531	2.361	0.988	0.687	0.975
	40	Std	40S	0.203	2.469	4.79	1.704	0.753	0.646	5.793	2.076	1.530	1.064	0.947
	80	XS	80S	0.276	2.323	4.24	2.254	0.753	0.608	7.661	1.837	1.925	1.339	0.924
	160			0.375	2.125	3.55	2.945	0.753	0.556	10.01	1.535	2.353	1.637	0.894
	......	XXS	......	0.552	1.771	2.464	4.03	0.753	0.464	13.70	1.067	2.872	1.998	0.844
	......	......	......	0.675	1.525	1.826	4.663	0.753	0.399	15.860	0.792	3.0890	2.1490	0.8140
	......	......	......	0.800	1.275	1.276	5.212	0.753	0.334	17.729	0.554	3.2250	2.2430	0.7860
3 3.500	......	......	5S	0.083	3.334	8.73	0.891	0.916	0.873	3.03	3.78	1.301	0.744	1.208
	......	......	10S	0.120	3.260	8.35	1.274	0.916	0.853	4.33	3.61	1.822	1.041	1.196
	40	Std	40S	0.216	3.068	7.39	2.228	0.916	0.803	7.58	3.20	3.02	1.724	1.164
	80	XS	80S	0.300	2.900	6.61	3.02	0.916	0.759	10.25	2.864	3.90	2.226	1.136
	160			0.437	2.626	5.42	4.21	0.916	0.687	14.32	2.348	5.03	2.876	1.094
	......	XXS	......	0.600	2.300	4.15	5.47	0.916	0.602	18.58	1.801	5.99	3.43	1.047
	......	......	......	0.725	2.050	3.299	6.317	0.916	0.537	21.487	1.431	6.5010	3.7150	1.0140
	......	......	......	0.850	1.800	2.543	7.073	0.916	0.471	24.057	1.103	6.8530	3.9160	0.9840
3½ 4.000	......	......	5S	0.083	3.834	11.55	1.021	1.047	1.004	3.47	5.01	1.960	0.980	1.385
	......	......	10S	0.120	3.760	11.10	1.463	1.047	0.984	4.97	4.81	2.756	1.378	1.372
	40	Std	40S	0.226	3.548	9.89	2.680	1.047	0.929	9.11	4.28	4.79	2.394	1.337
	80	XS	80S	0.318	3.364	8.89	3.68	1.047	0.881	12.51	3.85	6.28	3.14	1.307
	......	XXS	......	0.636	2.728	5.845	6.721	1.047	0.716	22.850	2.530	9.8480	4.9240	1.2100

TABLE 54. (Continued)

nominal pipe size outside diameter, in.	schedule number* a	b	c	wall thickness, in.	inside diameter, in.	inside area, sq. in.	metal area, sq. in.	sq ft outside surface, per ft	sq ft inside surface, per ft	weight per ft, lb†	weight of water per ft, lb	moment of inertia, in.⁴	section modulus, in.³	radius gyration, in.
			5S	0.083	4.334	14.75	1.152	1.178	1.135	3.92	6.40	2.811	1.249	1.562
			10S	0.120	4.260	14.25	1.651	1.178	1.115	5.61	6.17	3.96	1.762	1.549
				0.188	4.124	13.357	2.547	1.178	1.082	8.560	5.800	5.8500	2.6000	1.5250
4	40	Std	40S	0.237	4.026	12.73	3.17	1.178	1.054	10.79	5.51	7.23	3.21	1.510
4.500	80	XS	80S	0.337	3.826	11.50	4.41	1.178	1.002	14.98	4.98	9.61	4.27	1.477
	120			0.437	3.626	10.33	5.58	1.178	0.949	18.96	4.48	11.65	5.18	1.445
				0.500	3.500	9.621	6.283	1.178	0.916	21.360	4.160	12.7710	5.6760	1.4250
	160			0.531	3.438	9.28	6.62	1.178	0.900	22.51	4.02	13.27	5.90	1.416
		XXS		0.674	3.152	7.80	8.10	1.178	0.825	27.54	3.38	15.29	6.79	1.374
				0.800	2.900	6.602	9.294	1.178	0.759	31.613	2.864	16.6610	7.4050	1.3380
				0.925	2.650	5.513	10.384	1.178	0.694	35.318	2.391	17.7130	7.8720	1.3060
			5S	0.109	5.345	22.44	1.868	1.456	1.399	6.35	9.73	6.95	2.498	1.929
			10S	0.134	5.295	22.02	2.285	1.456	1.386	7.77	9.53	8.43	3.03	1.920
5	40	Std	40S	0.258	5.047	20.01	4.30	1.456	1.321	14.62	8.66	15.17	5.45	1.878
5.563	80	XS	80S	0.375	4.813	18.19	6.11	1.456	1.260	20.78	7.89	20.68	7.43	1.839
	120			0.500	4.563	16.35	7.95	1.456	1.195	27.04	7.09	25.74	9.25	1.799
	160			0.625	4.313	14.61	9.70	1.456	1.129	32.96	6.33	30.0	10.80	1.760
		XXS		0.750	4.063	12.97	11.34	1.456	1.064	38.55	5.62	33.6	12.10	1.722
				0.875	3.813	11.413	12.880	1.456	0.998	43.810	4.951	36.6450	13.1750	1.6860
				1.000	3.563	9.966	14.328	1.456	0.933	47.734	4.232	39.1110	14.0610	1.6520
			5S	0.109	6.407	32.2	2.231	1.734	1.677	5.37	13.98	11.85	3.58	2.304
			10S	0.134	6.357	31.7	2.733	1.734	1.664	9.29	13.74	14.40	4.35	2.295
				0.219	6.187	30.100	4.410	1.734	1.620	15.020	13.100	22.6600	6.8400	2.2700
6	40	Std	40S	0.280	6.065	28.89	5.58	1.734	1.588	18.97	12.51	28.14	8.50	2.245
6.625	80	XS	80S	0.432	5.761	26.07	8.40	1.734	1.508	28.57	11.29	40.5	12.23	2.195
	120			0.562	5.501	23.77	10.70	1.734	1.440	36.39	10.30	49.6	14.98	2.153
	160			0.718	5.189	21.15	13.33	1.734	1.358	45.30	9.16	59.0	17.81	2.104
		XXS		0.864	4.897	18.83	15.64	1.734	1.282	53.16	8.17	66.3	20.03	2.060
				1.000	4.625	16.792	17.662	1.734	1.211	60.076	7.284	72.1190	21.7720	2.0200
				1.125	4.375	15.025	19.429	1.734	1.145	66.084	6.517	76.5970	23.1240	1.9850
			5S	0.109	8.407	55.5	2.916	2.258	2.201	9.91	24.07	26.45	6.13	3.01
			10S	0.148	8.329	54.5	3.94	2.258	2.180	13.40	23.59	35.4	8.21	3.00
				0.219	8.187	52.630	5.800	2.258	2.150	19.640	22.900	51.3200	11.9000	2.9700
8	20			0.250	8.125	51.8	6.58	2.258	2.127	22.36	22.48	57.7	13.39	2.962
8.625	30			0.277	8.071	51.2	7.26	2.258	2.113	24.70	22.18	63.4	14.69	2.953
	40	Std	40S	0.322	7.981	50.0	8.40	2.258	2.089	28.55	21.69	72.5	16.81	2.938
	60			0.406	7.813	47.9	10.48	2.258	2.045	35.64	20.79	88.8	20.58	2.909
	80	XS	80S	0.500	7.625	45.7	12.76	2.258	1.996	43.39	19.80	105.7	24.52	2.878
	100			0.593	7.439	43.5	14.96	2.258	1.948	50.87	18.84	121.4	28.14	2.847
	120			0.718	7.189	40.6	17.84	2.258	1.882	60.63	17.60	140.6	32.6	2.807
8	140			0.812	7.001	38.5	19.93	2.258	1.833	67.76	16.69	153.8	35.7	2.777
8.625	160			0.906	6.813	36.5	21.97	2.258	1.784	74.69	15.80	165.9	38.5	2.748
				1.000	6.625	34.454	23.942	2.258	1.734	81.437	14.945	177.1320	41.0740	2.7190
				1.125	6.375	31.903	26.494	2.258	1.669	90.114	13.838	190.6210	44.2020	2.6810
			5S	0.134	10.482	86.3	4.52	2.815	2.744	15.15	37.4	63.7	11.85	3.75
			10S	0.165	10.420	85.3	5.49	2.815	2.728	18.70	36.9	76.9	14.30	3.74
				0.219	10.312	83.52	7.24	2.815	2.70	24.63	36.2	100.46	18.69	3.72
	20			0.250	10.250	82.5	8.26	2.815	2.683	28.04	35.8	113.7	21.16	3.71
	30			0.307	10.136	80.7	10.07	2.815	2.654	34.24	35.0	137.5	25.57	3.69
10	40	Std	40S	0.365	10.020	78.9	11.91	2.815	2.623	40.48	34.1	160.8	29.90	3.67
10.750	60	XS	80S	0.500	9.750	74.7	16.10	2.815	2.553	54.74	32.3	212.0	39.4	3.63
	80			0.593	9.564	71.8	18.92	2.815	2.504	64.33	31.1	244.9	45.6	3.60
	100			0.718	9.314	68.1	22.63	2.815	2.438	76.93	29.5	286.2	53.2	3.56
	120			0.843	9.064	64.5	26.24	2.815	2.373	89.20	28.0	324	60.3	3.52
				0.875	9.000	63.62	27.14	2.815	2.36	92.28	27.6	333.46	62.04	3.50
	140			1.000	8.750	60.1	30.6	2.815	2.291	104.13	26.1	368	68.4	3.47
	160			1.125	8.500	56.7	34.0	2.815	2.225	115.65	24.6	399	74.3	3.43
				1.250	8.250	53.45	37.31	2.815	2.16	126.82	23.2	428.17	79.66	3.39
				1.500	7.750	47.15	43.57	2.815	2.03	148.19	20.5	478.59	89.04	3.31
			5S	0.156	12.438	121.4	6.17	3.34	3.26	20.99	52.7	122.2	19.20	4.45
			10S	0.180	12.390	120.6	7.11	3.34	3.24	24.20	52.2	140.5	22.03	4.44
	20			0.250	12.250	117.9	9.84	3.34	3.21	33.38	51.1	191.9	30.1	4.42
	30			0.330	12.090	114.8	12.88	3.34	3.17	43.77	49.7	248.5	39.0	4.39

TABLE 54. (Continued)

nominal pipe size outside diameter, in.	schedule number* a	b	c	wall thick-ness, in.	inside diam-eter, in.	inside area, sq. in.	metal area, sq. in.	sq ft outside surface, per ft	sq ft inside surface, per ft	weight per ft, lb†	weight of water per ft, lb	moment of inertia, in.⁴	section modu-lus, in.³	radius gyra-tion, in.
		Std	40S	0.375	12.000	113.1	14.58	3.34	3.14	49.56	49.0	279.3	43.8	4.38
	40			0.406	11.938	111.3	15.74	3.34	3.13	53.53	48.5	300	47.1	4.37
		XS	80S	0.500	11.750	108.4	19.24	3.34	3.08	65.42	47.0	362	56.7	4.33
12	60			0.562	11.626	106.2	21.52	3.34	3.04	73.16	46.0	401	62.8	4.31
12.750	80			0.687	11.376	101.6	26.04	3.34	2.978	88.51	44.0	475	74.5	4.27
				0.750	11.250	99.40	28.27	3.34	2.94	96.2	43.1	510.7	80.1	4.25
	100			0.843	11.064	96.1	31.5	3.34	2.897	107.20	41.6	562	88.1	4.22
				0.875	11.000	95.00	32.64	3.34	2.88	110.9	41.1	578.5	90.7	4.21
	120			1.000	10.750	90.8	36.9	3.34	2.814	125.49	39.3	642	100.7	4.17
	140			1.125	10.500	86.6	41.1	3.34	2.749	139.68	37.5	701	109.9	4.13
				1.250	10.250	82.50	45.16	3.34	2.68	153.6	35.8	755.5	118.5	4.09
	160			1.312	10.126	80.5	47.1	3.34	2.651	160.27	34.9	781	122.6	4.07
			5S	0.156	13.688	147.20	6.78	3.67	3.58	23.0	63.7	162.6	23.2	4.90
			10S	0.188	13.624	145.80	8.16	3.67	3.57	27.7	63.1	194.6	27.8	4.88
				0.210	13.580	144.80	9.10	3.67	3.55	30.9	62.8	216.2	30.9	4.87
				0.219	13.562	144.50	9.48	3.67	3.55	32.2	62.6	225.1	32.2	4.87
	10			0.250	13.500	143.1	10.80	3.67	3.53	36.71	62.1	255.4	36.5	4.86
				0.281	13.438	141.80	11.21	3.67	3.52	41.2	61.5	285.2	40.7	4.85
	20			0.312	13.376	140.5	13.63	3.67	3.50	45.68	60.9	314	44.9	4.84
				0.344	13.312	139.20	14.76	3.67	3.48	50.2	60.3	344.3	49.2	4.83
14	30	Std		0.375	13.250	137.9	16.05	3.67	3.47	54.57	59.7	373	53.3	4.82
14.000	40			0.437	13.126	135.3	18.62	3.67	3.44	63.37	58.7	429	61.2	4.80
				0.469	13.062	134.00	19.94	3.67	3.42	67.8	58.0	456.8	65.3	4.79
		XS		0.500	13.000	132.7	21.21	3.67	3.40	72.09	57.5	484	69.1	4.78
	60			0.593	12.814	129.0	24.98	3.67	3.35	84.91	55.9	562	80.3	4.74
				0.625	12.750	127.7	26.26	3.67	3.34	89.28	55.3	589	84.1	4.73
	80			0.750	12.500	122.7	31.2	3.67	3.27	106.13	53.2	687	98.2	4.69
	100			0.937	12.126	115.5	38.5	3.67	3.17	130.73	50.0	825	117.8	4.63
	120			1.093	11.814	109.6	44.3	3.67	3.09	150.67	47.5	930	132.8	4.58
	140			1.250	11.500	103.9	50.1	3.67	3.01	170.22	45.0	1127	146.8	4.53
	160			1.406	11.188	98.3	55.6	3.67	2.929	189.12	42.6	1017	159.6	4.48
			5S	0.165	15.670	192.90	8.21	4.19	4.10	28	83.5	257	32.2	5.60
			10S	0.188	15.624	191.70	9.34	4.19	4.09	32	83.0	292	36.5	5.59
	10			0.250	15.500	188.7	12.37	4.19	4.06	42.05	81.8	384	48.0	5.57
	20			0.312	15.376	185.7	15.38	4.19	4.03	52.36	80.5	473	59.2	5.55
16	30	Std		0.375	15.250	182.6	18.41	4.19	3.99	62.58	79.1	562	70.3	5.53
16.000	40	XS		0.500	15.000	176.7	24.35	4.19	3.93	82.77	76.5	732	91.5	5.48
	60			0.656	14.688	169.4	31.6	4.19	3.85	107.50	73.4	933	116.6	5.43
				0.843	14.314	160.9	40.1	4.19	3.75	136.46	69.7	1157	144.6	5.37
	100			1.031	13.938	152.6	48.5	4.19	3.65	164.83	66.1	1365	170.6	5.30
	120			1.218	13.564	144.5	56.6	4.19	3.55	192.29	62.6	1556	194.5	5.24
	140			1.437	13.126	135.3	65.7	4.19	3.44	223.64	58.6	1760	220.0	5.17
	160			1.593	12.814	129.0	72.1	4.19	3.35	245.11	55.9	1894	236.7	5.12
			5S	0.165	17.670	245.20	9.24	4.71	4.63	31	106.2	368	40.8	6.31
			10S	0.188	17.624	243.90	10.52	4.71	4.61	36	105.7	417	46.4	6.30
	10			0.250	17.500	240.5	13.94	4.71	4.58	47.39	104.3	549	61.0	6.28
	20			0.312	17.376	237.1	17.34	4.71	4.55	59.03	102.8	678	75.5	6.25
		Std		0.375	17.250	233.7	20.76	4.71	4.52	70.59	101.2	807	89.6	6.23
18	30			0.437	17.126	230.4	24.11	4.71	4.48	82.06	99.9	931	103.4	6.21
18.000	40			0.562	16.876	223.7	30.8	4.71	4.42	104.75	97.0	1172	130.2	6.17
	60			0.750	16.500	213.8	40.6	4.71	4.32	138.17	92.7	1515	168.3	6.10
	80			0.937	16.126	204.2	50.2	4.71	4.22	170.75	88.5	1834	203.8	6.04
	100			1.156	15.688	193.3	61.2	4.71	4.11	207.96	83.7	2180	242.2	5.97
	120			1.375	15.250	182.6	71.8	4.71	3.99	244.14	79.2	2499	277.6	5.90
	140			1.562	14.876	173.8	80.7	4.71	3.89	274.23	75.3	2750	306	5.84
	160			1.781	14.438	163.7	90.7	4.71	3.78	308.51	71.0	3020	336	5.77
			5S	0.188	19.634	302.40	11.70	5.24	5.14	40	131.0	574	57.4	7.00
			10S	0.218	19.564	300.60	13.55	5.24	5.12	46	130.2	663	66.3	6.99
	10			0.250	19.500	298.6	15.51	5.24	5.11	52.73	129.5	757	75.7	6.98
20	20	Std		0.375	19.250	291.0	23.12	5.24	5.04	78.60	126.0	1114	111.4	6.94
20.000	30	XS		0.500	19.000	283.5	30.6	5.24	4.97	104.13	122.8	1457	145.7	6.90
	40			0.593	18.814	278.0	36.2	5.24	4.93	122.91	120.4	1704	170.4	6.86
	60			0.812	18.376	265.2	48.9	5.24	4.81	166.40	115.0	2257	225.7	6.79

TABLE 54. (Continued)

nominal pipe size outside diameter, in.	schedule number*			wall thick- ness, in.	inside diam- eter, in.	inside area, sq in.	metal area, sq in.	sq ft outside surface, per ft	sq ft inside surface, per ft	weight per ft, lb†	weight of water per ft, lb	moment of inertia, in.⁴	section modu- lus, in.³	radius gyra- tion, in.
	a	b	c											
20 20.000		...	...	0.875	18.250	261.6	52.6	5.24	4.78	178.73	113.4	2409	240.9	6.77
	80	...	...	1.031	17.938	252.7	61.4	5.24	4.70	208.87	109.4	2772	277.2	6.72
	100	...	...	1.281	17.438	238.8	75.3	5.24	4.57	256.10	103.4	3320	332	6.63
	120			1.500	17.000	227.0	87.2	5.24	4.45	296.37	98.3	3760	376	6.56
	140			1.750	16.500	213.8	100.3	5.24	4.32	341.10	92.6	4220	422	6.48
	160			1.968	16.064	202.7	111.5	5.24	4.21	379.01	87.9	4590	459	6.41
22 22.000			5S	0.188	21.624	367.3	12.88	5.76	5.66	44	159.1	766	69.7	7.71
			10S	0.218	21.564	365.2	14.92	5.76	5.65	51	158.2	885	80.4	7.70
	10	...		0.250	21.500	363.1	17.18	5.76	5.63	58	157.4	1010	91.8	7.69
	20	Std		0.375	21.250	354.7	25.48	5.76	5.56	87	153.7	1490	135.4	7.65
	30	XS		0.500	21.000	346.4	33.77	5.76	5.50	115	150.2	1953	177.5	7.61
				0.625	20.750	338.2	41.97	5.76	5.43	143	146.6	2400	218.2	7.56
				0.750	20.500	330.1	50.07	5.76	5.37	170	143.1	2829	257.2	7.52
	60			0.875	20.250	322.1	58.07	5.76	5.00	197	139.6	3245	295.0	7.47
	80			1.125	19.750	306.4	73.78	5.76	5.17	251	132.8	4029	366.3	7.39
	100			1.375	19.250	291.0	89.09	5.76	5.04	303	126.2	4758	432.6	7.31
	120			1.625	18.750	276.1	104.02	5.76	4.91	354	119.6	5432	493.8	7.23
	140			1.875	18.250	261.6	118.55	5.76	4.78	403	113.3	6054	550.3	7.15
	160			2.125	17.750	247.4	132.68	5.76	4.65	451	107.2	6626	602.4	7.07
24 24.000	10			0.250	23.500	434	18.65	6.28	6.15	63.41	188.0	1316	109.6	8.40
	20	Std		0.375	23.250	425	27.83	6.28	6.09	94.62	183.8	1943	161.9	8.35
		XS		0.500	23.000	415	36.9	6.28	6.02	125.49	180.1	2550	212.5	8.31
	30			0.562	22.876	411	41.4	6.28	5.99	140.80	178.1	2840	237.0	8.29
				0.625	22.750	406	45.9	6.28	5.96	156.03	176.2	3140	261.4	8.27
	40			0.687	22.626	402	50.3	6.28	5.92	171.17	174.3	3420	285.2	8.25
				0.750	22.500	398	54.8	6.28	5.89	186.24	172.4	3710	309	8.22
			5S	0.218	23.564	436.1	16.29	6.28	6.17	55	188.9	1152	96.0	8.41
				0.875	22.250	388.6	63.54	6.28	5.83	216	168.6	4256	354.7	8.18
	60			0.968	22.064	382	70.0	6.28	5.78	238.11	165.8	4650	388	8.15
	80			1.218	21.564	365	87.2	6.28	5.65	296.36	158.3	5670	473	8.07
	100			1.531	20.938	344	108.1	6.28	5.48	367.40	149.3	6850	571	7.96
	120			1.812	20.376	326	126.3	6.28	5.33	429.39	141.4	7830	652	7.87
	140			2.062	19.876	310	142.1	6.28	5.20	483.13	134.5	8630	719	7.79
	160			2.343	19.314	293	159.4	6.28	5.06	541.94	127.0	9460	788	7.70
26 26.000				0.250	25.500	510.7	19.85	6.81	6.68	67	221.4	1646	126.6	9.10
	10			0.312	25.376	505.8	25.18	6.81	6.64	86	219.2	2076	159.7	9.08
		Std		0.375	25.250	500.7	30.19	6.81	6.61	103	217.1	2478	190.6	9.06
	20	XS		0.500	25.000	490.9	40.06	6.81	6.54	136	212.8	3259	250.7	9.02
				0.625	24.750	481.1	49.82	6.81	6.48	169	208.6	4013	308.7	8.98
				0.750	24.500	471.4	59.49	6.81	6.41	202	204.4	4744	364.9	8.93
				0.875	24.250	461.9	69.07	6.81	6.35	235	200.2	5458	419.9	8.89
				1.000	24.000	452.4	78.54	6.81	6.28	267	196.1	6149	473.0	8.85
				1.125	23.750	443.0	87.91	6.81	6.22	299	192.1	6813	524.1	8.80
28 28.000				0.250	27.500	594.0	21.80	7.33	7.20	74	257.3	2098	149.8	9.81
	10			0.312	27.376	588.6	27.14	7.33	7.17	92	255.0	2601	185.8	9.79
		Std		0.375	27.250	583.2	32.54	7.33	7.13	111	252.6	3105	221.8	9.77
	20	XS		0.500	27.000	572.6	43.20	7.33	7.07	147	248.0	4085	291.8	9.72
	30			0.625	26.750	562.0	53.75	7.33	7.00	183	243.4	5038	359.8	9.68
				0.750	26.500	551.6	64.21	7.33	6.94	218	238.9	5964	426.0	9.64
				0.875	26.250	541.2	74.56	7.33	6.87	253	234.4	6865	490.3	9.60
				1.000	26.000	530.9	84.82	7.33	6.81	288	230.0	7740	552.8	9.55
				1.125	25.750	520.8	94.98	7.33	6.74	323	225.6	8590	613.6	9.51
30 30.000			5S	0.250	29.500	683.4	23.37	7.85	7.72	79	296.3	2585	172.3	10.52
	10		10S	0.312	29.376	677.8	29.19	7.85	7.69	99	293.7	3201	213.4	10.50
		Std		0.375	29.250	672.0	34.90	7.85	7.66	119	291.2	3823	254.8	10.48
	20	XS		0.500	29.000	660.5	46.34	7.85	7.59	158	286.2	5033	335.5	10.43
	30			0.625	28.750	649.2	57.68	7.85	7.53	196	281.3	6213	414.2	10.39

TABLE 54. (*Continued*)

nominal pipe size outside diameter, in.	schedule number*			wall thickness, in.	inside diameter, in.	inside area, sq. in.	metal area, sq. in.	sq ft outside surface, per ft	sq ft inside surface, per ft	weight per ft, lb†	weight of water per ft, lb	moment of inertia, in.⁴	section modulus, in.³	radius gyration, in.
	a	b	c											
30 30.000	40			0.750	28.500	637.9	68.92	7.85	7.46	234	276.6	7371	431.4	10.34
				0.875	28.250	620.7	80.06	7.85	7.39	272	271.8	8494	566.2	10.30
				1.000	28.000	615.7	91.11	7.85	7.33	310	267.0	9591	639.4	10.26
				1.125	27.750	604.7	102.05	7.85	7.26	347	262.2	10653	710.2	10.22
32 32.000				0.250	31.500	779.2	24.93	8.38	8.25	85	337.8	3141	196.3	11.22
	10			0.312	31.376	773.2	31.02	8.38	8.21	106	335.2	3891	243.2	11.20
		Std		0.375	31.250	766.9	37.25	8.38	8.18	127	332.5	4656	291.0	11.18
	20	XS		0.500	31.000	754.7	49.48	8.38	8.11	168	327.2	6140	383.8	11.14
	30			0.625	30.750	742.5	61.59	8.38	8.05	209	321.9	7578	473.6	11.09
	40			0.688	30.624	736.6	67.68	8.38	8.02	230	319.0	8298	518.6	11.07
				0.750	30.500	730.5	73.63	8.38	7.98	250	316.7	8990	561.9	11.05
				0.875	30.250	718.3	85.52	8.38	7.92	291	311.6	10372	648.2	11.01
				1.000	30.000	706.8	97.38	8.38	7.85	331	306.4	11680	730.0	10.95
				1.125	29.750	694.7	109.0	8.38	7.79	371	301.3	13023	814.0	10.92
34 34.000				0.250	33.500	881.2	26.50	8.90	8.77	90	382.0	3773	221.9	11.93
	10			0.312	33.376	874.9	32.99	8.90	8.74	112	379.3	4680	275.3	11.91
		Std		0.375	33.250	867.8	39.61	8.90	8.70	135	376.2	5597	329.2	11.89
	20	XS		0.500	33.000	855.3	52.62	8.90	8.64	179	370.8	7385	434.4	11.85
	30			0.625	32.750	841.9	65.53	8.90	8.57	223	365.0	9124	536.7	11.80
	40			0.688	32.624	835.9	72.00	8.90	8.54	245	362.1	9992	587.8	11.78
				0.750	32.500	829.3	78.34	8.90	8.51	266	359.5	10829	637.0	11.76
				0.875	32.250	816.4	91.01	8.90	8.44	310	354.1	12501	735.4	11.72
				1.000	32.000	804.2	103.67	8.90	8.38	353	348.6	14114	830.2	11.67
				1.125	31.750	791.3	116.13	8.90	8.31	395	343.2	15719	924.7	11.63
36 36.000				0.250	35.500	989.7	28.11	9.42	9.29	96	429.1	4491	249.5	12.64
	10			0.312	35.376	982.9	34.95	9.42	9.26	119	426.1	5565	309.1	12.62
		Std		0.375	35.250	975.8	42.01	9.42	9.23	143	423.1	6664	370.2	12.59
	20	XS		0.500	35.000	962.1	55.76	9.42	9.16	190	417.1	8785	488.1	12.55
	30			0.625	34.750	948.3	69.50	9.42	9.10	236	411.1	10872	604.0	12.51
	40			0.750	34.500	934.7	83.01	9.42	9.03	282	405.3	12898	716.5	12.46
				0.875	34.250	920.6	96.50	9.42	8.97	328	399.4	14903	827.9	12.42
				1.000	34.000	907.9	109.96	9.42	8.90	374	393.6	16851	936.2	12.38
				1.125	33.750	894.2	123.19	9.42	8.89	419	387.9	18763	1042.4	12.34
42 42.000				0.250	41.500	1352.6	32.82	10.99	10.86	112	586.4	7126	339.3	14.73
		Std		0.375	41.250	1336.3	49.08	10.99	10.80	167	579.3	10627	506.1	14.71
	20	XS		0.500	41.000	1320.2	65.18	10.99	10.73	222	572.3	14037	668.4	14.67
	30			0.625	40.750	1304.1	81.28	10.99	10.67	276	565.4	17373	827.3	14.62
	40			0.750	40.500	1288.2	97.23	10.99	10.60	330	558.4	20689	985.2	14.59
				1.000	40.000	1256.6	128.81	10.99	10.47	438	544.8	27080	1289.5	14.50
				1.250	39.500	1225.3	160.03	10.99	10.34	544	531.2	33233	1582.5	14.41
				1.500	39.000	1194.5	190.85	10.99	10.21	649	517.9	39181	1865.7	14.33

TABLE 55. Friction of Water in Pipes

Loss of Head in Feet Due to Friction, per 100 Feet of Ordinary Iron Pipe

Size Pipe, Inches

Gallons per Minute	¼ Vel.	¼ Fric.	⅜ Vel.	⅜ Fric.	½ Vel.	½ Fric.	¾ Vel.	¾ Fric.	1 Vel.	1 Fric.	1¼ Vel.	1¼ Fric.	1½ Vel.	1½ Fric.	2 Vel.	2 Fric.
1	3.08	28.0	1.67	6.4	1.05	2.1										
2	6.16	103.0	3.35	23.3	2.10	7.4	1.20	1.9								
3			5.02	49.0	3.16	15.8	1.80	4.1	1.12	1.26						
4			6.70	84.0	4.21	27.0	2.41	7.0	1.49	2.14	0.86	0.57	0.63	0.26		
5			8.37	126.0	5.26	41.0	3.01	10.5	1.86	3.25	1.07	0.84	0.79	0.39		
10					10.52	147.0	6.02	38.0	3.72	11.7	2.14	3.05	1.57	1.43	1.02	0.50
15							9.02	80.0	5.60	25.0	3.2	6.50	2.36	3.0	1.53	1.0
20							12.03	136.0	7.44	42.0	4.29	11.1	3.15	5.2	2.04	1.82
25									9.30	64.0	5.36	16.6	3.94	7.8	2.55	2.73
30									11.15	89.0	6.43	23.5	4.72	11.0	3.06	3.84
35									13.02	119.0	7.51	31.2	5.51	14.7	3.57	5.1
40									14.88	152.0	8.58	40.0	6.3	18.8	4.08	6.6
45											9.65	50	7.08	23.2	4.60	8.2
50											10.72	60	7.87	28.4	5.11	9.0
70											15.01	113	11.02	53.0	7.15	18.4
90													14.17	84.0	9.19	29.4
100													15.74	102.0	10.21	35.8
120													18.89	143.0	12.25	50.0
140													22.04	190.0	14.30	67.0
160															16.34	86.0
180															18.38	107.0
200															20.42	129.0
220															22.47	154.0
240															24.51	182.0
260															26.55	211.0

Size Pipe, Inches

Gallons per Minute	2½ Vel.	2½ Fric.	3 Vel.	3 Fric.	4 Vel.	4 Fric.	5 Vel.	5 Fric.	6 Vel.	6 Fric.	8 Vel.	8 Fric.	10 Vel.	10 Fric.	12 Vel.	12 Fric.
10	0.65	0.17	0.45	0.07												
15	0.98	0.36	0.68	0.15												
20	1.31	0.61	0.91	0.25												
25	1.63	0.92	1.13	0.38												
30	1.96	1.29	1.36	0.54												
35	2.29	1.72	1.59	0.71												
40	2.61	2.20	1.82	0.91	1.02	0.22										
45	2.94	2.80	2.05	1.15	1.17	0.29										
50	3.27	3.32	2.27	1.38	1.28	0.34										
70	4.58	6.2	3.18	2.57	1.79	0.63	1.14	0.21								
75					1.92	0.73	1.22	0.24								
90	5.88	9.8	4.09	4.08												

Flow	P1 Vol	P1 Fric	P2 Vol	P2 Fric	P3 Vol	P3 Fric	P4 Vol	P4 Fric	P5 Vol	P5 Fric	P6 Vol	P6 Fric	P7 Vol	P7 Fric	P8 Vol	P8 Fric
100	6.54	12.0	4.54	4.96	2.55	1.23	1.63	0.39	1.14	0.14						
120	7.84	16.8	5.45	7.0	3.06	1.71	1.96	0.57								
125					3.19	1.81	2.04	0.64	1.42	0.25						
140	9.15	22.3	6.35	9.2					1.48	0.28						
150					3.84	2.55	2.45	0.88	1.71	0.32						
160	10.46	29.0	7.26	11.8												
175					4.45	3.36	2.86	1.18	2.00	0.48						
180	11.76	35.7	8.17	14.8												
200	13.07	43.1	9.08	17.8	5.11	4.37	3.27	1.48	2.28	0.62						
220	14.38	52.0	9.99	21.3												
225							3.67	1.86	2.57	0.74						
240	15.69	61.0	10.89	25.1												
250	16.09	70.0			6.32	5.51	4.08	2.24	2.80	0.92	1.60	0.22				
260			11.80	29.1	6.40	6.72					1.70	0.25				
270					6.90	7.70	4.42	2.60	3.03	1.13	1.73	0.27				
275					7.03	7.99	4.50	2.72	3.06	1.15						
280	18.30	81.0	12.71	33.4												
300	19.61	92.0	13.62	38.0	7.66	9.38	4.90	3.15	3.40	1.29	1.90	0.36				
350					8.90	12.32	5.72	4.19	3.98	1.69	2.20	0.41				
400					10.20	15.82	6.54	5.33	4.54	2.21	2.60	0.56				
450					11.50	19.74	7.35	6.65	5.12	2.74	2.92	0.64	1.80	0.21		
470					12.16	22.40			5.49	3.12	3.07	0.77	1.92	0.24		
475					12.30	22.96	7.78	7.42	5.55	3.21	3.10	0.79	1.94	0.25		
500					12.77	24.08	8.17	8.12	5.60	3.20	3.20	0.81	2.04	0.28	1.42	0.11
550							8.99	9.66	6.16	3.93	3.52	0.98	2.25	0.33	1.57	0.14
600							9.80	11.34	6.72	4.70	3.84	1.16	2.46	0.39	1.71	0.15
650							10.62	13.16	7.28	5.50	4.16	1.34	2.66	0.46	1.85	0.19
700							11.44	15.12	7.84	6.38	4.46	1.54	2.80	0.52	2.00	0.22
750							12.26	17.22	8.50	7.00	4.80	1.74	3.06	0.59	2.13	0.24
800									9.08	7.90	5.12	1.97	3.28	0.67	2.27	0.27
850									9.58	8.75	5.48	2.28	3.48	0.75	2.41	0.31
900									10.30	10.11	5.75	2.46	3.63	0.83	2.56	0.34
950									10.72	10.71	6.06	2.87	3.88	0.91	2.70	0.35
1000									11.32	12.01	6.40	3.02	4.08	1.01	2.84	0.41
1050									11.90	13.30	6.70	3.21	4.29	1.09	3.08	0.48
1100									12.50	14.31	7.03	3.51	4.50	1.20	3.13	0.49
1150									12.95	15.34	7.35	3.84	4.71	1.34	3.27	0.53
1200									13.52	16.09	7.67	4.26	4.91	1.46	3.41	0.57
1250									14.10	18.20	8.00	4.45	5.11	1.51	3.55	0.62
1500											9.00	6.27	6.10	2.09	4.20	0.85
2000											12.70	10.71			5.60	1.43
2500													10.10	5.33	7.00	2.18
3000													12.10	7.42	8.40	3.39
3500													14.10	10.08	9.80	3.92
4000															11.35	5.32
4200															11.93	6.30
4500															12.78	6.75
5000															14.20	8.15

Vol.—velocity foot per second.   Fric.—friction head in foot.

TABLE 56. Properties of Saturated Steam

Pressure Lbs. per Sq. In. Absolute	Pressure Lbs. per Sq. In. Gage	Temperature Degrees F.	Heat of the Liquid Btu/lb.	Latent Heat of Evaporation Btu/lb.	Total Heat of Steam Btu/lb.	Specific Volume Cu. Ft. per Lb.	Weight of 1 Cu. Ft. Steam, lb.
14.696	0.0	212.00	180.07	970.3	1150.4	26.80	.03794
20.0	5.3	227.96	196.16	960.1	1156.3	20.089	.05070
25.0	10.3	240.07	208.42	952.1	1160.6	16.303	.06253
30.0	15.3	250.33	218.82	945.3	1164.1	13.746	.07420
35.0	20.3	259.28	227.91	939.2	1167.1	11.898	.08576
40.0	25.3	267.25	236.03	933.7	1169.7	10.498	.09721
45.0	30.3	274.44	243.36	928.6	1172.0	9.401	.1086
50.0	35.3	281.01	250.09	924.0	1174.1	8.515	.1198
55.0	40.3	287.07	256.30	919.6	1175.9	7.787	.1311
60.0	45.3	292.71	262.09	915.5	1177.6	7.175	.1422
65.0	50.3	297.97	267.50	911.6	1179.1	6.655	.1533
70.0	55.3	302.92	272.61	907.9	1180.6	6.206	.1643
75.0	60.3	307.60	277.43	904.5	1181.9	5.816	.1753
80.0	65.3	312.03	282.02	901.1	1183.1	5.472	.1862
85.0	70.3	316.25	286.39	897.8	1184.2	5.168	.1971
90.0	75.3	320.27	290.56	894.7	1185.3	4.896	.2080
95.0	80.3	324.12	294.56	891.7	1186.2	4.652	.2188
100.0	85.3	327.81	298.40	888.8	1187.2	4.432	.2296
105.0	90.3	331.36	302.10	886.0	1188.1	4.232	.2403
110.0	95.3	334.77	305.66	883.2	1188.9	4.049	.2510
115.0	100.3	338.07	309.11	880.6	1189.7	3.882	.2617
120.0	105.3	341.25	312.44	877.9	1190.4	3.728	.2724
125.0	110.3	344.33	315.68	875.4	1191.1	3.587	.2830
130.0	115.3	347.32	318.81	872.9	1191.7	3.455	.2936
135.0	120.3	350.21	321.85	870.6	1192.4	3.333	.3042
140.0	125.3	353.02	324.82	868.2	1193.0	3.220	.3147
145.0	130.3	355.76	327.70	865.8	1193.5	3.114	.3253
150.0	135.3	358.42	330.51	863.6	1194.1	3.015	.3358
160.0	145.3	363.53	335.93	859.2	1195.1	2.834	.3567
170.0	155.3	368.41	341.09	854.9	1196.0	2.675	.3775
180.0	165.3	373.06	346.03	850.8	1196.9	2.532	.3983
190.0	175.3	377.51	350.79	846.8	1197.6	2.404	.4191
200.0	185.3	381.79	355.36	843.0	1198.4	2.288	.4400
225.0	210.3	391.79	366.09	833.8	1199.9	2.0422	.4897
250.0	235.3	400.95	376.00	825.1	1201.1	1.8438	.5478
275.0	260.3	409.43	385.21	816.9	1202.1	1.6804	.5951
300.0	285.3	417.33	393.84	809.0	1202.8	1.5433	.6515
360.0	345.3	434.40	412.67	791.4	1204.1	1.2895	.7755
400.0	385.3	444.59	424.00	780.5	1204.5	1.1613	.8572
460.0	445.3	458.50	439.70	764.9	1204.6	1.0094	.9907
500.0	485.3	467.01	449.40	755.0	1204.4	0.9278	1.062
560.0	545.3	478.85	463.00	740.8	1203.8	0.8265	1.210
600.0	585.3	486.21	471.60	731.6	1203.2	0.7698	1.266
660.0	645.3	496.58	483.80	718.3	1202.1	0.6971	1.435
700.0	685.3	503.10	491.50	709.7	1201.2	0.6554	1.470
760.0	745.3	512.36	502.60	697.1	1199.7	0.6007	1.665

TABLE 57. Steam Flow in Pipes
Reasonable Velocities for Fluid Flow Through Pipes

Fluid	Service	Pressure PSI. (Gage)	Velocities FPM.
Saturated Steam	Heating Mains	0–15	4000–6000
Saturated Steam	Miscellaneous	50 and Up	6000–8000
Superheated Steam	Turbine and Boiler Leads	200 and Up	10000–15000
Water	City Service	25–40	120–300
Water	General Service	50–150	300–600
Water	Boiler Feed	150	600

Pounds per Hour of Saturated Steam at 6000 Feet per Minute Velocity

In Iron or Steel Pipe

Pipe Size In.	SATURATED STEAM, LB. PER HOUR — PRESSURE, PSI. (GAGE)											
	5	25	50	75	100	150	200	250	300	400	500	600
½	30	55	90	120	150	210	270	330	390	495	660	740
¾	55	100	160	220	280	390	510	620	730	930	1150	1390
1	90	170	270	390	460	650	840	1020	1200	1530	1900	2280
1¼	160	300	480	650	820	1150	1490	1830	2160	2760	3410	4100
1½	220	470	660	900	1100	1600	2060	2550	3000	3820	4750	5700
2	370	700	1100	1500	1900	2680	3450	4200	4980	6350	7900	9450
2½	525	1000	1600	2175	2750	3850	4950	6050	7150	9120	11300	13600
3	800	1600	2500	3350	4250	6000	7700	9450	11200	14200	17700	21300
3½	1100	2100	3400	4550	5700	8050	10200	12700	15000	19100	23700	28500
4	1450	2750	4300	5850	7400	10450	13450	16400	19200	24400	30400	36500
5	2300	4400	6800	9300	11700	16500	21200	26000	30800	39300	48890	58500
6	3200	6200	9800	13200	16800	23700	30800	38900	44000	56000	69700	83500
8	5700	10800	17100	23300	29300	41300	53100	65200	76800	98000	121000	145500
10	9300	17800	28100	38000	48100	67500	87100	106500	126000	160000	200000	240000
12	13500	25800	40700	55300	69900	98500	126500	154500	183100	233000	250000	347000

Comparative Capacities of Different Sizes of Pipe

Pipe Size .... inches	½	¾	1	1¼	1½	2	2½	3	3½	4	4½	5	6	7	8	9	10
Capacity Factor	2	3.5	5.5	10	13.5	22.5	31.5	48.5	65	84	105	131.5	190	255	329	430	539

EXAMPLE: To get size of pipe to serve a ½-inch and a ¾-inch pipe, add factors thus: ½-inch factor (2) + ¾-inch factor (3.5) = 5.5 (1-inch factor).

TABLE 58. Approximate Viscosities and Specific Gravities of Common Liquids

Liquid	Specific Gravity Value	Specific Gravity °F	Viscosity SSU Value	Viscosity SSU °F
Asphalt:				
*Virgin	1.3	60	7500	250
			2000	300
Blended				
*RC-1, MC-1 or SC-1	1.0	60	3700	100
			1100	122
*RC-3, MC-3 or SC-3	1.0	60	9000	122
			3700	140
*RC-5, MC-5 or SC-5	1.0	60	55000	140
			4500	180
Gasoline	.71	70	31	70
*Glucose	1.4	60	70000	100
			7500	150
Glycerine	1.25	70	3000	70
			800	100
Glycol:				
Propylene	1.04	70	240	70
Triethylene	1.13	70	190	70
Diethylene	1.12	70	150	70
Ethylene	1.13	70	90	70
Milk	1.03	70	33	70
Molasses:				
*A	1.43	60	12000	100
			4500	130
*B	1.45	60	33000	100
			9000	130
*C (Blackstrap)	1.48	60	130000	100
			40000	130
Oils:				
Petroleum				
*Crude (Pennsylvania)	.82	60	130	60
			60	100
*Crude (Texas, Okla.)	.85	60	400	60
			120	100
*Crude (Wyo., Mont.)	.87	60	650	60
			180	100
*Crude (California)	.85	60	2600	60
			380	100
*No. 1 Fuel Oil	.88	60	37	70
			34	100
*No. 2 Fuel Oil	.88	60	43	70
			37	100

Liquid	Specific Gravity Value	Specific Gravity °F	Viscosity SSU Value	Viscosity SSU °F
*No. 3 Fuel Oil	.88	60	40	100
			36	130
*No. 5A Fuel Oil	.88	60	90	100
			60	130
*No. 5B Fuel Oil	.88	0	250	100
			175	130
*No. 6 Fuel Oil	.88	0	1700	122
			500	160
*SAE No. 10	.91	60	200	100
			105	130
*SAE No. 30	.91	60	490	100
			220	130
*SAE No. 50	.91	60	1300	100
			90	210
*SAE No. 70	.91	60	2700	100
			140	210
*SAE No. 90 (Trans.)	.91	60	1200	100
			400	130
*SAE No. 140 (Trans.)	.91	60	1600	100
			160	210
*SAE No. 250 (Trans.)	.91	60	Over 2300	130
			Over 200	210
Vegetable:				
Castor	.97	60	1300	100
			500	130
China Wood	.94	160	1400	100
			600	130
Coconut	.93	60	140	100
			80	130
Corn	.92	60	140	130
			50	212
Cotton Seed	.90	60	170	100
			100	130
Linseed, Raw	.93	60	140	100
			90	130
Olive	.92	60	200	100
			110	130
Palm	.92	60	220	100
			125	130
Peanut	.92	60	200	100
			110	130
Rosin	.98	60	1500	130
			600	130

Liquid	Specific Gravity Value	Specific Gravity °F	Viscosity SSU Value	Viscosity SSU °F
Sesame	.92	60	190	100
			110	130
Soya Bean	.94	60	170	100
			100	130
Turpentine	.86	60	33	60
			32	100
Syrups:				
*Corn	1.43	100	250000	100
			30000	130
Sugar:				
	1.29	60 Brix	230	70
			90	100
	1.30	62 Brix	300	70
			110	100
	1.31	64 Brix	450	70
			150	100
	1.32	66 Brix	650	70
			200	100
	1.34	68 Brix	1000	70
			280	100
	1.35	70 Brix	1700	70
			400	100
	1.36	72 Brix	2700	70
			650	100
	1.38	74 Brix	5500	70
			1150	100
	1.39	76 Brix	10000	70
			2000	100
Tar:				
*Coke Oven	1.12	60	5000	70
			1000	100
*Gas House	1.24	60	150000	70
			11000	100
Road				
*RT-2	1.07	60	250	122
			60	212
*RT-6	1.09	60	1500	122
			110	212
*RT-10	1.14	60	40000	122
			300	212
Water	1.0	60	32	70

*Values given are average values and the actual viscosity may be greater or less than the value given.

TABLE 59. Friction of Air in Pipes
Loss of pressure, in pounds per square inch,
in 1000-foot lengths of pipe

Cu. Ft. Free Air per Min- ute	Equiv. Cu. Ft. Com- pres- sed Air per Min.	At 100 Pounds Gauge —Loss of Pressure, PSI—— ——Nominal Diameter, Inches——									
		½	¾	1	1¼	1½	2	2½	3	3½	4
10	1.28	4.5	1.0	.28	.....	.....	.....	.....	....	....	....
20	2.56	18.2	4.1	1.11	.25	.11	.....	.....	....	....	....
30	3.84	40.8	9.2	2.51	.57	.26	.....	.....	....	....	....
40	5.12	72.6	16.3	4.45	1.03	.46	.....	.....	....	....	....
50	6.41	....	25.5	6.96	1.61	.71	.19	.....	....	....	....
60	7.68	....	36.7	10.0	2.32	1.02	.28	.....	....	....	....
70	8.96	....	50.0	13.7	3.16	1.40	.37	.....	....	....	....
80	10.24	....	65.3	17.8	4.14	1.83	.49	.19	....	....	....
90	11.52	....	82.6	22.6	5.23	2.39	.62	.24	....	....	....
100	12.81	....	....	27.9	6.47	2.86	.77	.30	....	....	....
125	15.82	....	....	48.6	10.2	4.49	1.19	.46	....	....	....
150	19.23	....	....	62.8	14.6	6.43	1.72	.66	.21	....	....
175	22.40	....	....	.....	19.8	8.72	2.36	.91	.28	....	....
200	25.62	....	....	.....	25.9	11.4	3.06	1.19	.37	.17	....
250	31.64	....	....	.....	40.4	17.9	4.78	1.85	.58	.27	....
300	38.44	....	....	.....	58.2	25.8	6.85	2.67	.84	.39	.20
350	44.80	....	....	.....	.....	35.1	9.36	3.64	1.14	.53	.27
400	51.24	....	....	.....	.....	45.8	12.1	4.75	1.50	.69	.35
450	57.65	....	....	.....	.....	58.0	15.4	5.98	1.89	.88	.46
500	63.28	....	....	.....	.....	71.6	19.2	7.42	2.34	1.09	.55
600	76.88	....	....	.....	.....	.....	27.6	10.7	3.36	1.56	.79
700	89.60	....	....	.....	.....	.....	37.7	14.5	4.55	2.13	1.09
800	102.5	....	....	.....	.....	.....	49.0	19.0	5.89	2.77	1.42
900	115.3	....	....	.....	.....	.....	62.3	24.1	7.6	3.51	1.80
1000	126.6	....	....	.....	.....	.....	76.9	29.8	9.3	4.35	2.21

TABLE 60. Wire and Sheet Metal Gauges
(diameters and thicknesses in decimal parts of an inch)

gauge no.	American wire gauge, or Brown and Sharpe (for copper wire)	steel wire gauge, or Washburn and Moen or Roebling (for steel wire)	Birmingham wire gauge (B.W.G.) (for steel wire or sheets)	stubs steel wire gauge	manufacturers' standard gage for steel sheet thickness[a]
0000000	----	0.4900	----	----	----
000000	----	0.4615	----	----	----
00000	----	0.4305	----	----	----
0000	0.460	0.3938	0.454	----	----
000	0.410	0.3625	0.425	----	----
00	0.365	0.3310	0.380	----	----
0	0.325	0.3065	0.340	----	----
1	0.289	0.2830	0.300	0.227	----
2	0.258	0.2625	0.284	0.219	----
3	0.229	0.2437	0.259	0.212	0.2391
4	0.204	0.2253	0.238	0.207	0.2242
5	0.182	0.2070	0.220	0.204	0.2092
6	0.162	0.1920	0.203	0.201	0.1943
7	0.144	0.1770	0.180	0.199	0.1793
8	0.128	0.1620	0.165	0.197	0.1644
9	0.114	0.1483	0.148	0.194	0.1495
10	0.102	0.1350	0.134	0.191	0.1345
11	0.091	0.1205	0.120	0.188	0.1196
12	0.081	0.1055	0.109	0.185	0.1046
13	0.072	0.0915	0.095	0.182	0.0897
14	0.064	0.0800	0.083	0.180	0.0747
15	0.057	0.0720	0.072	0.178	0.0673
16	0.051	0.0625	0.065	0.175	0.0598
17	0.045	0.0540	0.058	0.172	0.0538
18	0.040	0.0475	0.049	0.168	0.0478
19	0.036	0.0410	0.042	0.164	0.0418
20	0.032	0.0348	0.035	0.161	0.0359

21	0.0329	0.157	0.032	0.0317	0.0285
22	0.0299	0.155	0.028	0.0286	0.0253
23	0.0269	0.153	0.025	0.0258	0.0226
24	0.0239	0.151	0.022	0.0230	0.0201
25	0.0209	0.148	0.020	0.0204	0.0179
26	0.0179	0.145	0.018	0.0181	0.0159
27	0.0164	0.143	0.016	0.0173	0.0142
28	0.0149	0.139	0.014	0.0162	0.0126
29	0.0135	0.134	0.013	0.0150	0.0113
30	0.0120	0.127	0.012	0.0140	0.0100
31	0.0105	0.120	0.010	0.0132	0.0089
32	0.0097	0.115	0.009	0.0128	0.0080
33	0.0090	0.112	0.008	0.0118	0.0071
34	0.0082	0.110	0.007	0.0104	0.0063
35	0.0075	0.108	0.005	0.0095	0.0056
36	0.0067	0.106	0.004	0.0090	0.0050
37	0.0064	0.103	-----	0.0085	0.0045
38	0.0060	0.101	-----	0.0080	0.0040
39	-----	0.099	-----	0.0075	0.0035
40	-----	0.097	-----	0.0070	0.0031
41	-----	0.095	-----	0.0066	-----
42	-----	0.092	-----	0.0062	-----
43	-----	0.088	-----	0.0060	-----
44	-----	0.085	-----	0.0058	-----
45	-----	0.081	-----	0.0055	-----
46	-----	0.079	-----	0.0052	-----
47	-----	0.077	-----	0.0050	-----
48	-----	0.075	-----	0.0048	-----
49	-----	0.072	-----	0.0046	-----
50	-----	0.069	-----	0.0044	-----

*Replaces U. S. standard (revised) gauge.

INDEX

INDEX

Towers, cooling. *See* Cooling towers
Transfer of heat. *See* Heat

"U" factor, 11

Valves, 319
  ball and butterfly, 322
  check, 321
  gate, 319
  globe, angle and "Y", 319
  installation costs of. *See* Estimating labor costs
  needle, 323
  plug, 322
  refrigerant, 323
Ventilating load, computation of, 22

Ventilation, 13, 20, 22, 29
  bypassed air in, 38
  quantity of air required for, 396, 397
Volume, specific, 31

Walls
  heat travel through, 375
  solar heat gains in, 385
  surface films on, 380
Water system, 46, 66
Water treatment, 326, 327
  equipment, installation costs of. *See* Estimating labor costs
Water vapor, 4
Watt density, 83
Weather data analysis, 41
  methods for, 41, 42
  general characteristics of, 42